Zoology Essentials

Zoology Essentials

Edited by Toby Simpson

MURPHY & MOORE
www.murphy-moorepublishing.com

Published by Murphy & Moore Publishing,
1 Rockefeller Plaza,
New York City, NY 10020, USA
www.murphy-moorepublishing.com

Zoology Essentials
Edited by Toby Simpson

International Standard Book Number: 978-1-63987-577-1 (Hardback)

Cataloging-in-Publication Data

Zoology essentials / edited by Toby Simpson.
p. cm.
Includes bibliographical references and index.
ISBN 978-1-63987-577-1
1. Zoology. 2. Animals. I. Simpson, Toby.
QL45.2 .Z66 2022
590--dc23

TABLE OF CONTENTS

PREFACE

It is often said that books are a boon to mankind. They document every progress and pass on the knowledge from one generation to the other. They play a crucial role in our lives. Thus I was both excited and nervous while editing this book. I was pleased by the thought of being able to make a mark but I was also nervous to do it right because the future of students depends upon it. Hence, I took a few months to research further into the discipline, revise my knowledge and also explore some more aspects. Post this process, I begun with the editing of this book.

The branch of biology that deals with the study of animal kingdom, including their structure, classification, habits, embryology, evolution, and distribution is termed as zoology. It also focuses on their interactions with their ecosystems. There are various branches of zoology such as zoography, comparative anatomy, animal physiology and behavioral ecology. Zoography is the branch of zoology which describes the animals and their habitats. Comparative anatomy deals with the structure of animals. The scientific study of functions and mechanisms in animals is known as animal physiology. Behavioral ecology focuses on evolution of behavior of animals due to ecological pressure. This book provides significant information of this discipline to help develop a good understanding of zoology and related fields. It will also provide interesting topics for research which interested readers can take up. The extensive content of this book provides the readers with a thorough understanding of the subject.

I thank my publisher with all my heart for considering me worthy of this unparalleled opportunity and for showing unwavering faith in my skills. I would also like to thank the editorial team who worked closely with me at every step and contributed immensely towards the successful completion of this book. Last but not the least, I wish to thank my friends and colleagues for their support.

Editor

1

A new 'horned' *Stenocercus* from the highlands of southeastern Brazil, and redescription of *Stenocercus tricristatus* (Reptilia: Tropiduridae)

Teresa C. S. Ávila-Pires[1], Cristiano de C. Nogueira[2], Marcio Martins[2]

[1]*Museu Paraense Emílio Goeldi, Coordenação de Zoologia, Caixa Postal 399, 66017-970 Belém, PA Brazil.*
[2]*Departamento de Ecologia, Instituto de Biociências, Universidade de São Paulo. 05508-090 São Paulo, SP, Brazil.*
Corresponding author. Teresa C. S. Ávila-Pires (avilapires@museu-goeldi.br)

ABSTRACT. A new species of the lizard genus *Stenocercus* Duméril & Bibron, 1837 is described based on six specimens from Serra da Canastra, Minas Gerais, southeastern Brazil. Similar to *S. tricristatus* (Duméril, 1851), *S. dumerilii* (Steindachner, 1867), *S. quinarius* Nogueira & Rodrigues, 2006, and *S. squarrosus* Nogueira & Rodrigues, 2006, the new species has a pyramidal head and an enlarged, dorsally projected post-supraciliary scale, like a small horn. It is most similar to *S. tricristatus*, from which it differs in having a more elongated head in ventral view; a larger number of scales around midbody; a distinctly enlarged, smooth preauricular scale; and probably a larger body size. Moreover, the new species has distinctly keeled scales, with phylloid and mucronate dorsals; no enlarged supraoculars; no neck folds; a vertebral, and a pair of dorsolateral and (weak) lateral crests. A redescription of *S. tricristatus* is presented for comparison with the new species.

KEY WORDS. Minas Gerais, new species, Serra da Canastra, Squamata, taxonomy.

INTRODUCTION

The enigmatic *Stenocercus tricristatus* was originally described by A. Duméril (Duméril and Duméril 1851) as the type-species of *Ophryoessoides*. Genus and species were based on a single specimen from "Brésil", collected by M. Claussen. A short characterization of the genus and species also was given in Duméril et al. (1854). Subsequently, Duméril (1856) complemented the description of *S. tricristatus* with a color drawing of the animal on a rocky substrate, thereby illustrating its color pattern in detail (Fig. 1). *Ophryoessoides dumerilii* Steindachner, 1867 was likewise described on the basis of a single specimen from Pará, northern Brazil, but Etheridge (in Peters and Donoso-Barros 1970) referred it as a synonym of *O. tricristatus*. The former taxon remained poorly known until Cunha (1981) published a study based on several specimens from eastern Pará. Avila-Pires (1995), however, examining the holotypes of both species, demonstrated that these two taxa, by then referred to *Stenocercus* Duméril & Bibron, 1837 (Frost 1992), corresponded to distinct species. Taking into account where M. Claussen lived and traveled, Avila-Pires (1995) suggested that *S. tricristatus* probably would occur in Minas Gerais, southeastern Brazil. Torres-Carvajal (2007a), in a taxonomic revision of the genus, compared *S. tricristatus* with all other congeners and presented a description based on the literature and on data provided by R. Etheridge.

During field trips to Serra da Canastra, Minas Gerais, six lizards were collected that at first we thought belonged to *S. tricristatus* (Figs 2, 3). Considering that Nogueira and Rodrigues (2006) described two other species, *S. quinarius* Nogueira & Rodrigues, 2006 and *S. squarrosus* Nogueira & Rodrigues, 2006, from central and northeastern Brazil, with the characteristic pyramidal head and enlarged, dorsally projected post-supraciliaries present in *S. dumerilii* and *S. tricristatus*, we performed a comparison with these four species (from here on referred to as the 'horned *Stenocercus*'). Our data showed that, although the specimens from Serra da Canastra are most similar to *S. tricristatus*, they also show some significant differences and represent clearly a distinct species, described here. In order to facilitate comparisons, we also present a redescription of the holotype of *S. tricristatus*.

MATERIAL AND METHODS

Specimens were collected during field trips to Serra da Canastra National Park (20°04'32.73" to 20°39'50.91"S, 46°12'17.87" to 46°59'56.54"W), state of Minas Gerais, southeastern Brazil (collection permit IBAMA 02027.008026/03-68,

Figures 1–3. (1) Drawing of *Stenocercus tricristatus* from Plate XXII in Duméril (1851); (2) photograph (by O.A.V. Marques) of a live adult specimen of *Stenocercus canastra* sp. nov. (holotype, MZUSP 88873); (3) photograph (by CCN) of a live juvenile specimen of *S. canatra* sp. nov. (MZUSP 94456).

03/2003 to CCN). Direct comparisons were made with other four species of horned *Stenocercus*: *S. tricristatus*, *S. dumerilii*, *S. quinarius*, and *S. squarrosus*. The holotype and only known specimen of *S. tricristatus* was examined by TCSAP and CCN. Data on *S. dumerilii* are based on MPEG 6031–6037, MPEG 7376–7377, MPEG 7382–7383, MPEG 7386, RMNH 28047, 28048, 28049 (former MPEG 6089, 7324, 7330, respectively) and on Avila-Pires (1995). Data on *S. quinarius* and *S. squarrosus* refer to the type series. For comparisons with other *Stenocercus* species, data were taken from the literature (Cadle 1991, 2001, Torres-Carvajal 2005, 2007a). We used the museum acronyms listed in Sabaj-Perez (2013).

To define the species we applied the 15 sets of characters and the character terminology used by Cadle (1991). Additionally to these 15 sets, we also used the following two sets of features: 1) number of scales around midbody and number of ventrals between the anterior margin of forelimbs and the anterior margin of hind limbs, and 2) shape of the scale bordering the anterior portion of the ear opening. Description follows Avila-Pires (1995), with some additional characters added from Cadle (1991; internasals counted immediately posterior to postrostrals, number of supraoculars in a transverse line across the greatest width of one of the orbits, angulate temporals, number of gular scales between ventral edges of ear-openings). All measurements were taken from the specimens after preservation. Snout–vent length (SVL) and tail length were measured with a ruler to the nearest 1 mm, whereas all other measurements were taken with an electronic caliper, to the nearest 0.1 mm. Head length was measured from the tip of the snout to the anterior border of the ear-opening; head width at the widest point of the head; head height at highest mid-dorsal point (prominent scales on canthal-supraciliary crests not considered). Limb length was measured from axil/groin to tip of claw of longest digit. Lamellae under fourth finger and fourth toe were counted on left side.

A Principal Components Analysis (PCA) was performed in PAST (version 3.04 for Mac; Hammer et al. 2001) to compare the new species with other horned *Stenocercus*. Of the 27 meristic and morphometric characters obtained, we used the only 11 that had a normal or almost normal distribution ($p > 0.025$ in Shapiro-Wilk normality test). In order to eliminate the influence of size on the comparisons, for the eight morphometric characters we used the residuals of regressions of the variable and head length (for head measurements) or SVL (for all other measurements). Characters included were relative tail length, relative head width, relative head height, relative leg length, relative body height, relative body width, relative tail height, relative tail width, scales around midbody, longitudinal ventrals, and infradigital lamellae under fourth toe. As juveniles tend to show differences in morphology in relation to adults, we used a different symbol for the only young specimen from Serra da Canastra in the PCA biplot. Since our variables showed different scales, the PCA was based on a correlation matrix (after normalizing all variables by dividing them by their standard deviations – Hammer et al. 2001).

TAXONOMY

Comparison between the specimens from Serra da Canastra and the four other horned *Stenocercus* species showed several differences. Among them were body shape, tail shape and length, distinctions on some of the scales on head and body, and color pattern (Table 1). The first four axes of the PCA explained 71.0% of the variance in the data. Axes 1 and 2 explained 49.3% of total variation and were able to almost completely separate the groups, except for *S. quinarius* and *S. squarrosus* that showed a larger overlap (Fig. 4). However, these two species differ in a number of characters not included in the PCA (Table 1). The most important variables in component 1 were relative tail length, scales around midbody, relative tail height, longitudinal ventrals, and relative body width, whereas in component 2, they were relative body height, lamellae under fourth toe, relative tail width, relative tail height, and relative head width (Table 2). Along axis 1, the group formed by the Serra da Canastra specimens is well separated from *S. quinarius*, *S. squarrosus*, and *S. dumerilii*, whereas the single specimen of *S. tricristatus* appears well separated from the Serra da Canastra specimens, on the highest values of axes 1 and 2, especially the former. These comparisons (Tables 1, 2 and Fig. 4) demonstrate that the specimens from Serra da Canastra cannot be attributed to any of the known taxa and represent a new species, which we describe below.

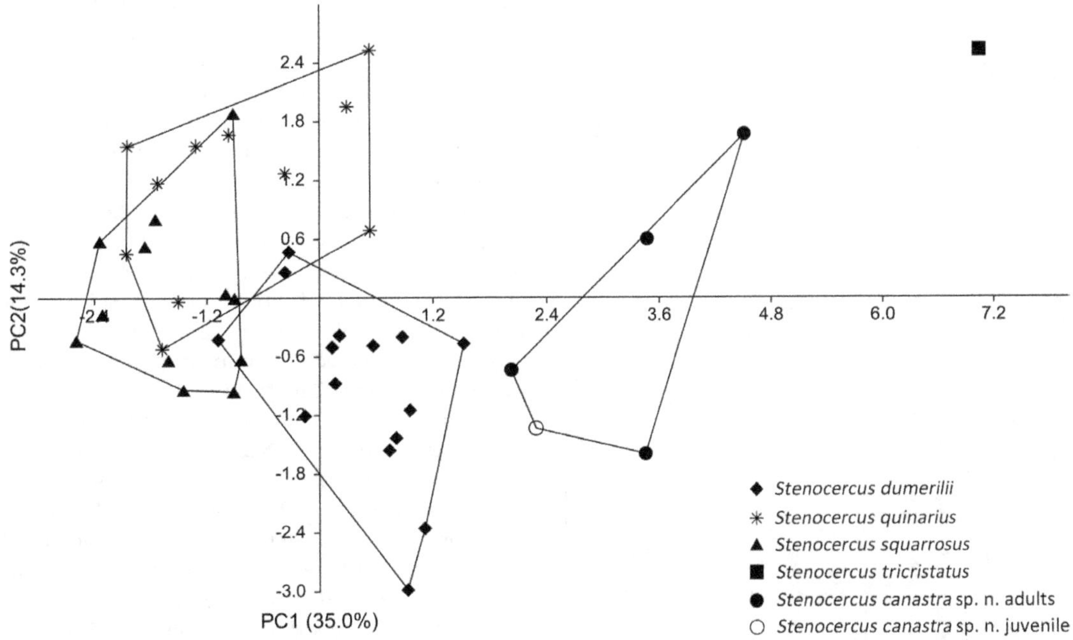

Figure 4. Plot of axis 1 versus axis 2 from Principal Component Analysis comparing the five horned *Stenocercus* species.

Table 1. Comparison between the five horned *Stenocercus* species (n = number of specimens). *A low crest on flanks, between dorsolateral and lateral crests.

	S. canastra (n = 6)	*S. tricristatus* (n = 1)	*S. dumerilii* (n = 15)	*S. quinarius* (n = 11)	*S. squarrosus* (n = 12)
SVL in adult males	70–77 mm	58 mm	84–104 mm	65–75 mm	76–88 mm
Body shape	Slightly depressed	Slightly compressed	Depressed	Depressed	Strongly depressed
Tail shape	Nearly cylindrical to slightly compressed	Compressed near base, cylindrical distally	Depressed near base, cylindrical distally	Depressed near base, cylindrical distally	Depressed, blunt and thick toward end
Tail length/SVL	1.4–1.8	1.7	1.2–1.4	1.0–1.1	0.8–0.9
Head shape ventrally	Elongated, neck distinct	Broad, neck distinct	Elongated, neck poorly distinct	Elongated, neck poorly distinct	Elongated, neck poorly distinct
Interparietal	Large	Large	Small	Large	Large
Supraciliaries in dorsal view	2 (rarely 3)	2	4 (rarely 3)	4	4
Preauricular scale	Enlarged, projecting, smooth	Not enlarged, projecting, keeled	Not enlarged, not projecting	Not enlarged, only slitghtly projecting	Not enlarged, only slitghtly projecting
Mental scales	Subequal	Subequal	Anterior ones distinctly smaller	Anterior ones distinctly smaller	Anterior ones distinctly smaller
Vertebral crest	Prominent, serrate, to proximal part of tail	Prominent, serrate, tail?	Low, not on tail	Low (sliglhtly serrated ant.), to base of tail	Low, to base of tail
Dorsolateral crest	Prominent, serrate, to base of tail	Prominent, serrate	Low, not on tail	Low or serrate, to base of tail; stronger on neck	Prominent, serrate, to base of tail; stronger on neck
Lateral crests	Visible, but less conspicuous in 2 out of 6 specimens	Uncertain, but apparently absent	Low, may be rather inconspicuous	Serrate, stronger on neck	Serrate, very strong on neck
Mid-lateral crest*	Absent	Absent	Present, variable in extension	Absent	Absent
Dorsals	High keels, mucronate	High keels, mucronate	Low keels, not or hardly mucronate	Low keels, not or hardly mucronate	Low keels, not or hardly mucronate
Dorsals at midbody	8–9	7–8	11–14	13–15	13–15
Scales at midbody	39–41	33	41–50	47–56	47–53
Color pattern	Back and flanks densely covered by bands/blotches	Back and flanks densely covered by bands/blotches	Back with sparse blotches or uniform	Uniform or some sparse blotches on back	Back with sparse blotches or uniform

Table 2. Results from Principal Component Analysis comparing the five horned *Stenocercus* species, including factor loadings, eigenvalues, and the percent variance explained by the five first components. Highest factor loadings in components 1 and 2 are shown in bold.

	PC 1	PC 2	PC 3	PC 4	PC 5
Relative tail length	**0.464910**	-0.043799	0.1314000	-0.024299	0.177340
Relative head width	0.221170	0.285550	0.2435400	0.548100	-0.025339
Relative head height	0.058938	0.248490	0.5807400	-0.121060	-0.646790
Relative leg length	0.285790	-0.060514	0.2099100	-0.359920	0.073845
Relative body height	-0.043074	**0.480880**	0.4335200	-0.056876	0.587490
Relative body width	**-0.345650**	0.229540	-0.0059336	0.047379	0.320150
Relative tail height	**0.365020**	**0.381700**	-0.2947500	-0.080645	-0.043359
Relative tail width	0.218990	**0.414650**	-0.4548700	0.315760	-0.124840
Scales around midbody	**-0.440310**	0.135830	-0.0236180	-0.026822	-0.030189
Longitudinal ventrals	**-0.364530**	0.119990	0.0792200	0.329860	-0.226500
Lamellae under 4th toe	0.143220	**-0.46554**	0.2313100	0.577570	0.163700
Eigenvalues	3.846870	1.575790	1.3274500	1.065050	0.800367
% variance explained	34.97200	14.32500	12.0680000	9.682300	7.276100

Stenocercus canastra sp. nov.

Stenocercus cf. *tricristatus*; Teixeira et al. 2016: 411, 413.

Type material. Holotype. MZUSP 88873 (Fig. 2), an adult male, from Brazil, Minas Gerais: São Roque de Minas municipality, Parque Nacional da Serra da Canastra (PNSC), at the beginning of the road to Casca D'Anta waterfall, 20°14'38" S, 46°32'56" W, ca 1410 m. Paratypes. Four specimens from the type-locality: MZUSP 88874, a half-grown male; MZUSP 88875, a female; MPEG 31738, a male; and MNHN 2014.0061, a male. One juvenile, MZUSP 94456, also from PNSC, Chapadão da Zagaia, 20°09'58"S, 46°41'54" W, ca 1350 m (Fig. 3).

Diagnosis. *Stenocercus canastra* sp. nov. is characterized by the following combination of features: (1) Dorsal head scales keeled. (2) Interparietal distinct, moderately enlarged; parietals about as large as, or smaller than interparietal; postparietals large, with a prominent keel; other posterior head scales variable in size. (3) Internasals six. (4) No distinctly enlarged supraoculars, but medial ones larger than lateral ones; all with pronounced keels. (5) An enlarged canthal on each side, in contact anteriorly with two elongate scales that form a double canthal ridge. (6) An enlarged, prominent, pointed scale immediately posterior to supraciliaries; no projecting, blade-like, angulate temporal scales. (7) Gulars and ventrals distinctly keeled. (8) Parietal eye distinct. (9) Neck folds absent. (10) Dorsals phylloid, keeled, mucronate, and imbricate; scales on flanks similar to dorsals. (11) A prominent serrate vertebral crest, a slightly less prominent dorsolateral crest, and an even less prominent lateral crest. (12) Mite pockets absent. (13) Scales on posterior surface of thighs imbricate, keeled. (14) Tail nearly cylindrical to moderately compressed, verticils absent. (15) Dorsal coloration with numerous, bold, dark brown rhomboid marks forming a longitudinal series on back, and dark brown vertical bands on flanks; head with at least a dark brown spot between nostrils, and a large, rhomboid spot between eyes; no sexual dimorphism in color pattern. (16) Scales around midbody 39–41, ventrals between anterior margin of forelimbs and anterior margin of hind limbs 25–28. (17) Ear opening bordered anteriorly by a distinctly enlarged, smooth scale. *Stenocercus canastra* sp. nov. is distinguished from all other *Stenocercus* except *S. dumerilii*, *S. tricristatus*, *S. quinarius* and *S. squarrosus* by the presence of an enlarged, prominent post-supraciliary scale (Figs 5–9); from all others except *S. tricristatus*, *S. quinarius* and *S. squarrosus* by a moderately enlarged interparietal (although not as large as in the Tropidurini; Figs 10–11). It is distinguished from *S. dumerilii*, *S. quinarius* and *S. squarrosus* (character states in parentheses) by the presence of a prominent, serrate vertebral crest (in contrast to a low vertebral crest); two, rarely 3, supraciliaries (4, rarely 3); ear-opening bordered anteriorly by a distinctly enlarged scale projecting over tympanum (no enlarged scale projecting over tympanum; Figs 12–17); dorsals distinctly keeled and mucronate (dorsals with a low keel, not or hardly mucronate); 9–10 scales across midbody from one dorsolateral row to the other (11–13 in *S. dumerilii*, 13–15 in *S. quinarius* and *S. squarrosus*); ventrals between anterior level of fore- and hind limbs 24–28 (28–32 in *S. dumerilii*, 30–34 in *S. quinarius*, 28–34 in *S. squarrosus*); scales on chin subequal and imbricate, or only most anterior ones smaller and subimbricate (scales on chin smaller, polygonal, and subimbricate anteriorly, grading into larger, pointed, and imbricate posteriorly; Figs 18–22); tail 1.4–1.8 times SVL (1.2–1.4 times in *S. dumerilii*, 1.0–1.1 in *S. quinarius* and 0.8–0.9 in *S. squarrosus*); and color pattern (Table 1). From *S. tricristatus* it differs in presenting head width 0.78–0.89 times head length (0.96); 39–41 scales around midbody (33); and ear-opening bordered anteriorly by a distinctly enlarged, smooth scale (scale not larger than adjacent temporal scales, keeled). Moreover, it probably reaches a larger body size (adult males ≥ 70 mm in *S. canastra* sp. nov. versus 58 mm in the holotype of *S. tricristatus*).

Description. Tropidurid with a known maximum SVL in males of 77 mm (MZUSP 88873, holotype), and 65 mm in the only female known (MZUSP 88875). Head 24–25% of SVL, 1.1–1.2 times as long as wide, 1.2–1.3 times as wide as high. Snout bluntly pointed. Canthus rostralis well defined, continuous with supraciliaries, which end in an enlarged, prominent scale, so that the head has the shape of a four-sided pyramid. Neck narrower than head and body. Body roughly cylindrical. Limbs well developed, forelimbs 42–46% of SVL, hind limbs 58–65% of SVL, tibia 15–17% of SVL. Tail nearly cylindrical to slightly compressed, 1.7–1.8 times SVL in the largest specimens (SVL 71–77 mm), 1.4–1.5 in the smallest specimens (SVL 42–65 mm).

Rostral roughly triangular, largest width three to four times the medial height, only barely visible from above. Postrostrals 4–5. Snout, supraocular, and interorbital regions covered with irregularly polygonal, subimbricate to juxtaposed, strongly

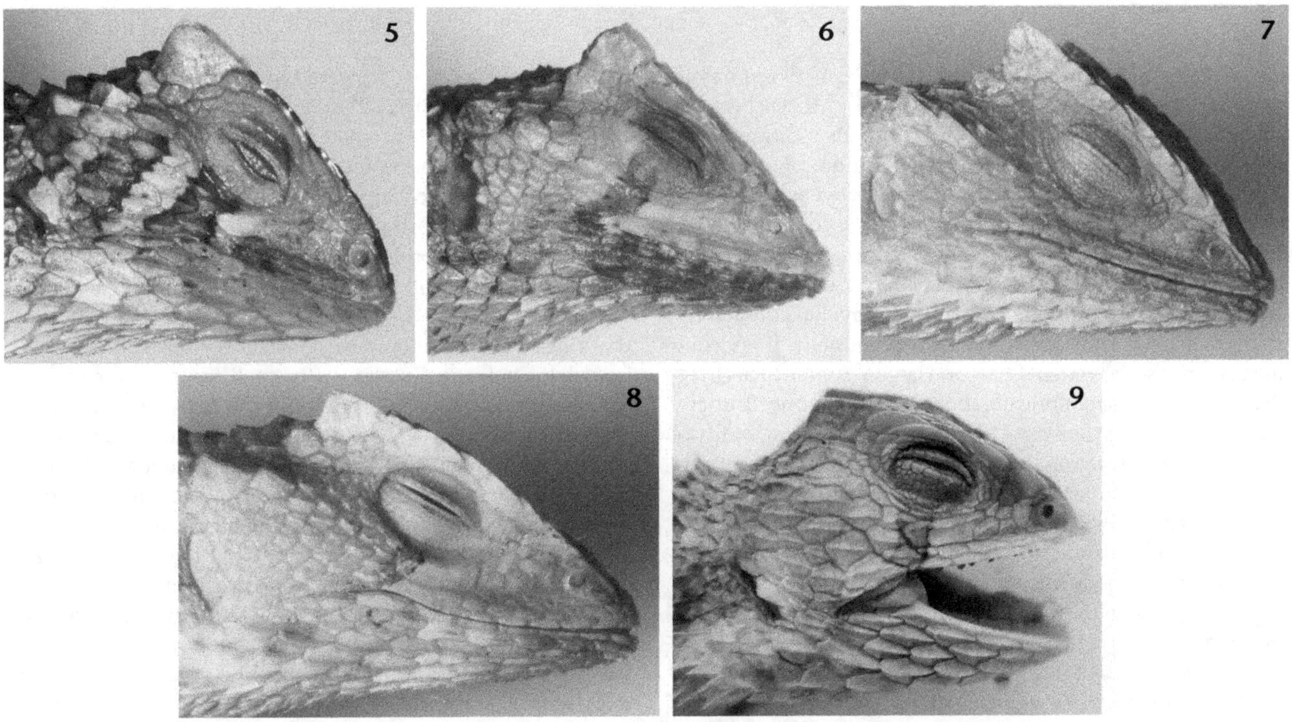

Figures 5–9. Lateral view of head of: (5) *Stenocercus canastra* sp. nov., MZUSP 88873, holotype; (6) *S. dumerilii*, MPEG 6036; (7) *S. quinarius*, MZUSP 94069, holotype; (8) *S. squarrosus*, MZUSP 94056, holotype; (9) *S. tricristatus*, MNHN 6825, holotype.

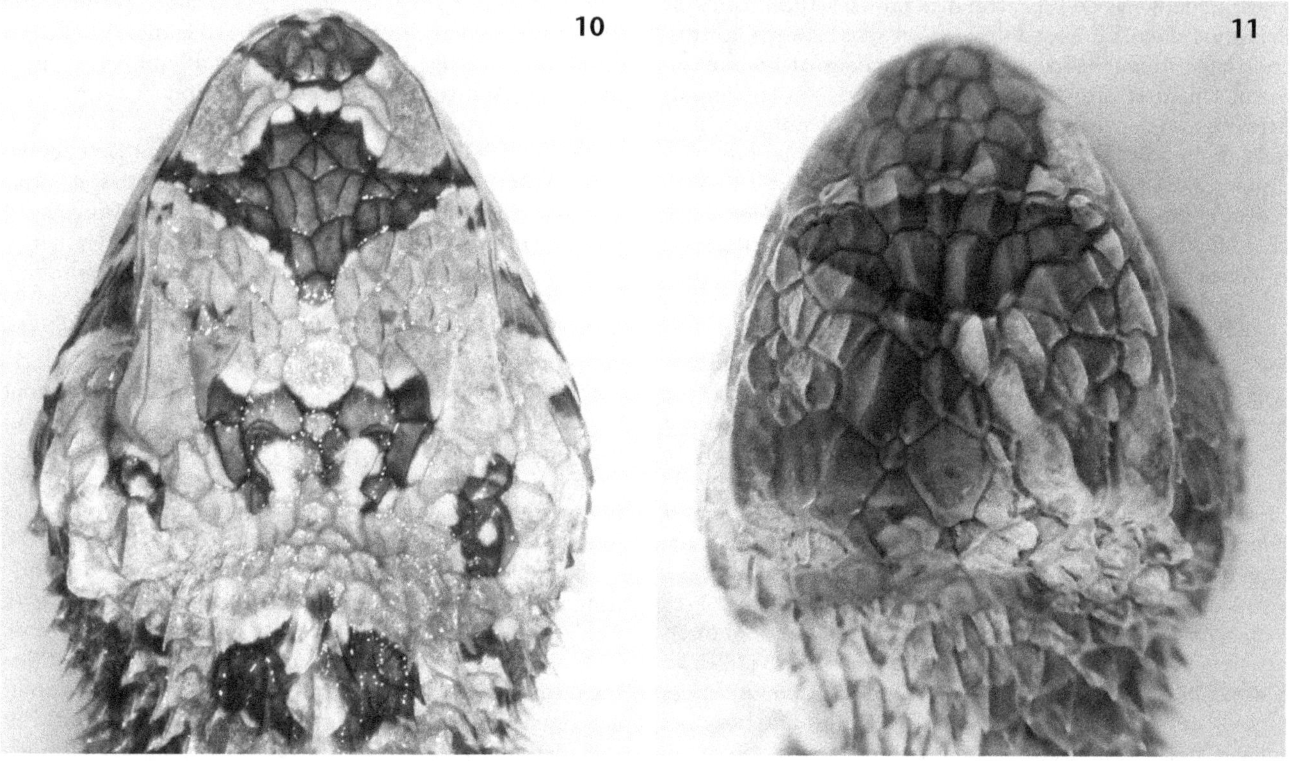

Figures 10–11. Dorsal view of head in (10) *Stenocercus canastra* sp. nov., MZUSP 88873, holotype; (11) *Stenocercus tricristatus*, MNHN 6825, holotype.

keeled scales, heterogeneous in size. Six scales between nasals, of which one supranasal at each side and four scales between them. Scales across the snout between canthals 3–5 (minimum number in a transverse row). One large canthal on each side, anteriorly in contact with two elongate, keeled scales (one of which is the supranasal); canthal posteriorly in contact with the first supraciliary. Supraorbital semicircle formed by 7–8 scales (between canthal ridge and post-supraciliary scale), of which the third and fourth, or only the fourth, are in medial contact; the posterior scale is the largest and most prominent, the second most anterior scale is often the second largest. Supraoculars irregularly polygonal, subimbricate, strongly keeled, none distinctly enlarged; 3–5 in a transverse line across the greatest width of each orbit. Two (three on the left side of MPEG 31738) elongate supraciliaries in contact with each other (narrowly overlapping), with a third supraciliary lateral to, and in contact with, the other two. Canthals and supraciliaries form a distinct ridge that ends in a very prominent pointed scale immediately posterior to supraciliaries; in the juvenile, MZUSP 94456, this scale is less prominent. This ridge delimits a sharp angle between dorsal and lateral surfaces of head. Interparietal moderately enlarged, roughly rhomboid or hexagonal, parietal eye distinct. Parietals about as large as, or smaller than interparietal; in some specimens separated from interparietal by small scales. Postparietals large with a prominent keel, separated medially by small scales (in medial contact in MNHN 2014.0061). Other posterior head scales heterogeneous in size. Angulate temporals distinctly keeled, but not especially large or prominent (Fig. 10).

Nasals lateral to canthal ridge, undivided, separated from rostral by the postrostrals; nostril in posterior part of nasal, directed latero-posteriorly. Loreal region with a distinct row of lorilabials, anteriorly and posteriorly forming a single row of scales, medially a double row; between lorilabials and canthal ridge a few large (and in some cases a few small as well), irregularly polygonal scales, with small keels and/or variably rugose surface. Scales in a transverse row between canthal and supralabials, 4–5. An elongate subocular, preceded by a distinctly shorter pre-subocular, both keeled, the keels aligned and closer to the upper margin of the scales. Elongate subocular separated from supralabials by the row of lorilabials. Supralabials 5–6, very narrow and mostly with a keel on their lower margin. Temporal region with polygonal (mostly rhomboid to hexagonal), strongly keeled, imbricate scales, in approximately oblique rows; upper scales larger and distinctly more prominent (these scales are ventral to what was called by Cadle 1991 "angulate temporals"). Ear opening vertically oval, largely covered by an enlarged, smooth scale on its anterior margin; tympanum slightly recessed (Figs 5, 12, 13).

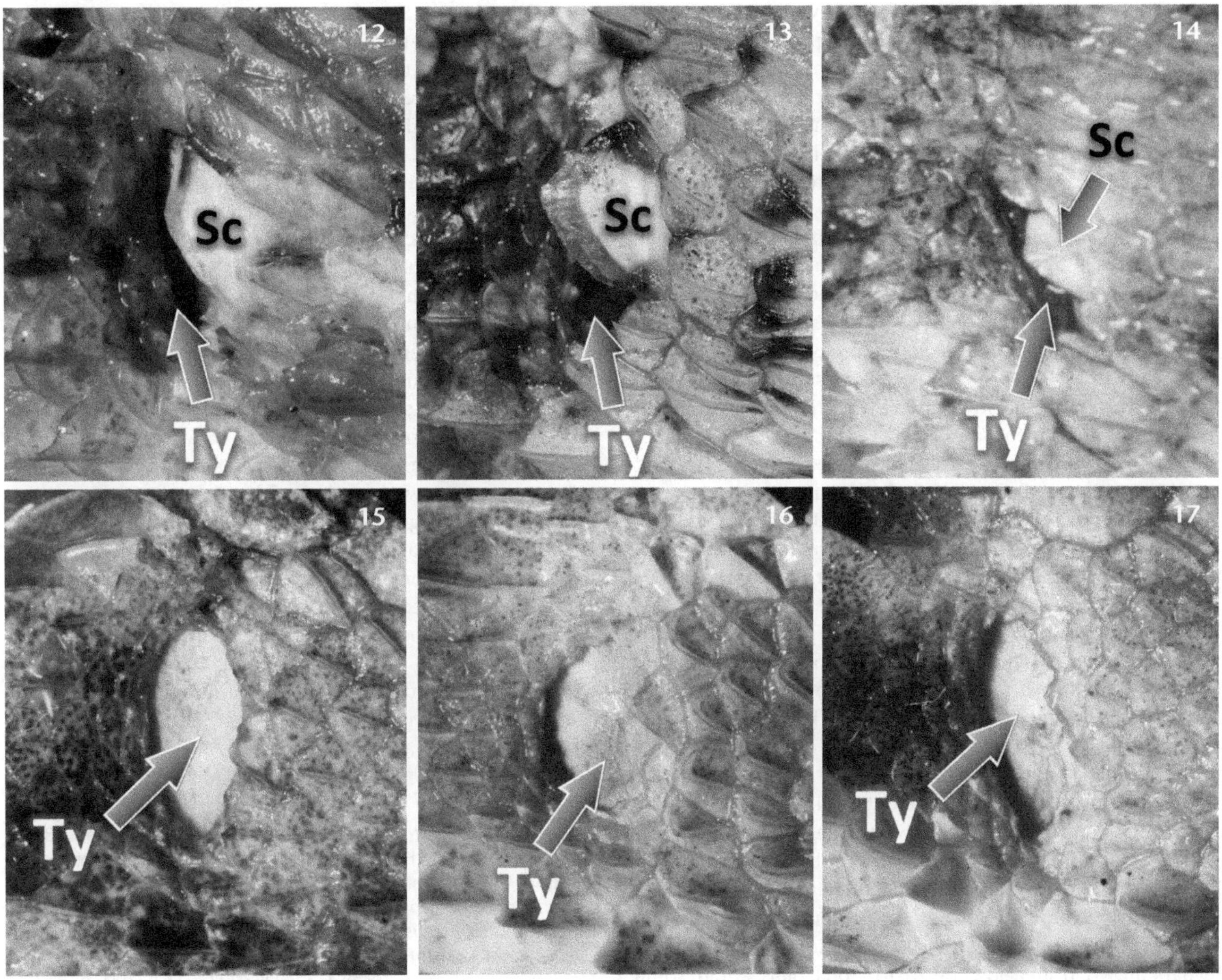

Figures 12–17. Ear region of: two specimens of *Stenocercus canastra* sp. nov.: (12) MPEG 31738, (13) MZUSP 88873, holotype; (14) *S. tricristatus*, MNHN 6825, holotype; (15) *S. dumerilii*, MPEG 6036; (16) *S. quinarius*, MZUSP 94069, holotype; (17) *S. squarrosus*, MZUSP 94056, holotype. (Ty) Tympanic membrane, (Sc) enlarged ear scale. Note the evident enlarged preauricular scale in *S. canastra* sp. nov., the relatively small and keeled preauricular scale in *S. tricristatus* and the virtual absence of this scale in *S. dumerilii*, *S. quinarius*, and *S. squarrosus*.

Mental relatively small, but larger than adjacent infralabials and the two or four post-mentals. Infralabials 5–7 (mostly 6), very narrow and keeled; followed by a relatively large, elongate, keeled scale. Scales on chin anteriorly polygonal, keeled, subimbricate, posteriorly (including gulars) mostly phylloid, distinctly keeled, mucronate, and imbricate, longer and with higher keels laterally. The keels of the phylloid scales are mostly aligned in longitudinal rows, which also continue along neck and belly; anteriorly on the chin these rows converge towards the mental. Gulars between ventral edges of ear-openings 11–14 (12.7 ± 1.0). No gular or lateral folds (Fig. 18).

Scales on nape relatively small, subimbricate, and keeled medially, dorsolaterally similar to dorsals but shorter, some of them with a very prominent mucron. On sides of neck the mucron is similarly well developed, giving the neck a spinose appearance. Dorsals phylloid, keeled, mucronate, and imbricate, in approximately longitudinal rows. A prominent serrate vertebral crest of moderately enlarged, strongly keeled, and mucronate scales is present from nape to base of tail. A dorsolateral row of similar, but slightly smaller, scales on each side forms a lower, serrate crest also from nape to base of tail. A third, less prominent longitudinal crest, is present on each side, between fore- and hind limbs. Vertebral and dorsolateral crests usually separated by three scales in a transverse line at midbody, totaling nine scales from one dorsolateral crest to the other (10 scales in MZUSP 88875); dorsolateral and lateral crests are separated by 4–5 rows of scales. Vertebral crest with 25–28 (26.2 ± 1.1) scales from occiput to posterior margin of hind limbs. Paravertebral scales between the same points 36–42 (39.7 ± 2.3). Ventrals similar to dorsals, but flatter and slightly larger, forming distinct longitudinal rows (the keels forming low longitudinal ridges); 24–28 (25.9 ± 1.5) ventral scales along a midventral line between anterior level of forelimbs and anterior level of hind limbs. Scales around midbody 39–41 (39.7 ± 0.8). Scales on preanal plate similar to dorsals, but slightly smaller. Mite pockets absent.

Scales at base of tail similar to dorsals and ventrals. Distally the scales grade into polygonal shapes and become arranged in transverse (slightly oblique) rows; the crests disappear, but the ridges formed by the keels, on the ventral surface, continue to the tip of the tail. No distinct verticils (which probably indicates that autotomic segments are absent).

Limbs with phylloid, strongly keeled, mucronate, and imbricate scales, similar to, or smaller and more elongate than, dorsals. The mucron is especially well developed on the upper aspects of hind limbs. Subdigital lamellae single, mostly tricarinate; 13–15 (14.0 ± 0.8) under fourth finger, 17–20 (19.0 ± 1.2) under fourth toe.

In life, ground color predominantly light brown, with several dark brown blotches. Borders of most of blotches deep dark brown followed by white to cream areas (Fig. 2). Main features are: (1) an irregular dark brown blotch on anterior part of snout, present or absent; (2) a very conspicuous, large, rhomboid dark brown blotch on posterior part of snout and anterior to supraocular region, present in all individuals; it may be irregularly bordered by a thin white line; (3) posteriorly, on post-parietal and supratemporal areas, either three relatively large, irregular dark brown blotches (medial one may have a shape grossly reminding of a butterfly), or four small dark brown blotches; (4) sides of head with a brown oblique band, bordered by dark brown, from below the eye to supralabials, and in some specimens one or two other similar bands posteriorly; (5) a series of 6–9 roughly triangular or rhomboid (a few irregular) dark brown spots along the vertebral region, from nape to level of hind limbs; intermediate dorsolateral spots are usually present; (6) sides of neck and flanks with irregular brown bands bordered by dark brown, usually separated by cream to light brown areas; (7) brown to dark brown spots form irregular bands across both limbs and tail, separated by wider (tail) or narrower (limbs) light brown spaces. Ventral region immaculate, dirty white to tan. In preservative, all colors tend to become darker.

Holotype main features. SVL 77 mm, tail length 132 mm, head length 19.0 mm, head width 16.7 mm, head height 13.6 mm, fore limb length 33 mm, hind limb length 45 mm, tibia length 12.5 mm. Scales around midbody 40. Vertebrals (from occiput to posterior margin of hind limbs) 27. Ventrals (between anterior margin of forelimbs and that of hind limbs) 28. Lamellae under fourth finger 15, under fourth toe 20. See also Figs 2, 5, 10, 13, 18.

Distribution. *Stenocercus canastra* sp. nov. is only known from Serra da Canastra National Park, Minas Gerais, Brazil (Figs 23–24). Besides the type series, two additional individuals of *S. canastra* sp. nov. were observed, but not collected, in an area very close to (less than 1 km N) the type-locality (O. Marini, pers. comm.).

Etymology. The specific epithet *canastra* is a noun in apposition and refers to the mountains ('Serra da Canastra') where all specimens were collected.

Habitat and natural history. These lizards were found in two localities 18 km apart, at Serra da Canastra National Park, in Minas Gerais, southeastern Brazil. Elevation varies from 800 to 1507 m. The vegetation is characterized by savanna-like 'cerrado', grass/shrubland on deep, well-drained soil ('campo limpo' and 'campo sujo') and on rocky, shallow soils ('campos rupestres'), and narrow strips of riparian forests. Dietz (1984) gives a general description of the vegetation in the area. *Stenocercus canastra* sp. nov. was only found in open cerrado formation with dense grass vegetation on deep, well-drained soil, between 1350–1410 m. It was not found in areas with rocky quartzitic soil/sandy soil and on rock outcrops, which cover a large portion of the park. Six specimens were on or inside termite mounds – all six found in about one hour, after a fire. The juvenile was obtained in a pitfall trap in campo limpo. Scales of *S. canastra* sp. nov. were found in scats of the omnivorous maned wolf (Canidae), *Chrysocyon brachyurus* (Illiger, 1815) (Queirolo and Motta-Junior 2007).

Stenocercus tricristatus (A. Duméril, 1851)

Ophryoessoides tricristatus A. Duméril (in Duméril and Duméril 1851): 66 (holotype MNHN 6825, type-locality: 'Brésil'); Duméril et al. 1854: 267; A. Duméril, 1856: 531; Etheridge 1966: 89; Peters and Donoso-Barros 1970: 215 (part); Fritts 1974: 35 (part).
Liocephalus tricristatus; Boulenger 1885: 170.
[*Stenocercus*] *tricristatus*; Frost 1992: 43 (part).
Stenocercus tricristatus; Torres-Carvajal 2007a: 161, 2007b: 317.
non *Ophryoessoides tricristatus*; Cunha 1981: 4.

Material examined. Holotype, MNHN 6825, M, Brazil, leg. Claussen.

Diagnosis. *Stenocercus tricristatus* is characterized by the following combination of features: (1) Dorsal head scales keeled. (2) Interparietal distinct, moderately enlarged; posterior head scales variable in size. (3) Internasals six. (4) No distinctly enlarged supraoculars. (5) An enlarged canthal at each side, in contact anteriorly with two elongate scales that form a double canthal ridge. (6) An enlarged, prominent, obtusely pointed scale immediately posterior to supraciliaries; no projecting, blade-like, angulate temporal scales. (7) Gulars and ventrals distinctly keeled. (8) Parietal eye distinct. (9) Neck folds absent. (10) Dorsals phylloid, keeled, mucronate and imbricate; scales on flanks similar to dorsals. (11) A moderately prominent, serrate vertebral crest, and a slightly less prominent dorsolateral crest (a less conspicuous lateral crest may be present, but it is not very clear from the present condition of the specimen). (12) Mite

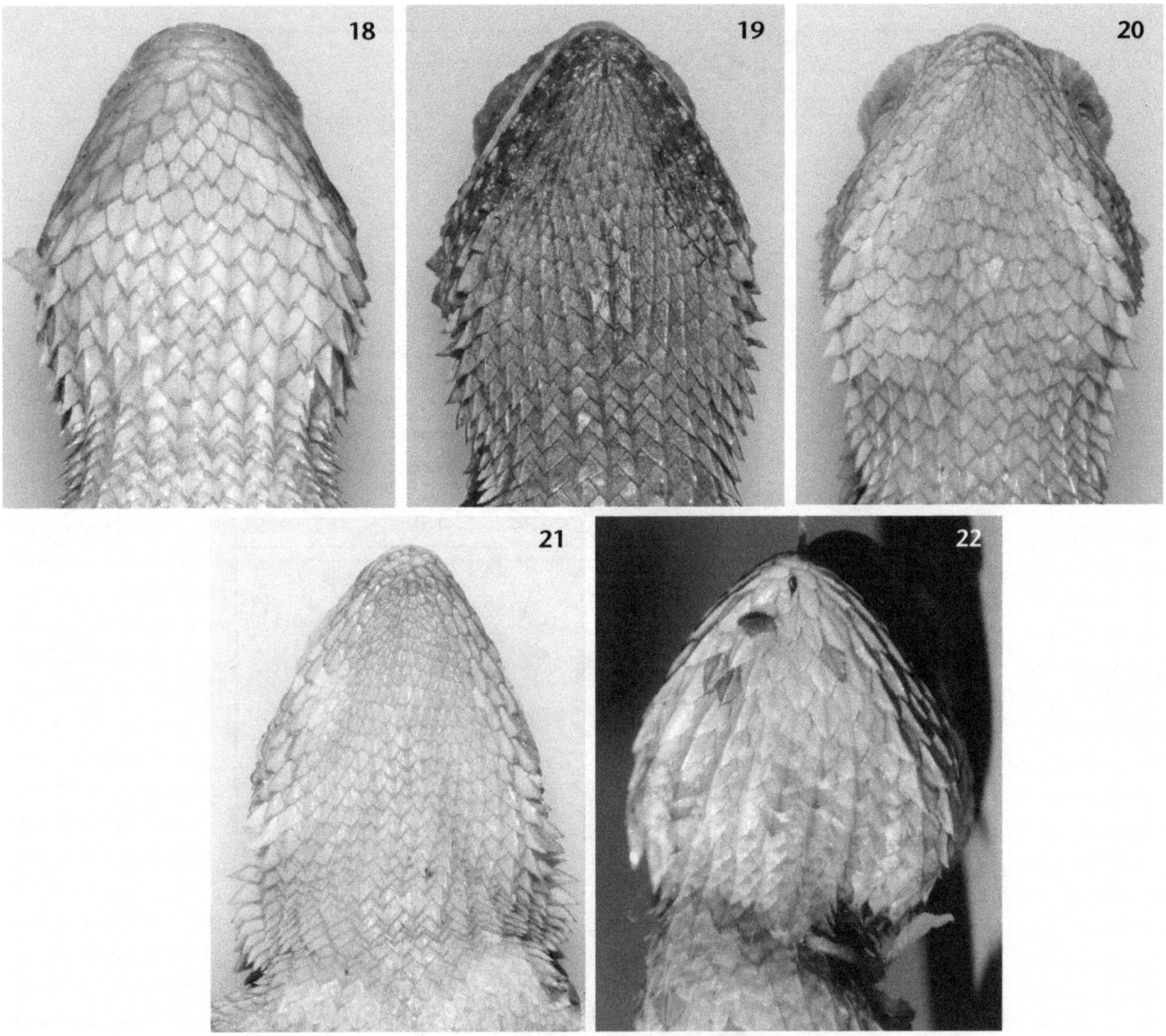

Figures 18–22. Ventral view of head of: (18) *Stenocercus canastra* sp. nov., MZUSP 88873, holotype; (19) *S. dumerilii*, MPEG 6036; (20) *S. quinarius*, MZUSP 94069, holotype; (21) *S. squarrosus*, MZUSP 94056, holotype; (22) *S. tricristatus*, MNHN 6825, holotype.

pockets absent. (13) Scales on posterior surface of thighs imbricate, keeled. (14) Tail moderately compressed, verticils absent. (15) Dorsal coloration probably with numerous dark brown spots on back and flanks; head with at least a large, triangular spot on posterior part of snout. (16) Scales around midbody 33, ventrals between anterior margin of forelimbs and anterior margin of hind limbs 23. (17) No distinctly enlarged scale on anterior margin of ear-opening.

Stenocercus tricristatus is distinguished from all other *Stenocercus* except *S. canastra* sp. nov., *S. quinarius*, *S. squarrosus* and *S. dumerilii* by the presence of an enlarged, prominent post-supraciliary scale; from all others except *S. canastra* sp. nov., *S. quinarius* and *S. squarrosus* by a moderately enlarged interparietal (although not as large as in the Tropidurini). It is distinguished from *S. dumerilii*, *S. quinarius* and *S. squarrosus* (character states in parentheses) by the presence of a prominent, serrate vertebral crest (in contrast to a low vertebral crest); two supraciliaries (4, rarely 3); absence of two distinctly enlarged upper temporals (two distinctly enlarged upper temporals); preauricular scale projecting over the tympanum, keeled (preauricular scale not or only slightly projecting over the tympanum, smooth); dorsals distinctly keeled and mucronate (dorsals with a low keel,

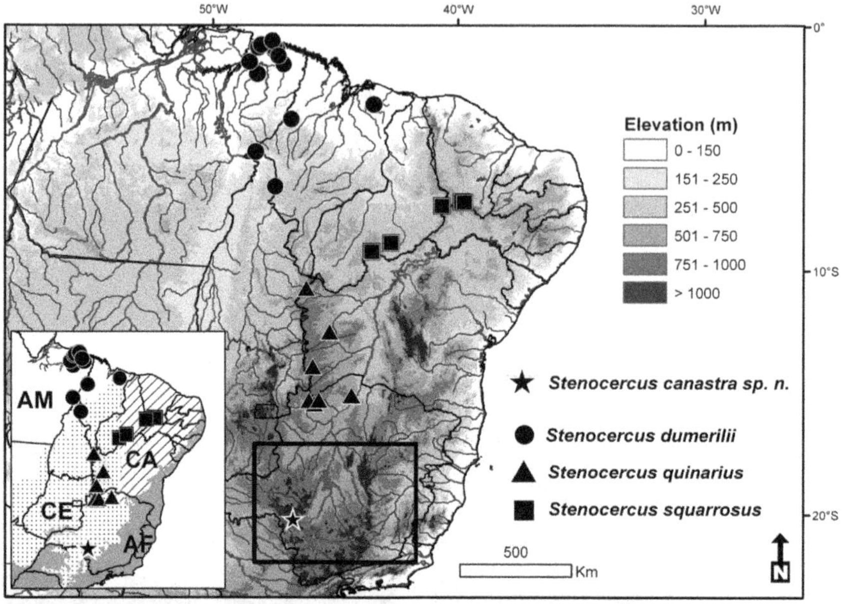

23

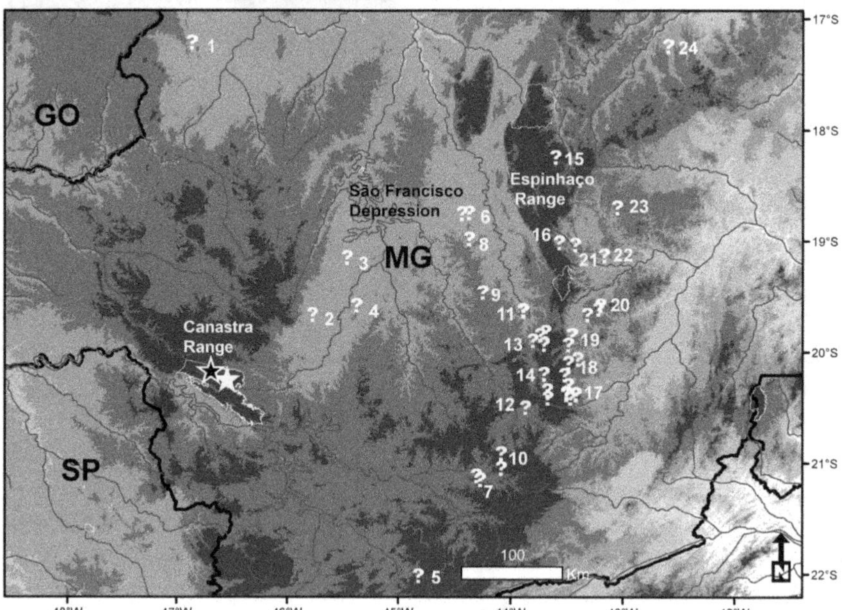

24

Figures 23–24. Distribution of the new species and related horned *Stenocercus*: (23) Distribution according to altitudinal variation. Inset – Distribution in major Neotropical phytogeographic domains. (AM) Amazonia; (CA) Caatinga; (CE) Cerrado; (AF) Atlantic Forest. (24) Enlarged area around the type-locality of *Stenocercus canastra* sp. nov. (white star = type-locality, black star = Chapadão da Zagaia), showing major massifs (above 1000m, in dark grey) and the São Francisco depression (light grey). White "?" represent localities visited or mentioned in Claussen (1841), Castelnau (1949) and Luna (2007). Localities are numbered from West to East, with groups of clustered localities represented by a single number: (1) Paracatu; (2) Baú; (3) Abaeté; (4) Extrema; (5) Bananal; (6) Curvelo, Fazenda Porteirinhas (farm owned by Claussen), and Piranga; (7) Brumado and São João Del Rei; (8) Maquiné; (9) Sete Lagoas, Fazenda Melancias (farm owned by Claussen); (10) Prados and Lagoa Dourada; (11) Santa Luzia and Lagoa Santa; (12) Congonhas do Campo; (13) Sabará, Cuiabá, Morro Vermelho and Serra da Piedade; (14) Cachoeira do Campo (includes farm owned by Claussen), Gongosco and São Vicente; (15) Serro do Frio; (16) Itacolomi de Mariana; (17) Ouro Preto, Sumidouro and São Sebastião; (18) Mariana, Bento Rodrigues, Caraça and Catas Altas; (19) Brucutu; (20) Itabira, Chapada and Santana; (21) Ouro Fino; (22) Santa Rita; (23) Pinheiro; (24) Minas Novas. Protected areas are shown as light polygons. (MG) Minas Gerais state; (SP) São Paulo state; (GO = Goiás state.

not or hardly mucronate); 7–8 scales across midbody from one dorsolateral row to the other (11–13 in *S. dumerilii*, 13–15 in *S. quinarius* and *S. squarrosus*); ventrals between anterior level of fore- and hind limbs 22 (28–32 in *S. dumerilii*, 30–34 in *S. quinarius*, 28–34 in *S. squarrosus*); scales on chin subequal and imbricate (scales on chin smaller, poligonal and subimbricate anteriorly, grading into larger, pointed, and imbricate posteriorly); tail 1.7 times SVL, slightly depressed near base (at most 1.4 times, compressed near base); and color pattern (Table 1). From *S. canastra* sp. nov. it differs by its wider head when seen from the ventral side (0.96 times as wide as long versus 0.78–0.89 times); 33 scales around midbody (39–41); preauricular scale about as large as adjacent temporals, keeled (distinctly larger, smooth); and no pre-subocular larger than adjacent scales (a larger pre-subocular in *S. canastra* sp. nov.). Besides, it reaches probably a smaller adult size (57 mm versus ≥ 70 mm in adult males of *S. canastra* sp. nov.).

Redescription. Holotype. The holotype and only known specimen of *S. tricristatus* is an adult male (as evidenced by the distinctly swollen base of the tail) with the following measurements: SVL 58 mm, tail length 100 mm (1.7 times SVL), head length 14.1 mm (24% of SVL), head width 13.6 mm, head height 10.6 mm (head 1.04 times as long as wide, 1.28 times as wide as high), forelimb 28 mm (48% of SVL), hind limb 39 mm (67% of SVL), tibia 9.4 mm (16% of SVL). Snout bluntly pointed. Canthus rostralis well defined, continuous with supraciliaries, and followed by a prominent post-supraciliary, so that the head has the shape of a four-sided pyramid. Neck narrower than head and body. Body and proximal part of tail slightly compressed, tail distally cylindrical.

Rostral roughly triangular, largest width approximately four times the medial height, only barely visible from above. Postrostrals 4. Snout, supraocular, and interorbital regions covered with irregularly polygonal, subimbricate to juxtaposed, strongly keeled scales, variable in size. Six scales between nasals, of which one supranasal at each side and four scales between them. Scales across the snout between canthals 4 (including one small, asymmetrical). One large canthal at each side, anteriorly in contact with two elongate, keeled scales (one of which is the supranasal); canthal posteriorly in contact with the first supraciliary. Supraorbital semicircle formed by 6–7 scales (between canthal ridge and post-supraciliary scale), posterior scale largest; the third supraorbital scale on the left side is in medial contact with the fourth on the right side. Supraoculars irregularly polygonal, subimbricate, strongly keeled, none distinctly enlarged; three in a transverse line across the greatest width of each orbit. Two elongate supraciliaries in contact with each other (narrowly overlapping), with a third supraciliary lateral to, and in contact with, the other two. Canthals and supraciliaries form a distinct ridge that ends in a prominent, obtusely pointed scale immediately posterior to supraciliaries. This ridge delimits a sharp angle between dorsal and lateral surfaces of head. Interparietal moderately enlarged, roughly rhomboid, parietal eye distinct; at each side a smaller parietal. Post-parietal region with irregularly polygonal and strongly keeled scales, heterogeneous in size: two relatively small scales immediately posterior to the interparietal, three moderately large (but smaller than the interparietal) scales on the left side, and two scales on the right side, of which the posterior is about as large as, or slightly larger than, the interparietal. No projecting angulate temporals (as described by Cadle, 1991; Fig. 11).

Nasals lateral to canthal ridge, undivided, separated from the rostral by the postrostrals; nostril in posterior part of nasal, directed latero-posteriorly. Loreal region with a distinct row of lorilabials, anteriorly and posteriorly forming a single row of scales, medially a double row; between lorilabials and canthal ridge a few scales of variable size, irregularly polygonal, with small keels and/or variably rugose surface. Scales in a transverse row between canthal and supralabials, 3–4. An elongate subocular, preceded by a small pre-subocular, both keeled, the keels closer to the upper margin of the scales, approximately aligned. Elongate subocular separated from supralabials by the row of lorilabials. Supralabials 5, very narrow and mostly keeled on their lower margin. Temporal region with polygonal (mostly rhomboid to hexagonal), keeled, imbricate scales, in approximately oblique rows, upper scales larger. The upper three scales in a row between the lateral side of the post-suparciliary and the posterior margin of the head (in direction to, but not reaching, the ear-opening) are the largest and have a prominent keel. Ear-opening obliquely oval, anteriorly bordered by two main scales projecting over tympanum; the mid-anterior one is keeled and slightly larger than the one dorsal to it, but not larger than adjacent temporal scales; tympanum slightly recessed (Figs 9, 14).

Mental relatively small, only slightly larger than medial post-mentals. Four post-mentals, two of which very narrow, in contact with first and second infralabials at each side. Infralabials 6, very narrow and keeled. Scales on chin anteriorly polygonal, keeled and slightly imbricate, posteriorly (including gulars) phylloid, distinctly keeled, mucronate, and imbricate, the scales on the lateral rows slightly larger. The keels of the phylloid scales are mostly aligned in longitudinal rows, which also continue along neck and belly; anteriorly on the chin these rows converge toward the mental. Gulars between ventral edges of ear-openings 12. No gular or lateral folds (Fig. 22).

Scales on nape medially relatively small, subimbricate, and keeled, dorsolaterally similar to dorsals but shorter, some of them with a prominent mucron. Sides of neck with relatively large scales, each with a well-developed mucron. Dorsals and scales on flanks phylloid, keeled, mucronate, and imbricate, mostly in approximately longitudinal rows. A serrate vertebral crest of enlarged, strongly keeled, and mucronate scales is present from nape to base of tail. A dorsolateral row of similar, but slightly smaller, scales on each side forms a lower, serrate crest also from nape to base of tail. The presence of a third, less prominent longitudinal crest, on each side, between fore- and hind limbs, is dubious (throughout most of the body the external cuticle is missing, which makes some characteristics less distinct, but the scales on this lateral row seem to be slightly larger than the scales on neighboring rows). Vertebral and dorsolateral crests are separated by 2–3 scales in a transverse line at midbody (7–8 scales from one dorsolateral crest to the other). Dorsolateral crest and lateral row of slightly enlarged scales are separated by four rows of scales. Vertebral crest with 27 scales from occiput to posterior margin of hind limbs. Paravertebral scales between the same points 34. Ventrals similar to dorsals, forming distinct longitudinal rows; 22 ventral scales along a midventral line between anterior level of forelimbs and anterior level of hind limbs. Scales around midbody 33. Scales on preanal plate similar to dorsals, but slightly smaller. Mite pockets absent.

Scales at base of tail similar to dorsals and ventrals. Distally the scales grade into polygonal shapes and become arranged in transverse (slightly oblique) rows; the crests disappear, but the ridges formed by the keels, on the ventral surface, continue to the tip of the tail. No distinct verticils (which probably indicates that autotomic segments are absent).

Limbs with phylloid, strongly keeled, mucronate, and imbricate scales, similar to, or smaller and more elongate than, dorsals. The mucron is better developed on the upper aspects of hind limbs. Subdigital lamellae single, mostly tricarinate; 14–15 under fourth finger, 18–19 under fourth toe.

Color pattern in the holotype has largely faded out, but the description (p. 531) and drawing (plate XXII, fig. 1) of Duméril (1856), together with what is still visible in the specimen, indicates that the color pattern would be similar to that of *S. canastra* sp. nov. (Fig. 1). According to the original description, the animal had a fawn color dorsally, with darker brown bands finely bordered by yellowish-white. The specimen at present has a tan ground color. A dark brown spot, with a thin black margin, on posterior part of snout and anterior to supraocular region is still very conspicuous; unlike that of *S. canastra* sp. nov. (which is rhomboid), the shape is triangular, pointing posteriorly (Fig. 11). On the posterior part of head there are three relatively small dark brown spots, arranged in a symmetrical way; the three together remind a stylized butterfly, with the central one (on the two scales behind the interparietal) representing the body, and the two lateral ones the partial contour of the wings. A dark band runs from lower margin of eye to labials. Between the vertebral and dorsolateral crests there are a few irregular, oblique, dark brown lines, in alternating positions left and right to the vertebral crest. Flanks with alternating tan and whitish vertical bands. Dark brown bands across hind limbs are still well visible, less so on the forelimbs. Color pattern on tail well preserved, proximally with middorsal irregular dark brown spots, each bordered by a thin black line, and lateral vertically elongate spots of similar color. Distally the middorsal and lateral spots fuse into transverse bands. Ventral region immaculate white under head and body, dirty white under limbs and tail.

Distribution. *Stenocercus tricristatus* was described by A. Duméril as coming from Brazil, with no further specification of locality (Figs 23–24). However, the specimen had been collected by [Peter] Claussen (also spelled Clausen), a Danish who lived in Brazil for about 20 years. Even though he travelled to other places in his first years in the country, during most of his stay in Brazil he lived and owned farms in Minas Gerais, in Curvelo (Fazenda Porteirinhas; Luna Filho 2007) and in Cachoeira do Campo (Castelnau 1949), and all evidences indicate that his scientific contributions, including some geological publications and collections of fossils, plants and animals, refer to Minas Gerais, even though in many cases only "Brazil" is indicated as origin (British Museum 1904, 1906, Papavero 1971). Claussen visited and collected fossils at several of the caves where Peter W. Lund worked, mostly between Curvelo and Lagoa Santa, at the Rio das Velhas drainage, Minas Gerais (Luna Filho 2007). All these localities lie east of the São Francisco river, but he also visited areas (e.g., Abaeté) in the western margin of this river, for geological studies (Claussen 1841). In 1843 Claussen joined the expedition led by Count F. de Castelnau in Ouro Preto (Castelnau 1949) and left the expedition while still in Minas Gerais (according to a letter from Lund, in Luna Filho 2007). Although Warming (1880) indicates that Claussen accompanied Castelnau from Rio de Janeiro to Sabará, Castelnau's report of the expedition in Brazil clearly states that they met in Ouro Preto, and Claussen is mentioned only while the expedition was travelling from Ouro Preto to Sabará, in the central portions of Minas Gerais (Castelnau 1949). No mention of Claussen is made later, while Castelnau travelled west, on his way from Sabará to Catalão, Goiás, central Brazil (Castelnau 1949). Papavero (1971: 91) mentioned the following localities visited by Claussen, all in the State of Minas Gerais: Ouro Preto, Cachoeira do Campo, Curvelo, Itabira, Itacolomi, Serra do Caraça, and Rio São Francisco. These data corroborate the idea that *S. tricristatus* comes from the State of Minas Gerais, most likely from its central, upland areas in or near the Espinhaço range, where most localities cited by Claussen are located, including the two farms where he lived while collecting in Brazil (Luna Filho 2007). Among the many localities visited or mentioned by Claussen, no localities are found in upland areas near Serra da Canastra, and all localities west of the São Francisco River are found in low lying areas, below 500 m.

Habitat. Unknown. Considering the possibility of its occurrence in Minas Gerais (see below) and data on other horned *Stenocercus*, it could be expected to occur either in cerrado vegetation or in shrublands on rock outcrops, both of which are widespread in these areas.

DISCUSSION

Stenocercus is a complex and heterogeneous genus, with 68 species known at present (Uetz et al. 2017), of which ten occur in Brazil (Costa and Bérnils 2015). The five species studied here, however, differ from all other *Stenocercus* by the presence of a pyramidal head and enlarged, dorsally projected post-supraciliary. Torres-Carvajal 2007b, in a phylogeny of the genus based on molecular and morphological characters, recovered *S. dumerilii* and *S. tricristatus* as sister taxa, but *S. quinarius* and *S. squarrosus* were not included in the analysis. Teixeira et al. (2016) constructed a new molecular phylogeny based on the same mitochondrial genes studied by Torres-Carvajal (2007b), where they included *S. dumerilii*, *S. quinarius*, *S. squarrosus* and also *S. canastra* sp. nov. (named *Stenocercus* cf. *tricristatus*). They recovered this latter species as sister to the clade (*S. dumerilii* (*S. quinarius*, *S. squarrosus*)), dating the divergence between *S. canastra* sp. nov. and the other three species in the Miocene, and among these three in the Pliocene/Pleistocene. In agreement with Teixeira et al. (2016), our data suggest that horned *Stenocercus* form two groups, one including *S. tricristatus* and *S. canastra* sp. nov., with body not or only slightly depressed, longer tails, and a serrated vertebral crest, among other characteristics, that occupies the southern portion of the horned *Stenocercus* distribution. The second group includes *S. dumerilii*, *S. quinarius* and *S. squarrosus*, with depressed body, shorter tails, and a low vertebral crest, and occupies the northern portion of the distribution.

Stenocercus canastra sp. nov. is most similar to *S. tricristatus*, which is also probably its geographically closest congener. In spite of the fact that *S. tricristatus* is only known from the holotype, the consistent differences between the series of *S. canastra* sp. nov. specimens and the holotype of *S. tricristatus*, especially in body size, shape of preauricular scales, head shape and number of scales around midbody (see Fig. 4, Table 1), indicate that they do not belong to the same species. Midbody scale counts in *S. tricristatus*, for instance, falls far outside the observed range in *S. canastra* sp. nov. Differences in snout-vent length and shape of the head could be explained if the holotype of *S. tricristatus*, with 57 mm SVL, would be a juvenile. However, it has a distinctly swollen base of tail, indicating well-developed hemipenes, which was confirmed by a small dissection at the base of the tail. Besides, the specimens of *S. canastra* sp. nov. varies from 42 mm to 77 mm SVL, among which the only female is 65 mm SVL. There is a male of 61 mm SVL which is a half-grown, and the remaining adult males are 71 mm SVL or larger. Finally, the only specimen from Serra da Canastra that has characteristics of a juvenile (SVL 42 mm), with disproportionate head and limbs in relation to body length, appeared far from the single specimen of *S. tricristatus* in our PCA (Fig. 4). Moreover, this juvenile specimen presents dorsal scales with less developed keels and longitudinal crests than adults of the same species, as well as in relation to the holotype of *S. tricristatus*, where dorsals are strongly keeled and mucronate (see Figs 1–3). Thus, the holotype of *S. tricristatus*, despite its small size, clearly differs from the juvenile of *S. canastra* sp. nov., indicating that differences between *S. tricristatus* and *S. canastra* sp. nov. are not due to ontogenetic variation. All other differences, in scale shape and numbers, are not known to vary with age or size. We also considered the possibility that some of the differences are due to the loss of the Oberhäutchen layer of the epidermis in the holotype of *S. tricristatus*. This may have affected in some way the shape of the post-supraciliary scale and the lateral crest, but other scale differences, such as the larger size of the preauricular scale in *S. canastra* sp. nov., are not a result of such a loss. Two other characteristics may differ between the two species, but needs confirmation with the examination of additional (not yet available) specimens: the shape/size of scales on chin (as observed in Figs 18 and 22) and the shape, number and position of dark spots on head (Figs 10 and 11). Given all the evidence above, we conclude that *S. canastra* sp. nov. and *S. tricristatus* are two distinct species, diagnosed by a suite of meristic and morphometric characters.

Stenocercus tricristatus has not yet been assessed by the IUCN global red list (IUCN 2016). In Brazil, it was recently assessed as Data Deficient (DD) because its distribution is poorly known (ICMBio 2015). This assessment wrongly assigned specimens of *S. canastra* sp. nov. from Serra da Canastra deposited at the University of Brasília collection (CHUNB) to *S. tricristatus*; however, if corrected, this change would not affect its categorization. In relation to *S. canastra* sp. nov., the vegetation and soil types where the species was found covers at least 200 km^2 of the northern part of the Serra da Canastra National Park (SCNP). If this area were assumed as its Area of Occupancy (IUCN 2012), it would be a candidate for a threat category (Endangered) under criterion B. However, its known distribution is inside a well-protected national park and thus continuing decline in its population, distribution or habitat is not expected. Although bushfires are common at the SCNP, we found five out of the six specimens of the type series of *S. canastra* sp. nov. just after a fire; all individuals were hidden inside burrows and termite mounds, indicating that the species is not severely affected by bushfires (that are a natural feature of its habitat). Thus, at present, *S. canastra* sp. nov. should be assessed as Least concern.

Although *S. canastra* sp. nov. is known only from the northern portion of the park (Chapadão da Canastra), considering that the species is hard to find, it is possible that it occurs also in its southern part (Chapadão da Babilônia), that comprises 65% of the area of the SCNP (approximately 200,000 ha) and is still poorly known regarding lizard diversity. On the other hand, land tenure is regularized only in the northern part of the SCNP, while the southern portion is still under legal dispute. If this latter area is removed from the park, *S. canastra* sp. nov. might be affected.

The distribution in upland areas and plateus is typical in Stenocercini (see Torres-Carvajal 2007b), and except for *S. dumerilii*, all other species of horned *Stenocercus* are found on uplands or plateaus with at least 500 m.a.s.l. *Stenocercus squarrosus* is typical of isolated upland areas (Serra das Confusões, Araripe and Ibiapaba complex, see Fig. 23) scattered in the predominantly lowland Caatinga, while *S. quinarius* is largely confined to the Serra Geral plateau, a large sandstone tabletop in the eastern portion of the Cerrado (Nogueira and Rodrigues 2006, see Fig. 23). Minas Gerais is a state with a very complex topography and geomorphology (see Castelnau 1949), with a number of hills and isolated plateaus with elevations above 800 m.a.s.l. and some peaks above 1800 m.a.s.l. Two upland blocks are largely separated by the São Francisco depression. Serra da Canastra, where *S. canastra* sp. nov. is found, lies to the west of the São Francisco depression, and Serra do Espinhaço, to the east. Although the exact provenance of the type specimen of *S. tricristatus* remains unknown, most localities mentioned by Claussen are clustered in a small upland portion of Minas Gerais, along Serra do Espinhaço (near Caraça, Ouro Preto, Cachoeira do Campo, Itacolomi), most above 1000 m.a.s.l. The few localities visited by Claussen west of the São Francisco river are in low lying areas, all below 500 m.a.s.l. (see Fig. 24). One possibility therefore is that these two species occupy different upland areas, separated by the São Francisco depression, where only a few localities are cited by Claussen and where the presence of horned *Stenocercus* species is unlikely. Such a pattern of upland open areas harboring a complex and diverse herpetofauna, mostly with allopatric ranges, has already been detected for Squamates in the Cerrado (Nogueira et al. 2011) and for snakes in the Caatinga region (Guedes et al. 2014). The pattern here inferred for Stenocercini in Minas Gerais is probably similar to that observed in closely related amphibians, that also show narrow ranged endemics in different plateus in Minas Gerais and neighboring states, like the Hylidae *Ololygon tripui* (Lourenço, Nascimento & Pires, 2009), known from Ouro Preto, Serra do Espinhaço (Lourenço et al. 2009), and *Ololygon pombali* (Lourenço, Carvalho, Baêta, Pezzuti & Leite, 2013), from Serra da Canastra (Lourenço et al. 2013); and the Phyllomedusidae *Pithecopus ayeaye* (Lutz, 1966), from Serra da Canastra, Poços de Caldas, Minas Gerais, and Pedregulho, São Paulo (Araujo et al. 2007), and *P. megacephalus* (Miranda-Ribeiro, 1926), from Serra do Espinhaço, in Minas Gerais and Bahia (Brandão et al. 2012). Our findings, with two closely related species possibly found within a relatively small but topographically complex portion of south-central Minas Gerais, reinforce the importance of targeted, spatially accurate conservation strategies to safeguard the conservation of these isolated, easternmost allopatric components of Stenocercini, and highlight the relevance of taxonomic and distributional knowledge as basis for the conservation of the Neotropical herpetofauna, especially for its rare, restricted elements.

ACKNOWLEDGMENTS

We thank Otávio A.V. Marques for sharing with us information and photos of the species, and for critically commenting on the manuscript; O. Torres-Carvajal for comments on an earlier version of this manuscript; and the curators of MZUSP (H. Zaher), CHUNB (G. Colli), and MHNH (I. Ineich and L. Pierre-Huyet) for allowing access to specimens under their care. This study was partially funded by grants from Fundação de Amparo à Pesquisa do Estado de São Paulo (1995/09642-5, 2011/50206-9 and 2015/20215-7). Conselho Nacional de Desenvolvimento Científico e Tecnológico provided fellowships to TCSAP (312674/2013-9) and MM (306961/2015-6).

LITERATURE CITED

Araujo CO, Condez TH, Haddad CFB (2007) Amphibia, Anura, *Phyllomedusa ayeaye* (B. Lutz, 1966): distribution extension, new state record, and geographic distribution map. Check List 3: 156–158.

Avila-Pires TCS (1995) Lizards of Brazilian Amazonia (Reptilia: Squamata). Zoologische Verhandelingen Leiden 299: 1–706. Available online at http://www.repository.naturalis.nl/document/149074

Boulenger GA (1885) Catalogue of the Lizards in the British Museum (Natural History). London, Trustees of the British Museum, vol. 2, 497 pp. Available online at https://www.biodiversitylibrary.org/bibliography/21097

Brandão RA, Leite FSF, Françoso RD, Faivovich J (2012) *Phyllomedusa megacephala* (Miranda-Ribeiro 1926) (Amphibia, Anura, Hylidae, Phyllomedusinae): Distribution extension, new state record and map. Herpetology Notes 5: 535–537.

British Museum (1904) The history of the collections contained in the Natural history departments of the British museum. London, Trustees of the British Museum, vol. 1, 1442 pp. Available online at https://www.biodiversitylibrary.org/bibliography/20926 [Accessed: 26/03/2015]

British Museum (1906) The history of the collections contained in the natural history departments of the British Museum. London, Trustees of the British Museum, vol. 2, 782 pp. Available online at https://www.biodiversitylibrary.org/bibliography/20926 [Accessed: 26/03/2015]

Cadle JE (1991) Systematics of lizards of the genus *Stenocercus* (Iguania: Tropiduridae) from Northern Peru: new species and comments on relationships and distribution patterns. Proceedings of the Academy of Natural Sciences of Philadelphia 143: 1–96.

Cadle JE (2001) A new species of lizard related to *Stenocercus caducus* (Cope) (Squamata: Iguanidae) from Peru and Bolivia, with a key to the "*Ophryoessoides* Group". Bulletin of the Museum of Comparative Zoology 157: 183–221.

Castelnau F (1949) Expedição às regiões centrais da América do Sul. São Paulo, Companhia Editora Nacional, vol. 1, 383 pp. Available online at http://www.brasiliana.com.br/brasiliana/colecao/obras/133/Expedicao-as-regioes-centrais-da-America-do-Sul-v1

Claussen P (1841) Notes Géologiques sur la Province de Minas Gerais au Brésil. Bulletin de l' Académie Royale de Bruxelles 8: 1–22. Available online at https://books.google.com.br/books?id=UjxYAAAAcAAJ&pg=PA1&dq=claussen+notes+géologiques&hl=en&sa=X&ved=0CC4Q6AEwBWoVChMIwML20q6xyAIVBqweCh3jQQJe#v=onepage&q=claussen%20notes%20géologiques&f=false [Accessed 06/10/2015]

Costa HC, Bérnils RS (2015) Répteis brasileiros: Lista de espécies 2015. Herpetologia Brasileira 4: 75–93.

Cunha OR (1981) Lacertílios da Amazônia. VIII – Sobre *Ophryoessoides tricristatus* Duméril, 1851, com redescrição da espécie e notas sobre ecologia e distribuição na região leste do Pará (Lacertilia, Iguanidae). Boletim do Museu Paraense Emilio Goeldi, Zoologia, 108: 1–23. Available online at http://repositorio.museu-goeldi.br/handle/mgoeldi/673

Dietz JM (1984) Ecology and social organization of the maned wolf (*Chrysocyon brachyurus*). Smithsonian Contributions to Zoology 392: 1–51.

Duméril A (1856) Description des Reptiles nouveaux ou imparfaitement connus de la collection du Muséum d'Histoire Naturelle et remarques sur la classification et les caractères des Reptiles. Deuxième Mémoire. Archives du Muséum National d'Histoire Naturelle Paris 8: 437–588. Available online at https://www.biodiversitylibrary.org/bibliography/5035

Duméril AMC, Duméril A (1851) Catalogue méthodique de la collection des Reptiles du Muséum d'Histoire Naturelle de Paris. Paris, Gide et Baudry, 224 pp. Available online at http://gallica.bnf.fr/ark:/12148/bpt6k883907t

Duméril AMC, Bibron G, Duméril A (1854) Erpétologie générale ou Histoire naturelle complète des Reptiles. Paris, Librarie Encyclopédique de Roret, vol. 9, 440 pp. Available online at https://www.biodiversitylibrary.org/bibliography/45973

Etheridge R (1966) The systematic relationships of West Indian and South American lizards referred to the iguanid genus *Leiocephalus*. Copeia 1966: 79–91.

Fritts TH (1974) A multivariate evolutionary analysis of the Andean iguanid lizards of the genus *Stenocercus*. Memoirs San Diego Society of Natural History 7: 1–89.

Frost DR (1992) Phylogenetic analysis and taxonomy of the *Tropidurus* Group of lizards (Iguania: Tropiduridae). American Museum Novitates 3033: 1–68.

Guedes TB, Sawaya RJ, Nogueira CC (2014) Biogeography, vicariance, and conservation of snakes of the neglected and endangered Caatinga region, northeastern Brazil. Journal of Biogeography 41: 919–931. http://dx.doi.org/10.1111/jbi.12272

Hammer Ø, Harper DAT, Ryan PD (2001) PAST: Paleontological statistics software package for education and data analysis. Palaeontologia Electronica 4: 9 pp. Available online at: http://palaeo-electronica.org/2001_1/past/issue1_01.htm [Acessed: 01/10/2015]

ICMBio (2015) Lista das espécies consideradas com Dados Insuficientes (DD). Available online at: http://www.icmbio.gov.br/portal/faunabrasileira/lista-de-especies-dados-insuficientes [Acessed: 21/03/2017]

IUCN (2012) IUCN Red List Categories and Criteria. Gland, International Union for Conservation of Nature, v. 3.1, 2nd ed., 32 pp. Available online at: http://www.iucnredlist.org/technical-documents/categories-and-criteria/2001-categories-criteria [Acessed: 21/03/2017]

IUCN (2016) The IUCN Red List of Threatened Species – *Stenocercus*. Gland, International Union for Conservation of Nature, v. 2016.3. Available online at: http://www.iucnredlist.org/search [Acessed: 23/03/2017]

Lourenço ACC, Nascimento LB, Pires MRS (2009) A new species of the *Scinax catharinae* species group (Anura: Hylidae) from Minas Gerais, southeastern Brazil. Herpetologica 65: 468–479. https://doi.org/10.1655/07-088.1

Lourenço ACC, Carvalho ALG, Baêta D, Pezzuti TL, Leite FSF (2013) A new species of the *Scinax catharinae* group (Anura, Hylidae) from Serra da Canastra, southwestern state of Minas Gerais, Brazil. Zootaxa 3613: 573–588. http://dx.doi.org/10.11646/zootaxa.3613.6.4

Luna Filho PE (2007) Peter Wilhelm Lund: o auge das suas investigações científicas e a razão para o término das suas pesquisas. Doctoral Dissertation, Faculdade de Filosofia, Letras e Ciências Humanas, Universidade de São Paulo, São Paulo, 465 pp. https://doi.org/10.11606/T.8.2007.tde-09102007-142632

Miranda-Ribeiro A (1926) Notas para servirem ao estudo dos Gymnobatrachios (Anura) brasileiros. Arquivos do Museu Nacional 27: 1–227.

Nogueira C, Rodrigues MT (2006) The genus *Stenocercus* (Squamata: Tropiduridae) in extra-Amazonian Brazil, with the description of two new species. South American Journal of Herpetology 1: 149–165. http://dx.doi.org/10.2994/1808-9798(2006)1%5b149:TGSSTI%5d2.0.CO;2

Nogueira C, Ribeiro SR, Costa GC, Colli GR (2011) Vicariance and endemism in a Neotropical savanna hotspot: distribution patterns of Cerrado squamate reptiles. Journal of Biogeography 38: 1907–1922. https://doi.org/10.1111/j.1365-2699.2011.02538.x

Papavero N (1971) Essays on the history of neotropical Dipterology, with special reference to collectors (1750-1905). São Paulo, Museu de Zoologia, Universidade de São Paulo, 260 pp.

Peters JA, Donoso-Barros R (1970) Catalogue of the Neotropical Squamata. Part II. Lizards and Amphisbaenians. U.S. National Museum Bulletin 297: 1–293.

Queirolo D, Motta-Junior JC (2007) Prey availability and diet of maned wolf in Serra da Canastra National Park, southeastern Brazil. Acta Theriologica 52: 391–402.

Sabaj-Perez MH (2013) Standard symbolic codes for institutional resource collections in herpetology and ichthyology: an online reference. USA, American Society of Ichthyologists and Herpetologists, v. 4.0. Available online at http://www.asih.org/resources [Acessed: 11/07/2014, archived by WebCite® at http://www.webcitation.org/6Qzfl6ZjN]

Teixeira Jr M, Prates I, Nisa C, Silva-Martins NSC, Strüssmann C, Rodrigues MT (2016) Molecular data reveal spatial and temporal patterns of diversification and a cryptic new species of lowland *Stenocercus* Duméril and Bibron, 1837 (Squamata: Tropiduridae). Molecular Phylogenetics and Evolution 94: 410–423. https://doi.org/10.1016/j.ympev.2015.09.010

Torres-Carvajal O (2005) A new species of *Stenocercus* (Squamata, Iguania) from central-western Brazil with a key to

Brazilian *Stenocercus*. Phyllomedusa 4: 123–132. https://doi.org/10.11606/issn.2316-9079.v4i2p123-132

Torres-Carvajal O (2007a) A taxonomic revision of South American *Stenocercus* (Squamata: Iguania) lizards. Herpetological Monographs 21: 76–178. https://doi.org/10.1655/06-001.1

Torres-Carvajal O (2007b) Phylogeny and biogeography of a large radiation of Andean lizards (Iguania, *Stenocercus*). Zoologica Scripta 36: 311–326. https://doi.org/10.1111/j.1463-6409.2006.00284.x

Uetz P, Freed PHallermann J, Hošek J (2017) The Reptile Database – Tropiduridae. Available online at: http://reptile-database.reptarium.cz/advanced_search?taxon=Tropiduridae&submit=Search [Accessed: 07/12/2017]

Warming E (1880) Den Danske Botaniske Literatur Fra De Ældste Tider Til 1880. Botanisk tidsskrift 12: 42–247.

Author Contributions: TCSAP, CCN and MM examined and compared the specimens. MM made the statistical analyses, CCN elaborated the maps, and both were responsible for the ecological data. TCSAP wrote the first draft of the manuscript, which was revised and complemented by CCN and MM.
Competing Interests: The authors have declared that no competing interests exist.

2

Changes in trophic characteristics of two fish species of *Astyanax* (Teleostei: Characidae) in response to aquatic pollution

Mirella B. Alonso[1], Débora R. de Carvalho[1], Carlos B. M. Alves[2], Marcelo Z. Moreira[3], Paulo S. Pompeu[1]

[1] *Laboratório de Ecologia de Peixes, Setor de Ecologia, Departamento de Biologia, Universidade Federal de Lavras. Campus Universitário, 37200-000 Lavras, MG, Brazil.*
[2] *Laboratório Nuvelhas, Projeto Manuelzão, Universidade Federal de Minas Gerais. Avenida Antônio Carlos 6627, 31270-901 Belo Horizonte, MG, Brazil.*
[3] *Laboratório de Ecologia Isotópica, Centro de Energia Nuclear na Agricultura, Universidade de São Paulo. Avenida Centenário 303, 13416-000 Piracicaba, SP, Brazil.*
Corresponding author: Paulo Santos Pompeu (pompeu@ufla.br)

ABSTRACT. The trophic plasticity of most fish species of *Astyanax* Baird & Girard, 1854 in response to environmental changes and resource availability is high. This work evaluates the differences in the trophic characteristics of two congeneric species, *Astyanax taeniatus* (Jenyns, 1842) and *Astyanax lacustris* (Lütken, 1875), in Rio das Velhas Basin, which is highly impacted by the discharge of sewage from the Metropolitan Region of Belo Horizonte (MRBH). Eight sites were sampled and grouped into three regions: upper course (two sites upstream of the MRBH); middle course (three sites located in the middle portion of the Rio das Velhas, region with greater influence of the MRBH), and lower course (three sites downstream of the MRBH). Samples of fish and food resources were collected from all sites to obtain the isotopic composition of nitrogen ($\delta^{15}N$) and carbon ($\delta^{13}C$), and the stomach contents of the two species was analized. The most common items in the stomach of *A. lacustris* and *A. taeniatus*, respectively, were from plants and insects, followed by algae/periphyton (especially at the low course of Rio das Velhas). In contrast, stable isotope analyses indicated that algae (in polluted sites) and periphyton (in least-disturbed sites) were best assimilated both species. Both analyses indicated that the trophic niches of the two species overlap more in more polluted sites relative to less polluted sites. *Astyanax taeniatus* and *A. lacustris* only presented different isotopic composition of carbon and nitrogen in the upper course of the Rio das Velhas, probably in response to the greater diversity of food items consumed by each species. In the other regions, the species presented similar isotopic signatures, with $\delta^{15}N$ and $\delta^{13}C$ notably enriched in the most polluted regions (middle and low course). Our results suggest that pollution acts by increasing trophic niche overlap of these species, altering the type of resources most assimilated, and promoting a greater enrichment of $\delta^{15}N$ in fish and resources.

KEY WORDS. *Astyanax lacustris, Astyanax taeniatus*; carbon, nitrogen enrichment, stable isotopes, stomach contents.

INTRODUCTION

In many developing countries, a large proportion of untreated raw sewage is released into aquatic environments (Macedo and Sipaúba-Tavares 2010), increasing the load of organic matter and pollutants in rivers, which are considered as the main drivers of artificial eutrophication in these environments (Tundisi and Tundisi 2008). Nutrients from human activities, when released into the water, contribute to the rapid growth of algal blooms and aquatic plants, altering the physico-chemical and ecological conditions of aquatic systems (Pereira and Mercante 2005, Hicks et al. 2016, Baeta et al. 2017). Among the consequences of an increase in primary productivity is the rapid reduction in water oxygen levels, which has drastic effects on fish and invertebrate communities (Macedo and Sipaúba-Tavares 2010, Baeta et al. 2017). In addition, changes in primary productivity in response to pollutants affect directly the diets of these aquatic consumers (Cabana and Rasmussen 1996, Esteves and Aranha 1999), and may also be responsible for promoting the local extinction of specialist and less tolerant species.

By favoring primary productivity, environmental pollution of aquatic systems may homogenize the type of resources available to organisms in higher trophic levels. This process of homogenization in aquatic systems has been also described in several taxonomic groups such as diatoms, zooplankton and macroinvertebrates (Lougheed et al. 2008). Such changes in the balance of available resources may consequently affect the food web since changes in nutritional composition or abundances of basal food sources can induce shifts in primary consumers or even their exclusion (Hall 2004). However, the effects of this homogenization of producer communities on upper trophic levels remains unclear.

According to ecological theory, generalist species are less sensitive to environmental change than specialists as they have the capacity of varing their diet according to the availability of resources present in their respective habitats (Tundisi and Tundisi 2008). Therefore, trophic plasticity is an important strategy to allow species to tolerate changes in environment condition (Gerking 1994, Wootton 1999). Species of the American fish *Astyanax* Baird & Girard, 1854 are well known for their broad geographic distribution and their ability to inhabit environments with differing levels of preservation, including highly polluted environments (eg., Souza and Lima-Júnior 2013, Carvalho et al. 2015). *Astyanax* is composed of approximately 100 species that are distributed from the southern United States to northern Argentina (Eigenmann 1921, Géry 1997, Weitzman and Fink 1983). Most species have omnivorous feeding habits, with diets composed of animal and vegetable items, of both autochthonous and allochthonous origins (e.g., Esteves 1996, Vilela et al. 2002, Cassemiro et al. 2002, Bennemann et al. 2005). In addition, some species of this genus present generalist feeding habits and high trophic plasticity in response to environmental changes and resource availability (Lobón-Cerviá and Bennemann 2000, Carvalho et al. 2015), which increases their chance of survival in disturbed habitats (Menezes et al. 2007). However, congeneric species (species of the same genus) may respond differently to changes in the aquatic environment and the availability of resources.

One way to identify how distinct species respond to changes in the environment is by comparing their feeding habits in regions under differing levels of human disturbance (e.g, Carvalho et al. 2015). Accordingly, analyses of stomach contents and stable isotopes (carbon and nitrogen), can be used simultaneously for robust and reliable assessment of feeding habits (e.g., Carassou et al. 2017, Connan 2017). Stomach contents analyses provide useful taxonomic information on consumed prey items (Beaudoin et al. 1999). However, there are often uncertainties in the identification of such items, due to the different stages of digestion of food items, and that not all ingested items are in fact assimilated into biomass (Manetta and Benedito-Cecílio 2003). Stable isotopes analyses, on the other hand, provide information on the energy flow in food chains (Peterson and Fry 1987, Kling et al. 1992). The nitrogen ($\delta^{15}N$) isotope is consistently fractionated throughout the trophic web, allowing researchers to make inferences about the trophic relationships of consumers with their diet (Vander Zanden et al. 1997, Post 2002, Vanderklift and Ponsard 2003). The carbon ($\delta^{13}C$) isotope, in turn, allows to delineate the energy flow in environments that present several foods with different carbon values (Manetta and Benedito-Cecílio 2003).

Based on this information, we aimed to evaluate how trophic characteristics of two congeneric species, *Astyanax lacustris* (Lütken, 1875) and *Astyanax taeniatus* (Jenyns, 1842), change across an environmental pollution gradient. The diet and the trophic niches occupied by these two species were evaluated in different regions of a highly disturbed Neotropical river basin, the Rio das Velhas, south east Brazil. The main source of disturbance in this river basin is the discharge of untreated domestic and industrial sewage from a large nearby urban conurbation. We tested the following hypotheses: 1) Under natural (undisturbed) conditions, the congeneric species occupy different trophic niches, and consequently present little food overlap; 2) However, along a gradient of pollution, due to the simplification (homogenization) of the available resources, and due to their high trophic plasticity, both species will increase their food overlap and will present more similar isotopic signatures.

MATERIAL AND METHODS

The study was conducted in the Rio das Velhas Basin, southeast Brazil, with sampling sites located in the main channel of the Rio das Velhas. The Rio das Velhas is the largest tributary of the São Francisco river Basin (Alves and Pompeu 2001), and is located entirely in the territory of Minas Gerais state (CETEC 1983), covering 51 municipalities. The basin, with a drainage area of 29,173 km^2, has an average annual flow rate at its mouth of 300 m^3/s and average width of 38.3m (CETEC 1983). The Rio das Velhas is of significant economic and social importance. Its upper course is located at the Metropolitan Region of Belo Horizonte (MRBH); the third largest urban conurbation in Brazil, with almost six million inhabitants, and is the main water supply.

Eight sites were sampled along the Rio das Velhas channel (RV-01 to RV-08), which were divided into three regions (upper, middle and lower course). The upper course of the Rio das Velhas (Upper RV) corresponds to the region with the best water quality (RV-01 and RV-02). The middle course (Middle RV) is in the region with the greatest influence of the MRBH, characterized by the discharge of large amounts of domestic and industrial sewage (RV-03, RV-04, and RV-05). The lower course (Low RV), in turn, is the most distant region from the MRBH and is close to the river mouth (RV-06, RV-07, and RV-08). In this region the river partly recovers its quality, due to the presence of numerous well preserved tributaries (Alves and Pompeu 2001) (Table 1, Fig. 1).

Two sewage treatment plants (STP), Arrudas and Onça, were also sampled to obtain complementary samples of the suspended material to obtain the isotopic composition of the raw

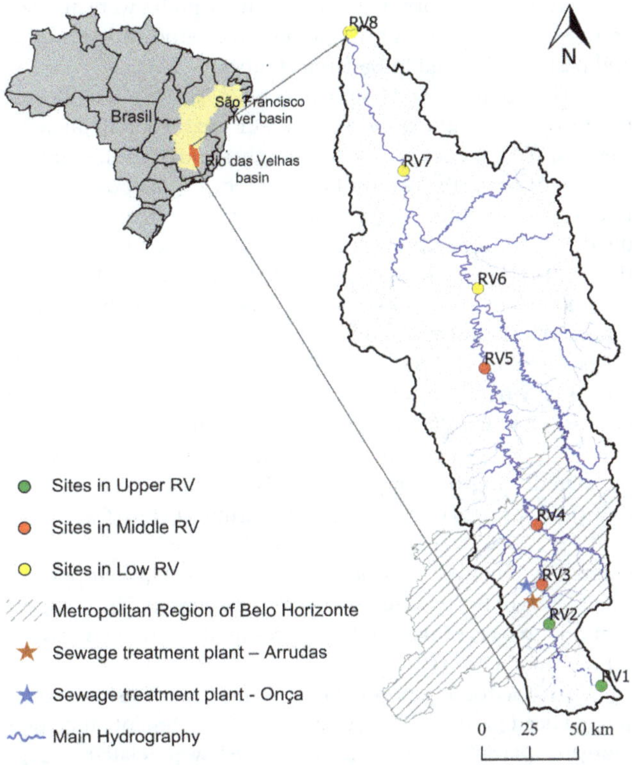

Figure 1. Sampling network in the Rio das Velhas Basin, Minas Gerais, Brasil.

sewage. All the sites were sampled between the years 2015 and 2016, in the dry (May to August) and wet (October to January) seasons (Table 1, Fig. 1).

The information about degradation level of sampling sites was obtained through data from literature (Feio et al. 2015). The sites RV08, RV11, RV10, RV12, RV13, RV14, RV15 and RV16 (Feio et al. 2015) were considered as the correspondents of RV-01, RV-02, RV-03, RV-04, RV-05, RV-06, RV-07 and RV-08, respectively (Table 1). Degradation levels range from I (preserved) to IV (degraded).

Data about water quality, hypereutrophic condition, toxic contamination and pressure factors in the study sites were accessed through IGAM's website (http://portalinfohidro.igam.mg.gov.br), which monitors water quality quarterly at several points across the Rio das Velhas Basin. The IGAM monitoring sites: BV001, BV139, BV105; BV137; BV141, BV150, BV151 and BV149, were considered as sampling points: RV-01, RV-02, RV-03, RV-04, RV-05, RV-06, RV-07 and RV-08, respectively (Table 1). Mean values of conductivity, dissolved oxygen, total ammoniacal nitrogen and total phosphorous were obtained from IGAM measurements carried out in the years 2015 and 2016. The hypereutrophic condition, toxic contamination and pressure factors acting in the study sites were extracted from the quarterly report of the year 2017.

Table 1. Geographic location (in degrees/minutes/seconds and UTM, date, altitude and municipality) and water quality of the sampling sites sampled in the main channel of Rio das Velhas. Cond.: Condutivity (µS/cm), D.O.: Dissolved oxygen (mg/l), Am. nitr.: Ammoniacal nitrogen (mg/l), Phosp.: Total phosphorus (mg/l), Tox. contam.: Toxic contamination, Deg. level: degradation level ranging from I to IV (Feio et al. 2015). *Sites with hypereutrophic condition according IGAM.

Characteristics of sampling sites						Water Quality						
Regions	Sampling points	Date of sampling	Coordinates	Altitude (m)	Municipality	Cond.	D.O.	Am. nitr.	Phosp.	Tox. contam.	Deg. level	Pressure factors
Upper RV	RV-01	20/08/2015 21/01/2016 10/06/2016	20°18'42.8"S 43°34'01.5"W 23K 649606 E 7753356 W	1010	Ouro Preto	26.09	8.10	0.11	0.06		I	
Upper RV	RV-02	19/08/2015 20/01/2016 9/06/2016	20°01'10.7"S 43°49'45.4"W 23K 622454 E 7785916 W	729	Nova Lima	73.21	7.54	0.12	0.08		II	
Middle RV	RV-03	17/08/2015 19/01/2016 7/06/2016	19°49'54.8"S 43°51'56.2"W 23K 618796 E 7806723 W	674	Santa Luzia	345.15	2.95	5.40	0.69	Total ammoniacal nitrogen	III	Discharge of domestic sewage
Middle RV	RV-04	18/08/2015 18/01/2016 8/06/2016	19°32'56.7"S 43°53'33.3"W 23K 616174 E 7838041 W	658	Lagoa Santa	330.92	4.29	4.96	0.49	Total ammoniacal nitrogen	III	Discharge of domestic sewage
Middle RV	RV-05	10/08/2015 11/01/2016 31/05/2016	18°48'19.2"S 44°09'09.2"W 23K 589298 E 7920498 W	567	Curvelo	287.20	7.25	0.92	0.41	Arsenic and total ammoniacal nitrogen	III*	Gold metallurgy and discharge of domestic sewage
Low RV	RV-06	11/08/2015 12/01/2016 1/06/2016	18°25'33.2"S 44°11'10.9"W 23K 585926 E 7962502 W	552	Corinto	203.23	7.30	0.23	0.22	Arsenic	II	Agriculture
Low RV	RV-07	13/08/2015 13/01/2016 3/06/2016	17°51'55.4"S 44°32'57.4"W 23 K 547752 E 8024649 W	495	Lassance	162.21	7.63	0.21	0.17	Arsenic	II	Discharge of domestic sewage and agriculture (sugar cane)
Low RV	RV-08	12/08/2015 14/01/2016 2/06/2016	17°12'25.9"S 44°48'49.8"W 23 K 519793 E 8097515 W	464	Várzea da Palma	153.00	8.35	0.14	0.11	Arsenic	II	Discharge of domestic sewage
Sewage MRBH	STP Arrudas	20/07/2016, 25/01/2017			Sabará							
Sewage MRBH	STP Onça	20/07/2016, 18/01/2017			Belo Horizonte							

Captures of specimens of *A. lacustris* and *A. taeniatus* were carried out with gillnets with mesh sizes of 2.4, 3.0 and 4.0 cm between opposing nodes and with cast nets, seines and sieves. A total of 137 individuals of *A. lacustris* and 103 individuals of *A. taeniatus* was sampled in the three regions. The captures with gillnets represented 63% of sampling. For the stable isotope analyses, we collected at least five samples of each species at each sampling site (whenever possible). In the field, dorsal muscle was removed for large specimens and for small the whole fish was used removing the digestive tract. All samples were kept frozen until laboratory processing to avoid decomposition and deterioration of the material. In the laboratory, the fish samples were lyophilized for 24 hours, ground to fine and homogeneous powder using mortar and pestle and stored in eppendorf tubes.

The individuals that were not selected to stable isotopes analyses were fixed in formalin (10%) in the field, washed in water after fixation and transferred to alcohol (70%) in laboratory. Individuals predated or in high stage of decomposition were discarded. The remain individuals were used to stomach contents analyses in laboratory, where they had their stomach contents carefully removed. The same individuals were not used for both isotopic and stomach contents analyses because the stomach contents were analyzed following the results of stable isotope analyses, when we detect the need for complementary information.

Whenever possible, we collected five samples of all basal food resources available at each sampling site: periphyton, filamentous algae, suspended matter, fine particulate organic matter (FPOM) from sediments, vegetation (grasses and riparian vegetation), coarse particulate organic matter (CPOM), and aquatic macrophytes. Complementary samples of the suspended material were made at the sewage treatment plants to obtain the isotope signature of the raw sewage.

Samples of algae, aquatic macrophytes, vegetation and CPOM were collected at all sites where they were present, stored in plastic bottles and kept frozen until laboratory processing. Filamentous algae and aquatic macrophytes were collected manually in each site where they were present. Leaves from pasture (grasses) and from the natural riparian vegetation were manually collected along river banks in each site, with the most common species being prioritized at the site. The CPOM was randomly collected from leaf litter deposits in the streams.

Liquid samples, like periphyton, suspended matter (including sewage samples) and sediment, were collected at each site and kept frozen until processing in laboratory, where they were filtered using a filtration device attached to a vacuum pump using calcined quartz fiber filters (Whatman® QMA quartz filters). The periphyton was collected by scraping rocks with a brush and placing the material in a plastic bottle with distilled water. FPOM samples were collected from sediment deposits revolving in each sampling site and stored in plastic bottles. The suspended matter presented in the sampling sites and at STPs were collected with a phytoplankton net (0.45 mm mesh) deployed for a period of three minutes at each sampling site.

In the laboratory, all basal resource samples were dried in an oven at 60 °C for 48 hours and then ground with a mortar and pestle and stored in Eppendorf tubes.

The contents of 44 stomachs of *A. lacustris*, and 31 stomachs of *A. taeniatus* were analyzed in total. Food items were weighed (0.001 g accuracy/ wet weight) and identified under stereomicroscope to the lowest taxonomic category possible. The frequency of occurrence (Fi = number of times item i occurred, divided by the total number of stomachs) and the relative weight (Pi = sum of the weight of item i divided by the sum of the weight of all items) of each item were obtained. The food index (IA), proposed by Kawakami and Vazzoler (1980), was then calculated for each species and region, according to the formula: $IAi = (Fi.Pi)/\Sigma Fi.Pi$, where, IAi = food index of item i; Fi = frequency of occurrence of item i, and Pi = weight of item i.

The degree of overlap in food items between species was calculated using the simplified Morisita index (Morisita-Horn index) (Krebs 2014), according to the formula below: $C_H = 2\Sigma Pij.Pik/\Sigma P^2ij + \Sigma P^2ik$, where, C_H = Simplified Morisita Index of overlap (Horn 1966) between species j and species k; Pij = Proportion resource i is of the total resources used by species j; Pik = Proportion resource i is of the total resources used by species k, and n = Total number of resource states (I = 1, 2, 3, ... n).

For the food items characterization, "detritus" was considered dead particulate organic material, "sediment" included inorganic particles, and "plant remnants" were related to fragment of terrestrial vegetation.

A total of 42 samples of *A. lacustris*, 47 samples of *A. taeniatus* and 703 basal resources samples were sent to the Center for Nuclear Energy in Agriculture (CENA) at University of São Paulo (USP) for isotopic analysis. About 2–5 mg of dry animal tissue material and approximately 5–10 mg of basal resources samples were selected for analysis.

To determine the isotopic ratio, a mass spectrometer system in the Continuous-flow (CF-IRMS) mode was used with a Carlo Erba elemental analyzer (CHN 1110) coupled to a Delta Plus mass spectrometer (Thermo Scientific). Results were expressed as relative difference of international reference standards, in the delta notation (δ ‰), and calculated using the following formula: $\delta X = [(R_{sample}/R_{standard})^{-1}] \times 10^3$, where X is ^{13}C or ^{15}N and R represents the isotopic ratio $^{13}C/^{12}C$ or $^{15}N/^{14}N$ (Barrie and Prosser 1996).

Differences in isotopic ratios of $\delta^{13}C$ and $\delta^{15}N$ of consumers and resources between the three regions were tested using one-way analysis of variance (ANOVAs) when the normality and homoscedasticity assumptions were met. The nonparametric Kruskal-Wallis test was used for data with non-normal distribution. When significant differences ($p < 0.05$) were observed, means were compared using the post-hoc Tukey's test. We also tested if isotopic signatures of *A. lacustris* and *A. taeniatus* presented variation between the dry and wet season, using t-tests (normal distribution) and Mann-Whitney tests (non-parametric). These analyses were performed in the software Statistica 6.0 (Statsoft 2004).

To evaluate the trophic structure of *A. lacustris* and *A. taeniatus* populations, individuals of the two species were plotted in the bi-plot space according to the isotopic values of carbon (x-axis) and nitrogen (y-axis) in each region (Fig. S1). Source contributions to the *A. lacustris* and *A. taeniatus* diet were estimated for the three regions based on stable isotope data analyzed through Bayesian stable isotope mixed models (Moore and Semmens 2008, Parnell et al. 2010), specifically using the MixSIAR package in R (Stock and Semmens 2016a). For both analyses, only the autochtonous sources, algae and periphyton, and the allochtounous sources, leaves of riparian vegetation and grasses were taken into account. The samples of sewage were also considered to sites "Middle RV" and "Low RV" since they are located in the area under influence of pollution. We used Markov chain Monte Carlo sampling based on the following parameters: number of chains = 3; chain length = 100,000; burn in = 50,000; thin = 50 and model 4 (Resid*Process) error structure (Stock and Semmens 2016b). Diagnostic tests (Gelmin-Rubin, Heidelberger-Welch and Geweke) and trace plots were examined for model convergence. The fractionation values used for consumers were 0.4 ± 1.3 ‰ for Carbon and 2.54 ± 1.27‰ for Nitrogen (Vanderklift and Ponsard 2003, Post 2002). Both the graphical representation and the partition analysis were done using the MixSIAR package in the R programming environment (Stock and Semmens 2016a).

The isotopic niches of *A. lacustris* and *A. taeniatus* in both regions (Upper RV, Middle RV and Low RV) were quantified based on standard ellipse areas (SEA – expressed in ‰^{2}) through use of the Stable Isotope Bayesian Ellipses package in R (SIBER, Jackson et al. 2011). The standard ellipse area (SEA) represents the core isotopic niche space and it is a proxy of the richness and evenness of resources consumed by the population (Bearhop et al. 2004). All measures were "bootstrapped" (n = 10,000, indicated by the letter "b") to compare groups with different sample sizes. A small sample size correction (indicated by the subscript letter "c") was applied to SEA to increase the accuracy of the comparisons, enabling the comparison of niches of populations with different sample sizes (Jackson et al. 2011). The SEAc allows to calculate the degree of niche overlap (in percentage, where 100% indicates total overlap) and can be used as a quantitative measure of diet similarity among different species (Hill et al. 2015).

RESULTS

Stomach contents

Only two stomachs were found empty, both of *A. taeniatus* sampled in Upper RV. Plant and insect remnants were the predominant items in the stomachs of *A. lacustris* and *A. taeniatus*, respectively (Table 2). However, both species presented variations in the type and proportion of ingested food items in each study region. In Upper RV, *A. lacustris* feed more on plant remnants, aquatic insects and detritus, while *A. taeniatus* feed on sediments and insect remnants. In the middle RV, *A. lacustris* maintained its diet based on plant remnants, however there was an increase of insect remnants. In this region *A. taeniatus* feed on insect remnants and aquatic insects. In the low RV, the most consumed item by *A. lacustris* was algae/periphyton, a pattern also observed for *A. taeniatus*, albeit to a lesser extent (Table 2).

Variation in resources used by *A. lacustris* and *A. taeniatus* was reflected in the food overlap of the two species in each region. The lowest food overlap was observed in the Upper RV (0%), followed by middle RV (34%) and low RV(83%).

Table 2. Food index (AI), frequency of occurrence (Freq.) and weight of each food item found in the stomachs of the species *A. lacustris* and *A. taeniatus* in each sampled region of the Rio das Velhas Basin.

	Item	Upper RV			Middle RV			Low RV		
		IA	Freq.	Weight	IA	Freq.	Weight	IA	Freq.	Weight
Astyanax lacustris	Algae/ Periphyton	0.00	0.00	0.00	0.00	0.08	0.00	0.80	0.32	0.85
	Aq. Macrophytes	0.00	0.00	0.00	0.00	0.00	0.00	0.00	0.04	0.01
	Sediment	0.00	0.00	0.00	0.00	0.15	0.00	0.02	0.24	0.03
	Detritus	0.20	0.17	0.33	0.00	0.00	0.00	0.00	0.00	0.00
	Plant remnants	0.40	0.33	0.33	0.53	0.54	0.44	0.00	0.16	0.00
	Aquatic insects	0.40	0.67	0.33	0.00	0.23	0.01	0.00	0.16	0.00
	Terrestrial insects	0.00	0.00	0.00	0.02	0.08	0.12	0.00	0.04	0.00
	Insects remnants	0.00	0.67	0.00	0.44	0.46	0.43	0.17	0.56	0.10
Astyanax taeniatus	Algae/ Periphyton	0.00	0.00	0.00	0.00	0.00	0.00	0.46	0.40	0.27
	Aq. Macrophytes	0.00	0.00	0.00	0.00	0.00	0.00	0.00	0.00	0.00
	Sediment	0.43	0.50	0.47	0.00	0.00	0.00	0.00	0.13	0.00
	Detritus	0.00	0.00	0.00	0.00	0.00	0.00	0.02	0.07	0.06
	Plant remnants	0.00	0.10	0.00	0.12	0.17	0.14	0.15	0.20	0.18
	Aquatic insects	0.00	0.20	0.00	0.64	0.17	0.77	0.00	0.00	0.00
	Terrestrial insects	0.00	0.00	0.00	0.02	0.17	0.03	0.10	0.07	0.34
	Insects remnants	0.57	0.60	0.52	0.22	0.67	0.07	0.28	0.40	0.16

Stable isotopes

The $\delta^{13}C$ and $\delta^{15}N$ of *A. lacustris* and *A. taeniatus* is different among the three regions (Figs 2–5). The $\delta^{13}C$ of *A. lacustris* is significantly different between Upper and Low RV (Fig. 2), while the $\delta^{13}C$ of *A. taeniatus* in Low RV is different of all other regions (Fig. 3). The $\delta^{15}N$ values of both species in Upper RV are different to the $\delta^{15}N$ values in Middle and Low RV (Figs 4, 5). When the comparison was made between the species, it was possible to observe that the isotopic composition of *A. lacustris* and *A. taeniatus* were different only in the upper RV for both carbon ($p < 0.01$) and nitrogen ($p = 0.02$). In other regions – middle RV ($\delta^{13}C$: $p = 0.41$ and $\delta^{15}N$: $p = 0.83$) and low RV ($\delta^{13}C$: $p = 0.61$ and $\delta^{15}N$: $p = 0.23$) – there was no variation between species. In addition, only the *A. taeniatus* presented variation in the $\delta^{15}N$ values between seasons ($p = 0.02$), with more enriched values being observed in the dry season.

Basal resources presented extensive variation in their isotopic composition, except for riparian vegetation and grasses, that did not vary in $\delta^{13}C$ between the three sampled regions (Table 3). Strikingly, autotrophic resources (algae, periphyton and aquatic

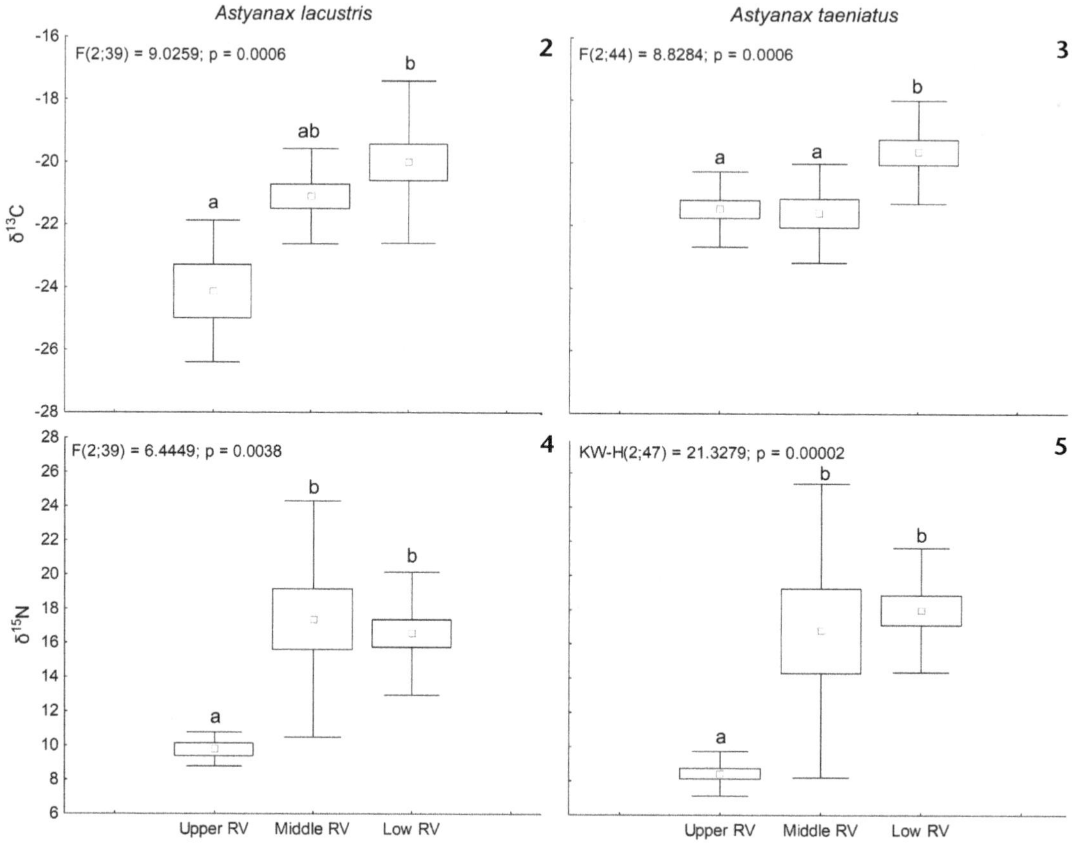

Figures 2–5. Variation in the isotopic composition of carbon (2–3) and nitrogen (4–5) in the species *A. lacustris* (2, 4) and *A. taeniatus* (3, 5) among the studied regions. Mean (small box), standard error (bars) and standard deviation (large box). Letters (a and b) indicate significant differences according to post-hoc Tukey's test.

Table 3. Variation in the carbon and nitrogen isotopic composition of the resources sampled in the three regions of the Rio das Velhas Basin. Letters (a, b and c) indicate significant differences according to post-hoc Tukey's test. AL: filamentous algae, CPOM: coarse particulate organic matter, FPOM: fine particulate organic matter, GR: grasses, MA: macrophytes, PE: periphyton, RV: riparian vegetation, SM: suspended matter and SW = raw sewage.

	$\delta^{13}C$ (mean and SD)				$\delta^{15}N$ (mean and SD)			
	Upper RV	Middle RV	Low RV	p	Upper RV	Middle RV	Low RV	p
AL	-27.74±4,66a	-22.84±6,35a	-5.52±2,47b	<0.01	4.67±2,99a	8.38±16,38ab	15.85±4,81b	<0.01
CPOM	-30.01±1,17a	-29.00±1,46b	-28.98±1,53b	<0.01	0.94±2,72a	4.48±3,86b	7.53±3,52c	<0.01
FPOM	-25.08±1,75a	-23.55±1,33b	-21.33±3,07c	<0.01	5.03±1,70a	6.66±7,10a	14.26±3,75b	<0.01
GR	-13.98±0,87a	-17.21±6,45a	-16.27±5,67a	0.56	0.28±1,81a	4.56±3,66b	8.00±3,89c	<0.01
MA	-30.54±1,08a	-25.94±5,91ac	-29.00±0,75c	<0.01	8.46±1,93a	19.55±14,18b	15.47±3,37b	<0.01
PE	-25.07±2,20a	-23.62±1,61a	-19.19±3,69b	<0.01	5.26±1,20a	9.15±10,72a	15.36±4,08b	<0.01
RV	-30.20±1,78a	-29.99±1,86a	-29.51±1,30a	0.54	0.64±1,72a	4.78±2,43b	7.28±2,39c	<0.01
SM	-25.88±1,67a	-24.56±1,75b	-20.45±4,62c	<0.01	4.47±1,39a	5.59±7,24a	14.15±4,96b	<0.01

macrophytes) showed highly enriched nitrogen isotopic values in the most polluted regions (middle and low RV) (Figs 6, 7). The range of carbon values of basal resources and fish species was higher in the low RV region and similar in other regions (Fig. 6). The range of nitrogen isotopic values, in turn, was higher in middle and low RV regions (Fig. 7).

According to the partition analysis, in the upper RV, periphyton was the most assimilated basal resource, followed

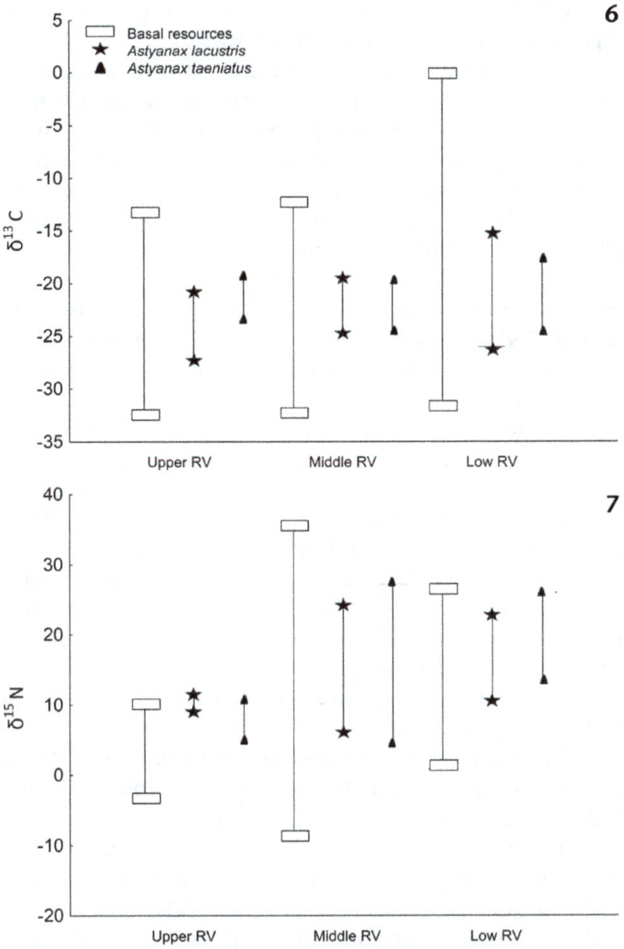

Figures 6–7. Isotopic values range for $\delta^{13}C$ (6) and $\delta^{15}N$ (7) of basal resources, *A. lacustris* and *A. taeniatus* sampled in three regions of Rio das Velhas basin.

by filamentous algae and grasses (mainly for *A. taeniatus*). In the middle course, both species assimilated more carbon from filamentous algae and the other resources had similar contributions. In the lower course, the periphyton was again the most assimilated resource by *A. lacustris* and *A. taeniatus*. However, riparian vegetation had a greater contribution in this site than in other sites, being the second most consumed resource by both species (Table 4).

Results of isotopic niche overlap were similar to those observed in the stomach contents analyses. We again observed a slight overlap of trophic niches in the upper RV (23%) (Fig. 8, see also Fig. S1). In this region, with no influence of the sewage from the MRBH, the two species presented little overlap in assimilated carbon sources and appeared to occupy the same trophic level (Fig. 8). In the middle course of the Rio das Velhas, where the discharge of sewage is high, the carbon and nitrogen values of the two species were very similar, presenting high overlap (71%). In addition, a large variation in nitrogen signatures for both species was observed in this region, with an amplitude of 4.68 to 27.22 ‰ (Figs 7, 9). In the low RV it was also observed a high niche overlap (62%) especially in the carbon source (Fig. 10). However, there was a decrease in variation of nitrogen isotopic composition (10.97 to 25.96 ‰) (Fig. 7).

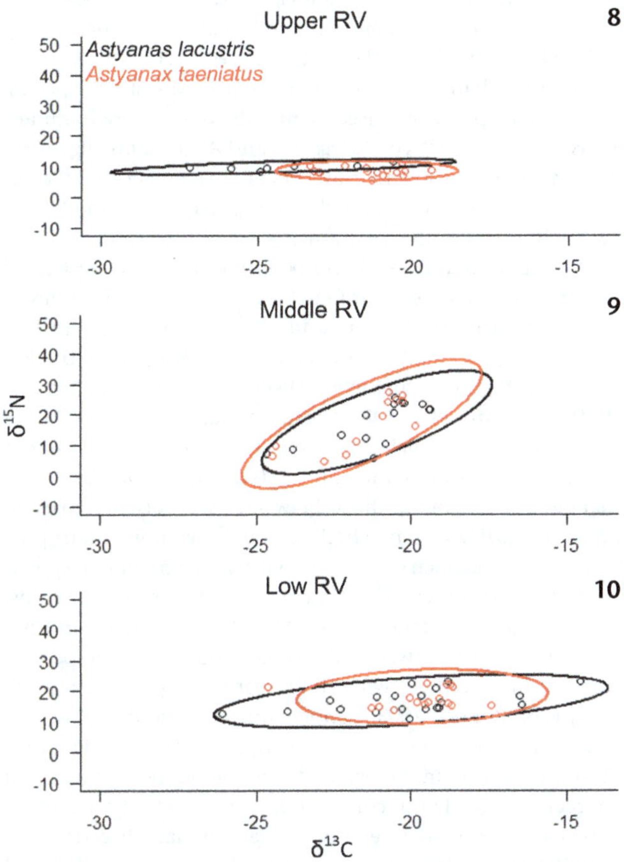

Figures 8–10. Trophic niche of *A. lacustris* and *A. taeniatus* (evaluated by the ellipse area with 95% confidence interval) in the Upper (8), Middle (9) and Low (10) regions of the Rio das Velhas Basin.

Table 4. Mean proportion of each basal resource assimilated by *Astyanax lacustris* and *A. taeniatus* at each sampling site. AL: filamentous algae, SW: raw sewage, GR: grasses, RV: riparian vegetation, PE: periphyton.

	Rio das Velhas regions	Basal resources				
		AL	SW	GR	RV	PE
A. lacustris	Upper	0.31		0.06	0.04	0.59
	Middle	0.58	0.09	0.08	0.08	0.18
	Low	0.04	0.04	0.05	0.14	0.74
A. taeniatus	Upper	0.19		0.22	0.02	0.57
	Middle	0.62	0.07	0.05	0.08	0.19
	Low	0.03	0.02	0.03	0.11	0.81

DISCUSSION

Food overlap between the two congeneric species was low in the least-disturbed region (upper Rio das Velhas), confirming our first hypothesis, that closely-related sympatric species diverge in their trophic niche to allow coexistence. In this study, the species *A. lacustris* and *A. taeniatus* presented high trophic plasticity in response to pollutants, increasing their food overlap and presenting similar isotopic signatures in the heavily polluted areas. Such aspect confirm our second hypothesis, that human disturbance promotes homogenization of fish species' diets. Despite plant and insect remnants were the predominant items in the stomachs of *A. lacustris* and *A. taeniatus*, algae and periphyton were also important food items (especially in lower sites). The importance of autochthonous resources as food items was highlighted in the partition analysis, which indicated that algae (in polluted regions) and periphyton (in least-disturbed region) were the most assimilated resources for both species.

The variation in $\delta^{13}C$ and $\delta^{15}N$ compositions and in stomach contents of *A. lacustris* and *A. taeniatus* along the Rio das Velhas highlight their generalistic habits and high trophic plasticity (Manna et al. 2012, Carvalho et al. 2015) probably as a result of resource availability found in the aquatic environment (Lobón-Cerviá and Bennemann 2000). The *A. taeniatus* also changes its $\delta^{15}N$ compositions between seasons (with enriched values in the dry season) which can be due changes on trophic levels (Vander Zanden et al. 1997), but also can occur in response to a higher enrichment in $\delta^{15}N$ values of resources in dry seasons.

The predominance of insects and plant remains in their stomach contents as well as the consumption of algae/periphyton are in agreement with the literature (e.g. Andrian et al. 2001, Casatti et al. 2003, Bennemann et al. 2005, Souza and Lima-Júnior 2013). It is likely that periphyton (in least-disturbed sites) and algae (in more degraded sites) are also being consumed indirectly through the consumption of insects. Castro et al. (2016) also observed a trend of changes in macroinvertebrates assimilation of algae and periphyton between degraded and preserved environments, which reinforces our statement. These changes in the type and proportion of autochthonous resources that sustain the two species are probably due to changes in environmental conditions as a result of pollution. Periphyton (or biofilm) is defined as an integral and independent micro-ecosystem in aquatic ecosystems, harboring biotic components (like algae, fungi, bacteria, protozoans, metazoans) and abiotic components (like substrata, extracellular polymeric substance and detritus) (Wu 2016). These organisms occur on the surface of rocks and submerged vegetation (Tundisi and Tundisi 2008), in environments with good water quality and greater presence of rocks and wood that favor the proliferation of periphyton. On the other hand, an increase of nutrients in the aquatic environment triggers a marked increase in algae (Tundisi and Tundisi 2008). Therefore, the expected greater abundance of algae in areas under the influence of pollution and of periphyton in sites with better environmental conditions, explain the shifts on basal resources assimilated by fish species.

In this study, stomach contents and stable isotopes analyses showed that there is a tendency to niche overlapping in *A. lacustris* and *A. taeniatus* in the presence of pollutants. The percentage of niche overlap observed by stomach contents and stable isotopes analyses were not the same, which is expected since not all items found in fish stomachs are assimilated (Manetta and Benedito-Cecílio 2003). In addition, the items consumed only occasionally or accidentally by individuals are observed on stomach contents, but will not be reflected on isotopic composition of fish. The greater overlap observed in the middle and lower course of Rio das Velhas could be due to the lower heterogeneity and resource availability in impacted sites (Gutiérrez-Cánovas et al. 2015). Fish tend to exhibit greater selectivity and specialization in the resources consumed in heterogeneous aquatic ecosystems, while in environments with few resources (or predominance of a single resource), fish tend to share the same food items (Knoppel 1970, Hurlbert 1978). Although we did not measure algae abundance, it is expected that in polluted sites populations reach high densities (e.g. Lata Dora et al. 2010, Macedo and Sipaúba-Tavares 2010), becoming an important food source consumed either directly or indirectly by generalist species, which can explain the higher niche overlap between *A. lacustris* and *A. taeniatus* in the middle and lower regions.

Trophic niche amplitude differed between regions. In the undisturbed region (Upper RV), both species had a broader trophic niche on the horizontal axis (niche with great carbon range). This trend is expected in food webs in which there are multiple basal resources with varying $\delta^{13}C$ values, enabling niche diversification at the base of a food web (Layman et al. 2007), which indicates that *A. lacustris* and *A. taeniatus* feed on a greater range of resources under natural conditions. On the other hand, in the most polluted region (Middle RV), the two species presented a narrow carbon range and a large nitrogen range (more vertical trophic niche). The narrow carbon range may be occurring in response to the restriction and homogenization of available food resources (the opposite of what has been observed in preserved sites). A larger range in $\delta^{15}N$ sometimes suggests more trophic levels and thus a greater degree of trophic diversity (Layman et al. 2007), however, probably this is not the explanation to the niche verticalization observed in this study, but the greater enrichment observed in the basal resources. This verticalization of the trophic niche has been found in fish (De Carvalho et al. 2017) and macroinvertebrates (Castro et al. 2016) in environments impacted by other anthropogenic activities (sugarcane).

The enriched nitrogen values of fish and resources especially in the middle section are probably related to the influence of sewage effluents, since $\delta^{15}N$ values of domestic wastes ranges between 7‰ to 38‰ (Dailer et al. 2010). Domestic wastes are nitrogen enriched especially because of isotopic fractionation during nitrification and volatilization in the case of ammonium, or denitrification in the case of nitrate (Nikolenko et al. 2018).

Therefore, the uptake of enriched $\delta^{15}N$ by primary producers are reflected in the entire food web (McClelland et al. 1997). Changes in $\delta^{15}N$ values along the pollution gradient, with a particularly large increase in regions affected by sewage effluent, were similar to those reported in studies of macroinvertebrates (e.g. Morrissey et al. 2013, Pastor et al. 2014, Baumgartner and Robinson 2016), and primary producers (e.g. MCclelland and Valiela 1998, Cole et al. 2004, Wang et al. 2016). Together, results of these studies, including ours, support the finding that high $\delta^{15}N$ values are good indicators of anthropogenic stress in aquatic systems.

Therefore, stomach contents and stable isotope analyses were very useful to evaluate the effects of the presence of pollutants in the trophic ecology of two congeneric species. It was possible to observe that even where species originally present different feeding habits (verified through the analysis of the stomach contents), food webs were mainly based on autochthonous items, such as algae and periphyton (verified through the isotopic analysis), assimilated directly and indirectly through aquatic insects. The presence of pollution, besides triggering increased food overlap between *A. lacustris* and *A. taeniatus*, also promoted an enrichment in $\delta^{15}N$ values of fish and resources. The $\delta^{15}N$ values of fish seems to be an effective means to detect anthropogenic impacts in aquatic ecosystems. In addition to providing important information on species biology, our work contributes to elucidate one of the 100 key ecological issues (Sutherland et al. 2013): How do resource pulses affect resource use and interactions between organisms?

ACKNOWLEDGMENTS

We thank the Agência Peixe Vivo and the Comitê de Bacia Hidrográfica do Rio das Velhas (CBH, Rio das Velhas) for the project financial support and the Projeto Manuelzão for the logistical support. Thanks to Aline J. Grossi (Universidade Federal de Lavras, UFLA) and Luiza Hoehne (Universidade Federal de Minas Gerais, UFMG) for the support on processing of samples. Thanks also to the Benthos Ecology Laboratory (UFMG) and the Laboratory of Fish Ecology (UFLA) that allowed processing of samples and infrastructure, and to the Centro de Energia Nuclear na Agricultura (CENA) for their support and partnership in the isotopic analysis. PSP received a research grant and a research fellowship from the Conselho Nacional de Desenvolvimento Científico e Tecnológico (CNPq 303548/2017-7) and from the Fundação de Amparo à Pesquisa do Estado de Minas Gerais (FAPEMIG PPM-00608/15).This study was financed in part by the Coordenação de Aperfeiçoamento Pessoal de Nível Superior (CAPES) – Finance code 32004010017P3. The manuscript underwent grammar revision by Alistair Campbell.

LITERATURE CITED

Alves CBM, Pompeu PS (2001) A Fauna de Peixes da Bacia do Rio das Velhas no Final do Século XX. In: Alves CBM, Pompeu PS (Eds) Peixes do rio das Velhas: passado e presente. Belo Horizonte, SEGRAC, 165–187.

Andrian IF, Silva HB, Peretti D (2001) Dieta de *Astyanax bimaculatus* (Linnaeus, 1758) (Characiformes, Characidae), da área de influência do reservatório de Corumbá. Estado de Goiás, Brasil. Acta Scientiarum 23(2): 435–440. http://dx.doi.org/10.4025/actascibiolsci.v23i0.2735

Baeta A, Vieira LR, Lírio AV, Canhoto C, Marques JC, Guilhermino L (2017) Use of stable isotope ratios of fish larvae as indicators to assess diets and patterns of anthropogenic nitrogen pollution in estuarine ecosystems. Ecological Indicators 83: 112–121. https://doi.org/10.1016/j.ecolind.2017.07.062

Barrie A, Prosser SJ (1996) Automated analysis of light-element stable isotopes by isotope ratio mass spectrometry. In: Boutton TW, Yamasaki S (Eds) Mass spectrometry of soils. New York, Marcel Dekker, 1–46.

Baumgartner SD, Robinson CT (2016) Changes in macroinvertebrate trophic structure along a land-use gradient within a lowland stream network. Aquatic Sciences 79(2): 407–418. https://doi.org/10.1007/s00027-016-0506-z

Bearhop SC, Adams E, Waldron S, Fuller AA, MacLeod H (2004) Determining trophic niche width: a novel approach using stable isotope analysis. Journal of Animal Ecology 73: 1007–1012. https://doi.org/10.1111/j.0021-8790.2004.00861.x

Beaudoin CP, Tonn WML, Prepas EEL, Wassenaar LI (1999) Individual specialization and trophic adaptability of norther pike (*Esox Lucius*): an isotope and dietary analysis. Oecologia 120(3): 386–396. https://doi.org/10.1007/s004420050871

Bennemann ST, Gealh AM, Orsi ML, Souza LM (2005) Ocorrência e ecologia trófica de quatro espécies de *Astyanax* (Characidae) em diferentes rios da bacia do rio Tibagi, Paraná, Brasil. Iheringia, Série Zoologia 95(3): 247–254. https://doi.org/10.1590/S0073-47212005000300004

Cabana G, Rasmussen JB (1996) Comparison of aquatic food chains using nitrogen isotopes. Ecology 93(20): 10844–10847. https://doi.org/10.1073/pnas.93.20.10844

Carassou L, Whitfield AK, Moyo S, Richoux NB (2017) Dietary tracers and stomach contents reveal pronounced alimentary flexibility in the freshwater mullet (*Myxus capensis*, Mugilidae) concomitant with ontogenetic shifts in habitat use and seasonal food availability. Hydrobiologia 799(1): 327–348. https://doi.org/10.1007/s10750-017-3230-3.

Carvalho DR, Castro D, Callisto M, Moreira MZ, Pompeu PS (2015) Isotopic variation in five species of stream fish under the influence of different land uses. Journal of Fish Biology 87(3): 559–578. https://doi.org/10.1111/jfb.12734

Casatti L, Mendes HF, Ferreira KM (2003) Aquatic macrophytes as feeding site for small fishes in the Rosana reservoir, Paranapanema river, Southeastern Brazil. Brazilian Journal of Biology 63(2): 213–222. https://doi.org/10.1590/S1519-69842003000200006

Cassemiro FAS, Hahn NS, Fugi R (2002) Avaliação da dieta de *Astyanax altiparanae* Garutti and Britski, 2000 (Osteichthyes, Tetragonopterinae) antes e após a formação do reservatório de salto Caxias, Estado do Paraná, Brasil. Acta Scientiarum 24(2): 419–425. https://doi.org/10.4025/actascibiolsci.v24i0.2314

Castro DMP, De Carvalho DR, Pompeu PS, Moreira MZ, Nardoto GB, Callisto M (2016) Land Use Influences Niche Size and the Assimilation of Resources by Benthic Macroinvertebrates in Tropical Headwater Streams. PLos ONE 11: e0150527. https://doi.org/10.1371/journal.pone.0150527

CETEC (1983) Diagnóstico Ambiental do Estado de Minas Gerais. Belo Horizonte, Fundação Centro Tecnológico de Minas Gerais/CETEC, Série de Publicações Técnicas/SPT-010, 158 pp.

Cole ML, Valiela I, Kroeger KD, Tomasky GL, Cebrian J, Wigand C, McKinney RA, Grady SP, da Silva MHC (2004) Assessment of a $\delta^{15}N$ Isotopic Method to Indicate Anthropogenic Eutrophication in Aquatic Ecosystems. Journal of Environmental Quality 33(1): 124–132. https://doi.org/10.2134/jeq2004.1240

Connan M, Bonnevie BT, Hagen C, van der Lingen CD, McQuaid C (2017) Diet specialization in a colonial seabird studied using three complementary dietary techniques: effects of intrinsic and extrinsic factors, Marine Biology 164: 171. https://doi.org/10.1007/s00227-017-3201-2

Dailer ML, Knox RS, Smith JE, Napier M, Smith CM (2010) Using $\delta^{15}N$ values in algal tissue to map locations and potential sources of anthropogenic nutrient inputs on the island of Maui, Hawai'i, USA. Marine Pollution Bulletin 60(5): 655–671. https://doi.org/10.1016/j.marpolbul.2009.12.021

De Carvalho DR, Castro DMP, Callisto M, Moreira MZ, Pompeu PS (2017) The trophic structure of fish communities from streams in the Brazilian Cerrado under different land uses: an approach using stable isotopes. Hydrobiologia 795(1): 199–217. https://doi.org/10.1007/s10750-017-3130-6

Eigenmann CH (1921) The American Characidae. Memoirs of the Museum of Comparative Zoology 43(3): 209–310.

Esteves KE (1996) Feeding Ecology of three *Astyanax* species (Characidae, Tetragonopterinae) from a foodplain lake of Mogi-Guaçú River, Paraná River Basin, Brazil. Environmental Biology of Fishes 46(1): 83–101. https://doi.org/10.1007/BF00001701

Esteves KE, Aranha JMR (1999) Ecologia trófica de peixes de riachos. Oecologia Brasiliensis 6: 157–182.

Feio MJ, Ferreira WR, Macedo DR, Eller AP, Alves CBM, França JS, Callisto M. (2015) Defining and testing targets for the recovery of tropical streams based on macroinvertebrate communities and abiotic conditions. River Research and Applications 31(1): 70–84. https://doi.org/10.1002/rra.2716

Gerking SD (1994) Feeding ecology of fish. Califórnia, Academic Press, 416 pp.

Géry J (1977) Characoids of the world. Neptune City, T.F.H. Publications, 672 pp.

Gutiérrez-Cánovas C, Sánchez-Fernández D, Velasco J, Millán A, Bonada N (2015) Similarity in the difference: changes in community functional features along natural and anthropogenic stress gradients. Ecology 96(9): 2458–2466. https://www.jstor.org/stable/24702350

Hall SR (2004) Stoichiometrically Explicit Competition between Grazers: Species Replacement, Coexistence, and Priority Effects along Resource Supply Gradients. American Society of Naturalists 164(2): 157–172. https://doi.org/10.1086/422201

Hicks KA, Loomer HA, Fuzzen MLM, Kleywegt S, Tetreault GR, McMaster ME, Servos MR (2016) $\delta^{15}N$ tracks changes in the assimilation of sewage-derived nutrients into a riverine food web before and after major process alterations at two municipal wastewater treatment plants. Ecological Indicators 72: 747–758. https://doi.org/10.1016/j.ecolind.2016.09.011

Hill JM, Jones RW, Hill MP, Weyl OLF (2015) Comparisons of isotopic niche widths of some invasive and indigenous fauna in a South African river. Freshwater Biology 60(5): 893–902. https://doi.org/10.1111/fwb.12542

Horn HS (1966) Measurement of "overlap" in comparative ecological studies. The American Naturalist 100(914): 419-424.

Hurlbert SH (1978) The measurement of niche overlap and some relatives. Ecology 59(1): 67–77. https://doi.org/10.2307/1936632

Jackson AL, Inger R, Parnell AC, Bearhop S (2011) Comparing isotopic niche widths among and within communities: SIBER – Stable Isotope Bayesian Ellipses in R. Journal of Animal Ecology 80(3): 595–602. https://doi.org/10.1111/j.1365-2656.2011.01806.x

Kling GW, Fry B, O'Brien W (1992) Stable isotopes and planktonic trophic structure in arctic lakes. Ecology 73(2): 561–566. https://doi.org/10.2307/1940762

Knoppel HA (1970) Food of Central Amazonian fishes. Amazoniana 2: 257–352.

Kawakami E, Vazzoler G (1980) Método gráfico e estimativa de índice alimentar aplicado no estudo de alimentação de peixes. Boletim do Instituto Oceangráfico 29(2): 205–207. https://doi.org/10.1590/S0373-55241980000200043

Krebs CJ (2014) Niche measures and resource preferences. In: Krebs CJ (Ed.) Ecological methodology. New York, Harper & Row, 3rd ed., 597–653.

Lata Dora S, Maiti SK, Tiwary RK, Anshumali (2010) Algae as an indicator of river water pollution – A Review. The Bioscan 2: 413–422.

Layman CA, Arrington DA, Montana CG, Post DM (2007) Can stable isotope ratios provide for community wide measures of trophic structure? Ecology 88(1): 42–48. https://doi.org/10.1890/0012-9658(2007)88[42:CSIRPF]2.0.CO;2

Lobón-Cerviá J, Bennemann S (2000) Temporal trophic shifts and feeding diversity in two sympatric, neotropical, omnivorous fishes: *Astyanax bimaculatus* and *Pimelodus maculatus* in Rio Tibagi (Paraná, Southern Brazil). Archiv fuer Hydrobiologie 149(2): 285–306. https://doi.org/10.1127/archiv-hydrobiol/149/2000/285

Lougheed VL, Mcintosh MD, Parker CA, Stevenson RJ (2008) Wetland degradation leads to homogenization of the biota local and landscape scales. Freshwater Biology 53(12): 2402–2413. https://doi.org/10.1111/j.1365-2427.2008.02064.x

Macedo CF, Sipaúba-Tavares LH (2010) Eutrofização e qualidade da água na piscicultura: consequências e recomendações. Boletim do Instituto de Pesca 36(2): 149–163.

Manetta GI, Benedito-Cecílio E (2003) Aplicação da técnica de isótopos estáveis na estimativa da taxa de turnover em estudos ecológicos: uma síntese. Acta Scientiarum, Biological Sciences 25(1): 121–129. https://doi.org/10.4025/actascibiolsci.v25i1.2090

Manna LR, Rezende CF, Mazzoni R (2012) Plasticity in the diet of Astyanax taeniatus in a coastal stream from south-east Brazil. Brazilian Journal of Biology 72(4): 919–928. https://doi.org/10.1590/S1519-69842012000500020

McClelland JW, Valiela I, Michener RH (1997) Nitrogen-stable isotope signatures in estuarine food webs: A record of increasing urbanization in coastal watersheds. American Society of Limnology and Oceanography 42(5): 930–937. https://doi.org/10.4319/lo.1997.42.5.0930

McClelland JW, Valiela I (1998) Liking nitrogen in estuarine producers to land-derived sources. American Society of Limnology and Oceanography 43(4): 577–585. https://doi.org/10.4319/lo.1998.43.4.0577

Menezes NA, Weitzman S, Oyakawa OT, Lima F, Castro R, Weitzman M (2007) Peixes de água doce da Mata Atlântica. São Paulo, Museu de Zoologia/USP, Conservação Internacional, FAPESP, CNPq, 407 pp.

Morrissey CA, Boldt A, Mapstone A, Newton J, Ormerod SJ (2013) Stable isotopes as indicators of wastewater effects on the macroinvertebrates of urban rivers. Hydrobiologia 700(1): 231–244. https://doi.org/10.1007/s10750-012-1233-7

Moore JW, Semmens BX (2008) Incorporating uncertainty and prior information into stable isotope mixing models. Ecology Letters 11(5): 470–480. https://doi.org/10.1111/j.1461-0248.2008.01163.x

Nikolenko O, Jurado A, Borges AV, Knöller K, Brouyère S (2018) Isotopic composition of nitrogen species in groundwater under agricultural areas: A review. Science of The Total Environment 621: 1415–1432. https://doi.org/10.1016/j.scitotenv.2017.10.086

Parnell AC, Inger R, Bearhop S, Jackson AL (2010) Source partitioning using stable isotopes: coping with too much variation. PLoS One 5: e09672. https://doi.org/10.1371/journal.pone.0009672

Pastor A, Peipoch M, Cañas L, Chappuuis E, Ribot M, Gacia E, Riera JL, Martí E, Sabater F (2014) Nitrogen stable isotopes in primary uptake compartments across streams differing in nutrient availability. Enviranmental Science & Technology 47(18): 10155–10162. https://doi.org/10.1007/s00027-013-0330-7

Pereira LPF, Mercante CTJ (2005) A amônia no sistema de criação de peixes e seus efeitos sobre a qualidade de água. Uma revisão. Boletim do Instituto de Pesca 31(1): 81–88.

Peterson BJ, Fry B (1987) Stable isotopes in ecosystem studies. Annual Review of Ecology and Systematics 18: 293–320. https://doi.org/10.1146/annurev.es.18.110187.001453

Post DM (2002) Using stable isotopes to estimate trophic position: models, methods, and assumptions. Ecology 83(3): 703–718. https://doi.org/10.1890/0012-9658(2002)083[0703:USITET]2.0.CO;2

Souza RG, Lima-Junior SE (2013) Influence of environmental quality on the diet of *Astyanax* in a microbasin of central western Brazil. Acta Scientiarum, Biological Sciences 35(2): 177–184. https://doi.org/10.4025/actascibiolsci.v35i2.15570

StatSoft (2004). STATISTICA: data analysis software system. Tulsa, StatSoft Inc, v. 7. Available online at: http://www.statsoft.com

Stock BC, Semmens BX (2016a) MixSIAR GUI. User Manual, v. 3.1. Available online at: https://github.com/brianstock/MixSIAR

Stock BC, Semmens BX (2016b) Unifying error structures in commonly used biotracer mixing models. Ecology 97(10): 2562–2569. https://doi.org/10.1002/ecy.1517

Sutherland WJ, Freckleton RP, Godfray HCJ, Beissinger SR, Benton T, Cameron DD, Hails RS (2013) Identification of 100 fundamental ecological questions. Journal of Ecology 101(1): 58–67. https://doi.org/10.1111/1365-2745.12025

Tundisi JG, Tundisi TM (2008) Limnologia. São Paulo, Oficina de Texto, 631 pp.

Vanderklift MA, Ponsard S (2003) Sources of variation in consumer-diet $\delta^{15}N$ enrichment: a meta-analysis. Oecologia 136(2): 169–182. https://doi.org/10.1007/s00442-003-1270-z

Vander Zanden M, Cabana G, Rasmussen JB (1997) Comparing trophic position of freshwater fish calculated using stable nitrogen isotope ratios ($\delta^{15}N$) and literature dietary data. Canadian Journal of Fisheries and Aquatic Sciences 54(5): 1142–1158. https://doi.org/10.1139/f97-016

Vilela FS, Becker FG, Hartz SM (2002) Diet of *Astyanax* species (Teleostei, Characidae) in an Atlantic forest river in Southern Brazil. Brazilian Archives of Biology and Technology 43(2): 223–232. https://doi.org/10.1590/S1516-89132002000200015

Wang Y, Liu D, Richard P, Di B (2016) Selection of effective macroalgal species and tracing nitrogen sources on the different part of Yantai coast, China indicated by macroalgal $\delta^{15}N$ values. Science of The Total Environment 542(Part A): 306–314. https://doi.org/10.1016/j.scitotenv.2015.10.059

Weitzman SH, Fink W (1983) Relationships of the Neon Tetras, a group of South American freshwater fishes (Teleostei, Characidae) with comments on the phylogeny of new world Characiforms. Bulletin of the Museum of Comparative Zoology 150(6): 339–395.

Wootton RJ (1999) Ecology of teleost fishes. The Netherlands, Kluwer Academic Publishers, 2nd ed., 386 pp.

Wu Y (2016) Periphyton: functions and application in environmental remediation. Amsterdam, Elsevier, 402pp.

Author Contributions: PSP, DRC and CBMA designed the experiments; DRC, PSP, and MBA conducted the experiments; MZM, PSP, MBA and DRC analyzed the data; MBA, DRC, CBMA, MZM and PSP wrote the paper.

Competing Interests: The authors have declared that no competing interests exist.

3

Multiple nesting attempts and long breeding seasons of *Mimus gilvus* (Aves: Mimidae) in southeastern Brazil

Rodrigo Morais[1], **Luiz Carlos de Araújo**[1], **Gleidson Ramos da Silva**[1], **Charles Duca**[2]

[1]*Programa de Pós-graduação em Ecologia de Ecossistemas, Universidade Vila Velha. Campus Boa Vista, 29102-920 Vila Velha, ES, Brazil.*
[2]*Laboratório de Ecologia de Populações e Conservação, Universidade Vila Velha. Campus Nossa Senhora da Penha, 29102-623 Vila Velha, ES, Brazil.*
*Corresponding author: Charles Duca (*cduca@uvv.br*)*

ABSTRACT. This study describes aspects of the life history of the Tropical Mockingbird *Mimus gilvus* (Vieillot, 1808), including the breeding period, clutch size, nestlings and a list of plants used for nesting. Nests were monitored in an area of Restinga (sand-coastal plain) habitat in a protected area in southeastern Brazil. The data from 181 nests during five breeding seasons (2010-2014) showed that the Tropical Mockingbird has a long breeding season (26.1 ± 2.6 weeks) with up to two peaks of active nests from August to March. The breeding pairs made up to four nesting attempts in the same breeding season. The mean (± SD) clutch size was 2.4 ± 0.6 eggs (n = 169). The mean (± SD) incubation period was 14 ± 0.6 days, and the nestling remained in the nest for 14.5 ± 2.2 days. The nests were built on thirty plant species, and *Protium icicariba* (DC.) Marchand. was the plant species most commonly used for nesting. The breeding parameters of the Tropical Mockingbird are similar to those of other Mimidae species. The knowledge gained from this study makes the Tropical Mockingbird a good choice for future studies, particularly for testing ecological and evolutionary hypotheses regarding life history attributes, habitat selection and parental investment.

KEY WORDS. Breeding biology, clutch size, re-nesting, Restinga, Tropical Mockingbird.

INTRODUCTION

The natural history of organisms involves a balance between energy spent in survival and reproductive functions (Bennett and Owens 2002). Reproduction invariably entails risks and results in energy loss, which often affects the behavior and other selected attributes of the breeder (Ricklefs 2010). Although birds are a relatively well-known group, little is known about the breeding biology of tropical birds (Stutchbury and Morton 2008), particularly endemic species with restricted distribution in the Neotropical region (Mason 1985, Stutchbury and Morton 2001, Xiao et al. 2016). Studies about natural history try to ascertain the breeding parameters and are an important source of information about a species (Bartholomew 1986). However, the incubation and nestling period of less than 1/10 of passerine species are known (Pienaar et al. 2013).

Variations in nesting duration and clutch size were recorded for species in the highly seasonal Neotropical savanna (Duca and Marini 2011) and semi-arid ecosystems (Cavalcanti et al. 2016). These studies demonstrated that birds use rainfall as an environmental cue to adjust the timing of their breeding activities. However, little is known about the breeding flexibility of these species and their ability to adapt to changes in the precipitation patterns of a less variable ecosystem such as the Restinga (sand-coastal plain). In this paper, we addressed hypotheses related to differences in clutch size between breeding seasons and the influence of the weather on the nesting activity of a typical species from the Restinga, where precipitation is less variable than in the savanna and semi-arid ecosystems of the Neotropical region.

The Tropical Mockingbird *Mimus gilvus* (Vieillot, 1808) is distributed from northern Mexico to southern Brazil (Sick 1997). Despite its wide latitudinal geographic distribution, its occurrence in the northeastern and southeastern coast of Brazil is associated with the physiognomies in the sandy coastal plains (here after Restinga), which are threatened habitats restricted to a narrow strip near the coast (< 4 km of wide) (Sick 1997). The

Tropical Mockingbird is an omnivorous species with frugivorous habits in association with Restinga plants, being a valuable seed disperser there (Gomes et al. 2007). Little is known about the breeding biology of the Tropical Mockingbird in the wild, except that it lives in cooperative breeding groups (Morton et al. 2004, Botero et al. 2007).

The Tropical Mockingbird is currently considered endangered on the Red List of the states of Espírito Santo and Rio de Janeiro (Alves et al. 2000, Passamani and Mendes 2007), and the subspecies *Mimus gilvus antelius* Oberholser, 1919 is classified as "Near Threatened" to extinction in Brazil's Red List (Machado et al. 2005). The subspecies *M. gilvus antelius* occurs on the Brazilian coast from the northern portion of the state of Pará to south of Rio de Janeiro. It is considered endemic to the Atlantic Forest (Gonzaga et al. 2000), and some authors have advocated that this subspecies should be considered a distinct species (Cody 2005, Zanon et al. 2015). Due to its close association with coastal environments, the Tropical Mockingbird was once the most common bird in the Restinga (Sick 1997). However, the occurrence of the subspecies *M. gilvus antelius* overlaps with the most threatened Restinga area in Brazil (Rocha et al. 2007). The populations of the Tropical Mockingbird have been disappearing in the southern limit of its distribution in recent decades (Teixeira and Nacinovic 1992, Araujo and Maciel 1998, Argel-de-Oliveira and Pacheco 1998, Gonzaga et al. 2000), confirming its status as an Endangered species (Zanon et al. 2015). If the subspecies *M. gilvus antelius* is validated as a distinct species, when it is described it will already be a threatened species (Zanon et al. 2015).

This study aimed to characterize the breeding attributes of the Tropical Mockingbird, such as breeding period, clutch size, incubation and nestling periods, fledgling characteristics and the plants used for nest support.

MATERIAL AND METHODS

This study was conducted in the Setiba Environmental Protection Area (Área de Proteção Ambiental de Setiba, hereafter APA-Setiba), which includes the Paulo César Vinha State Park (hereafter PEPCV). The APA-Setiba has an area of 12,960 ha. Inside the APA-Setiba is the PEPCV, which consists of a sand-coastal plain of 1,500 ha, located in the municipality of Guarapari, state of Espírito Santo, southeastern Brazil (20°33′ to 20°38′S; 40°23′ to 40°26′W). The study area is within the Atlantic Forest biome, Restinga vegetation (i.e., sand-coastal plain; Pereira 2003). The study sites were considered priority areas for biodiversity conservation falling into the category of high biological importance (MMA 2000). According to the Köppen's classification system, the region has a monsoon climate (Am) (Alvares et al. 2013), with mean temperature of 23.3 °C and a mean annual rainfall of 1,307 mm^3 (Fabris and Cesar 1996).

The study was carried out in the phytophysiognomies of open shrubby/non-flooded vegetation and shrubland vegetation (Pereira 2003). This vegetation, consisting primarily of *Clusia hilariana* Schltdl. and *Protium icicariba* (DC.) Marchand., occurs on sandy soil arranged in thickets and sandbanks at sites where the water table is far from the surface (CEPEMAR 2007).

Data sampling was from August 2010 to March 2015 (five breeding seasons). Based on previous observations, we searched for nests mainly from August to March. We did not search for nests during other months, but the population was monitored weekly during the entire year, and no evidence of breeding activity was found.

Capture and banding of individuals were carried out using mist nets 12 m in length and 3 m in height. The captured birds were individually marked with metal bands provided by the National Center for Bird Research and Conservation/Chico Mendes Institute for Biodiversity Conservation (CEMAVE/ICMBIO, license #3138/7), and unique combinations of three plastic color-bands were used. Individuals captured in the breeding season had their incubation patches and/or cloacal protuberances analyzed to confirm their sex. As the cloacal protuberance is exclusive to males and we never observed banded males with incubation patches or incubating the eggs, we assumed that all individuals with incubation patches and/or recorded incubating eggs were females. Observations of individuals' activities were made with binoculars using the focal sampling method (Martin and Baetson 1993). The breeding period was based on the use of nests and the presence of active nests (with eggs or nestlings).

Nests were monitored at intervals of three to four days. Nests were checked every two days just prior to hatching and fledgling, to estimate time intervals more precisely. In each occasion, nests were checked and classified as empty, eggs, or nestlings. The following data were collected from the occupied nests: date of egg laying and hatching, date of departure of the fledglings from the nest, and the results of breeding (successful, depredated, and abandoned). Data about nestling development was collected by visually inspecting the nestlings, noting the skin and bill color, the presence of yellow gape, the eyes, and the stage of development and color of feathers. This monitoring provided information for estimating clutch size, incubation period and nestling period. We also collected data on the development of the nestlings such as hatching order, growth and plumage color. When a nest became inactive, the exact location was marked for the identification of the plant species in which the nest was built. A total of 89 nests were used for this analysis.

Only nests found before first egg was laid and which survived until at least one egg hatched were used to estimate the incubation period. Only successful nests with known hatching dates were used to estimate the nestling period. Successful nests were those in which at least one nestling fledged. Empty nests that had previously harbored fledglings for at least 15 days were considered successful. Empty nests found during the incubation period or that contained nestlings for less than 15 days, usually presenting traces of predation such as structural damage, traces of blood and/or feathers, were considered depredated. A nest was considered abandoned when the eggs remained in it for 18 days without visits from the breeding pair. This number of days exceeds the observed incubation period of the Tropical Mockingbird (see results). Also, nests with dead nestlings without any signs of aggression were considered abandoned.

Month rainfall was obtained from the closest weather station (approximately 25 km) to the study area (APA-Setiba). These data were then correlated with the number of active nests.

To assess the normality of the data, we used a Kolmogorov-Smirnov test, and the clutch size data did not have a normal distribution. Therefore, we used a Kruskal-Wallis test (*H* test) to assess differences in clutch size among the five breeding seasons. When the difference was significant, we compared the means using a Wilcoxon-Mann-Whitney test (*U* test) with independent samples to verify which pairs of breeding seasons were significantly different. We set the significance level at $\alpha = 0.05$ and performed statistical analyses using the statistical packages PAST (Hammer et al. 2001) and BioEstat (Ayres et al. 2007). The results are reported as the means ± standard deviation (SD).

RESULTS

A total of 43 breeding groups were monitored in the study area. Of these, 38 groups were composed of one pair, three groups of three individuals, one group of four individuals and one group of six individuals. Forty-one adults from 21 different groups were banded. In total, 181 nests were monitored: 32 between 2010 and 2011, 38 between 2011 and 2012, 40 between 2012 and 2013, 25 between 2013 and 2014 and 46 between 2014 and 2015.

Breeding season and nest building

The Tropical Mockingbirds bred from August to March (n = 181 nests, Fig. 1). The mean duration of the breeding season was 26.1 ± 2.6 weeks (Table 1). The peaks of breeding activity (most active nests) were in December 2010/2011, 2012/2013 and 2014/2015, and in January 2011/2012 and 2013/2014 (Fig. 1).

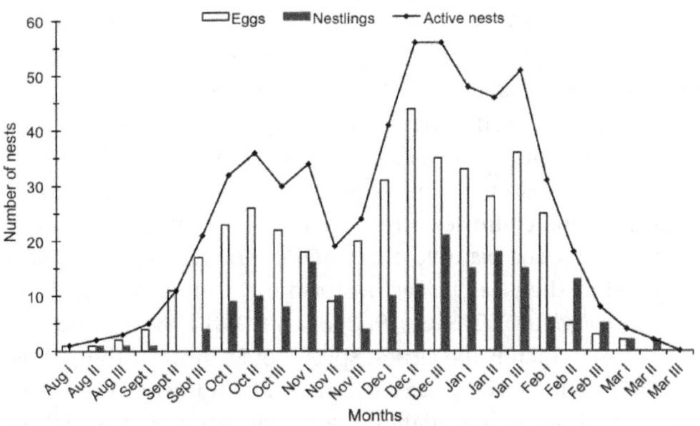

Figure 1. Number of active nests with eggs or nestlings of *Mimus gilvus* during the breeding seasons from 2010 to 2015 in a Restinga habitat (sand-coastal plain), southeastern Brazil. Roman numerals mean tens: I = 1-10 days; II = 11-20 days; III 21-30 (or 31) days.

In two seasons (2011/2012 and 2014/2015), we observed two peaks of breeding activity in the same season, the first being in mid-October and the second in mid-December (Fig. 1). The first active nests were found during the driest season of the five sampling years (Figs 1–2). However, most of the fledglings left their nests from October to January when the first peak of annual rainfall occurred (Figs 1–2).

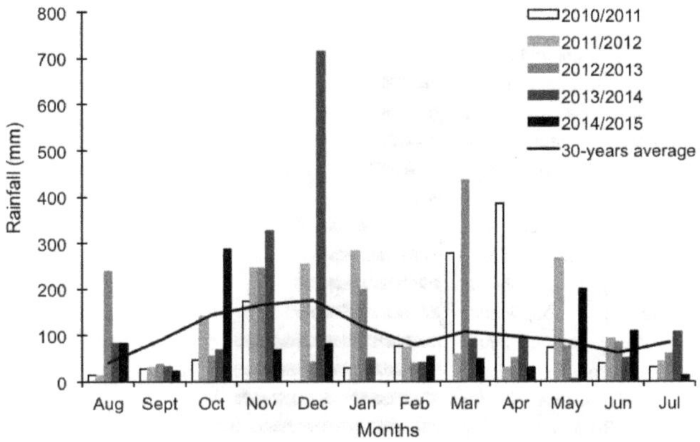

Figure 2. Average monthly rainfall in millimeters during the years 2010 to 2015 in the Municipality of Guarapari, southeastern Brazil.

Table 1. Period and duration of the breeding seasons of *Mimus gilvus* in a Restinga habitat (sand-coastal plain), southeastern Brazil.

Breeding season	First date of active nest	Last date of fledgling	Breeding season duration (weeks)
2010/2011	12/Sep/2010	11/Feb/2011	21.8
2011/2012	01/Aug/2011	19/Feb/2012	29.0
2112/2013	16/Sep/2012	21/Mar/2013	26.7
2013/2014	22/Aug/2013	24/Feb/2014	26.7
2014/2015	10/Aug/2014	08/Feb/2015	26.1

The breeding pairs made up to four nesting attempts in the same breeding season. The average number of nesting attempts was 2.6 ± 1.0 (n = 14), but we did not record couples that had success in one nesting attempt making another attempt in the same breeding season.

The average time spent on nest building was 6.7 ± 1.8 days (n = 15). The remaining nests were found in an advanced stage of construction. Both sexes were observed building nests (n = 19). New nests were built at each reproductive attempt without using material from previous nests. Nests were not repaired after egg laying.

Nest description

Tropical Mockingbird nests were cup-shaped and were placed on some branches. Nest exteriors were constructed with thicker sticks, and in the study area, *Eugenia cyclophylla* (Myrtaceae) and *Guapira* sp. (Nyctaginaceae) were commonly used. Nests interiors were lined with a layer of thin plant materials such as grasses, roots, and fragments of *Cassytha filiformis* (Lauraceae).

The plant species most commonly used to support the nests were *Protium icicariba* (Burseraceae, n = 14; 15.7% of nests) and *Byrsonima sericea* (Malpighiaceae, n = 8; 9.0% of nests). Additional species were *Clusia hilariana* (Clusiaceae), *Jacquinia armiralys* (Theophrastaceae) and *Neomitranthes obtuse* (Myrtaceae), which were used for seven nests (Fig. 3).

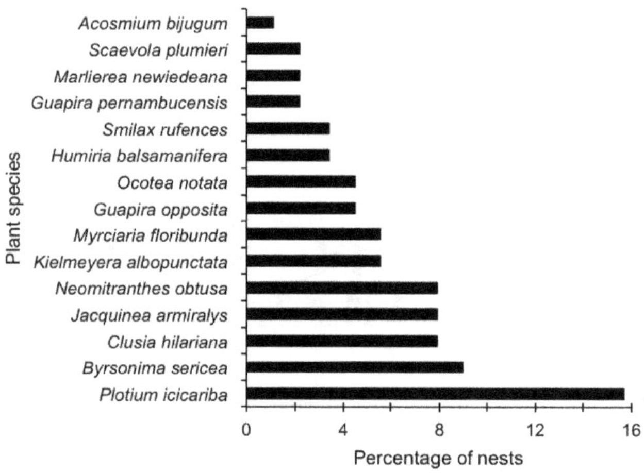

Figure 3. Plant species used for nest construction by *Mimus gilvus* in a Restinga habitat (sand-coastal plain), southeastern Brazil.

Clutch size

The mean clutch size was 2.4 ± 0.6 eggs (range: 1 to 3 eggs, n = 169) (Fig. 4). There was a significant difference in clutch size across breeding seasons ($H = 18.8$, df = 4, $p < 0.001$). However, the difference only occurred between the first and the last breeding seasons (2010/2011 and 2014/2015) ($U = 301$, $p < 0.001$) (Fig. 4).

Incubation and nestling period

Egg laying occurred on consecutive days with incubation beginning after the laying of the first egg, and the nestlings hatched asynchronously. The mean incubation period was 14 days ± 0.6 (range: 13–15 days; n = 19), and the mean nestling period was 14.5 ± 2.2 days (range: 12–18 days; n = 13). Only females participated in incubation. We recorded males playing a sentinel role during incubation by alerting the female when a potential predator approached the nest (n = 19). While caring for nestling, the couple took turns watching for potential predators and feeding the nestlings, and we did not record parental care behaviors from the other group members, contrasting with the idea there is cooperative breeding behavior in this species. Parental care continued up to 35 days after the fledglings left the nest (n = 22 couples).

Development of nestlings

At birth, the nestling presented grayish fuzz on the back and head, whereas the belly and wings were featherless (n = 77

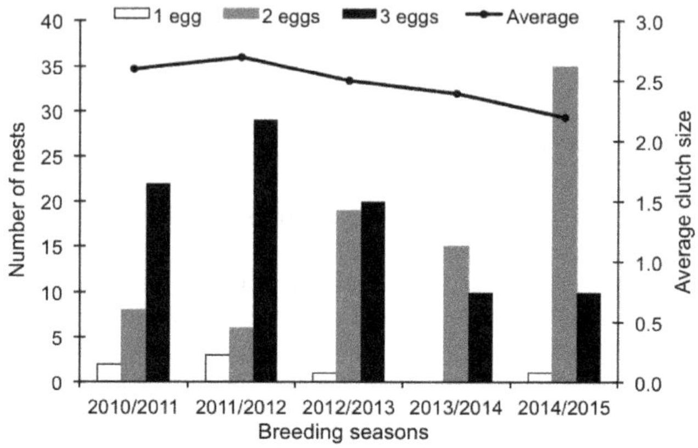

Figure 4. Number of nests and clutch size of *Mimus gilvus* from 2010 to 2015 in a Restinga habitat (sand-coastal plain), southeastern Brazil.

nestlings). The bare skin was pinkish and the beak a strongly yellow. A small calamus of feathers began to appear from the fifth day, first in the wings and with the head and body covered with feathers only in the final days of the nestling period. The nestlings were born with closed eyes, which became fully opened only at the end of the first week. When leaving the nest, the fledglings possessed plumage similar to the adults, except for black spots on the feathers of the breast and flanks, and rectrices that were half the size of the adult rectrices. The fledglings also showed the presence of a yellow gape, and the length of wings and tarsus was similar to adults.

DISCUSSION

Breeding Season and Nest Building. The breeding activities aligned with the rainy season in the study site. This species had a longer breeding season (6 to 7 months) than some Neotropical species from the Atlantic Forest (Aguilar et al. 2000, Duca and Marini 2004, Hoffmann and Rodrigues 2011). The breeding season of the Tropical Mockingbird was shorter than the breeding season of the subspecies *M. gilvus malnopterus* in Venezuela (Paredes et al. 2001). It is noteworthy that the breeding season of the congener *M. gilvus tobagensis*, also lasts six months (Hayes 2005). Variation in breeding season among populations is expected, especially in continental countries, since each region may have different climate types. For example, this variation in breeding period was recorded in the Nearctic region for the Northern Mockingbird (*Mimus polyglottos*) (Laskey 1962, Fischer 1981, Means and Goertz 1983), which is the closest species of *M. gilvus* (Arbosgast et al. 2006). The breeding season can also be influenced by other factors such as food availability, seasonal patterns of predation risk to the nest, and opportunities to renest. Birds can breed early or later each year (Wikelski et al. 2000, Langen and Berg 2016).

Long breeding seasons are expected for tropical species compared with species from temperate regions (Russell et al.

2004, Diniz et al. 2013, Langen and Berg 2016). A long breeding season and many nesting attempts per season may be an adaptive response to a high nest predation rate (Martin 1996, Roper 2005), but some species occasionally re-nest after fledgling a brood (Langen and Berg 2016). The breeding period of the Tropical Mockingbird began a few days before the onset of the rainy season in our study region, and the first fledglings left the nest at the beginning of this season. The relationship between rainfall and the nesting period has been shown by several studies of tropical species in most tropical biomes (Cruz and Andrews 1989, Aguilar and Marini 2007, Marini et al. 2012, Daros et al. 2018). The influence of the rainfall regimen on the breeding period of tropical species is secondarily related to the cascade effects that it triggers, for example, in the abundance of food (Boag and Grant 1984, Sick 1997, Langen and Berg 2016). The nesting period of the Chalk-browed Mockingbird, *Mimus saturninus* (Lichtenstein, 1823), extends from late August to December following the rainy season of central Brazil (Rodrigues et al. 2017).

The two peaks of active nests in the same breeding season observed in two breeding seasons for the Tropical Mockingbird (Fig. 1) are probably related to the re-nesting ability of this species. Biannual breeding cycles may be associated with two wet seasons in a year; these are particularly common in equatorial latitudes (Schondube et al. 2003). Two peaks of active nests within the same season can be a response to predation risk (Roper 2005), and both rainfall and prey availability may contribute (Diniz et al. 2013). It is possible that these two peaks of breeding activity of the Tropical Mockingbird also evolved coinciding with the rainfall pattern. However, it is different from the biannual breeding cycles near the equatorial latitudes because these two peaks of active nests occurred in the same breeding season. In Venezuela, the Tropical Mockingbirds are multi-brooded, and their peaks in breeding activity coincide with the rainy seasons (Paredes et al. 2001). This relationship between precipitation and breeding activities is widely discussed in the literature (e.g., Ricklefs and Bloom 1977, Stutchbury and Morton 2001) and is apparently more representative of semi-arid environments (Cavalcanti et al. 2016). Furthermore, the predation pressure may induce the Tropical Mockingbird to have many nesting attempts in the same breeding season. Studies about timing effects on breeding success and predation rates could elucidate which factor is most important to determine the pattern recorded in this study. It is noteworthy that the Tropical Mockingbird rarely attempted another nesting after success with the previous brood. This observation should be more fully investigated in future studies, but the predation pressure appears to be the main explanatory factor for several nesting attempts and the long breeding season of the Tropical Mockingbird.

Possibly, the long breeding season of Tropical Mockingbird is also associated with the stability of the environmental conditions in the region over the year, allowing the opportunity to renest many times in a single season. A breeding pair of Tropical Mockingbirds may make up to five nesting attempts in the same season (Hayes 2005). The Northern Mockingbirds also make several breeding attempts in a season with the number of attempts influenced by predation and the success of breeders (Laskey 1962). According to Roper (2005), sometimes only a high number of nesting attempts can ensure the seasonal breeding success in tropical regions. In contrast, the breeding phenology of tropical birds can also be related to other factors such as food availability and climate conditions after fledgling, the opportunity to re-nest, and timing of molt (Stutchbury and Morton 2001, Langen and Berg 2016).

The participation of both sexes of the Tropical Mockingbird in nest construction was also observed by Paredes et al. (2001). These authors showed that males select the nest site and are subsequently assisted by the female in nest building. Nest abandonment after construction may be related to the inability of inexperienced adults to perform tasks related to reproduction (Morbey and Ydenberg 2000). However, the energy cost of building a nest that will not be successful is high, and nest abandonment is costly for younger females. Therefore, the adaptive significance of the commitment of young or newly arrived females to the effective population may be the optimization of the number of offspring produced during their lifetime through learning from previous years (Curio 1983).

Plants Used to Support Nests. Tropical Mockingbirds nested on 30 plants species, and most nests were found on *P. icicariba* that has a relative frequency of 4.7% (CEPEMAR 2007) and low dominance in the study site (Ferreira et al. 2010). *Protium icicariba* is a shrub species that has favorable plant architecture and is the most commonly used plant for nesting by Mimidae species in open physiognomies (Argel-de-Oliveira 1994). The use of a common plant as support in the study area, such as *P. icicariba*, supports the potential prey hypothesis (Martin 1993). Nests located in the shrubs of species that are abundant are safer from predators due to the increased availability of potential sites for nest construction, thus reducing the efficiency of predators trying to locate them. The relative frequency of *P. icicariba* is the same for *Guapira pernambucensis* (Casar.) Lundell, but the latter harbored only two nests. Therefore, the preference of Tropical Mockingbird for *P. icicariba* over *G. pernambucensis* may be related to the possible protection that the first offers with respect to nest predation.

Clutch Size. The clutch size of the Tropical Mockingbirds in the study area was similar to another population in Venezuela (Paredes et al. 2001) for which a mean of 2.2 eggs (range, 2–3 eggs) was reported. For the tropical congener Chalk-browed Mockingbird, the mean clutch size was 2.9 eggs (Rodrigues et al. 2017). These low numbers of eggs are typical for tropical species (Stutchbury and Morton 2008). The high predation rate that occurs in the tropics favors smaller clutch sizes because adults invest in smaller numbers of eggs and many nesting attempts to ensure greater success in the breeding season (Slagsvold 1982). Regardless the regulatory mechanism of clutch size, the Tropical Mockingbird has a small clutch size and makes up to four nesting attempts in the same breeding season, which is consistent with expectations for other tropical species.

Incubation and Nestling Periods. The 14 days of incubation of the Tropical Mockingbird in our study was similar to that of the other two subspecies, *M. gilvus melanopterus* (13 days; Paredes et al. 2001) and *M. gilvus tobagensis* (13 days; Hayes 2005), and to the Mimidae species Long-tailed Mockingbird (*Mimus longicaudatus*) (12 to 13 days; Marchant 1960), Chalk-browed Mockingbird (14.2 days; Rodrigues et al. 2017), and White-banded Mockingbird (*Mimus triurus*) (13 days; Mezquida and Marone 2001). According to Sick (1997), the mean incubation period of Atlantic Forest passerines is 15 days, close to what has been observed in the present study. However, slight variations may occur due to environmental conditions and food availability (Murphy 1986, Rotenberry and Wiens 1991).

The nestling period of the Tropical Mockingbird (14.5 days) is similar to that of other Mimidae species (Marchant 1960, Rodrigues et al. 2017). Short incubation and nestling periods can be related to the type of and vulnerability to nest predators (Alves and Cavalcanti 1990). These two factors reduce the exposure time to nest predators (Martin 1987). However, leaving the nest too soon without sufficient time to develop significant features for flight can also expose the fledglings to predation. Thus, each species may adopt a different strategy to increase their breeding success.

The roles of each sex in reproduction appear to follow a common pattern for Mimidae species, in which only females incubate the eggs and males defend the nests against predators. This behavior was also observed for Chalk-browed Mockingbirds (Rodrigues et al. 2017) and Northern Mockingbirds (Laskey 1962). Although evidence of facultative cooperative breeding behavior has been documented for the Tropical Mockingbird (Morton et al. 2004), we did not record behaviors that suggest cooperative breeding in our study.

Nestling Development. The nestlings of the Tropical Mockingbird are typical altricial birds that usually hatch featherless and with the eyes closed, completely dependent on the care of adults (Starck and Ricklefs 1998). The offspring resemble those of the Chalk-browed Mockingbird (Rodrigues et al. 2017) and Northern Mockingbird (Laskey 1962). The similar length of the wings and tarsus of fledglings relative to adults is explained by the need for movement on the ground during the first days out of the nest, thus helping them to escape from predators (Sick 1997).

In this study, we highlight that six to seven continuous months of breeding season with up to two peaks of active nests per season are uncommon for a non-equatorial neotropical bird species, even compared with Neotropical birds from Cerrado, Pantanal and other physiognomies of the Atlantic Forest. The knowledge gained from this study makes the Tropical Mockingbird a good choice for future studies, particularly for testing ecological and evolutionary hypotheses regarding life history attributes, habitat selection and parental investment.

ACKNOWLEDGMENTS

We thank Rodrigo B. Ferreira and an anonymous referee for comments on the manuscript; IEMA for a sampling permit. W.B. Dutra, H. Daros, M. Percilios, U. Sabino, T.E.C. Tosta, and F. Eller for fieldwork support. R. Morais received a scholarship from Fundação de Amparo a Pesquisa e Inovação do Espírito Santo (FAPES). This study was funded by Fundação Nacional de Desenvolvimento do Ensino Superior Particular (FUNADESP) and the Conselho Nacional de Desenvolvimento Científico e Tecnológico (CNPq, process 45.6446/2014-1).

LITERATURE CITED

Aguilar TM, Marini MÂ (2007) Nest and nest-site reuse within and between breeding seasons by three Neotropical flycatchers (Tyrannidae). Brazilian Journal of Biology 67: 537–540. https://doi.org/10.1590/S1519-69842007000300020

Aguilar TM, Coelho MM, Marini MÂ (2000) Nesting biology of the Gray-hooded Flycatcher (*Mionectes rufiventris*). Ornitologia Neotropical 11: 223–230.

Alvares CA, Stable JL, Sentelhas PC, Gonçalves JLM, Sparovek G (2013) Köppen's climate classification map for Brazil. Meteorologische Zeitschrift 22: 711–728. https://doi.org/10.1127/0941-2948/2013/0507

Alves MAS, Cavalcanti RB (1990) Ninhos, ovos e crescimento de filhotes de *Neothraupis fasciata*. Ararajuba 1: 91–94.

Alves, MAS, Pacheco JF, Gonzaga LAP, Cavalcanti RB, Raposo M, Yamashita C, Maciel NC, Castanheira M. (2000) Aves. In: Bergallo HG, Duarte-da-Rocha CF, Alves MAS, Van Sluys M (Orgs) A fauna ameaçada de extinção do estado do Rio de Janeiro. Rio de Janeiro, Editora UERJ, 113–124.

Araujo DSD, Maciel NC (1998) Restingas fluminenses: biodiversidade e preservação. Boletim FBCN 25: 25–51.

Arbogast BS, Drovetski SV, Curry RL, Boag PT, Seutin G, Grant PR, Grant BR, Anderson DJ (2006) The origin and diversification of Galapagos mockingbirds. Evolution 60: 370–382. https://doi.org/10.1554/03-749.1

Argel de Oliveira MM (1994) Familia Mimidae. Boletim CEO 10: 3–14. Available onlie at: http://www.ceo.org.br/bolet/bolceo10.pdf

Argel-de-Oliveira MM, Pacheco JF (1998) Um resumo da situação: *Mimus saturninus* e *M. gilvus* no litoral sudeste brasileiro. Boletim FBCN 25: 53–69.

Ayres M, Ayres-Junior M, Ayres DL, Santos AS (2007) Bioestat 5.0: aplicações estatísticas nas áreas das ciências biomédicas. Belém, Sociedade Civil de Mamirauá, CNPq, 364 pp.

Bartholomew GA (1986) The role of natural history in contemporary biology. BioScience 36: 324–329. https://www.jstor.org/stable/1310237

Bennett PM, Owens IPE (2002) Evolutionary ecology of birds: life history, mating system and extinction. NewYork, Oxford University Press.

Boag PT, Grant PR (1984) Darwin's Finches (Geospiza) on Isla Daphne Major, Galapagos: breeding and feeding ecology in a climatically variable environment. Ecological Monographs 54: 463–489. https://doi.org/10.2307/1942596

Botero CA, Riveros JM, Vehrencamp SL (2007) Relative threat and recognition ability in the responses of tropical mockingbirds to song playback. Animal Behaviour 73: 661–669. https://doi.org/10.1016/j.anbehav.2006.10.007

Cavalcanti LMP, Paiva LV, França LF (2016) Effects of rain fall on bird reproduction in a semi-arid Neotropical region. Zoologia 33: 1–6. https://doi.org/10.1590/s1984-4689zool-20160018

CEPEMAR (2007) Plano de Manejo do Parque Estadual Paulo César Vinha. Relatório Técnico COM RT. Vitória, CEPEMAR Ambiental.

Cody M (2005) Family Mimidae (Mockingbirds and Thrashers) 448-495. In: Hoyo JD, Elliott A, Christie AD (Eds) Handbook of the birds of the world. Barcelona, Lynx Editions, vol. 10, 895 pp.

Cruz A, Andrews RW (1989) Observations on the breeding biology of Passerines in a seasonally flooded savanna in Venezuela. Wilson Bulletin 101: 62–76.

Curio E (1983) Why do young birds reproduce less well? Ibis 125: 400–404. https://doi.org/10.1111/j.1474-919X.1983.tb03130.x

Daros H, Dutra WB, Duca C (2018) Breeding biology of *Tyrannus melancholicus* (Aves: Tyrannidae) in a resting reserve of southeastern Brazil. Zoologia 35: e24569. https://doi.org/10.3897/zoologia.35.e24569

Diniz P, Ramos DM, Chemin N (2013) Breeding biology of Coal-crested Finches. The Wilson Journal of Ornithology 125: 592–599. https://doi.org/10.1676/12-166.1

Duca C, Marini MÂ (2004) Aspectos da nidificação de *Cacicus haemorrhous* (Passeriformes, Icterinae) no sudeste do Brasil. Ararajuba 12: 23–30.

Duca C, Marini MÂ (2011) Variation in breeding of the Shrike-Like Tanager in Central Brazil. The Wilson Journal of Ornithology 123: 259–265. https://doi.org/10.1676/10-116.1

Fabris LC, César O (1996) Estudos florísticos em uma mata litorânea no sul do estado do Espírito Santo. Boletim do Museu de Biologia Mello Leitão, Nova Série, 5: 15–46.

Ferreira LC, Thomazi RD, Oliveira DAC, Silva AG (2010) Estrutura populacional e padrão espacial de *Protium icicariba* (DC.) Marchand na Área de Proteção Ambiental de Setiba, Espírito Santo, sudeste do Brasil. Natureza on line 8: 39–45.

Fischer DH (1981) Factors affecting the reproductive success of the northern mockingbird in South Texas. The South western Naturalist 26: 289–293. https://doi.org/10.2307/3670907

Gomes VSM, Correia MCRH, Lima AL, Alvez MAS (2007) Potential role of frugivorous birds (Passeriformes) on seed dispersal of six plant species of a restinga habitat, southeastern Brazil. Revista de Biologia Tropical 56: 205–216. https://doi.org/10.15517/rbt.v56i1.5519

Gonzaga LP, Castiglioni GDA, Reis HBR (2000) Avifauna das restingas do Sudeste: estado do conhecimento e potencial para futuros estudos. In: Esteves FA, Lacerda LD (Eds) Ecologia de restingas e lagoas costeiras. Macaé, NUPEM/UFRJ, 179–188.

Hammer ØD, Harper AT, Ryan PD (2001) PAST: Paleontological Statistics software package for education and data analysis. Paleontologia Electronica 4: 1–9.

Hayes FE (2005) Apartial double-layered eggshell in the Tropical Mockingbird (*Mimus gilvus*). Ornitologia Neotropical 16: 263–266.

Hoffmann D, Rodrigues M (2011) Breeding biology and reproductive success of *Polystictus superciliaris* (Aves: Tyrannidae), an uncommon tyrant-flycatcher endemic to the highlands of eastern Brazil. Zoologia 28: 305–311. https://doi.org/10.1590/S1984-46702011000300000

Langen TA, Berg EC (2016) What determines the timing and duration of the nesting season for a tropical dry forest bird, the White-throated magpie-jay (*Calocitta formosa*)? The Wilson Journal of Ornithology 128: 32–42. https://doi.org/10.1676/wils-128-01-32-42.1

Laskey AR (1962) Breeding biology of Mockingbirds. The Auk 79: 596–606. https://doi.org/10.2307/4082641

Machado ÂBM, Martins CS, Drummond GM (2005) Lista da fauna brasileira ameaçada de extinção: incluindo as listas das espécies quase ameaças e deficientes em dados. Belo Horizonte, Fundação Biodiversitas, 160 pp.

Marchant S (1960) The breeding of some S.W. Ecuadorian Birds. Ibis, 102(3): 349–382. https://doi.org/10.1111/j.1474-919X.1960.tb08415.x

Marini MÂ, Borges FJA, Lopes LE, Sousa NOM, Gressler DT, Santos LR, Paiva LV, Duca C, Manica LT, Rodrigues SS, França LF, Costa PM, França LC, Heming NM, Silveira MB, Pereira ZP, Lobo Y, Medeiros RCS, Roper JJ (2012) Breeding biology of birds in the Cerrado of central Brazil. Ornitologia Neotropical 23: 385–405.

Martin TE (1987) Food as a limit on breeding birds: alife-history perspective. Annual Review of Ecology and Systematic 18: 453–487. https://doi.org/10.1146/annurev.es.18.110187.002321

Martin TE (1993) Nest predation among vegetation layers and habitat types: revising the dogmas. American Naturalist 141: 897–913. https://doi.org/10.1086/285515

Martin TE (1996) Life history evolution in tropical and south temperate birds: What do we really know? Journal of Avian Biology 27: 263–272. https://doi.org/10.2307/3677257

Martin P, Baetson P (1993) Measuring Behaviour: An introductory guide. Cambridge, Cambridge University Press, 2nd ed., 199 pp.

Mason P (1985) The nesting biology of some passerines of Buenos Aires, Argentina. In: Buckley PA, Foster MS, Morton ES, Ridgely RS, Buckley FG (Eds) Neotropical Ornithology. Ornithological Monographs 36. Washington, DC, American Ornithologist's Union, 954-972. https://doi.org/10.2307/40168328

Means LL, Goertz JW (1983) Nesting activities of Northern Mockingbirds in northern Louisiana. The Southwestern Naturalist 28: 61–70. https://doi.org/10.2307/3670593

Mezquida ET, Marone L (2001) Factors affecting nesting success of a bird assembly in the Central Monte Desert, Argentina. Journal of Avian Biology 32: 287- 296. https://doi.org/10.1111/j.0908-8857.2001.320401.x

MMA (2000) Avaliação e ações prioritárias para a conservação da biodiversidade da Mata Atlântica e Campos sulinos. Brasília, Secretaria de Biodiversidade e Florestas (SBF), Ministério do Meio Ambiente.

Morbey YE, Ydenberg RC (2000) Season al decline in nestling growth: support for the parental-quality hypothesis in Cassin's Auklets. The Auk 117: 1065–1068. https://doi.org/10.1642/0004-8038(2000)117[1065:SDINGS]2.0.CO;2

Morton ES, Stutchbury BJM, Piper WH (2004) Cooperative breeding in the Tropical Mockingbird (*Mimus gilvus*) in the Panama canal zone. Ornitologia Neotropical 15: 417–421.

Murphy MT (1986) Temporal components of reproductive variability in Eastern Kingbirds (*Tyrannus tyrannus*). Ecology 67: 1483–1492. https://doi.org/10.2307/1939079

Paredes M,Weir E, Gil K (2001) Reproduccíon del ave *Mimus gilvus* (Passeriformes: Mimidae) em Maracaibo,Venezuela. Revista Biologia Tropical 49: 1143–1146.

Passamani M, Mendes SL (2007) Espécies da fauna ameaçadas de extinção no Estado do Espírito Santo. Vitória, Instituto de Pesquisas da Mata Atlântica.

Pereira OJ (2003) Restinga: origem, estrutura e diversidade. In: Jardin MAG, Bastos NNC, Santos JUM (Eds) Desafios da botânica brasileira no novo milênio: Inventário, sistematização e conservação da diversidade vegetal. Belém, UFRA, Embrapa, Museu Paraense Emílio Goeldi, 177–179.

Pienaar J, Ilany A, Geffen E, Yom-Tov Y (2013) Macroevolution of life-history traits in passerine birds: adptation and phylogenetic inertia. Ecology Letters 16: 571–576. https://doi.org/10.1111/ele.12077

Ricklefs RE (2010) Economia da natureza. Rio de Janeiro, Guanabara koogan, 6th ed., 546 pp.

Ricklefs RE, Bloom G (1977) Components of avian breeding productivity. Auk 94: 86–96.

Rocha CFD, Bergallo HG, Van Sluys M, Alves M, Jamel CE (2007) The remnants of restinga habitats in the Brazilian Atlantic Forest of Rio de Janeiro state, Brazil: habitat loss and risk of disappearance. Brazilian Journal of Biology 67: 263–273. https://doi.org/10.1590/S1519-69842007000200011

Rodrigues SS, Lopes LE, Marini MM (2017) Breeding biology of Chalk-browed Mockinbird *Mimus saturninus* in a natural savana of central Brazil. Revista Brasileira de Ornitologia 25: 237–244.

Roper JJ (2005)Tryandtryagain: Nest predation favors persistence in neotropical birds. Ornitologia Neotropical 16: 253–262.

Rotenberry JT, Wiens JA (1991) Weather and reproductive variation in shrubsteppe sparrows: a hierarchical analysis. Ecology 72: 1325–1335. https://doi.org/10.2307/1941105

Russell EM, Yom-Tov Y, Geffen E (2004) Extended parental care and delayed dispersal: northern, tropical, and southern passerines compared. Behavioral Ecology 15: 831–838. https://doi.org/10.1093/beheco/arh088

Schondube JE, Castellon ES, Ruan-Tejeda I (2003) Biannual cycles of the Cinnamon bellied Flower piercer. Biotropica 35: 250–261. https://doi.org/10.1646/03026

Sick H (1997) Ornitologia Brasileira. Rio de Janeiro, Editora Nova Fronteira, 912 pp.

Slagsvold T (1982) Clutch size variation in passerine birds: the nest predation hypothesis. Oecologia 54: 159–169. https://doi.org/10.1007/BF00378388

Starck JM, Ricklefs RE (1998) Avian Growth and Development, Evolution within the Altricial-Precocial Spectrum. New York, Oxford University Press.

Stutchbury BJM, Morton ES (2001) Behavioral Ecology of Tropical Birds. San Diego, Academic Press, 165 pp.

Stutchbury BJM, Morton ES (2008) Recent advances in the behavioral ecology of tropical birds. The Wilson Journal of Ornithology 120: 26–37. https://doi.org/10.1676/07-018.1

Teixeira DE, Nacinovic JB (1992) Aves da Barra da Tijuca, 133–145. In: Carvalho-Hosken (Ed.) Parque da Gleba E, Rio de Janeiro: a entidade. Rio de Janeiro, Carvalho Hosken SA, 152 pp.

Wikelski M, Hau M, Wingfield JC (2000) Seasonality of reproduction in a neotropical rain forest bird. Ecology 81: 2458–2472. https://doi.org/10.1890/0012-9658(2000)081[2458:SORIAN]2.0.CO;2

Xiao H, Hu Y, Lang Z, Fang B, Guo W, Zhang Q, Pan X, Lu X (2016) How much do we know about breeding biology of bird species in the world? Journal of Avian Biology 48: 513–518. https://doi.org/10.1111/jav.00934

Zanon MS, Vale MM, Alves MAS (2015) Missing for the last twenty years: the case of the southernmost populations of the Tropical Mockingbird *Mimus gilvus* (Passeriformes: Mimidae). Zoologia 32: 1–8. https://doi.org/10.1590/S1984-46702015000100001

Author Contributions: RM and CD designed the experiments; RM, LCA and GRS conducted the experiments; RM, LCA and CD analyzed the data and wrote the paper.

Competing Interests: The authors have declared that no competing interests exist.

4

Association between deep-water scale-worms (Annelida: Polynoidae) and black corals (Cnidaria: Antipatharia) in the Southwestern Atlantic

José Eriberto De Assis[1], **José Roberto B. de Souza**[2], **Manuela M. de Lima**[2], **Gislaine V. de Lima**[2], **Ralf T.S. Cordeiro**[2], **Carlos D. Pérez**[3]

[1]*Departamento de Educação, Secretaria Municipal de Educação, Prefeitura Municipal de Bayeux. 50670-420 Bayeux, PB, Brazil.*
[2]*Departamento de Zoologia, Centro de Biociências, Universidade Federal de Pernambuco. 50670-420 Recife, PE, Brazil.*
[3]*Centro Acadêmico de Vitória, Universidade Federal de Pernambuco. 55608-680 Vitória de Santo Antão, PE, Brazil.*
Corresponding author: José Eriberto De Assis (eri.assis@gmail.com)

ABSTRACT. Polynoid scale-worms have been found living as commensals with deep-water antipatharians (commonly known as black corals) in the Potiguar Basin, off Rio Grande do Norte State, Northeastern Brazil. In this paper two polychaete species and four black corals species are redescribed. *Benhamipolynoe* cf. *antipathicola* and *Parahololepidella* cf. *greeffi*, and the black coral *Stylopathes adinocrada* Opresko, 2006 are recorded for the Southwestern Atlantic. *Benhamipolynoe* cf. *antipathicola* was first described from off New Zealand and the Malay Archipelago, as symbiont with the black coral *Stylopathes tenuispina* (Silberfeld, 1909). It was later reported for the North Atlantic, off Florida, associated with *Stylopathes columnaris* (Duchassaing, 1870). In our study, *B.* cf. *antipathicola* was found in association with the black coral *S. adinocrada*. *Parahololepidella* cf. *greeffi* was first described as a free-living from shallow waters off São Tomé and Cabo Verde Islands, West Africa, and later reported as symbiont with the black coral *Tanacetipathes* cf. *spinescens* in the same location. Our data expand both the geographical distribution and the host range of this species which is reported for the first time as symbiont with *Tanacetipathes barbadensis* (Brook, 1889), *T. tanacetum* (Pourtalès, 1880) and *T. thamnea* (Warner, 1981) in Brazil. The aim of this study is to discuss commensal associations between two species of scale-worm polynoids and black corals found in the Southwestern Atlantic, and also reporting their global distribution. Finally, we provided an updated list of the commensal polynoids and their black coral hosts.

KEY WORDS. Brazil, Hexacorallia, polynoids, South America, symbiosis.

INTRODUCTION

Specialized symbiotic associations involving polychaetes are ubiquitous in all oceans (Britayev and Antokhina 2012). The Polynoidae, also known as polynoids or scale-worms, includes the largest number and the most commonly reported species living as commensals, representing more than 45% of all known reports of symbiont polychaetes (Martin and Britayev 1998, 2018). More than 200 species are involved in about 600 relationships, representing about 25% of all known polynoid species (Britayev et al. 2014, Martin and Britayev 2018, Serpetti et al. 2017).

Polynoids are often found associated with other invertebrates, especially corals, echinoderms, mollusks, and other polychaetes (Pettibone 1991, Martin and Britayev 1998, 2018, Eckelbarger et al. 2005). They may either construct their own refuges on the host's surface or stimulate their hosts to build protective structures around them. Some species may live inside tunnels or gall-like cavities formed by coenenchymal walls of gorgonian or hydrocoral hosts (Martin and Britayev 1998, 2018, Williams and López-González 2005, Wagner et al. 2012).

Cnidarians are known to host abundant and diverse symbiotic fauna, including polychaetes, with records of 281 species of cnidarian hosts involved in 324 relationships with their hosts (Molodtsova et al. 2016). Antipatharia (Cnidaria: Anthozoa) is the least studied group of corals (Cairns 2007), particularly in the South Atlantic (Loiola 2007). Their colonies consist of chitinous skeletons covered with numerous tiny spines. Some families have developed close relationships with certain scale worms, forming tunnels by bending and anastomosing its branches or pinnules, providing ideal refuges (Molodtsova and Budaeva

2007). Black corals are generally slow-growing and long-living organisms, generally found in areas with hard substrates, low light levels and strong currents; with longevity ranging from 12 to 4265 years, in *Antipathes griggi* Opresko, 2009, known formerly as *Antipathes dichotoma* Pallas, 1766, and *Leiopathes* sp., respectively (Molodtsova and Budaeva 2007, Wagner et al. 2012, Brugler et al. 2013). Along with octocorals, antipatharians are habitual hosts not only for polychaetes, but also for a myriad of associated organisms including barnacles, ophiuroids, copepods, crabs, shrimps, anemones, zoanthids, hydroids, crinoids, bryozoans, snails, bivalves, tunicates and fishes (Buhl-Mortensen and Mortensen 2004, Wagner et al. 2012). The aim of this study is to discuss commensal associations between two species of scale-worm polynoids and black corals found in the Southwestern Atlantic. We redescribe and illustrate the specimens found, also reporting their global distribution and providing an updated list of the commensal polynoids and their black coral hosts.

MATERIAL AND METHODS

The specimens of scale-worms and their antipatharian hosts were collected in the Potiguar Basin, Rio Grande do Norte State, Northeastern Brazil (Fig. 1), by trawling during the surveys of the Program for megafauna characterization of the Potiguar Basin (funded by PETROBRAS), on board of the RV Seward Johnson, between 101 and 461 m depth. Specimens were fixed in a 4% formaldehyde-seawater solution, then rinsed with fresh water and preserved in 70% ethanol. The hosts were observed under a Zeiss stereomicroscope.

Polychaete identifications were based on chaetae, aciculae, and elytra observed under an Olympus BX41 compound microscope. Optical microscope Images were obtained with a Leica M205A. We also provide detailed scanning electron microscope (SEM) images of the parapodial structures and elytra. Measurements are given in mm. The polychaete structures were analyzed according to Pettibone (1989, 1991) and Britayev et al. 2014.

Black corals were identified and described following Loiola and Castro (2005) and Opresko (2006), mainly based on pinnule and subpinnule number, order and size, branching pattern, average distance between spines, and subpinnule number per pinnule using a Leica DM 300 compound microscope. Branche microstructures were also illustrated by SEM.

Voucher specimens were deposited in the Museu de Oceanografia Prof. Petrônio Alves Coelho (MOUFPE-CNI), Universidade Federal de Pernambuco, Brazil.

TAXONOMY

An updated list of the polychaetes associated to black corals, including the new records found in our study is presented in Table 1. The symbiotic partnerships found in Potiguar Basin are reported below.

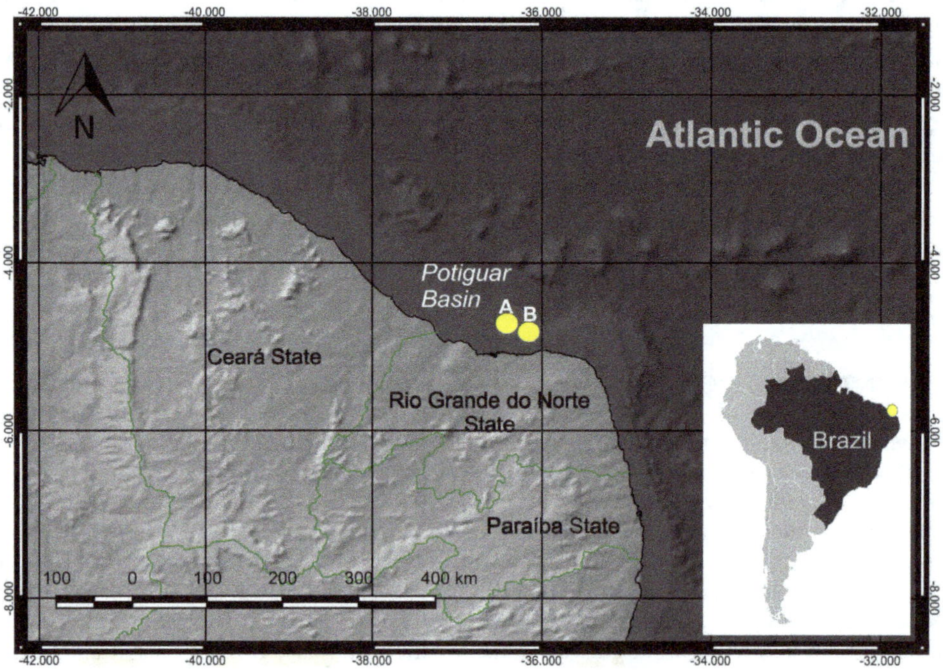

Figure 1. Records of association of scale-worm polynoids and black corals from deep-water in South America, Northeastern Brazil: (A) record of *Benhamypolynoe* cf. *antipathicola,* and its antipatharian hosts *Stylopathes adinocrada* in Potiguar Basin; (B) records of association of *Parahololepidella* cf. *greeffi* and its antipatharian hosts (*Tanacetipathes barbadensis, T. tanacetum* and *T. thamnea*) in Potiguar Basin, Brazil.

Table 1. List of known polychaete species associated with antipatharian hosts: 1: Pettibone (1970), 2: Wagner et al. (2012), 3: Britayev et al. (2014), 4: Pettibone (1991), 5: Martin and Britayev (1998), 6: Hanley and Burke (1991), 7: Barnich et al. (2013), 8: Molodtsova and Budaeva (2007), 9: Hartmann-Schröder and Zibrowius (1998), 10: Molodtsova et al. (2016), 11: Glasby (1994), O: Obligatory, F: Facultative.

Symbiont	Host antipatharian	Type of association	References
Polynoidae			
Benhamipolynoe cf. *antipathicola* (Benham, 1927)	Aphanithidae: *Asteriopathes arachniformis* Opresko, 2004	O	8, 10
	Stylopathidae: *Stylopathes tenuispina* Silberfeld, 1909	O	1
	Stylopathes columnaris (Duchassaing, 1870)	O	1, 2
	Stylopathes adinocrada Opresko, 2006, New host	O	This paper
Parahololepidella cf. *greeffi* (Augener, 1918)	Myriopathidae: *Tanacetipathes* cf. *spinescens* (Gray, 1857)	F	3
	Tanacetipathes barbadensis (Brook, 1889), New host	F	This paper
	Tanacetipathes tanacetum (Pourtalès, 1880), New host	F	This paper
	Tanacetipathes thamnea (Wagner, 1981), New host	F	This paper
Antipathypolyeunoa nuttingi Pettibone, 1991	*Tanacetipathes tanacetum* (Pourtalès, 1880)	F	2, 4
Bayerpolynoe floridensis Pettibone, 1991	Stylopathidae: *Stylopathes columnaris* (Duchassaing, 1870)	O	2, 5
Brychionoe karenae Hanley and Burke, 1991	Leiopathidae: *Leiopathes* sp.	O	6
Eunoe purpurea Treadwell, 1936	Schizopathidae: *Bathypathes* cf. *alternata* Brook, 1889	F	7
Neohololepidella antipathicola Hartmann-Schröder & Zibrowius, 1998	Aphanipathidae: *Elatopathes abietina* (Pourtalès, 1874)	F	8, 9
	Distichopathes filix (Pourtalès, 1867)		
Tottonpolynoe symantipatharia Pettibone, 1991	Schizopathidae: *Parantipathes* sp.	F	4
Eunicidae			
Eunice antipathum (Pourtalès, 1867)	Myriopathidae: *Distichopathes filix* (Pourtalès, 1867)	F	8, 9
	Elatopathes abietina (Pourtalès, 1874)		
Eunice kristiani (Hartmann-Schröder & Zibrowius, 1998)	*Anthipates cylindrical* Brook, 1889	F	8
Eunice marianae (Hartmann-Schröder & Zibrowius, 1998)	*Anthipates cylindrical* Brook, 1889	F	8
Syllidae			
Bollandiella antipathicola (Glasby, 1994)	Myriopathidae: *Anthipates* sp.	F	11

Benhamipolynoe cf. *antipaticola* and *Stylopathes adinocrada* Symbiotic scale-worm

Polynoidae

Benhamipolynoe Petttibone, 1970

Benhamipolynoe cf. *antipaticola* (Benham, 1927)

Figs 2–7

Lapidasthenia antipaticola Benham, 1927: 7–182, pl. 1–6.

Description. Body brownish, with continuous middorsal longitudinal reddish-brown band, transverse bands from elytra near parapodial bases. Body dorsoventrally flattened, elongate, slender, with nearly parallel sides, tapering posteriorly, subrectangular in cross-section. Seventeen symmetrical pairs of elytra; elytra colorless, transparent with median crescent-shaped dark pigmentation; on elytrophores of segments 2, 4, 5, 7, alternate segments to 29, 32, and 35; elytra continue to middle body, irregularly distributed, may be asymmetrical, with an elytron and cirrus on same segment, or from one to five cirri in succession (Figs 2–3). Elytra lacking fringes, papillae and tubercles, except for some scattered micro-tubercles. Prostomium bilobed, as wide as long, without cephalic peaks; with three antennae and two short, tapering, smooth palps. Ceratophore of median antenna short, cylindrical, inserted in anterior notch of prostomium; lateral antennae inserted terminally or subterminally on anterior part of prostomium, at the same level or slightly ventral to median antenna (without distinct ceratophores); antennae longer than prostomium, tapered, smooth; two pairs of eyes moderate in size; anterior pair on dorsolateral prostomial width, posterior pair near to posterior border. Tentacular segment anterolateral to prostomium, bearing two pairs of long tentacular cirri, upper pairs much longer than lower pairs; basis of tentacular cirri with aciculum, but lacking chaetae. Buccal segment lacking distinct nuchal fold; supporting first pair of dorsal elytra and long ventral buccal cirri (Fig. 4). Parapodia sub-biramous, with small notopodia on anterodorsal lobe; prechaetal lobe quadrangular or subtriangular, postchaetal short and subtriangular, both with acicula. Notopodia short, conical, achaetous acicular lobe. Neuropodia elongate, with a ventral and dorsal deep incision; with subequal, rounded prechaetal and postchaetal lobes (Fig. 5); with 5–11 neurochaetae, rather stout, smooth or with slight indications of spinous rows and falcate tips (Figs 6–7). Ventral cirri short, subulate, extending slightly beyond neuropodial lobe tips. Dorsal cirri with short cylindrical cirrophores; styles smooth, tapering gradually to filiform tips, extending beyond neurochaetae tips. Nephridial papillae short, starting at chaetiger 6 (Fig. 5). Pygidium short, with one pair of long anal cirri.

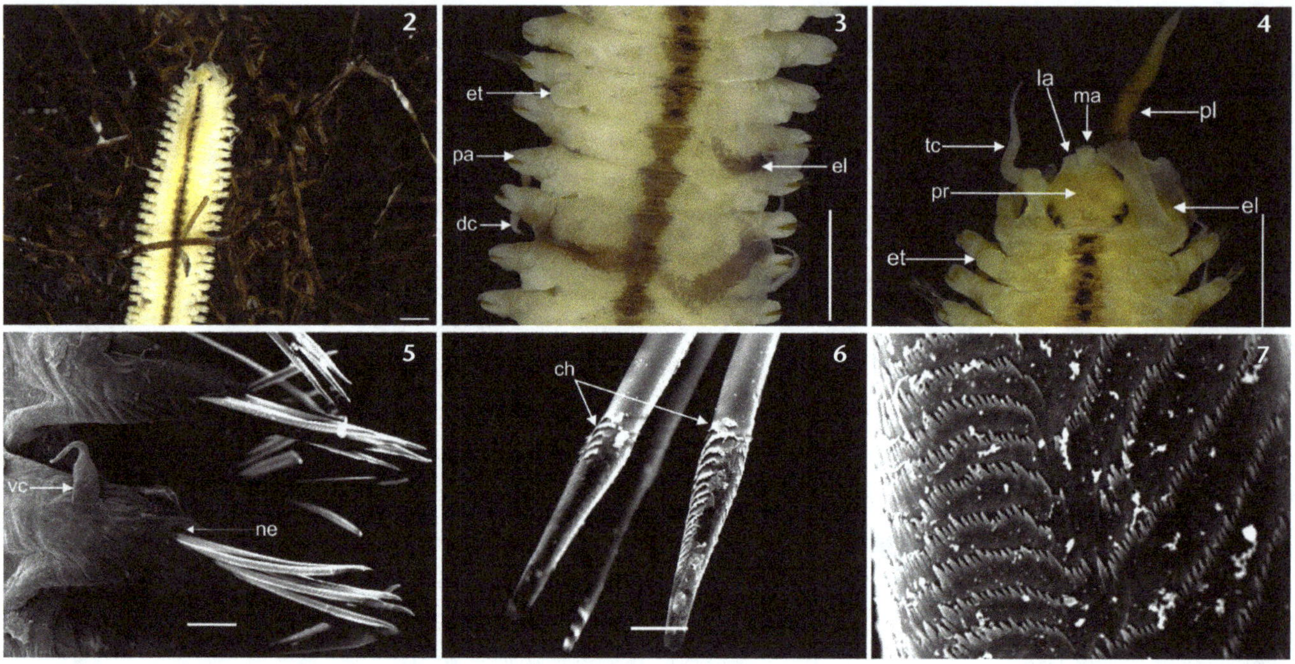

Figures 2–7. (2) *Benhamypolynoe* cf. *antipathicola* on the branches of *Stylopathes adinocrada*: (3) median chaetigers of *B. antipathicola* showing elytrophores, elytra and dorsal cirri; (4) head of *B. antipathicola* showing the prostomium, palps and antennae; (5) posterior chaetigers of *B. antipathicola* showing the ventral cirrus and neuropodia; (6) neurochaetae from posterior chaetigers, showing a faint serration; (7) detail of distal tip of neurochaetae. (ch) chaetae, (dc) dorsal cirrus, (el) elytra, (et) elytrophore, (ct) cirratophore, (la) lateral antennae; (ma) median antennae, (ne) neuropodium, (pa) parapodium, (pl) palp, (pr) prostomium, (tc) tentacular cirrus. Scale bars: 2–4: 1 mm, 5 = 200 μm, 6 = 100 μm, 7 = 10 μm.

Material examined. A complete specimen with 136 chaetigerous segments, with 103 mm in length, and 4 mm wide. MOUFPE-CNI 113; 04°47′50″S; 036°11′01″W (Potiguar Basin), 423–461 m depth (1 specimen).

Distribution. Pacific: Off New Zealand, 128 m depth (Benham 1927), Queensland, Australia, 398 m depth (Pettibone 1989), off New Caledonia and adjacent waters, French Polynesia, Wallis and Futuna, Solomon Islands, Indonesia, and Madagascar, from 300–600 m depth (Molodtsova and Budaeva 2007), Malay Archipelago; Atlantic: off Key West, Florida, 237 m depth (Pettibone 1970) and Northeastern Brazil, 423–461 m depth (present study), first report for the Southwestern Atlantic.

Remarks. *Benhamipolynoe antipaticola* was originally described as *Lapidasthenia antipaticola* Benham, 1927, from 128 m depth off New Zealand, in association with *Stylopathes tenuispina* (Silberfeld, 1909) (= *Parantipathes tenuispina*) (Opresko 2006). The morphological features of the Brazilian specimens agree with the original description, except in the absence of pigmentation in the holotype, according to Pettibone (1989).

Ecology. Found in association with *Stylopathes adinocrada* Opresko, 2006, a new host for this species. It is also known to live in association with *Stylopathes columnaris* (Duchassaing, 1870) (Pettibone 1970, Britayev et al. 2014).

Host black coral

Stylopathidae Opresko, 2006

Stylopathes adinocrada (Opresko, 2006)

Figs 8–10

Antipathes columnaris; Opresko, 1974: 101–115 (part).
Stylopathes adinocrada Opresko, 2006: 123.

Description. Colony highly pinnulated, monopodial, 19.5 cm long, with a central formation of fused pinnules forming a "worm run" harboring a polychaete (Figs 8–9). Primary pinnules with subpinnulation and a dense pinnulation pattern, arranged in three rows, 10–20 mm long, with ca. 11 secondary pinnules per posterior primary one. Colonies pinnulated to fourth order. Subpinnules slightly more abundant on abpolypar side of posterior primary pinnules; proximal end of primary pinnules with denser pinnulation pattern. Spines small, 0.024 to 0.06 mm long, arranged in ca. 5 rows in lateral view (Fig. 10), separated ca. 0.28 mm each other within same row. Polyps not seen (lost tissue).

Material examined. MOUFPE-CNI 113; 04°47′50″S; 036°11′01″W (Potiguar Basin), 423–461 m depth (1 colony).

Distribution. Bahamas, 134–708 m depth (Opresko 2006), Surinam (Opresko 2006) and northeastern Brazil (423–461 m

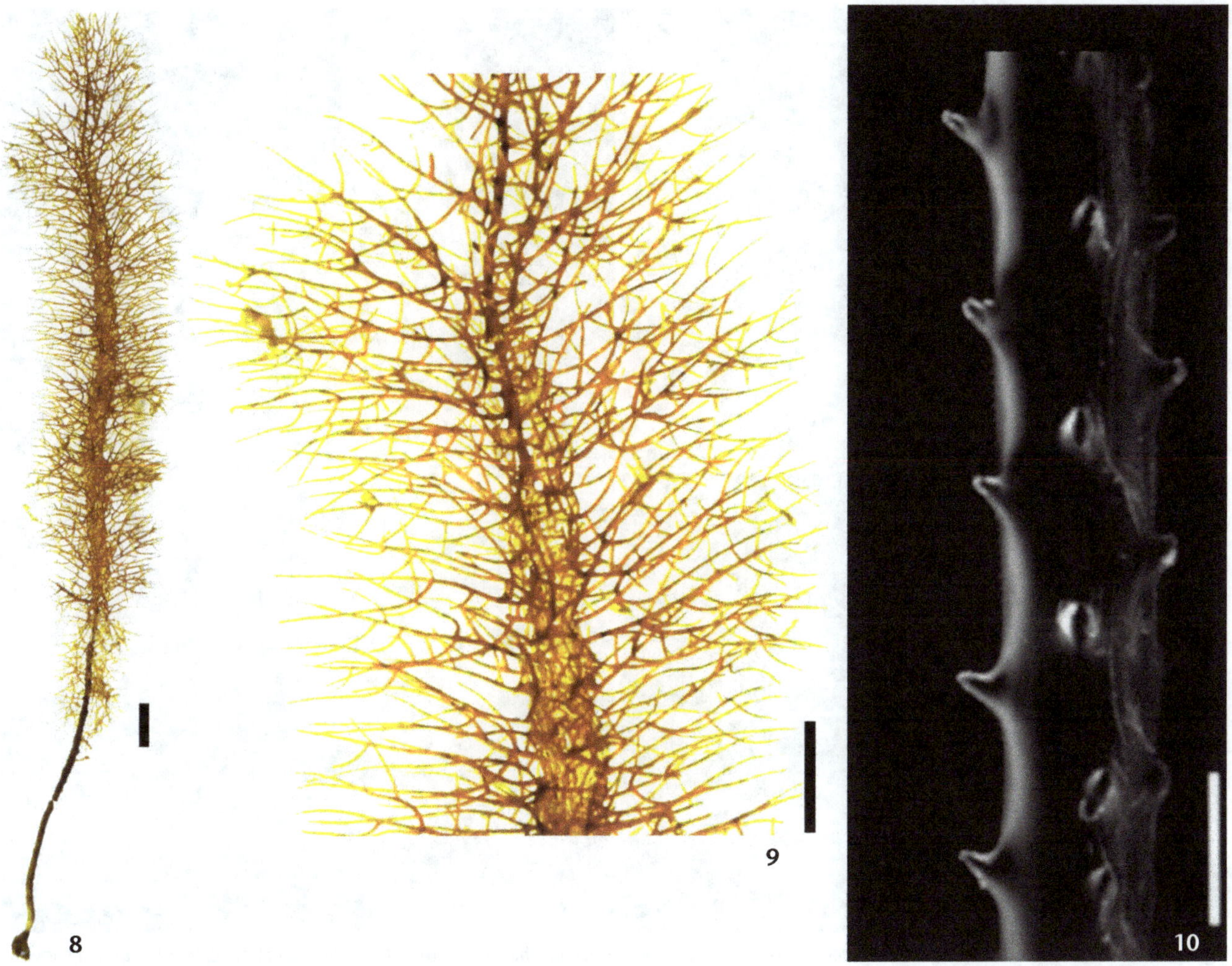

Figures 8–10. (8) Corallum morphology of *Stylopathes adinocrada*; (9) pinnulation pattern of *S. adinocrada*, showing the "worm run"; (10) organization of spines in *S. adinocrada*. Scale bars: 8–9 = 10 mm, 10: 100 µm.

depth, present study). *Stylopathes* (as *S. columnaris*) was known to occur in the coast of the state of Pará (Opresko 1974, Castro et al. 2006, Opresko 2006), but this is the first record of *S. adinocrada* for the Southwestern Atlantic.

Remarks. The studied material resemble the holotype, but was slightly larger, ca. 20 cm versus, ca. 17 cm, respectively. The pinnulation pattern was also similar but the spines were slightly smaller (up to 0.048 mm, rarely 0.06 mm) and more separated within a given row (i.e., up to 0.28 mm versus 0.1–0.2 mm the holotype). The type species of *Stylopathes* is *S. columnaris*, but two more species were included within the genus based on two morphotypes erroneously identified as *S. columnaris* (Opresko 2006). One of them, *S. adinocrada*, differs from *S. columnaris* mainly due to its scattered pinnulation pattern, although in both species the subpinules occur mainly in the distal portion of the primary pinnule, rarely covering the worm run (Opresko 2006).

Parahololepidella cf. *greeffi* and *Tanacetipathes* spp.

Symbiotic scale-worm

Polynoidae Kinberg, 1856

Parahololepidella Pettibone, 1969

Parahololepidella cf. *greeffi* (Augener, 1918)

Hololepidella greeffi Augener, 1918: 148, pl. 2, figs 22–24, pl. 3, fig. 52, text-fig. 9.
Hololepidella fagei Rullier, 1964: 132, fig. 4.

Diagnosis. Body dorsoventrally flattened, elongate, with up to 140 segments (Figs 11–12). Elytra very small, leaving mid-dorsum and parapodia uncovered. Fifty or more pairs of elytra on elytrophores of segments 2, 4, 5, 7, 9, 11, 13, 15, 17, 19, 21, 23, 26, 29, 32, alternating on posterior segments. Prostomium bilobed, with distinct cephalic peaks, and two pairs of large eyes.

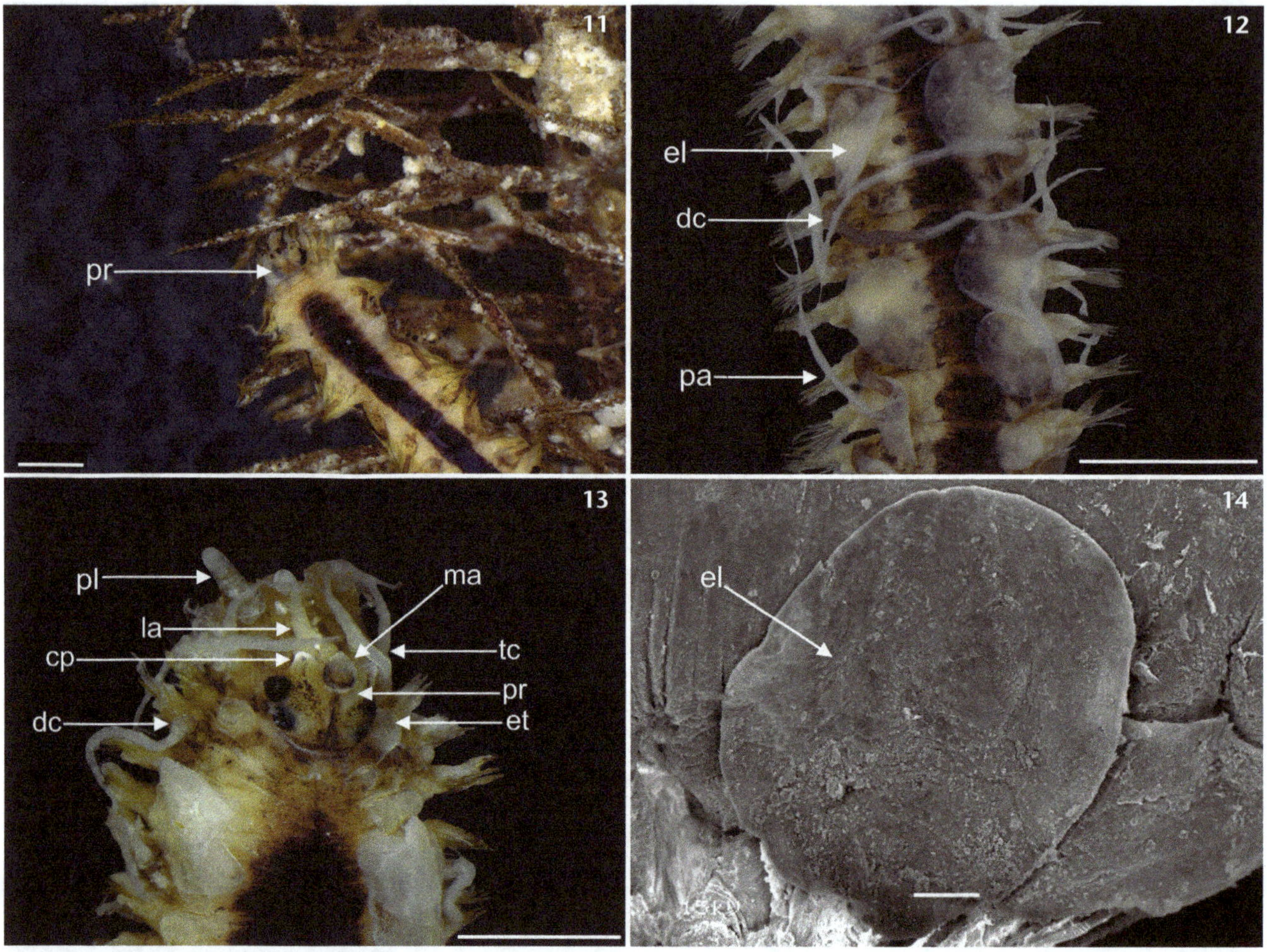

Figures 11–14. (11) *Paraholoiepidella* cf. *greeffi* on the branches of *Tanacetipathes tanacetum*; (12) median chaetigers of *P. greeffi* showing elytra and dorsal cirri; (13) head of *P. greeffi* showing the prostomium, palps and antennae; (14) elytrae from posterior chaetigers of *P. cf. greeffi* showing a smooth edge. (cp) Cephalic peak, (dc) dorsal cirrus, (el) elytra, (et) elythophore, (la) lateral antennae, (ma) median antennae, (pa) parapodium, (pl) palp, (pr) prostomium, (tc) tentacular cirrus. Scale bars: 11–13 = 1 mm, 14 = 100 μm.

Ceratophores of median antenna in anterior notch, smooth, tapering, longer than palps. Lateral antennae inserted ventrally to median antenna, smooth, tapering (Fig. 13). Dorsal cirri smooth, with cylindrical long cirrophores (Fig. 15). Parapodia sub-biramous. Notopodia small, digitiform. Neuropodia with longer rounded prechaetal lobe, with subacicular digitiform acicular lobe. Notopodia with reduced number 0–5 notochaetae, unidentate, with faint serration blade (Figs 16–17). Neuropodia with 16–17 notochaetae, unidentate, with faint serration (Fig. 18). Pygidium conical, with two long cirri.

For complete description, see Britayev et al. (2014).

Variability. Our complete specimens varied in number of segments and elytra after chaetiger 23: specimen 1 (33 mm in length, 72 chaetigers, and 44 pairs of elytra), specimen 2 (44 mm in length, 128 chaetigers, and 70 pairs of elytra), specimen 3 (35 mm in length, 68 chaetigers, and 30 pairs of elytra) and specimen 4 (33 mm in length, 65 chaetigers, and 32 pairs of elytra. Our specimens have elytra on posterior most chaetigers.

Material examined. Two complete specimens, four anterior, and three posterior (i.e., mid-posterior body end) fragments, found on the branches of *T. barbadensis*. MOUFPE-CNI 351, 04°44"31'S, 36°26'19"W, 101–108 m depth (1 specimen from *T. tanacetum*); MOUFPE-CNI 350, same collection data (2 specimens from *T. thamnea*). MOUFPE-CNI 349 (same collection data 1 specimens from *T. barbadensis*).

Distribution. Tropical and Equatorial East Atlantic, Cabo Verde and São Tomé Archipelagos; Western Atlantic Brazil (this paper).

Remarks. *Paraholoiepidella greeffi* was originally described as *Hololepidella greeffi* Augener, 1918, as free-living from shallow

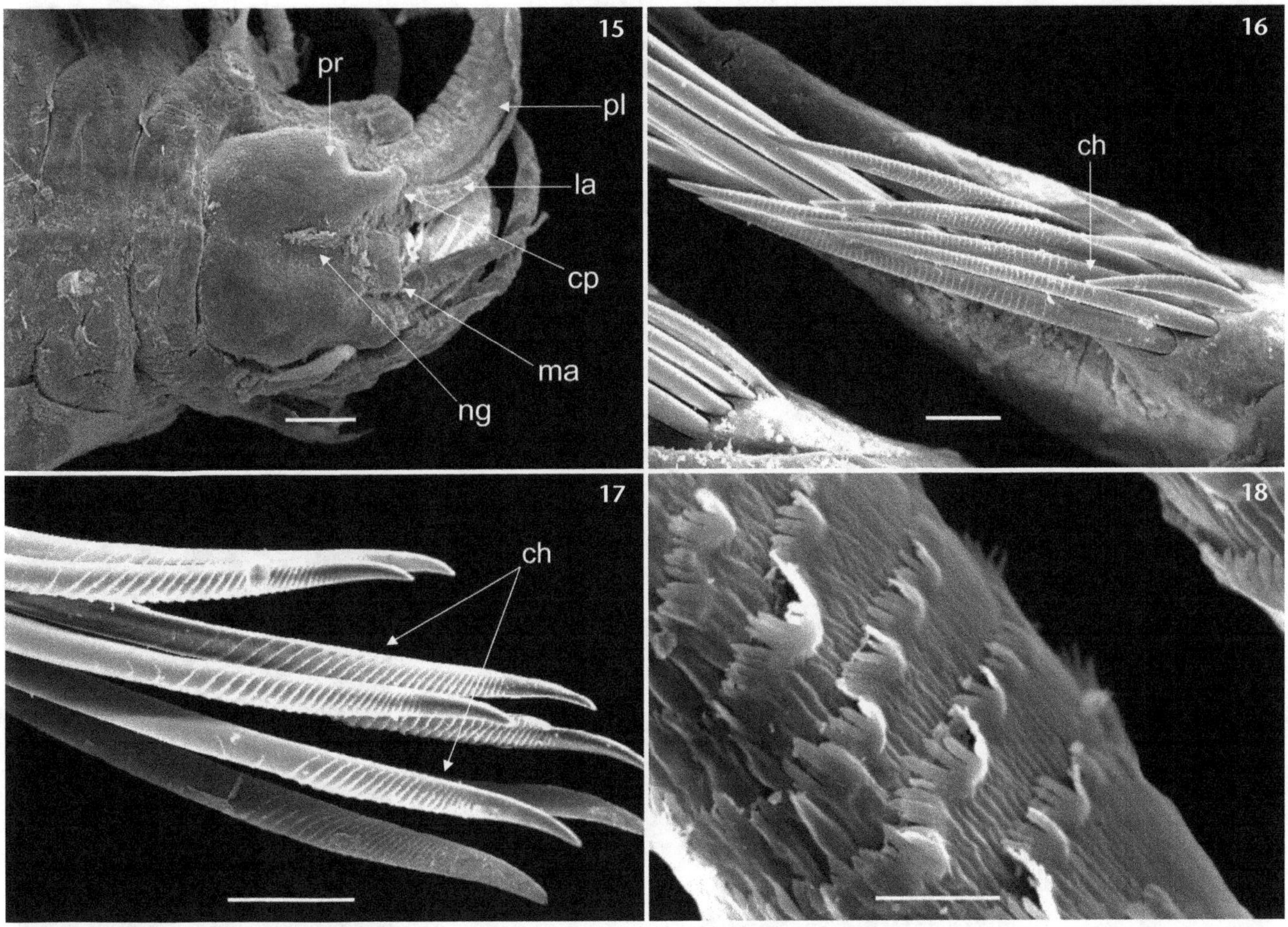

Figures 15–18. (15) Head of *P. cf. greeffi* showing the prostomium and their structures; (16) Parapodia of *P. cf. greeffi* showing notochaetae and neurochaetae; (17) Neurochaetae from median chaetigers of *P. cf. greeffi*; (18) Detail of distal tip of neurochaetae. (ch) Chaetae, (cp) cephalic peak, (la) lateral antennae, (ma) median antennae, (ng) nuchal groove, (pl) palp, (pr) prostomium. Scale bars: 15 = 1 mm, 16–17 = 100 µm, 18 = 10 µm.

waters off São Tomé and Cabo Verde Islands (West Africa). Posteriorly, Pettibone (1969) proposed *Parahololepidella* to include *Hololepidella*. The morphological features and cryptic color in preserved specimens from Brazil mostly agree with the currently known descriptions (Augener 1918, Hartman 1959, Pettibone 1969), including the most recent redescription (Britayev et al. 2014), but elytra are present till the end of the body.

Ecology. *Parahololepidella* cf. *greeffi* was first reported as a symbiont by Britayev et al. (2014), living in association with *Tanacetipathes* cf. *spinescens* (Gray, 1857) (= *Antipathes spinescens*) (Opresko 2001). Our results confirm the commensal character of the species, which appears to be a polyxenous black coral symbiont as, in addition to *T.* cf. *spinescens* it also lives in association *T. tanacetum*, *T. barbadensis* and *T. thamnea*. *Tanacetipathes tanacetum* has also been found too harbor *Antipathipolyeunoa nuttingi* Pettibone, 1991 for in North Atlantic Ocean and Caribbean Sea (i.e., Barbados) locations (Pettibone 1991, Molodtsova and Budaeva 2007).

Host black corals

Myriopathidae Opresko, 2001

Tanacetipathes Opresko, 2001

Tanacetipathes tanacetum (Pourtalès, 1880)

Figs 19–21

Antipathes tanacetum Pourtalès, 1880: 116, pl. 3, fig. 13.

Description. Corallum monopodial, with branches rarely up to 2nd order, in bottlebrush pattern, and primary pinnules usually in 4 main rows (Fig. 19). Primary posterior pinnules larger than primary anterior pinnules, 6–20 mm long, varying according to colony size. Anterior primary pinnules 5–7 mm long, cycle separated by ca. 1 mm. Angle between posterior and anterior primary pinnules (polypolar and abpolypar sides, respectively) of ca. 45°. Angle between two anterior or two posterior primary pinnules

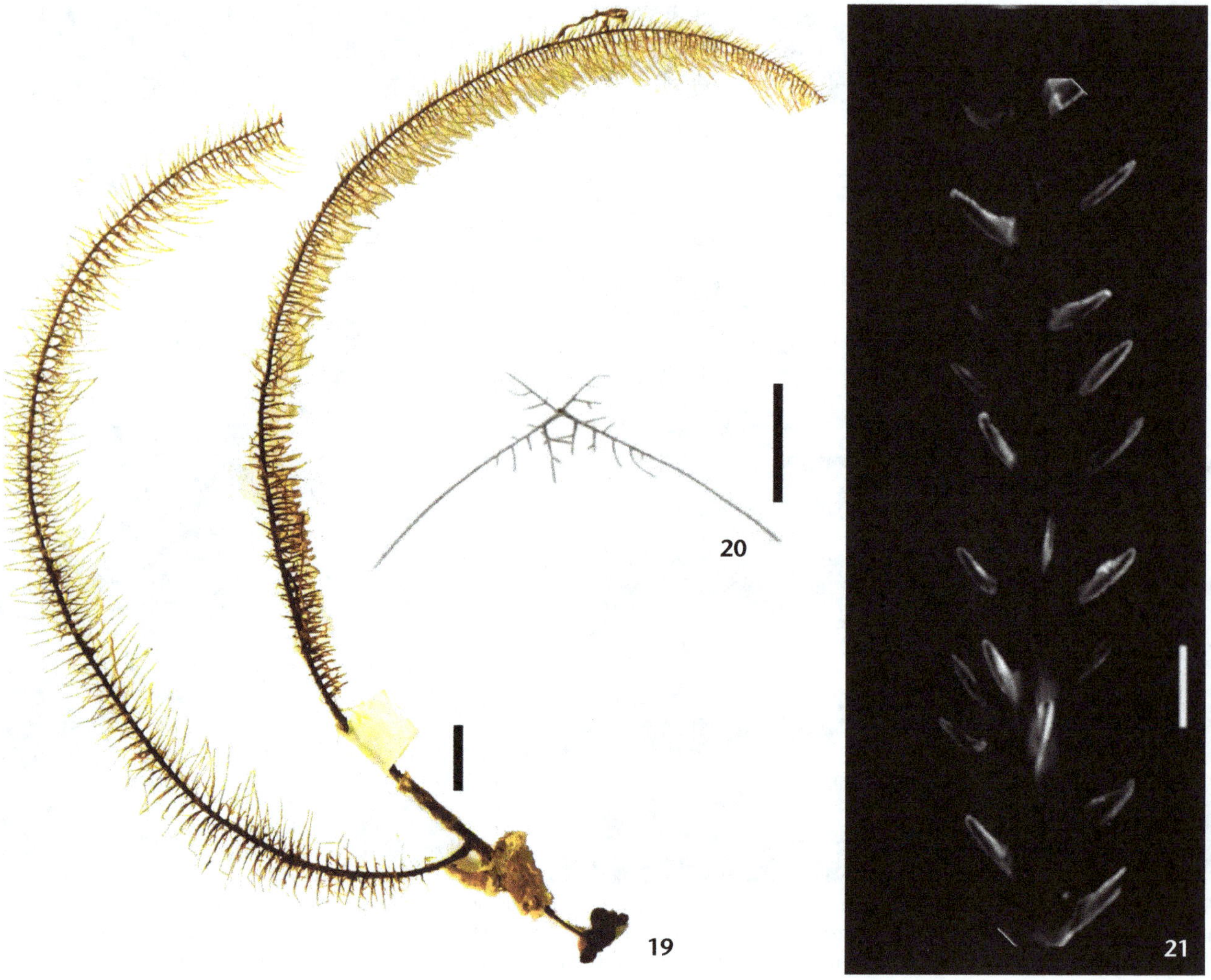

Figures 19–21. (19) Corallum morphology of *Tanacetipathes tanacetum*; (20) cross section showing pinnules cycle of *T. tanacetum*; (21) organization of spines in *T. tanacetum*. Sacale bars: 19, 20 = 1 cm, 21 = 200 μm.

(polypar/polypar sides or abpolypar/abpolypar, respectively) of ca. 160°. Secondary pinnules scarce, 3 – 7 (more common 4 – 5), usually near proximal end of posterior primary pinnules (Fig. 20). Tertiary pinnule rarely present, usually 1 or 2 per secondary pinnule when present. Spines with few ornaments, conical (Fig. 21). Polypar spines 0.048–0.1 mm long, separated by 0.12–0.18 mm. Abpolypar spines 0.036–0.06 mm long. Polyps not seen (lost tissue).

Material examined. MOUFPE-CNI 351, 04°44″31′S; 36°26′19″W, 101–108 m (Potiguar Basin).

Distribution. Bermuda (Cairns et al. 1986), Gulf of Mexico, Potiguar Basin (present study) and southern coast of Brazil (Pérez et al. 2005, Opresko 2009) 50 to 108 m. This is the first record for the Potiguar Basin, northeast Brazil.

Remarks. The lectotype of *T. tanacetum* designated by Pérez et al. (2005) has posterior primary pinnules 1.2 cm long versus the 2 cm of the Potiguar Basin material, which also lack the characteristic curved back pinnules of the lectotype. Nevertheless, not all specimens have these pinnules (Pérez et al. 2005). Our specimens resemble those from off Bahia, northeastern Brazil, which had polypar spines 0.04–0.27 mm long (Loiola and Castro 2005), in agreement with the 0.13 to 0.3 mm reported by Opresko and Sanchez (2005). In contrast, our colonies showed polypar spines 0.048–0.1 mm long, similar to those of the specimens from Bahia (Loiola and Castro 2005). The original description emphasizes the presence of a "parasitic worm" similar to that occurring in *S. columnaris*, but which did not produced changes in host growth (Pourtalès 1880). The symbiotic worm was described as *Antipathipolyeunoa nuttingi* Pettibone, 1991 in a paper that also provided an identification key based primarily on the associated species of black corals.

Tanacetipathes barbadensis (Brook, 1889)

Figs 22–27

Aphanipathes barbadensis Brook, 1889: 128, pl. II, fig. 10; pl. XI, fig. 4.

Description. Corallum monopodial or branched up to the first order with branches emerging close to base (Fig. 22). Colonies in bottlebrush pattern, with primary pinnules mainly arranged in four rows, reaching up to 6 rows. Secondary pinnules inclined towards distal end of primary posterior pinnule; anterior primary pinnules rarely with subpinnulation. 1–4 (more commonly 2–3) long secondary pinnule per posterior primary, usually at proximal end of primary pinnule (Fig. 23). No more than three tertiary pinnules, when present, near proximal end of secondary pinnule. Posterior primary pinnules 10–26 mm long. Anterior primary pinnules 3–12 mm long. Secondary pinnules 2–12 mm long (more commonly 6 mm). Smooth, conical spines, arranged in 5 to 6 rows visible under optical microscope (Fig. 26). Polypar spines 0.084–0.216 mm long (more commonly 0.144 mm). Abpolypar spines 0.048–0.084 mm long. Distance between spines mostly 0.12–0.18 mm, with ca. four spines per mm in each row. Polyps not seen (lost tissue).

Material examined. MOUFPE-CNI 349, 04°44"31'S; 36°26' 19"W, 101–108 m (Potiguar Basin).

Distribution. Barbados (Brook 1889); Ship Mouth, NW Trinidad (Warner 1981); Gulf of México (Opresko 2009), Brazilian coast, from Potiguar Basin, RN (present study) to the Abrolhos Bank (Loiola and Castro 2005), 30 to 108 m depth.

Remarks. *Tanacetipathes barbadensis* ressembles *T. tanacetum* (Pourtalès, 1880), but differs in having three or less secondary pinnules per subsequent primary pinnule (instead of 3–7 in *T. tanacetum*). Our specimens show posterior pinnules smaller than those previously reported for the species, up to 30 mm long versus 60 mm in Opresko and Sanchez (2005), while agreeing with those described by Brook (1889) and Loiola and Castro (2005). The latter, also from the Brazilian continental shelf, presented posterior primary pinnules up to 45 mm long. Therefore, we suggest that the presence of small sized pinnules may be an intraspecific variation typical of Brazilian populations.

Tanacetipathes thamnea (Warner, 1981)

Figs 25–27

Antipathes thamnea Warner, 1981: 148–151, figs. 2–4.
Tanacetipathes paula Perez, Costa and Opresko, 2005: 8–12, figs 5–8.

Description. Colonies 8–25 cm long, in bottlebrush pattern, monopodial, with branches absent or branching to second order, near colony base (Fig. 25). Primary pinnules usually arranged in four rows. Primary posterior pinnules 10–18 mm long (more common 14–16 mm), larger than the anterior ones (3–12 mm long, varying in juvenile colonies). Secondary pinnules tipping toward distal end of primary pinnule. 7–20 (more commonly 10–15) secondary pinnules per posterior primary pinnule (Fig. 26), 2–12 mm long. Tertiary pinnules scarce, usually more abundant on abpolipar side of posterior primary pinnule, 1–4 per posterior secondary pinnule. Angle between posterior and anterior primary pinnule (polypar and abpolypar sides, respectively) 55°–70°. Angle between two anterior or two posterior primary pinnules (polypar/polypar sides or abpolypar/abpolypar, respectively) 115–130°. Abpolypar side of posterior primary pinnules with more secondary pinnules than polypar side. Spines smooth, with few ornaments, slightly inclined toward distal end of pinnule or branch (Fig. 27). Distance between spines mainly 0.12–0.18 mm. Polypar spines 0.1–0.2 mm long; Abpolypar spines 0.036–0.084 mm long. Polyps not seen (lost tissue).

Material examined. MOUFPE-CNI 350, 04°44"31'S; 36°26' 19"W, 101–108 m (Potiguar Basin).

Distribution. Boca de Navios, NW Trinidad (Warner 1981), Gulf of México (Opresko 2009), and Brazilian coast (Loiola and Castro 2005, present study), 30 to 108 m depth. This is the first record of the species for the Potiguar Basin, Brazil.

Remarks. Our specimens differ from the syntype described by Warner (1981) in some characters. They were slightly shorter (up to 25 cm long versus 20–40 cm long). Some colonies had posterior primary pinnules more elongated and with more secondary pinnules than that described by Loiola and Castro (2005). Warner (1981) also noted that there may be 5–10 secondary pinnules per posterior primary in a10 mm primary pinnule, considering that a pinnule may reach up to 25 mm. Moreover, Loiola and Castro (2005) found 1–2 tertiary posterior pinnules per posterior secondary whereas our specimens had 1–4. Also, Warner (1981) mentioned a few distally inclined tertiary pinnules. Outnumber primary pinnules may occur especially in the more distal regions of the colonies as well as in larger colonies. Warner (1981) mentioned the presence of errant polychaetes in almost all colonies, living in a kind of tube formed by pinnules on the side of the colony turned toward the sea current.

DISCUSSION

Many authors have reported symbiotic relationships involving black corals since the 19th century, although most records were observations limited to strictly taxonomic papers (Brook 1889, Van Pesch 1914, Pourtalès 1874) and the symbiotic polychaetes, especially the Polynoidae, were not an exception (Beneden 1869, Paris 1955, Britayev 1989, Martin and Britayev 1998). The paucity of ecological information was strongly related to the difficultness of conducting observational studies in deep seas, but also to the low number of taxonomists with expertise in Antipatharia (Cairns 2007).

Black corals are key-species in deep-sea habitats, from sheltering quirky communities of microorganisms to nesting centers for several fish species (Tazioli et al. 2007, Wagner et al. 2012). Some decapod crustaceans, for example, house on black coral colonies to acquire a better position in the water column for filter-feeding (Wagner et al. 2012).

Figures 22–27. (22) Corallum morphology of *Tanacetipathes barbadensis*; (23) cross section showing pinnules cycle of *T. barbadensis*; (24) organization of spines in *T. barbadensis*; (25) corallum morphology of *Tanacetipathes thamnea*; (26) cross section showing pinnules cycle of *T. thamnea*; (27) organization of spines in *T. thamnea*. Scale bars: 22, 23, 25, 26 = 1 cm, 24 = 200 μm, 27 = 150 μm.

Molodtsova et al. (2016) listed around 19 black coral species hosting 14 species of symbiotic polychaetes worldwide, with the Polynoidae being the most diverse polychaete family associated with antipatharians, but also with gorgonians.

In many cases, the polychaetes associated with *Tanacetipathes* and *Stylopathes* changed the coral growth and caused modifications in specific morphological characters (Pettibone 1989, 1991, Molodtsova and Budaeva 2007, Wagner et al. 2012). Therefore, some taxonomically important morphological traits must be addressed with caution in antipatharians affected by symbiotic associations (Molodtsova and Budaeva 2007). The unbranched monopodial form in the examined colonies might be a physiological reaction of antipatharians to symbiotic polychaetes, as suggested by Molodtsova and Budaeva (2007).

In this study, most polychaetes associated with *Tanacetipathes* spp. were found on the main axis of the corallum, screwing on the secondary pinnules in the posterior side of black coral. In contrast, the polychaete associated with *S. adinocrada* was found inside a "worm run" formed by anastomosed secondary pinnules along the corallum main axis, typical of the genus.

Indo-Pacific and Northern Atlantic species of *Stylopathes* have been reported with *B. antipathicola* as obligatory commensals (Wagner et al. 2012). Although our data are insufficient to suggest coevolution, it is worth noting that the same relationship with *Stylopathes* seems to occur in two distinct ocean basins separated long ago. Since there is strong molecular evidence of vicariant speciation in deep-sea corals and polynoids between Pacific and Atlantic basins (e.g., Quattrini et al. 2013, Serpetti et al. 2017), the coevolution hypothesis deserves testing, particularly because the existence of cryptic diversity in *B. antipathicola* cannot be discarded. The morphological differences between specimens from the two basins seems to support the idea that they correspond to different taxa. However, more careful morphological studies based on new materials, in parallel with molecular studies, is certainly needed to clarify the situation of these populations living in association with deep-water corals.

We have recorded *Parahololepidella* cf. *greeffi* living as commensal with three species of *Tanacetipathes*: *T. thamnea*, *T. barbadensis* and *T. tanacetum*, which represents a new modification of the ecological status of the species, from free living to monoxenous symbiont in Britayev et al. (2014) to polyxenous symbiont according to our data. Our results also confirm that the worms did not cause drastic modifications to their hosts.

All 300 antipatharians from Myriopathidae and Antipathidae, collected in 75 Indo-Pacific stations showed symbiotic relationships with polynoid and eunicid polychaetes, including *B. antipathicola* forming latticed tunnels on *S. columnaris* (Molodtsova and Budaeva 2007). The tube was initially secreted by the polychaete and subsequently structured by anastomosis of pinules along the central axis of the colony. The same phenomenon was observed in our samples. Molodtsova and Budaeva (2007) also examined Caribbean samples of *Tanacetipathes* (*T. spinences*) with *P. greeffi* as symbiont of the colonies. In these case, corals also present anastomosed secondary pinules but, as in the Brazilian material, they did not form exactly a "worm run".

ACKNOWLEDGMENTS

We thank Daniel Martin (Centre d'Estudis Avançats de Blanes – CEAB/CSIC), Temir Britayev (Severtsov Institute of Ecology and Evolution RAS, Moscow, Russia), and the anonymous referees for suggestions to improve this manuscript. We heartily thank Paulo Lana (UFPR) for his suggestions to improve the manuscript. We also acknowledge Fundação de Amparo a Ciência e Tecnologia de Pernambuco for a post-doctoral scholarship to J.E. De Assis (DCR-0086-2.04/13).

LITERATURE CITED

Augener H (1918) Polychaeta. In: Beiträge zur Kenntnis der Meeresfauna Westafrikas. Hamburg, Herausgegeben von W. Michaelsen, 67–625.

Barnich R, Beuck L, Freiwald A (2013) Scale worms (Polychaeta: Aphroditiformia) associated with cold-water corals in the eastern Gulf of Mexico. Journal of the Marine Biological Association of the United Kingdom 93(8): 2129–2143. https://doi.org/10.1017/S002531541300088X

Beneden PJ (1869) Le commensalisme dans le règne animal. Bulletin de l'Académie royale de Belgique, Série 2, 28: 621–648.

Benham WB (1927). Polychaeta. British Antarctic 'Terra Nova' Expedition Natural History Reports, Zoology 7(2): 47–182.

Britayev TA (1989) The symbiotic polychaetes: morphology, ecology and distribution. In: Sveshnicov VA (Ed.) Symbiosis among marine animals. Moscow, A.N. Severtzov Institute, Russian Academy of Sciences, 60–74.

Britayev TA, Antokhina TI (2012) Symbiotic polychaetes from Nhatrang Bay, Vietnam. In: Britayev TA, Pavlov DS (Eds) Benthic fauna of the Bay of Nhatrang, Southern Vietnam. Moscow, KMK, vol. 2, 11–54.

Britayev TA, Gil J, Altuna Á, Calvo M, Martín D (2014) New symbiotic associations involving polynoids (Polychaeta, Polynoidae) from Atlantic waters, with redescriptions of *Parahololepidella greeffi* (Augener, 1918) and *Gorgoniapolynoe caeciliae* (Fauvel, 1913). Memoirs of the Museum Victoria 71: 27–43.

Brook G (1889) Report on the Antipatharia collected by HMS Challenger during the years 1873–1876. Report of the scientific results of the voyage of the HMS "Challenger". Zoology 32: 1–222.

Brugler MR, Opresko DM, France SC (2013) The evolutionary history of the order Antipatharia (Cnidaria: Anthozoa: Hexacorallia) as inferred from mitochondrial and nuclear DNA: implications for black coral taxonomy and systematics. Zoological Journal of the Linnean Society 169(2): 312–361. https://doi.org/10.1111/zoj.12060

Buhl-Mortensen L, Mortensen PB (2004) Symbiosis in deep-water corals. Symbiosis 37: 33–61.

Cairns SD (2007) Deep-water corals: an overview with special reference to diversity and distribution of deep-water scleractinian corals. Bulletin of Marine Science 81(3): 311–322.

Cairns SD, Hartog JC, Arneson C (1986) Marine fauna and flora of Bermuda. In: Sterrer W (Ed.) Marine fauna and flora of Bermuda: A systematic guide to the identification of marine organisms. New York, Wiley-Interscience Publications, 159–194.

Castro CB, Pires DO, Medeiros MS, Loiola LL, Arantes RCM, Thiago CM, Berman E (2006) Filo Cnidaria. Corais. In: Lavrado HP, Ignacio BL (Eds) Biodiversidade bentônica da região central da Zona Econômica Exclusiva brasileira. Rio de Janeiro, Museu Nacional, 147–192.

Eckelbarger KJ, Watling L, Fournier H (2005) Reproductive biology of the deep-sea polychaete Gorgoniapolynoe caeciliae (Polynoidae), a commensal species associated with octocorals. Journal of the Marine Biological Association of the United Kingdom 85(6): 1425–1433. https://doi.org/10.1017/S0025315405012609

Glasby CJ (1994) A new genus and species of polychaete, *Bollandia antipathicola* (Nereidoidea: Syllidae), from black coral. Proceedings of the Biological Society of Washington 107(4): 615–621.

Hanley JR, Burke M (1991) A new genus and species of scaleworm (Polychaeta, Polynoidae) from the Cascade Plateau, Tasman Sea. The Beagle, Records of the Northern Territory Museum of Arts and Sciences 8(1): 97–102.

Hartman O (1959) Catalogue of the polychaetous Annelids of the World, parts 1–2. Allan Hancock Foundation Publications, Occasional Papers 23: 1–628.

Hartmann-Schröder G, Zibrowius H (1998). Polychaeta associated with Antipatharia (Cnidaria: Anthozoa): description of Polynoidae and Eunicidae. Mitteilungen aus dem Hamburgischen zoologischen Museum und Institut 95: 29–44.

Loiola LL (2007) Black Corals (Cnidaria: Antipatharia) from Brazil: an overview. Conservation and Adaptive Management of Seamount and Deep-Sea Corals Ecosystems. Bulletin of Marine Science 81(Suppl. 1): 253–264.

Loiola LL, Castro CB (2005) *Tanacetipathes* Opresko, 2001 (Cnidaria: Antipatharia: Myriopathidae) from Brazil, including two new species. Zootaxa 1081: 1–31.

Martin D, Britayev TA (1998) Symbiotic polychaetes: Review of known species. Oceanography and Marine Biology: An Annual Review 36: 217–340.

Martin D, Britayev TA (2018) Symbiotic polychaetes revisited: an update of the known species and relationships (1998–2017). Oceanography and Marine Biology: An Annual Review 56: 367–446.

Molodtsova T, Britayev TA, Martin D (2016) Cnidarians and their polychaete symbionts. In: Goffredo S, Dubinsky Z (Eds) The Cnidaria, past, present and future. The world of Medusa and her sisters. Switzerland, Springer International Publishing, 387–413.

Molodtsova T, Budaeva N (2007) Modifications of corallum morphology in black corals as an effect of associated fauna. Bulletin of Marine Science 81: 469–480.

Opresko DM (1974) A study of the classification of the Antipatharia (Coelenterata: Anthozoa), with redescriptions of eleven species. PhD Thesis, University of Miami, Coral Gables, 194 pp [press in 1994 by University Microfilms, Ann Arbor].

Opresko DM (2001) Revision of the Antipatharia (Cnidaria: Anthozoa). Part I. Establishment of a new family, Myriopathidae. Zoologische Mededelingen Leiden 75(17): 343–370.

Opresko DM (2006) Revision of the Antipatharia (Cnidaria: Anthozoa). Part V. Establishment of a new family, Stylopathidae. Zoologische Mededelingen Leiden 80(4): 109–138.

Opresko DM (2009) Antipatharia (Cnidaria) of the Gulf of Mexico. *In:* Tunnell JW Jr, Felder DL, Earle SA (Eds) *Gulf of Mexico origin, waters, and Biota.* Corpus Christi, Texas A&M University Press, vol. 1, 359–363.

Opresko DM, Sanchez JA (2005) Caribbean Shallow-water Black Corals (Cnidaria: Anthozoa: Antipatharia). Caribbean Journal of Science 41(3): 492–507.

Paris J (1955) Commensalisme et parasitisme chez les annélides polychètes. Vie et Milieu 6: 525–536.

Pérez CD, Costa DL, Opresko DM (2005) A new species of *Tanacetipathes* from Brazil, with a redescription of the type species *T. tanacetum* (Pourtalès) (Cnidaria, Anthozoa, Antipatharia). Zootaxa 890: 1–12.

Pérez CD, Vila-Nova DA, Santos AM (2005) Associated community with the zoanthid *Palythoa caribaeorum* (Duchassaing & Michelotti, 1860) (Cnidaria, Anthozoa) from littoral of Pernambuco, Brazil. Hydrobiology 548: 207–215.

Pettibone MH (1969) The genera Polyeunoa McIntosh, *Hololepidella* Willey, and three new genera (Polychaeta, Polynoidae). Proceedings of the Biological Society of Washington 82(3): 43–62.

Pettibone MH (1970) Polychaeta Errantia of the Siboga Expedition. Part IV. Some additional polychaetes of the Polynoidae, Hesionidae, Nereidae, Goniadidae, Eunicidae, and Onuphidae, selected as new species by the late Dr. Hermann Augener with remarks on other related species. In: Siboga-Expeditie Uitkomsten op Zoologisch, Bonatisch, Oceanographisch en Geologisch gebied verzameld in Nederlandsch Oost-Indië. Leiden, E.J. Brill, 199–270.

Pettibone MH (1989) A new species of *Benhamipolynoe* (Polychaeta: Polynoidae: Lepidastheniinae) from Australia, associated with the unattached stylasterid coral *Conopora adeta.* Proceedings of the Biological Society of Washington 102: 300–304.

Pettibone MH (1991) Polynoid polychaetes commensal with antipatharian corals. Proceedings of the Biological Society of Washington 104: 714–726.

Pourtalès LF de (1874) Crinoids and Corals. Zoological results of the "Hassler" Expedition. – Ill. Catalogue of the Museum of Comparative of Zoology, Harvard, 8: 25–50.

Pourtalès LF de (1880) Zoological results of the "Blake" expedition to the Caribbean Sea. Bulletin of the Museum of Comparative Zoology Havard 6(4): 113–118.

Quattrini AM, Georgian SE, Byrnes L, Stevens A, Falco R, Cordes EE (2013) Niche divergence by deep-sea octocorals in the genus *Callogorgia* across the continental slope of the Gulf of Mexico. Molecular Ecology 22: 4123–4140. https://doi.org/10.1111/mec.12370

Rullier F (1964) Résultats scientifiques des campagnes de la Calypso – Campagne de la Calypso aux îles du Cap Vert (1959) (suite): 5. Annélides polychètes. Annales de l'Institut Océanographique 41(6): 113–218.

Serpetti S, Taylor ML, Green DH, Rogers AD, Paterson GLJ (2017) Ecological adaptations and commensal evolution of the Polynoidae (Polychaeta) in the Southwest Indian Ocean Ridge: A phylogenetic approach. Deep Sea Research Part II: Topical Studies in Oceanography 137: 273–281.

Tazioli S, Bo M, Boyer M, Rotinsulu H, Bavestrello G (2007) Ecological observations of some common antipatharian corals in the marine park of Bunaken (North Sulawesi, Indonesia). Zoological Studies 46: 227–241.

Van Pesch AJ (1914) The Antipatharia of the Siboga Expeditie. Monographe 17: 1–258.

Warner GF (1981) Species descriptions and ecological observations of black corals. (Antipatharia) from Trinidad. Bulletin of Marine Science 31(1): 147–163.

Wagner D, Luck DG, Toonen RJ (2012) The biology and ecology of black corals (Cnidaria: Anthozoa: Hexacorallia: Antipatharia). *Advances in Marine Biology* 63: 67–132. https://doi.org/10.1016/B978-0-12-394282-1.00002-8

Williams GC, López-González P (2005) A new genus and species of gorgonian octocoral (Anthozoa: Plexauridae) from Antarctic waters. Proceedings of the California Academy of Sciences 56(26): 379–390.

Author Contributions: JEA original idea, identified polynoids, composed figures, wrote the manuscript; JRBS wrote the manuscript; MML sorted the black corals, identified specimens, wrote the manuscript; GVL sorted specimens, coordinated the project, wrote the manuscript; RTSC wrote and helped with translation of the manuscript, composed figures; CDP lab head, original idea, wrote the manuscript.

Competing Interests: The authors have declared that no competing interests exist.

5

Parasites in feces of the endemic rattlesnake, *Crotalus triseriatus* (Serpentes: Viperidae), from Mexican highlands

Leroy Soria-Díaz[1], Jaime Luis Rábago-Castro[2], Hublester Domínguez-Vega[3], Yuriana Gómez-Ortiz[3], Javier Manjarrez[4], Lorena Garrido-Olvera[1]

[1]*Instituto de Ecología Aplicada, Universidad Autónoma de Tamaulipas. División del Golfo 356, La Libertad, 87019, Ciudad Victoria, Tamaulipas, Mexico.*
[2]*Facultad de Medicina Veterinaria y Zootecnia, Universidad Autónoma de Tamaulipas. Carretera Victoria-Mante km 5, 87000, Ciudad Victoria, Tamaulipas, Mexico.*
[3]*División de Desarrollo Sustentable, Universidad Intercultural del Estado de México. Libramiento Francisco Villa SN, 50640, San Felipe del Progreso, Estado de México, Mexico.*
[4]*Centro de Investigación en Recursos Bióticos, Facultad de Ciencias, Universidad Autónoma del Estado de México. Instituto Literario 100, 50000, Toluca, Estado de México, Mexico.*
Corresponding author: Lorena Garrido-Olvera (lolvera@docentes.uat.edu.mx)

ABSTRACT. There are few studies about parasitic infections in *Crotalus triseriatus* (Wagler, 1830), an endemic rattlesnake from the highlands of central Mexico. This species occupies several habitats, from conserved forested regions to heavily impacted landscapes. To increase the parasitological knowledge of this reptile species without impacting populations, we obtained fecal samples of 16 rattlesnakes between 2012 and 2014 from Toluca Valley, Mexico. By using flotation technique, we found oocysts of *Eimeria* sp. and eggs of Capillariidae sp. The most frequent parasite was *Eimeria* sp. (62.5%). This study provides the first records of occurrence of parasites in reptilian feces from Mexico. Our results may be important for wildlife conservation purposes, for example, they could indicate of the population health of this species during processes of translocation.

KEY WORDS. Coccidia, Mexico, Nematoda, reptile, urban wildlife.

Reptiles are hosts of a wide variety of protozoan and metazoan parasites (Wilson and Carpenter 1996). Most of these parasites have been recorded through necropsies of specimens, although the diagnosis could be achieved by revealing parasitic elements in feces (e.g., eggs, larvae, oocysts). Even though, the latter non-invasive method only allows detecting intestinal parasites, may be useful for conservation projects, which involve knowing the health status of a reptile population in wildlife (Jorge et al. 2013).

In Mexico, 864 reptile species have been recorded (i.e., 8.7% of the reptiles of the world), of which 57% are endemic (Flores-Villela and García-Vazquez 2014). However, the inventory of their parasites is far from complete, since only a few more than 200 Mexican reptile species have been recorded as hosts of helminths, acari and crustaceans (Paredes-León et al. 2008).

Mexico holds the highest species richness of rattlesnakes, with 42 species (Heimes 2016), but wild populations have declined due to direct hunting and habitat loss caused by the expansion of croplands and urban areas. Particularly, the Mexican dusky rattlesnake, *Crotalus triseriatus* (Wagler, 1830), is restricted to the Trans-Mexican Volcanic Belt, where occupies forested regions heavily impacted (Bryson et al. 2014). This species is considered in the category of "least concern" in the IUCN and it has not received any level of protection by Mexican government due probably to a lack of knowledge in demographic and ecological aspects (Bryson et al. 2014, Sunny et al. 2015).

Translocations have increased for wildlife conservation, including the herpetofauna. One concern associated with conservation translocations is that the released individuals may suffer from parasitic diseases resulting in establishment failure. However, infectious hazards may be mitigated when diseases and/or parasites of translocated species or species at the destination are known (Bobadilla Suarez et al. 2017). Unfortunately, for *C. triseriatus* little is known about its parasites (Paredes-León et al. 2008), even though this information is essential to know the population health of this species and potentially useful in future translocation processes of species from one population to another. Our objective was to investigate the occurrence of parasites in feces of *C. triseriatus* to increase parasitological knowledge without affecting wild populations.

This study was conducted in Toluca Valley located in the State of Mexico, a highly anthropized area and considered the fifth largest metropolis in Mexico, with 2.1 million inhabitants (SEDESOL, CONAPO, INEGI 2004). The region has a humid temperate climate, with an annual precipitation of 500–600 mm and a range of average annual temperature between 12 and 15 °C (García 2004, Gobierno del Estado de México 2012). The sampling sites were El Cerrillo located at 5.1 km north of metropolitan area of Toluca City (19°24'27"N; 99°41'40"W; 2615 masl) and the Sierra Morelos Park located on the edge of the metropolitan area of Toluca City (19°18'39"N; 99°41'33"W; 2630 masl).

To capture individuals of *C. triseriatus* and collect individual fecal samples, during 2012 to 2014 we made 24 monthly visits at each study site, where five observers simultaneously walked five linear transects of 800 m long and separated by 300 m in a schedule of 09:00 am–13:00 pm. We capture the rattlesnakes with herpetological hooks and once subjected, we recorded total length (snout-vent length) in centimeters and weight in grams only with descriptive purposes. After we obtained fecal samples of each specimen place them in Eppendorf tubes and we released the rattlesnakes at their capture sites.

Each fecal sample was conserved in 70% alcohol and examined in the laboratory using the flotation technique (Bowman et al. 2004). Approximately 0.5 g of each sample was macerated and mixed with 9.0 ml of 0.8% saline solution. Each homogenized sample was filtered and placed in a vial, which was centrifuged for three minutes at 1500 rpm. After centrifugation, the supernatant was discarded, and a saturated glucose solution was used to re-suspend the pellet. The suspension was centrifuged for five minutes at 1000 rpm and three subsamples of top part of the flotation were obtained and examined by light microscopy. Parasitic elements (i.e., eggs and oocysts) were counted in each subsample to obtain an average per sample and to determine the levels of infection according to Rodríguez-Vivas and Cob-Galera (2005). Eggs and oocysts were photographed and measured using a light microscope at 40 and 100X magnifications (Leica DFC 490) and identified in accordance with Mehlhorn et al. (1993) and Wolf et al. (2014).

A total 16 Mexican dusky rattlesnakes were caught, 6 in El Cerrillo with 46–66 cm (56.42 ± 6.56) and 71–270 g (190.17 ± 73.7), and 10 in Sierra Morelos Park with 43–67 cm (55.03 ± 7.47) and 70–350 g (202.78 ± 78.19). Ten of the 16 fecal samples showed the presence of parasitic elements: nematode eggs of Capillariidae sp. and coccidian oocysts of *Eimeria* sp., with total prevalence of 62.5% y 6.25%, respectively (Table 1, Figs 1–2). The most frequent parasite was *Eimeria* sp. with high infection levels in both localities. Only one double parasitic infection was found in El Cerrillo.

Knowledge about the parasites of Mexican rattlesnakes is still far from complete. For example, in *Crotalus* spp. only 21 taxa of metazoan parasites (1 Acanthocephala, 3 Cestoda, 9 Nematoda, 3 Acari, and 5 Crustacea) have been recorded through necropsies of bred or wild specimens (Table 2). Of them, *Ophiotaenia perspicua* La Rue, 1911, *Kalicephalus inermis coronellae* (Ortlepp, 1923), *Ixobioides* sp., and *Ophionyssus natricis* (Gervais, 1844) were recorded in *C. triseriatus*. Thus, findings of Capillariidae sp. and *Eimeria* sp. represent new host records and this study is the first on the parasites of wild reptilians in Mexico using fecal samples.

The eggs of Capillariidae sp. found in *C. triseriatus* were not identified at specific level, since capillariid eggs are morphologically almost identical even among species of different genera (see Moravec 1994). Therefore, we cannot know if the nematode is a parasite or a pseudoparasite for Mexican dusky rattlesnakes, i.e., a parasite of the rodents on which hosts feed (personal observation of hair follicles in all fecal samples). In this context, Fuehrer (2014) recognized that endo- and ectoparasites of potential prey animals can be recorded as transiting parasites in the intestinal tract of reptiles, for example eggs of *Capillaria hepaticum* (Bancroft, 1893), a parasite species in mammals, have been documented in fecal samples from reptiles fed with infected rodents. On the other, Šlapeta et al. (2018) pointed out that the nematodes of *Capillaria* (Zeder, 1800) are the only known trichurids affecting reptiles, infecting primarily the intestine and secondarily other organs as the liver and gonads. It is known that their life cycle is direct, unfortunately the life cycles of snakes capillarids have not been investigated (Anderson 2000).

An examination of fecal samples of captive reptiles, including specimens of *Crotalus* Linnaeus, 1758, from Campania region in southern Italy revealed the most frequent parasites were species of groups of Oxyurida, *Rhabdias* Stiles & Hassall, 1905, *Kalicephalus* Molin, 1861, *Capillaria* and Eimeriidae Minchin, 1903 (Rinaldi et al. 2012). Analysis of the fecal samples in the present work confirmed that capillariids and eimeriids are the most frequent parasites of the reptiles, but they did not show the presence of any of the helminths recorded of rattlesnakes in Mexico (Table 2).

The oocysts of *Eimeria* sp. were not identified at specific level, because we did not observe some taxonomic characteristics. But it is known that *Eimeria* Schneider, 1875 is the most numerous group of snake coccidian with more than 80 species, which differ in both the morphology of exogenous stages (oocysts) and in endogenous development (Modrý 2004). Most species of *Eimeria* inhabit the intestine of reptiles, which become infected after ingesting oocysts. Schizogony occurs in the epithelial lining of intestine, depending on parasite species. Oocysts are shed with the feces following gametogony. The pathologic effects of coccidian appear mild for most reptiles; however, more severe cases resulting in epithelial ulceration and fibrosis, and septicemia from the loss of epithelial integrity, may occur (Mitchell 2007).

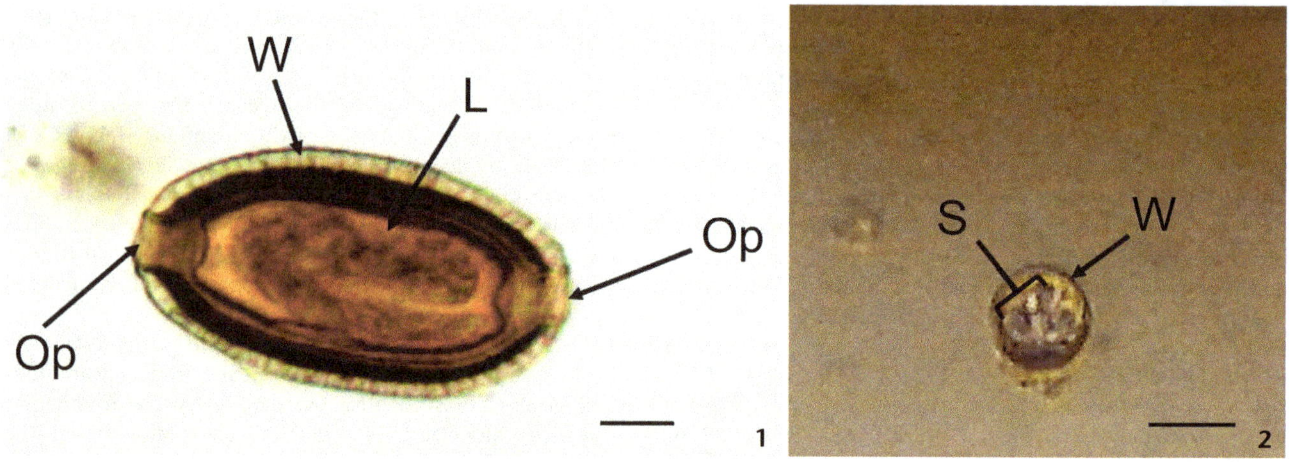

Figures 1–2. Parasitic elements in feces of *Crotalus triseriatus* from Mexico: (1) Egg of Capillariidae sp. with one larva (L), brown color, thick and rough wall (W), one operculum at each pole (Op), which are asymmetrical. (2) Sporulated oocyst of *Eimeria* sp., spheroidal type, rough outer wall (W), absence of micropyle, ovoid sporocysts (S). Scale bars: 10 µm.

Table 1. Infection levels of parasitic elements in feces of *Crotalus triseriatus* from two localities in the State of Mexico, Mexico. Range of oocysts/eggs per gram of feces and infection levels slightly modified of Rodríguez-Vivas and Cob-Galera (2005): 2–9 very low (+), 8–18 Low (++), 20–30/10–15 Medium (+++), 32–50 High (++++).

Localities	Host			Collection date	*Eimeria* sp.	Capillariidae sp.
	ID	Length (cm)	Weight (g)			
Sierra Morelos Park	SMP1	43.0	70	May 26, 2012	+	Negative
	SMP2	57.0	233	July 12, 2012	Negative	Negative
	SMP3	63.0	250	September 15, 2012	Negative	Negative
	SMP4	59.0	212	September 15, 2012	+++	Negative
	SMP5	53.0	Undetermined	April 27, 2014	++	Negative
	SMP6	53.0	185	September 20, 2014	Negative	Negative
	SMP7	51.8	200	October 18, 2014	++++	Negative
	SMP8	45.0	125	October 18, 2014	++++	Negative
	SMP9	67.0	350	November 01, 2014	Negative	Negative
	SMP10	58.5	200	November 01, 2014	Negative	Negative
El Cerrillo	CE1	66.0	250	October 20, 2012	++++	Negative
	CE2	54.5	140	December 08, 2012	+++	+++
	CE3	46.0	71	December 08, 2012	++	Negative
	CE4	55.0	200	August 30, 2014	+	Negative
	CE5	58.0	210	November 08, 2014	+++	Negative
	CE6	59.0	270	November 08, 2014	Negative	Negative

Jorge et al. (2013) found the detectability of nematodes was significantly lower in feces than in the intestine of reptiles, suggesting a large number of fecal samples and temporal replication to increase the likelihood of detection of intestinal parasites. Undoubtedly, parasitological studies based solely on feces most probably underestimate the true species richness present in a given host population and/or locality. However, such studies may be important for wildlife conservation purposes mainly when the translocations are employed, since the released individuals or individuals within the destination ecosystem may suffer from parasitic disease linked to the translocation process (Bobadilla Suarez et al. 2017). Specifically, it has been shown the vulnerability of populations of *C. triseriatus* in the Toluca Valley, due to the proximity to urban areas (Sunny et al. 2015), therefore, a complete parasitological record would aid wellbeing of snakes by receiving appropriate treatment plan and/or quarantine procedures. That allow to move the rattlesnakes outside the urban area, without the danger of infecting other populations with any diseased.

ACKNOWLEDGEMENTS

We thank Cruz Armida Olmeda Espinoza for the processing of the samples as laboratory technician in Facultad de Medicina Veterinaria y Zootecnia of the Universidad Autónoma de Tamaulipas.

Table 2. Records of metazoan parasites for *Crotalus* spp. in Mexico.

Parasite	Stage	Habitat	Host	State	Reference
Cestoda					
Mesocestoides sp.	Tetratiridium	Body cavity	*C. enyo* (Cope, 1861)	Baja California Sur	Goldberg et al. (2003)
			C. molossus Baird & Girard, 1853	Undetermined	Goldberg and Bursey (1999)
			C. pricei Van Denburgh, 1895		
			C. willardi Meek, 1905	Chihuahua, Sonora	Goldberg and Bursey (2000)
Oochoristica sp.	Adult	Intestine	*C. basiliscus* (Cope, 1864)	Colima	Paredes-León et al. (2008)
Ophiotaenia perspicua	Adult	Intestine	*C. cinerens*	Hidalgo	Flores-Barroeta et al. (1961)
Acanthocephala					
Oligacanthorhynchidae sp.	Cystacanth	Body cavity	*C. basiliscus*	Michoacán	Goldberg et al. (2006)
			C. enyo	Baja California Sur	Goldberg et al. (2003)
			C. lepidus (Kennicott, 1861)	Undetermined	Goldberg and Bursey (1999)
			C. tigris Kennicott, 1859		
			C. willardi	Chihuahua, Sonora	Goldberg and Bursey (2000)
Nematoda					
Hastospiculum onchocercum Chitwood, 1932	Adult	Stomach	*C. tzabcan* Klauber, 1952	Quintana Roo	Carbajal-Márquez et al. (2018)
Hexametra boddaertii (Baird, 1860)	Adult	Body cavity	*C. tzabcan*	Yucatán	Carbajal-Márquez et al. (2018)
			C. basiliscus	Sinaloa	Goldberg et al. (2006)
Kalicephalus inermis coronellae	Adult	Intestine	*C. pusillus* Klauber, 1952	Michoacán	Comroe (1948), Schad (1962)
			C. triseriatus		
Kalicephalus inermis macrovulvus (Caballero, 1954)	Adult	Intestine	*C. molossus*	Ciudad de México	Goldberg and Bursey (1999)
Macdonaldius oschei Chabaud & Frank, 1961	Adult	Post vena cava	*C. basiliscus*	Colima	Telford (1965)
Ophidascaris labiatopapillosa Walton, 1927	Adult		*C. molossus*	Durango	Klauber (1972)
Ozolaimus ctenosauri Caballero, 1938	Adult	Intestine	*C. polystictus* (Cope, 1865)	Ciudad de México	Caballero y Caballero (1939)
Physocephalus sp.	Larva	Mesentery	*C. mitchellii* (Cope, 1861)	Baja California	Goldberg et al. (2013)
Travassosascaris araujoi (Schneider, 1866)	Adult	Liver	*C. tzabcan*	Quintana Roo	Carbajal-Márquez et al. (2018)
Acari					
Amblyomma dissimile Koch, 1844	Adult	Body surface	*C. durissus* Linnaeus, 1758	Guerrero	Paredes-León et al. (2008)
Argasidae sp.		Body surface	*C. viridis viridis* (Rafinesque, 1818)	Chihuahua	Gatica-Colima et al. (2014)
Ixobioides sp.	Adult	Body surface	*C. triseriatus*	Ciudad de México	Paredes-León et al. (2008)
Ophionyssus natricis	Adult	Body surface	*C. durissus*	Ciudad de México	Paredes-León et al. (2008)
			C. triseriatus		
			C. cerastes cercobombus Savage & Cliff, 1953	Nuevo León	Rodríguez and Lazcano (1992)
			C. lepidus lepidus (Kennicott, 1861)		
			C. atrox Baird & Girard, 1853	Puebla	Paredes-León et al. (2008)
			C. ravus Cope, 1865		
Ornithodoros turicata (Dugès, 1876)	Larva		*C. mitchellii*	Baja California	Gutsche and Mutschmann (2011)
			C. ruber Cope, 1892	Baja California Sur	
Crustacea					
Porocephalus basiliscus Riley & Self, 1979	Adult		*C. basiliscus*	Colima	Riley and Self (1979)
Porocephalus crotali Humboldt, 1812	Adult		*C. atrox*	Nuevo León	Peláez and Julia (1983)
			C. basiliscus	Colima	Peláez and Julia (1983)
			C. catalinensis Cliff, 1954	Santa Catalina Island, Baja California Sur	Goldberg et al. (2003)
			C. culminatus Klauber, 1952	Guerrero	Pelaéz and Julia (1983)
			C. durissus	Tamaulipas	Peláez and Julia (1983)
			C. tzabcan	Campeche, Quintana Roo	Carbajal-Márquez et al. (2018)
Porocephalus tortugensis Riley and Self, 1979	Adult		*C. tortugensis* Van Denburgh & Slevin, 1921	Tortuga Island-Baja California Sur	Riley and Self (1979)
Raillietiella crotali Ali, Riley and Self, 1984	Adult	Lung	*C. ruber*	Pond Island, Baja California	Ali et al. (1984)
Raillietiella furcocercum (Diesing, 1836)	Adult	Lung	*C. atrox tortugensis* Baird & Girard, 1853	Tortuga Island-Baja California Sur	Ali et al. (1984), Klauber (1972)

LITERATURE CITED

Ali JH, Riley J, Self JT (1984) A revision of the taxonomy of pentastomid parasites (genus *Raillietiella* Sambon, 1910) from American snakes and amphisbaenians. Systematic Parasitology 6: 87–97. https://doi.org/10.1007/BF02185516

Anderson RC (2000) Nematode parasites of vertebrates. Their development and transmission. Wallingford, CABI Publishing, 650 pp.

Bobadilla Suarez M, Ewen JG, Groombridge JJ, Beckmann K, Shotton J, Masters N, Hopkins T, Sainsbury AW (2017) Using qualitative disease risk analysis for herpetofauna conservation translocations transgressing ecological and geographical barriers. EcoHealth 14: S47–S60. https://doi.org/10.1007/s10393-015-1086-4

Bowman D, Lynn RC, Eberhard ML (2004) Georgis parasitología para veterinarios. Madrid, Elsevier, 453 pp.

Bryson RW Jr, Linkem CW, Dorcas ME, Lathrop A, Jones JM, Alvarado-Díaz J, Grünwald CI, Murphy RW (2014) Multilocus species delimitation in the *Crotalus triseriatus* species group (Serpentes: Viperidae: Crotalinae), with the description of two new species. Zootaxa 3826: 475–496. https://doi.org/10.11646/zootaxa.3826.3.3

Caballero y Caballero E (1939) Nemátodos de los reptiles de México III. Anales del Instituto de Biología, Universidad Nacional Autónoma de México, Serie Zoología 10: 73–82.

Carbajal-Márquez RA, González-Solís D, Cedeño-Vázquez JR (2018) Endoparasites of *Crotalus tzabcan* (Serpentes: Viperidae), with a checklist in rattlesnakes. Journal of Parasitic Diseases 42: 303–314. https://doi.org/10.1007/s12639-018-1001-3

Comroe DB (1948) *Kalicephalus conoidus*, n. sp. (Strongylata), a nematode from the rattlesnake *Crotalus triseriatus*. Transactions of the American Microscopical Society 67: 280–284. https://www.jstor.org/stable/3223195

Flores-Barroeta L, Hidalgo-Escalante E, Montero-Gei F (1961) Céstodos de Vertebrados. VIII. Revista de Biología Tropical 9: 187–207.

Flores-Villela O, García-Vázquez UO (2014) Biodiversity of reptiles in Mexico. Revista Mexicana de Biodiversidad 85: S467–S475. https://doi.org/10.7550/rmb.43236

Fuehrer HP (2014) An overview of the host spectrum and distribution of *Calodium hepaticum* (syn. *Capillaria hepatica*): part 2 – Mammalia (excluding Muroidea). Parasitology Research 113: 641–651. https://doi.org/10.1007/s00436-013-3692-9

García E (2004) Modificaciones al sistema de clasificación climática de Köppen. Serie Libros Núm. 6. Instituto de Geografía, Mexico, Universidad Nacional Autónoma de México, 90 pp.

Gatica-Colima A, Macias-Rodríguez EF, Paredes-León R (2014) *Crotalus viridis viridis* (Prairie Rattlesnake). Ectoparasites. Herpetological Review 45: 143–144.

Gobierno del Estado de México (2012) Programa para Mejorar la Calidad del Aire del Valle de Toluca (2012-2017). Mexico, Gobierno del Estado de México, Secretaría del Medio Ambiente, 105 pp.

Goldberg SR, Bursey CR (1999) *Crotalus lepidus* (Rock Rattlesnake), *Crotalus molossus* (Blacktail Rattlesnake), *Crotalus pricei* (Twin-spotted Rattlesnake), *Crotalus tigris* (Tiger Rattlesnake). Endoparasites. Herpetological Review 30: 44–45.

Goldberg SR, Bursey CR (2000) *Crotalus mitchellii* (Speckled rattlesnake) and *Crotalus willardi* (Ridgenose rattlesnake). Endoparasites. Herpetological Review 31: 104.

Goldberg SR, Bursey CR, Beaman KR (2003) *Crotalus enyo* (Baja California Rattlesnake). Endoparasites. Herpetological Review 34: 64–65.

Goldberg SR, Bursey CR, Beaman KR, Dugan EA (2006) *Crotalus basiliscus* (Mexican West Coast Rattlesnake). Endoparasites. Herpetological Review 37: 94.

Goldberg SR, Bursey CR, Glaudas X (2013) Helminths of the Speckled Rattlesnake, *Crotalus mitchellii* (Squamata: Viperidae). Western North American Naturalist 73: 533–535. https://doi.org/10.3398/064.073.0407

Gutsche A, Mutschmann F (2011) *Crotalus mitchellIi* (Speckled Rattlesnake) and *Crotalus ruber* (Red Diamond Rattlesnake). Ectoparasites. Herpetological Review 42: 287–288.

Heimes P (2016) Herpetofauna Mexicana. Frankfurt am Main, Edition Chimaira, vol. 1, 572 pp.

Jorge F, Carretero MA, Roca V, Poulin R, Perera A (2013) What you get is what they have? Detectability of intestinal parasites in reptiles using faeces. Parasitology Research 112: 4001–4007. https://doi.org/10.1007/s00436-013-3588-8

Klauber LM (1972) Rattlesnakes: their habits, life histories, and influence on mankind. Berkeley, University of California Press, 1533 pp.

Mehlhorn H, Düwel D, Raether W (1993) Manual de Parasitología Veterinaria. Bogotá, Grass-Iatros, 436 pp.

Mitchell MA (2007) Parasites of reptiles. In: Baker DG (Ed) Flynn's parasites of laboratory animals. Iowa, Blackwell Publishing, 177–216.

Modrý D, Ne as P, Mazuch T, Kamler M (2004) *Eimeria atheridis* n. sp. (Apicomplexa: Eimeriidae), a new coccidium from the western bush viper *Atheris chlorechis* (Pel, 1851) from tropical Africa. Systematic Parasitology 59: 71–74. https://doi.org/10.1023/B:SYPA.0000038444.21384.93

Moravec F (1994) Parasitic Nematodes of Freshwater Fish of Europe. The Netherlands, Kluwer Academic Publishers, 473 pp.

Paredes-León R, García-Prieto L, Guzmán-Cornejo C, León-Règagnon V, Pérez TM (2008) Metazoan parasites of Mexican amphibians and reptiles. Zootaxa 1904: 1–166. http://www.realitat.com/websites/zoologia/zoo_08/tecnicos_s_853645/archivos/Metazoan%20parasites.pdf

Peláez D, Juliá J (1983) Pentastómidos de México. I. Porocefalosis en serpientes. (Comunicación preliminar). (Pentastom.: Poroceph. y Reptilia: Squamata). Anales de la Escuela Nacional de Ciencias Biológicas 27: 29–37.

Riley J, Self JT (1979) On the systematics of the pentastomid genus *Porocephalus* (Humboldt, 1811) with descriptions of two new species. Systematic Parasitology 1: 25–42. https://doi.org/10.1007/BF00009772

Rinaldi L, Mihalca AD, Cirillo R, Maurelli MP, Montesano M, Capasso M, Cringoli G (2012) FLOTAC can detect parasitic and pseudoparasitic elements in reptiles. Experimental Parasitology 130: 282–284. https://doi.org/10.1016/j.exppara.2012.01.011

Rodríguez ML, Lazcano D (1992) Primer reporte de ácaro *Ophionyssus natricis* (Acarina: Macronyssidae) para México. The Southwestern Naturalist 37: 426. https://www.jstor.org/stable/3671798

Rodríguez-Vivas RI, Cob-Galera LA (2005) Técnicas diagnósticas en parasitología veterinaria. Mérida, Universidad Autónoma de Yucatán, 306 pp.

Schad GA (1962) Studies on the genus *Kalicephalus* (Nematoda: Diaphanocephalidae). II. A taxonomic revision of the genus *Kalicephalus* Molin, 1861. Canadian Journal of Zoology 40:1035–1165. https://doi.org/10.1139/z62-090

SEDESOL, CONAPO, INEGI (2004) Delimitación de las zonas metropolitanas de México. Mexico, Secretaría de Desarrollo Social, Consejo Nacional de Población, Instituto Nacional de Estadística, Geografía e Informática, 169 pp.

Šlapeta J, Modrý D, Johnson R (2018) Reptile parasitology in health and disease. In: Doneley B, Monks D, Johnson R, Carmel B (Eds) Reptile medicine and surgery in clinical practice. Sussex, Wiley Blackwell, West 425–439.

Sunny A, Monroy-Vilchis O, Zarco-González MM, Mendoza-Martínez GD, Martínez-Gómez D (2015) Genetic diversity and genetic structure of an endemic Mexican Dusky Rattlesnake (*Crotalus triseriatus*) in a highly modified agricultural landscape: implications for conservation. Genetica 143: 705–716. https://doi.org/10.1007/s10709-015-9868-8

Telford SR (1965) A study of filariasis in Mexican snakes. Japanese Journal of Experimental Medicine 35: 565–586.

Wilson SC, Carpenter JW (1996) Endoparasitic diseases of reptiles. Seminars in Avian and Exotic Pet Medicine 5: 64–74. https://doi.org/10.1016/S1055-937X(96)80019-3

Wolf D, Vrhovec M, Failing K, Rossier C, Hermosilla C, Pantchev N (2014) Diagnosis of gastrointestinal parasites in reptiles: comparison of two coprological methods. Acta Veterinaria Scandinavica 56: 44. https://doi.org/10.1186/s13028-014-0044-4

Author Contributions: LSD and LGO contributed to the original idea. LGO JLRC and LSD wrote the draft of the manuscript. LGO and JLRC conducted laboratory analysis. LSD, YGO, HDV and JM contributed to field samplingt and provided revisions on the manuscript.

Competing Interests: The authors have declared that no competing interests exist.

6

Influences of the area, shape and connectivity of coastal lakes on the taxonomic and functional diversity of fish communities in Southern Brazil

Sandra M. Hartz[1,2], **Elise A. Rocha**[2], **Fernanda T. Brum**[3], **André L. Luza**[2], **Taís de F.R. Guimarães**[2], **Fernando G. Becker**[1,2]

[1] *Departamento de Ecologia, Universidade Federal do Rio Grande do Sul. Avenida Bento Gonçalves 9500, 91501-970 Porto Alegre, RS, Brazil.*

[2] *Programa de Pós-Graduação em Ecologia, Universidade Federal do Rio Grande do Sul. Avenida Bento Gonçalves 9500, 91501-970 Porto Alegre, RS, Brazil.*

[3] *Programa de Pós-Graduação em Ecologia e Conservação, Setor de Ciências Biológicas, Universidade Federal do Paraná. Avenida Coronel Francisco H. dos Santos 210, Jardim das Américas, 81531-980 Curitiba, PR, Brazil.*

*Corresponding author: Sandra Maria Hartz (*sandra.hartz@ufrgs.br*)*

ABSTRACT. In this study we investigated the influence of landscape variables on the alpha taxonomic and functional diversity of fish communities in coastal lakes. We built an analytical framework that included possible causal connections among variables, which we analyzed using path analysis. We obtained landscape metrics for the area, shape and connectivity (estuary connectivity and primary connectivity to neighboring lakes) of 37 coastal lakes in the Tramandaí River Basin. We collected fish data from 49 species using standardized sampling with gillnets and obtained a set of traits related to dispersal abilities and food acquisition. The model that best explained the taxonomic diversity and functional richness took into account the shape of the lakes. Functional richness was also explained by estuary connectivity. Functional evenness and dispersion were not predicted by area or connectivity, but they were influenced by the abundant freshwater species. This indicates that all lakes support most of the regional functional diversity. The results highlight the importance of the dispersal process in this lake system and allow the conclusion that considering multiple diversity dimensions can aid the conservation of local and regional fish communities.

KEY WORDS. Alpha diversity, environmental filters, limiting similarity, neutral paradigm, geographic information system.

INTRODUCTION

Coastal lakes are spatially and temporally dynamic ecosystems containing a substantial quantity of biodiversity (Esteves et al. 2008, Petry et al. 2016). Even though in these systems many factors influence biological communities, most previous studies have focused on the importance of water quality in determining alpha diversity. From a regional perspective, the local climate (temperature, precipitation and wind patterns) is the main driver of water quality and nutrient availability, which influence taxonomic diversity (Heino et al. 2007, Caliman et al. 2010). Besides these regional factors, local biotic interactions and anthropogenic disturbances can modify taxonomic diversity (Esteves et al. 2008, Petry et al. 2016). Additionally, landscape characteristics (position, morphometry and connectivity) of the lakes may also determine community structure in lakes, although their use is very recent (Quinlan et al. 2003).

Aquatic biodiversity has many facets that can be quantified, including taxonomic and functional diversity (Villéger et al. 2017). Functional diversity results from the functional trait space occupied by the species that compose a community (functional richness), from the distribution of abundance within the functional trait space (functional evenness) and from the functional dissimilarity of species across the functional trait space (functional dispersion) (Mason et al. 2005, Villéger et al. 2008, Laliberté and Legendre 2010, Mouchet et al. 2010). Functional redundancy indicates a high overlap of functional groups across the functional trait space (Laliberté and Legendre 2010). In contrast, functional complementarity indicates low functional overlap and that each species in the community has unique traits (Villéger et al. 2008, Laliberté and Legendre 2010, Leitão et al. 2016). Splitting the three independent functional diversity facets (richness, evenness and dispersion) can provide insights into the strength of the impact of environmental or biotic filters on community assembly (Mouchet et al. 2010, Leitão et al. 2016, Córdova-Tapia et al. 2017). Furthermore, incorporating the abundance of traits within the community may improve our understanding of functional diversity and community assembly, as some species may disproportionately occupy the functional space and, therefore, have a stronger effect on ecosystem functioning. The high abundance of some species may result from their tolerance to anthropogenic environmental changes (Villéger et al. 2008, Leitão et al. 2016).

The mechanisms that influence fish community diversity in lakes are mainly associated with the complexity of the area and the shoreline habitat (Tonn and Magnuson 1982, Eadie and Keast 2011, Petry et al. 2016), which are positively associated with primary productivity (Dodson et al. 2000), habitat heterogeneity (Dodson et al. 2009, Scheffer et al. 2006) and availability of refuges from large predators (Jennings et al. 1999). Particularly in coastal lakes, the salinity regimen is an important factor for fish communities at local and regional scales, in view of the changes it produces in species composition (Jones and West 2005, Pérez-Ruzafa et al. 2007, Camara et al. 2018). Regional dispersal can be an important determinant of alpha diversity, (Özkan et al. 2013, Spasojevic et al. 2014, Cardoso et al. 2017) since it is correlated with habitat isolation and landscape connectivity, which also affect fish species richness and composition (Olden et al. 2001, Bouvier et al. 2009, Guimarães et al. 2014). Connectivity, more than environmental heterogeneity, is directly linked to dispersal of organisms, which is one of the predominant mechanisms contributing to community diversity (Drakou et al. 2009). Recent studies have addressed the effects of landscape factors on the functional diversity of fish communities, mainly for lotic ecosystems (Pease et al. 2012, Dala-Corte et al 2016, among others). For lakes, in general, the functional richness and evenness decrease with habitat isolation (Miyazono et al. 2010, Cheng et al. 2014).

The coastal lakes of Southern Brazil harbor a considerable proportion of the Neotropical fish diversity in fresh and brackish waters. These lakes are influenced by both marine and continental processes and were subjected to events of sea retraction and expansion, which connected and disconnected lakes during their formation (Schwarzbold and Schäfer 1984). Azevedo et al. (2017) compared distinct types of coastal ecosystems in Southeast Brazil and observed that coastal lakes supported more functionally distinctive fish assemblages than beaches and bays. In highly connected lakes, communities may become functionally homogenized due to the frequent movement of vagile species (Guimarães et al. 2014). Thus, the ecological complexity of lakes (e.g. marginal marshes, grasslands, coastal 'restingas' and rock outcrops) and the historical seawater retractions and expansions that shaped the coastal landscapes of Southern Brazil offer a rare opportunity to evaluate how landscape factors such as lake connectivity, area and shape determine patterns of functional diversity in fish communities (Mouillot 2007, Miyazono et al. 2010).

We quantified the relative importance of landscape metrics to the alpha taxonomic and functional diversity of fish communities inhabiting lakes in the Tramandaí river basin in Southern Brazil. More specifically, we aimed to understand whether functional richness, evenness and dispersion are affected by lake characteristics (area and shape) and the connectivity of the lake with the surrounding lakes (primary connectivity) or the estuary zone (estuarine connectivity). For this, the causal relationships between these different variables were analyzed using a path analysis. We built an analytical framework that included all plausible causal connections between variables (Fig. 1). The effect of the area on taxonomic diversity is predicted to be mediated by how complex the shape of the lake is. Following to Tonn and Magnuson (1982), we assumed that the largest lakes are also the largest reservoirs of the species' pool in the system, therefore indicating the effect on both connectivity metrics. Assuming that functional diversity increases proportionally with taxonomic diversity (Mouillot et al. 2007), we predict that taxonomic diversity will have a positive effect on the functional richness, evenness and dispersion of fish communities (Villéger et al. 2010). Connectivity, mainly with the estuarine zone, should increase the effect of the area on taxonomic diversity by adding species from partially distinct habitats, increasing functional richness and dispersion in the lakes. On the other hand, the primary connectivity increases only the functional evenness, since it would facilitate and homogenize the effects of dispersal and abundance of the species.

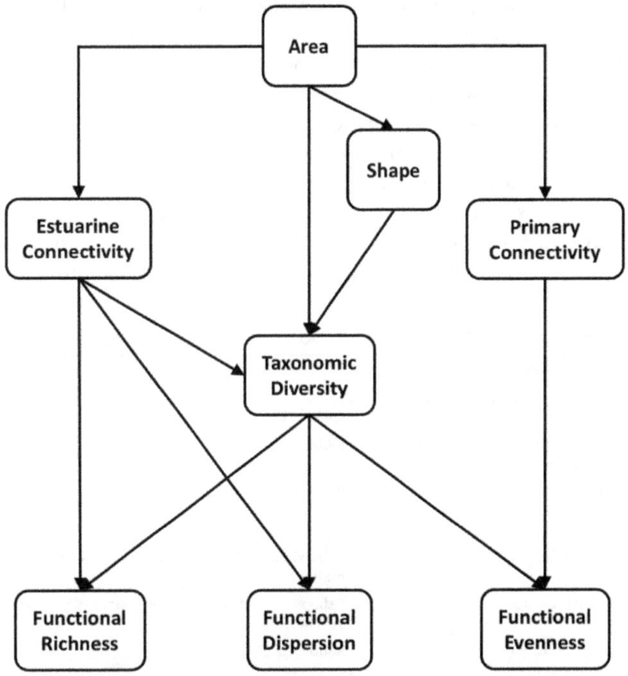

Figure 1. Hypothetical framework used to build the path model, with plausible causal connections between the landscape variables and both the taxonomic and functional diversity of the fish communities of the coastal lakes of the Tramandaí River Basin in Southern Brazil

MATERIAL AND METHODS

The Tramandaí River Basin (29°37'–30°30'S; 49°74'–50°24'W) is situated on the coast of the state of Rio Grande do Sul, Southern Brazil. It is a very representative coastal ecosystem containing 41 shallow coastal lakes formed recently in geological terms (Schwarzbold and Schäfer 1984), with varying degrees of interconnection and a single link to the ocean through the Tramandaí Estuary (see a map of the system in Guimarães et al. 2014). The formation of the system began in the Quaternary period through the deposition of sediments from marine transgressions and regressions, which produced sand barriers and isolated portions of water (Schwarzbold and Schäfer 1984). The region has a humid subtropical climate without a dry season (Hasenack and Ferraro 1989) and the mean temperature is 20 °C. February is the warmest month of the year.

We sampled 37 out of the 41 lakes from the river basin as it was not possible to access the other four lakes. The lakes have different degrees of connection between them, ranging from hydrologically isolated freshwater lakes to lakes that are directly connected to each other and/or to the estuary by channels. The estuary, which includes the Tramandaí Lagoon and its outlet channel to the sea, is a region of micro-tidal influence (Loitzenbauer and Mendes 2012) with seawater inflow and shorelines that flood during high tide. Consequently, three of the lakes have brackish waters (Würdig 1987).

Sampling in each lake consisted of the capture of fish using two sets of gillnets (in both the littoral and limnetic zone). Each gillnet set had an area of 180 square meters (mesh nodes of 15, 20, 25, 30, 35 and 40 mm) and was left for 24 hours in each lake. Samples were taken from May 2009 to April 2013. Since the activity periods of the species differ, we conducted between one and three sampling events in each lake per season (warm season, from October to April, and the cold season, from May to September). Larger lakes were sampled more often. The sampling effort was standardized by dividing the abundance of each species by the number of sampling events in the lake, which ranged from two to six events.

Morphological functional traits of fish were chosen based on the work of Gatz (1979), Sibbing and Nagelkerke (2001), Dumay et al. (2004), Mouillot et al. (2007), Villéger et al. (2010, 2017) and Córdova-Tapia et al. (2017). Traits of the largest possible number of adults from each species were measured using an electronic caliper (precision of 0.1mm, Supplementary file S1). We measured morphological traits that represent mobility (capacity to swim and dispersal), food acquisition and behavior (the ability to use resources in the environment) (Table 1). For rare species whose presence in the lakes was known but no adult individuals were found, measurements were collected from individuals in scientific collections. Functional traits were calculated from the ratios between these morphological measurements (Table 1). Trait values were obtained for each species from the mean of their individual measurements as it was assumed that intraspecific variation is lower than interspecific variation (Villéger et al. 2012). In addition to the continuous morphological trait data taken from sampled individuals, other categorical trait data (caudal fin shape and trophic category) were taken from specialized literature (Rodrigues and Hartz 2001, Nunes and Hartz 2006, Malabarba et al. 2013) and from FishBase (http://www.fishbase.de, Froese and Pauly 2016). We obtained a set of 18 traits (6 continuous, 12 categorical/binary), but retained only those attributes represented by at least four species (16 traits: 6 quantitative, 10 binaries; Table 1). None of the traits had a Spearman correlation coefficient greater than |0.7|. Gower index was used to build the matrix of functional distances among fish species (Pavoine et al. 2009).

Lake attributes (Table 2) were estimated using GIS Idrisi Taiga (Eastman 2009) applied to a spatial database of the Tramandaí River Basin, which was produced from Landsat -TM images. The two connectivity indices (named estuarine, linking each lake with the estuarine zone and primary, linking each lake with the surrounding neighbor lakes, Table 2) were calculated using the

Table 1. Functional traits measured for fish species sampled from May 2009 to April 2013 in 37 coastal lakes of the Tramandaí River Basin.

Morphological traits measured (mm) or categorized	Functional trait	Ecological meaning	References
Standard length (Ls) Body depth (Bd)	Ratio of Ls/Bd	Hydrodynamic ability	Sibbing and Nagelkerke 2001
Length (Pl) and depth (Pd) of the pectoral fin	Ratio of Pl/Pd	Swimming ability, maneuverability at slow speeds and locomotion efficiency	Bellwood et al. 2002, Dumay et al. 2004
Length (Cl) and depth (Cd) of the caudal fin	Ratio of Cl/Cd	Decreases as the swimming ability of the fish declines; benthic species tend to have higher ratios whereas rapid swimmers have lower ratios	Dumay et al. 2004, Mouillot et al. 2007
Total length (Lt)	Lt maximum (Standardized by log (x+1) transformation)	Species with longer bodies are more successful over long distance dispersers than species of smaller sizes	Lucas and Baras 2001
Caudal fin shape	Cshape (forked, emarginate, rounded, truncated*, semilunar*)	Influence the forces exerted on the water	Lauder 2000
Eye diameter (D)	Ratio D/Lt	Detection of food and visual acuity	Piet 1998, Boyle and Horn 2006
Mouth depth (M)	Ratio M/Lt	Maximum size of the prey ingested	Sibbing and Nagelkerke 2001
Trophic Category	CT (invertivorous, planktophagous, piscivorous, omnivorous, benthophagous, detritivorous, insectivorous)	Position in the food web and feeding behavior	Gerking 1994

* Removed because the trait was not present in more than four species.

Table 2. Patch-level metrics for 37 coastal lakes of the Tramandaí River Basin. For more details see Guimarães et al. (2014).

Variable	Equation	Description
Area (A)		Calculated in hectares.
Shape (S)	Shape = $(0.282 \times \text{perimeter})/\sqrt{\text{area}}$	Corrected relationship between perimeter/area. Ranges from 1 (a perfect circle) to infinite (a long narrow shape, Farina 1998).
Primary Connectivity (PC)	$PC_i = \Sigma[(RP)] \times [\log(\text{smallest CDis of the system}/\log10(CDis_{ij})]$	Related to recolonization potential (RP) and to the cost distance (Adriaensen et al. 2003, Guimarães et al. 2014). It considers the type of connection, lake area and the number and length of each connection. This index shows the degree of connection of each lake to its neighboring lakes.
	$RP = \log(Area_j)/\log(\text{largest lake of the system})$	RP assumes that the largest lake is also the largest reservoir of the species pool in the system. Therefore, the larger the lake the greater its potential to contribute to the recolonization of nearby lakes (Tonn and Magnuson 1982).
Estuarine Connectivity (EC)	$EC = 1/\log(CDis)$	Measure of functional connectivity based on the cost distance from each lake to the estuary, considering the friction value for each connection (Adriaensen et al. 2003, Guimarães et al. 2014).

concept of cost distance (CDis), which expresses the distance between two habitat patches weighted by the friction (resistance) imposed on the dispersal of an organism as it moves through a certain type of habitat (Adriansen et al. 2003). Friction values were designated for each connection between lakes in the following way: rivers (friction = 3), channels (friction = 5) and small man-made channels (friction = 7). To measure the connectivity of the lakes connected to the estuary region, the other lakes were scored with a friction value of 1to indicate that fish dispersal in these habitats is basically influenced by the distance they travel (Guimarães et al. 2014). None of the landscape covariates had a Spearman correlation coefficient greater than $|0.7|$.

Spatial autocorrelation may influence the patterns of functional similarity between assemblages (Devictor et al. 2010, Stuart-Smith et al. 2013), which may confound assessments of the trait-habitat relationship. To define the strength of the spatial autocorrelation on the functional patterns, we first calculated the mean pairwise functional distances between fish assemblages (Webb et al. 2008), which consist of the differences in species composition (Hellinger-transformed composition matrix, Legendre and Legendre 2012) weighted by the functional distance between species (based on the Gower distance according to the traits, Pavoine et al. 2009). The mean pairwise functional distance is a measure of functional beta diversity, which indicates the expected functional distance between two species randomly drawn from different communities (Webb et al. 2008). The functional beta diversity was calculated using the 'comdist' function of the picante package (R Core Team 2018). Then, we related the functional beta diversity to the spatial distances between lakes (Euclidean distance of Lat-Long coordinates) using a Co-Inertia analysis (Dolédec and Chessel 1994). Co-Inertia is a very general and flexible analysis as it allows us to relate two or more data sets that were composed using various distance measures and ordination methods. The Co-inertia analysis projects the position of points from two ordinations on to a common multivariate space, which was generated by an eigen-analysis of the cross-set covariance matrix (Dolédec and Chessel 1994, Legendre and Legendre 2012). We ran the Co-Inertia analysis using all the axes (i.e., the total variation from each dataset) of two different Principal Component Analyses: one for the distances between lakes and the another for the functional distances between lakes (two and 36 ordination axes, respectively). Overlaps in the position of objects (lakes) from each data set across the common multivariate space indicate the overall consistency of the relationship between the spatial and functional distances. The RV coefficient measures the degree of consistency in the relationship between ordinations, like a correlation coefficient (Legendre and Legendre 2012). To evaluate the consistency of the co-structure of the spatial and functional distances between lakes, we applied the Monte-Carlo Permutation Test to the sum of the eigenvalues of the Co-Inertia analysis (RV coefficient $\geq$ random RV coefficients, 999 permutations). This analysis was performed using the 'ade4' and 'picante' packages in the R environment (R Core Team 2018).

We used the complement of the Simpson Index (1-D) to access species taxonomic diversity (Magurran 2004) and three different functional indices (functional richness, functional evenness and functional dispersion) to access functional diversity (Villéger et al. 2008, Laliberté and Legendre 2010). We used different functional diversity components to evaluate the influence of lake area, size and connectivity on the overall functional space occupied by species (functional richness) and on how this space was occupied by fish species (functional evenness and dispersion). Functional richness, as measured by the FRic index, calculates the overall multivariate space occupied by coexistent species by linking the most extreme trait values, forming the vertices of a functional convex hull volume (Villéger et al. 2008). For this, we used nine PCoA axes (with 95% of total information retained). The number of PCoA axes was defined according to the number of species that was greater than the number of traits ("max" option in the 'dbFD' function, FD package, R Core Team 2018). The FRic index was standardized to obtain values between 0 and 1 (Laliberté et al. 2015). Functional evenness, as measured by the FEve index, represents the distribution of species abundance across the multivariate functional space (Villéger et al. 2008). Values will be low when some parts of the functional space are empty, while other parts are densely occupied. Neither FRic nor FEve estimate how species and abundance are spatially dispersed, as they only measure the position of the species in the functional space (convex hull, Laliberté et al. 2015). Thus, we used the Functional Dispersion index (FDis), which calculates the mean distance of each species from the functional-space centroid (Laliberté and Legendre 2010). For the FDis calculation, the centroid and species dispersion values were weighted by species abundance, so that the position of the centroid would be near to the most abundant species. If the functional space contained rare species located close to the abundant ones, we would expect low FDis values (Laliberté and Legendre 2010). We run the FRic, FEve and FDis analyses using the function 'dbFD' from the FD package (R Core Team 2018).

We assessed the influence of lake area, shape and connectivity on fish-community diversity using a path analysis (Shipley 2000), following the analytical approach proposed by Brum et al. (2012). First, we built a theoretical path model with all the plausible causal connections, based on our hypothesis (Fig. 1). Next, we found the generalized linear models that best explained each of the endogenous variables in the path model (i.e. a variable that is caused some other variable in the path model) by using the model selection procedure based on Akaike's information criterion (AIC, Burnham and Anderson 2010). We selected the generalized linear model with the lowest AIC value. Since our dependent variables were of different types, we choose the appropriate error distribution for each type in the model selection procedure. We used the beta distribution in the model selection for FRic, FEve, taxonomic diversity and estuarine connectivity (proportion data from Ferrari and Cribari-Neto 2004) and gamma distribution in the model selection for FDis, shape and primary connectivity (continuous positive data from Zuur et al. 2009). The path coefficients of the constructed model were the β regression coefficients from the best previous models. In order to validate the entire path model, we employed Fisher's test for composite probabilities (Sokal and Rohl 1994), called the C statistics by Shipley (2000). First, we compute the basis set by identifying the independence relationships between the variables in the path model (pairs of variables not connected by an arrow). For each independence test in the basis set, we evaluated the independence relationships by using partial correlation to control for the effects of all variables with direct ascendancy on both variables included in the independence relationship. Some of the variables in the path model presented correlated errors (CP and CE, FDis and FEve, FEve and FRic), and we could not establish a causal or unidirectional relationship between them. These relationships were incorporated to the path model by using a double-headed curve arrow between the variables, indicating the existence of correlated errors (Shipley 2000). The values on the double-headed curve arrows were the partial correlations after removing the effect of any covariates. After that, we computed the C statistics:

$$C = -2\sum_{i=1}^{k} \ln P_i$$

where k is the number of independence relationships in the basis set, Pi is the probability resulting from the partial multiple regression test for the independence relationship i. The C statistics follows a chi-square distribution with 2k degrees of freedom and the corresponding P value was used to validate the path model (Shipley 2000). The null hypothesis was that the independence relationships postulated in the model are valid, i.e. a P value below determined threshold would indicate that the independence assumptions of the basis set are false, and thus the model should be rejected. We used the 'stats' (R Core Team 2018) and 'betareg' (Cribari-Neto and Zeiles 2010) packages in R (R Core Team 2018) to run the model selection based on AIC and the partial correlations.

In addition to the direct causal effects (path coefficients) shown in the path model, we also computed the indirect effect and the total net effect of the predictors on the dependent variables in the model. The indirect effect was obtained from the product of the path coefficients of the sequence of arrows that lead from one variable to another (Shipley 2000). As we had more than one path connecting two variables, we obtained the overall indirect effect from all paths by summing all the indirect effects along all the paths between two variables. The net effect was then the sum of the direct and the overall indirect effect (Shipley 2000).

RESULTS

We captured a total of 7,870 individuals belonging to 49 fish species, 23 families and 9 orders (Supplementary file S1). The most frequent and abundant species (comprising more than 20% of the captured individuals) was *Cyphocharax voga* (Hensel, 1870) (Curimatidae) followed by *Geophagus brasiliensis* (Quoy and Gaimard, 1824) (Cichlidae) and *Oligosarcus jenynsii* (Günther, 1864) (Characidae). Functional richness ranged from 0.76 to 0.95 (mean ± SD = 0.88 ± 0.05), functional evenness from 0.61 to 0.90 (0.75 ± 0.69), functional dispersion from 1.60 to 2.41 (1.93 ± 0.22) and taxonomic diversity (Simpson index) from 0.17 to 28.62 (6.41 ± 5.38). The Co-Inertia analysis revealed a lack of co-structure between the functional beta diversity and the spatial distances between lakes (RV coefficient = 0.172; P = 0.124; Supplementary file S2), indicating that there was no spatial autocorrelation in the data.

The final path model, representing the significant best model connecting the variables according to our hypothetical model, is presented in Fig. 2 (C = 32.54, df = 36, P = 0.63). It shows that the complexity of the shape of the lake was the only landscape variable that significantly and positively explained the observed patterns of taxonomic diversity. Taxonomic diversity had a positive effect on the functional richness, but not on functional evenness or dispersion. The only connectivity value positively associated with functional richness was estuarine connectivity. The indirect effects of landscape variables on the taxonomic and functional diversity were very small, and only the shape of the lakes had a positive influence on the functional richness (Table 3).

Table 3. The direct, indirect and net effect values of the lake area, shape and connectivity on the taxonomic and functional diversity of the fish communities in the coastal lakes of the Tramandaí River Basin in Southern Brazil, based on the final path model. TD = taxonomic diversity; CP = primary connectivity; CE = estuarine connectivity; FRic = functional richness; FEve = functional eveness; FDis = functional dispersion.

Preditor	Response	Direct	Indirect	Net
Area	Shape	-0.0400	0.0000	-0.0400
Area	TD	0.0000	-0.0064	-0.0064
Area	FRic	0.0000	-0.0052	-0.0052
Area	FEve	0.0000	0.0002	0.0002
Area	FDis	0.0000	0.0001	0.0001
Area	CP	0.1230	0.0000	0.1230
Shape	TD	0.1610	0.0000	0.1610
Shape	FRic	0.0000	0.1288	0.1288
Shape	FEve	0.0000	-0.0048	-0.0048
Shape	FDis	0.0000	-0.0032	-0.0032
CE	FRic	0.2370	0.0000	0.2370
TD	FRic	0.8000	0.0000	0.8000
TD	FEve	-0.0200	0.0000	-0.0200
TD	FDis	-0.0300	0.0000	-0.0300

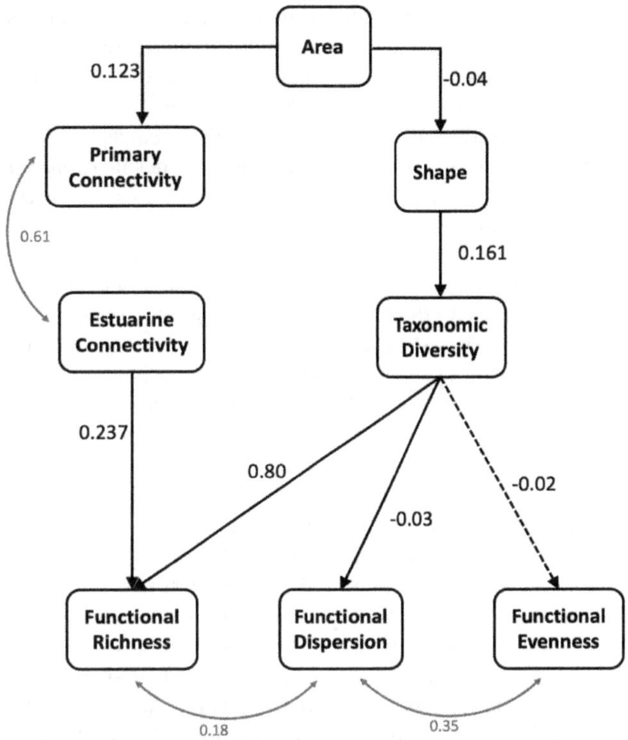

Figure 2. The final path model showing the causal relationships between the landscape variables and both the taxonomic and functional diversity of fish communities in the coastal lakes of the Tramandaí River Basin in Southern Brazil. Dotted line = not significant ($p > 0.05$). The curved double-headed arrows in grey depict correlated errors among variables.

DISCUSSION

The relationship between species richness and area has been established for a long time in ecology, particularly for freshwater ecosystems (Drakou et al. 2009, Petry et al. 2016). However, lake area was not the best predictor of taxonomic fish diversity in the Tramandaí River Basin. In our final model, area had a small negative influence on lake shape, indicating that larger lakes are less complex in form. Scheffer et al. (2006) discussed the small habitat size of shallow lakes and presented a framework suggesting that these lakes present a higher species richness than predicted by the classic island biogeography theory. In shallow lakes, the primary productivity occurs throughout the water column, resulting in greater biomass of macrophytes, adding resources and refuges to fish (Kruk et al. 2009). The lakes in the Tramandaí River Basin are shallow (mean maximum depth of around 3 m) and well connected (area had a positive influence on primary connectivity), allowing small lakes to support a high taxonomic diversity that is compatible with the diversity of larger lakes (lack of effect of area), in agreement with Scheffer et al. (2006).

The complexity of the shape of the lake had a direct positive influence on the taxonomic diversity and an indirect positive influence on the functional richness. This result may be related to the high proportion of different lake border habitats and the decreased available inner area (limnetic zone) of irregularly shaped lakes (i.e., lakes with higher shape index values, Farina 1998). The input of nutrients is favored in aquatic habitat patches with large perimeters (Pérez-Ruzafa et al. 2007), and therefore high productivity and availability of food resources would be expected. We only sampled fish communities with gillnets, which favor the capture of large species (Franco et al. 2012), even though the nets were placed in both the littoral and limnetic zones of the lakes. In the Tramandaí River Basin there are many small species (Lt max around 100 mm) living in the littoral zone (Malabarba et al. 2013) that were not detected by our sampling method. Despite of that, the effect of lake shape on functional richness was detected in our results. This reinforces the importance of the shape of the lake and consequently the importance of the littoral zones in shallow coastal lakes.

The positive association found in our study between taxonomic diversity and functional richness is not surprising because several other studies revealed a positive correlation between them (Pool et al. 2014, Carvalho and Tejerina-Garro 2015, among others). The volume of the functional space increases when functionally very distinct species are added in the community (Villéger et al. 2008). In our case, this occurred in the lakes near to the estuarine zone. Therefore, only the estuarine connectivity was important for (positively correlated to) functional richness in the Tramandaí River Basin, indicating that the lakes that are better connected with de estuarine zone have different species' composition, with functionally distinct species. Salinity is one of the main factors influencing the distribution of fish species in coastal zones (Barletta et al. 2005, Mouillot et al. 2007, Pérez-Ruzafa et al. 2007, Sosa-López et al. 2007, Moura et al. 2012). Estuarine habitats allow for the occurrence of rare ecomorphologically distinct species such as the sole *Citharichthys macrops* Dresel, 1885 (higher value of the Ls/Bd) and the ladyfish *Elops saurus* Linnaeus, 1766 (higher value of the Pl/Pd). Guimarães et al. (2014) found that, in the same study area, lake connectivity was more important than lake area for estuarine species richness. In the Tramandaí River Basin three of the lakes studied have brackish waters (strictly coastal lagoons), despite the fact that these lakes are highly interconnected with other lakes. Azevedo et al. (2017) also found a higher functional diversity in coastal lagoons compared to bays or beaches (a consequence of the salinity gradient) on the Southeastern coast of Brazil. However, it is important to note that in their study the lagoons evaluated were not interconnected.

On the other hand, taxonomic diversity had a very small and negative influence on functional evenness and dispersion. Furthermore, no measure of connectivity was retained in our final model for both metrics. The results obtained showed that even when functional richness increased, FEve and FDis remained unaltered or even decreased. This result may be due to the effect of the more dominant functionally similar species sampled in the lakes. These metrics are heavily influenced by the most abundant and dominant species (Villéger et al. 2008, Laliberté and Legendre 2010) and, according to Hillebrand et al. (2008), the dominance reflects the distribution of traits in a community. In the Tramandaí River Basin the species with the greatest numbers of individuals were also present in almost all lakes (the characins *Cyphocharax voga* and *Oligosarcus jenynsii* and the cichlid *Geophagus brasiliensis*). We recalculated FEve and FDis metrics with only presence data of the species in the lakes and we obtained higher values significantly different for both metrics (Supplementary file S3), indicating the effect of the abundance of the dominant species.

Initially we may suppose that lakes seem to have sufficiently favorable environments to support the regional functional pool of traits related to dispersal and resource use, since the same dominant species are in all lakes. In natural systems organisms are subject to directional dispersal to find suitable environmental situations (Moritz et al. 2013). Even though it does not occur frequently, for some lake systems, other studies have found no evidence that fish species have dispersal limitations (Magnuson et al. 1998, Mason et al. 2007). The coastal lakes of the Tramandaí River Basin are shallow, which facilitates wind action (Cenzano and Würdig 2006). Wind is a major ecological force in this region, producing waves that resuspend bottom sediments (Schwarzbold and Schäfer 1984). This wind-driven resuspension of nutrients increases the efficiency of energy flow through the food chain, providing particulate organic matter to the lower trophic levels which consequently benefits the higher trophic levels (Jeppesen et al. 1999). In general, shallow ponds and lakes have a greater tendency for eutrophication; however, the coastal lakes of the Tramandaí River Basin do not follow this pattern. The constant exposure to ocean winds produces both vertical and horizontal movements of the water, and since the lakes are shallow the sediment is well oxygenated, which means that there is a chemical fixation of phosphorus (Schäfer et al. 2013). This reduces the natural availability of this nutrient, decreasing the chances of eutrophication and preventing the oxygen reduction that these events generally cause (Schäfer et al. 2013). The efficiency of energy flow and the high levels of oxygenation make these lakes extremely productive ecosystems. This leads us to suppose that perhaps from a biological point of view, the lakes do not differ in their availability of resources.

Other environmental factors not measured on the lakes, like water quality and salinity (Schäfer 1992), could be important for FEve and FDis metrics since these variables influencing the local abundance of the species (Macedo-Soares et al. 2010, Souza et al. 2018). Similarly, if we had used other types of traits, such as habitat use, or traits related to life history strategies (Winemiller 1989), maybe these functional metrics could have been explained by the landscape variables used. Particularly the traits linked with habitat use (Villéger et al. 2017) should be included in future studies, since they reflect the abundance of species with degradation of shoreline habitats in the lakes (Jennings et al. 1999, Villéger et al. 2010).

In addition, since the shape and connectivity (estuarine) positively influenced the functional richness, as previously discussed, both niche filtering and dispersal processes are involved for the functional diversity in the coastal lakes of the Tramandaí River Basin. As FEve and FDis were influenced by the most abundant and dispersal-prone species, these different processes could affect different groups of species (Pandit et al. 2009, Leitão et al. 2016): the more functionally original species would be influenced by niche processes, while the less functionally abundant species would be influenced by neutral processes. In our study, the most abundant and frequent species in the lakes are close in the functional space (Supplementary file S4). On the other hand, the most distinctive species are either estuarine or they belong to the loricariid family (the catfish *Loricariichthys anus* (Valenciennes, 1835) and *Rineloricaria quadrensis* Reis, 1983), a group that has low dispersal capacity (Pagotto et al. 2011, Rodrigues-Filho et al. 2018). Accordingly, the study about the dispersal capacity of the species becomes another important point for understanding the local dynamics of fish communities in this region.

The local diversity of a habitat patch is determined by dispersal, spatial dynamics and processes of extinction and colonization. In the coastal lakes of the Tramandaí River Basin, lake shape and estuarine connectivity were good predictors of the taxonomic diversity and functional richness of fish communities, but not of their functional evenness and dispersion. There was a certain level of functional redundancy, produced by the spatially frequent and abundant species. Species coexistence could be explained by either niche and, for the traits analyzed, neutral dynamics. Coastal lakes are human-dominated ecosystems (Esteves et al. 2008) and according to our results, in terms of functional diversity conservation of fish (Mouillot 2007), it is important to maintain the diversity and the heterogeneity of the regional habitats among lakes to prevent biotic homogenization and to preserve the regional abundance of the species.

ACKNOWLEDGMENTS

We would like to thank Coordenação de Aperfeiçoamento de Pessoal de Nível Superior (CAPES) for the master scholarship to EAR, Phd scholarship to TFRG, postdoctoral fellowship to FTB, and PROEX program) and the Conselho Nacional de Desenvolvimento Científico e Tecnológico (CNPq, processes 304820/2014-8 and 483873/2007-1) for financial support. Additionally, we thank Ana Petry, João P. Vieira Sobrinho, Leandro Duarte, Luigi Naselli-Flores, David Hoeinghaus and the Subject Editor S. Adriano Melo and the anonymous reviewers for their valuable suggestions. We also extend special thanks to all the field assistants who made this research possible. All collection licenses (ICMBio – process 11174) and Ethics Committee approvals (UFRGS – process 15044) were obtained.

LITERATURE CITED

Adriaensen F, Chardon J, De Blust G, Swinnen E, Villalba S, Gulinck H, Matthysen E (2003) The application of 'least-cost' modelling as a functional landscape model. Landscape and Urban Planning 64: 233–247. https://doi.org/10.1016/S0169-2046(02)00242-6.

Azevedo M, Gomes-Gonçalves R, Mattos T, Uehara W, Guedes G, Araújo F (2017) Taxonomic and functional distinctness of the fish assemblages in three coastal environments (bays, coastal lagoons and oceanic beaches) in Southeastern Brazil. Marine Environmental Research 129: 180–188. https://doi.org/10.1016/j.marenvres.2017.05.007

Barletta M, Barletta-Bergan A, Saint-Paul U, Hubold G (2005) The role of salinity in structuring the fish assemblages in a tropical estuary. Journal of Fish Biology 66: 45–72. https://doi.org/10.1111/j.0022-1112.2005.00582.x

Bellwood D, Wainwright R, Fulton C, Hoey A (2002) Assembly rules and functional groups at global biogeographical scales. Functional Ecology 16: 557–562. https://doi.org/10.1046/j.1365-2435.2002.00655.x

Boyle K, Horn M (2006) Comparison of feeding guild structure and ecomorphology of intertidal fish assemblages from central California and central Chile. Marine Ecology Progress Series 319: 65–84. https://doi.org/10.3354/meps319065

Bouvier L, Cottenie K, Doka S (2009) Aquatic connectivity and fish metacommunities in wetlands of the lower Great Lakes. Canadian Journal of Fisheries and Aquatic Sciences 66: 933–948. https://doi.org/10.1139/F09-050

Brum FT, Kindel A, Hartz SM, Duarte L (2012) Spatial and phylogenetic structure drive frugivory in Tyrannidae birds across the range of Brazilian Araucaria forests. Oikos 121: 899–906. https://doi.org/10.1111/j.1600-0706.2011.19978.x

Burnham K, Anderson D (2010) Model Selection and Multimodel Inference. A Practical Information-Theoretic Approach. New York, Springer-Verlag, 2nd ed., 488 pp.

Caliman A, Carneiro L, Santangelo J, Guariento R, Pires A, Suhett A, Quesado L, Scofield V, Fonte E, Lopes P, Sanches L, Azevedo F, Marinho C, Bozelli R, Esteves E, Farjalla V (2010) Temporal coherence among tropical coastal lagoons: a search for patterns and mechanisms. Brazilian Journal of Biology 70: 803–814. https://doi.org/10.1590/S1519-69842010000400011

Camara E, Caramaschi E, Di Dario F, Petry A (2018) Short-term changes in two tropical coastal lagoons: effects of sandbar openings on fish assemblages. Journal of Coastal Research 34: 90–105. https://doi.org/10.2112/jcoastres-D-16-00026.1

Cardoso S, Nabout J, Farjalla V, Lopes P, Bozelli R, Huszar V, Roland F (2017) Environmental factors driving phytoplankton taxonomic and functional diversity in Amazonian floodplain lakes. Hydrobiologia 802: 115–130. https://doi.org/10.1007/s10750-017-3244-x

Carvalho R, Tejerina-Garro F (2015) Relationships between taxonomic and functional components of diversity: implications for conservation of tropical freshwater fishes. Freshwater Biology 60: 1854–1862. https://doi.org/10.1111/fwb.12616

Cenzano C, Würdig N (2006) Spatial and temporal variations of the benthic macrofauna in different habitats of a lagoon of the northern coastal system of Rio Grande do Sul State, Brazil. Acta Limnologica Brasiliensia 18:153–163. https://www.ablimno.org.br

Cheng L, Blanchet S, Loot G, Villéger S, Zhang T, Lek S, Lek-Ang S, Li Z (2014) Temporal changes in the taxonomic and functional diversity of fish communities in shallow Chinese lakes: the effects of river-lake connections and aquaculture. Aquatic Conservation: Marine and Freshwater Ecosystems 24: 23–34. https://doi.org/10.1002/aqc.2418

Córdova-Tapia F, Hernández-Marroquin V, Zambrano L (2017) The role of environmental filtering in the functional structure of fish communities in tropical wetlands. Ecology of Freshwater Fish 27: 522–532. https://doi.org/10.1111/eff.12366

Cribari-Neto F, Zeileis A (2010) Beta Regression in R. Journal of Statistical Software 34: 1–24. https://doi.org/10.18637/jss.v034.i02

Dala-Corte R, Giam X, Olden J, Becker F, Guimarães T, Melo A (2016) Revealing the pathways by which agricultural land-use affects stream fish communities in South Brazilian grasslands. Freshwater Biology 61: 1921–1934. https://doi.org/10.1111/fwb.12825

Devictor V, Mouillot D, Meynard C, Jiguet F, Thuiller W, Mouquet N (2010) Spatial mismatch and congruence between taxonomic, phylogenetic and functional diversity: the need for integrative conservation strategies in a changing world. Ecology Letters 13: 1030–1040. https://doi.org/10.1111/j.1461-0248.2010.01493.x

Dodson S, Arnott S, Cottingham K (2000) The relationship in lake communities between primary productivity and species richness. Ecology 81: 2662–2679. https://doi.org/10.1890/0012-9658(2000)081

Dodson S, Newman A, Will-Wolf S, Alexander M, Woodford M, Van Egeren S (2009) The relationship between zooplankton community structure and lake characteristics in temperate lakes (Northern Wisconsin, USA). Journal of Plankton Research 31: 93–100. https://doi.org/10.1093/plankt/fbn095

Dolédec S, Chessel D (1994) Co-inertia analysis: an alternative method for studying species-environment relationships. Freshwater Biology 31: 277–294. https://doi.org/10.1111/j.1365-2427.1994.tb01741.x

Drakou E, Bobori D, Kallimanis A, Mazaris A, Sgardelis S, Pantis J (2009) Freshwater fish community structured more by dispersal limitation than by enrironmental heterogeneity. Ecology of Freshwater Fish 18: 369–379. https://doi.org/10.1111/j.1600-0633.2009.00354.x

Dumay O, Tari P, Tomasini J, Mouillot D (2004) Functional groups of lagoon fish species in Languedoc Roussillon, southern France. Journal of Fish Biology 64: 970–983. https://doi.org/10.1111/j.1095-8649.2004.00365.x

Eadie J, Keast A (2011) Resource heterogeneity and fish species diversity in lakes. Canadian Journal of Zoology 62: 1689–1695. https://doi.org/10.1139/z84-248

Eastman JR (2009) IDRISI Taiga. Worcester, Clark University, 342 pp.

Esteves F, Caliman A, Santangelo J, Guariento R, Farjalla V, Bozelli R (2008) Neotropical coastal lagoons: an appraisal of their biodiversity, functioning, threats and conservation management. Brazilian Journal of Biology 68: 967–981. https://doi.org/10.1590/s1519-69842008000500006

Farina A (1998) Principles and methods in landscape ecology. London, Chapman & Hall, 412 pp.

Ferrari SLP, Cribari-Neto F (2004) Beta Regression for Modeling Rates and Proportions. Journal of Applied Statistics, 31: 799–815. https://doi.org/10.1080/0266476042000214501

Franco A, Pérez-Ruzafa A, Drouineau H, Franzoi P, Koutrakis E, Lepage M, Verdiell-Cubedo D, Bouchoucha M, López-Capel A, Riccato F, Sapounidis A, Marcos C, Oliva-Paterna F, Torralva-Forero M, Torricelli P (2012) Assessment of fish assemblages in coastal lagoon habitats: effect of sampling method. Estuarine, Coastal and Shelf Science 112: 115–125. https://doi.org/10.1016/j.ecss.2011.08.015

Froese R, Pauly D (2016) FishBase. World Wide Web electronic publication. https://www.fishbase.org.

Gatz A (1979) Community organization in fishes as indicated by morphological features. Ecology 60: 711–718. https://doi.org/10.2307/1936608

Gerking S (1994) Feeding Ecology of Fish. San Diego, Academic Press, 416 pp.

Guimarães T, Hartz SM, Becker F (2014) Lake connectivity and fish species richness in southern Brazilian coastal lakes. Hydrobiologia 740: 207–217. https://doi.org/10.1007/s10750-014-1954-x

Hasenack H, Ferraro L (1989) Considerações sobre o clima da região de Tramandaí, RS. Pesquisas 22: 53–70. https://seer.ufrgs.br/PesquisasemGeociencias/article/view/21456

Heino J, Mykrä H, Kotanen J, Muotka T (2007) Ecological filters and variability in stream macroinvertebrates communities: do taxonomic and functional structure follow the same path? Ecography 30: 217–230. https://doi.org/10.1111/j.2007.0906-7590.04894.x

Hillebrand H, Bennett D, Cadotte MW (2008) Consequences of dominance: a review of evenness effects on local and regional ecosystem processes. Ecology 89: 1510–1520. https://doi.org/10.1890/07-1053.1

Jennings M, Bozek M, Hatzenbeler G, Emmons E, Staggs M (1999) Cumulative effects of incremental shoreline habitat modification on fish assemblages in North Temperate Lakes. North American Journal of Fisheries Management 19: 18–27. https://doi.org/10.1577/1548-8675(1999)019<0018:-CEOISH>2.0.CO;2

Jeppesen E, Jensen J, Søndergaard M, Lauridsen T (1999) Trophic dynamics in turbid and clearwater lakes with special emphasis on the role of zooplankton for water clarity. Hydrobiologia 408/409: 217–231. https://doi.org/10.1007/978-94-017-2986-4_24

Jones M, West R (2005) Spatial and temporal variability of seagrass fishes in intermittently closed and open coastal lakes in southeastern Australia. Estuarine, Coastal and Shelf Science 64: 277–288. https://doi.org/10.1016/j.ecss.2005.02.021

Kruk C, Rodríguez-Gallego L, Meerhoff M, Quintans F, Lacerot G, Mazzeo N, Scasso F, Paggi J, Peeters E, Scheffer M (2009) Determinants of biodiversity in subtropical shallow lakes (Atlantic coast, Uruguay). Freshwater Biology 54: 2628–2641. https://doi.org/10.1111/j.1365-2427.2009.02274.x

Laliberté E, Legendre P (2010) A distance-based framework for measuring functional diversity from multiple traits. Ecology 91: 299–305. https://doi.org/10.1890/08-2244.1

Laliberté E, Legendre P, Shipley B, Laliberté M (2015) Measuring functional diversity (FD) from multiple traits, and other tools for functional ecology. R package version 1.0–12, 27pp. http://idg.pl/mirrors/CRAN/web/packages/FD/FD.pdf

Lauder G (2000) Function of the caudal fin during locomotion in fishes: kinematics, flow visualization, and evolutionary patterns. American Zoologist 40: 101–122. https://doi.org/10.1093/icb/40.1.101

Legendre P, Legendre L (2012) Numerical Ecology. Montreal, Elsevier, 1006 pp.

Leitão R, Zuanon J, Villéger S, Williams S, Baraloto C, Fortunel C, Mendonça F, Mouillot D (2016) Rare species contribute disproportionately to the functional structure of species assemblages. Proceedings Royal Society B 283: 20160084. https://doi.org/10.1098/rspb.2016.0084

Loitzenbauer E, Mendes CAB (2012) Salinity dynamics as a tool for water resources management in coastal zones: an application in the Tramandaí River basin, southern Brazil. Ocean and Coastal Management 55: 52–62. https://doi.org/10.1016/j.ocecoaman.2011.10.011

Lucas M, Baras E (2001) Migration of Freshwater Fishes. Oxford, Blackwell Science, 440 pp.

Macedo-Soares P, Petry A, Farjalla V, Caramaschi E (2010) Hydrological connectivity in coastal inland systems: lessons from a Neotropical fish metacommunity. Ecology of Freshwater Fish 19: 7–18. https://doi.org/10.1111/j.1600-0633.2009.00384.x

Magnuson J, Tonn W, Banerjee A, Toivonen J, Sanchez O, Rask M (1998) Isolation vs. extinction in the assembly of fishes in small northern lakes. Ecology 79: 2941–2956. https://doi.org/10.1890/0012-9658(1998)079

Magurran A (2004) Measuring biological diversity. Oxford, Blackwell Pub., 264 pp.

Malabarba L, Neto L, Bertaco V, Carvalho T, Santos J, Artioli L (2013) Guia de Identificação dos Peixes da Bacia do Rio Tramandaí. Porto Alegre, Via Sapiens, 143 pp.

Mason N, Mouillot D, Lee W, Wilson J (2005) Functional richness, functional evenness and functional divergence: the primary components of functional diversity. Oikos 111: 112–118. https://doi.org/10.1111/j.0030-1299.2005.13886.x

Mason N, Lanoiselée C, Mouillot D, Irz P, Argillier C (2007) Functional characters combined with null models reveal inconsistency in mechanisms of species turnover in lacustrine fish communities. Oecologia 153: 441–452. https://doi.org/10.1007/s00442-007-0727-x

Miyazono S, Aycock J, Miranda L, Tietjen T (2010) Assemblage patterns of fish functional groups relative to habitat connectivity and conditions in floodplain lakes. Ecology of Freshwater Fish 19: 578–585. https://doi.org/10.1111/j.1600-0633.2010.00438.x

Moritz C, Meynard C, Devictor V, Guizien K, Labrune C, Guarini J, Mouquet N (2013) Disentangling the role of connectivity, environmental filtering, and spatial structure on metacommunity dynamics. Oikos 122: 1401–1410. https://doi.org/10.1111/j.1600-0706.2013.00377.x

Mouchet M, Villéger S, Mason N, Mouillot D (2010) Functional diversity measures: an overview of their redundancy and their ability to discriminate community assembly rules. Functional Ecology 24: 867–876. https://doi.org/10.1111/j.1365-2435.2010.01695.x

Mouillot D (2007) Niche-assembly vs. dispersal-assembly rules in coastal fish metacommunities: implications for management of biodiversity in brackish lagoons. Journal of Applied Ecology 44: 760–767. https://doi.org/10.1111/j.1365-2664.2007.01330.x

Mouillot D, Dumay O, Tomasini J (2007) Limiting similarity, niche filtering and functional diversity in coastal lagoon fish communities. Estuarine, Coastal and Shelf Science 71: 443–456. https://doi.org/10.1016/j.ecss.2006.08.022

Moura P, Vieira J, Garcia A (2012) Fish abundance and species richness across an estuarine freshwater ecosystem in the Neotropics. Hydrobiologia 696: 107–122. https://doi.org/10.1007/s10750-012-1187-9

Nunes D, Hartz SM (2006) Feeding dynamics and ecomorphology of *Oligosarcus jenynsii* (Gunther, 1864) and *Oligosarcus robustus* (Menezes, 1969) in the Fortaleza Lake, southern Brazil. Brazilian Journal of Biology 66: 121–132. https://doi.org/10.1590/s1599-69842006000100016

Olden J, Jackson D, Peres-Neto P (2001) Spatial isolation and fish communities in drainage lakes. Oecologia 127: 572–585. https://doi.org/10.1007/s004420000620

Özkan K, Jeppesen E, Sondergaard M, Lauridsen T, Liboriussen L, Svenning J-C (2013) Contrasting roles of water chemistry, lake morphology, land-use, climate and spatial processes in driving phytoplankton richness in the Danish landscape. Hydrobiologia 710: 173–187. https://doi.org/10.1007/s10750-011-0996-6

Pagotto J, Goulart E, Oliveira E, Yamamura C (2011) Trophic ecomorphology of Siluriformes (Pisces, Osteichthyes) from a tropical stream. Brazilian Journal of Biology 71: 469–479. https://doi.org/10.1590/S1519-69842011000300017

Pandit S, Kolasa J, Cottenie K (2009) Contrasts between habitat generalists and specialists: an empirical extension to the basic metacommunity framework. Ecology 90: 2253–2262. https://doi.org/10.1890/08-0851.1

Pavoine S, Vallet J, Dufour A-B, Gachet S, Daniel H (2009) On the challenge of treating various types of variables: application for improving the measurement of functional diversity. Oikos 118: 391–402. https://doi.org/10.1111/j.1600-0706.2008.16668.x

Pease A, González-Díaz A, Rodiles-Hernández R, Winemmiller K (2012) Functional diversity and trait-environmental relationships of stream fish assemblages in a large tropical catchment. Freshwater Biology 57: 1060–1075. https://doi.org/10.1111/j.1365-2427.2012.02768.x

Pérez-Ruzafa A, Mompeán M, Marcos C (2007) Hydrographic, geomorphologic and fish assemblage relations in coastal lagoons. Hydrobiologia 577: 107–125. https://doi.org/10.1007/978-1-4020-6008-310

Petry A, Guimarães T, Vasconcellos F, Hartz SM, Becker F, Rosa R, Goyenola G, Caramaschi E, Díaz de Astarloa J, Sarmento-Soares L, Vieira J, Garcia A, Teixeira de Mello F, Melo F, Meerhoff M, Attayde J, Menezes R, Mazzeo N (2016) Fish composition and species richness in eastern South American coastal laggons: additional support for the freshwater ecoregions of the world. Journal of Fish Biology 89: 280–314. https://doi.org/10.1111/jfb.13011

Piet G (1998) Ecomorphology of a size-structured tropical freshwater fish community. Environmental Biology of Fishes 51: 67–86. https://doi.org/10.1023/A:100733853

Pool T, Grenouillet G, Villéger S (2014) Species contribute differently to the taxonomic, functional, and phylogenetic alpha and beta diversity of freshwater fish communities. Diversity and Distributions 20: 1235–1244. https://doi.org/10.1111/ddi.12231

Quinlan R, Paterson A, Hall R, Dillon P, Wilkinson A, Cumming B, Douglas M, Smol J (2003) A landscape approach to examining spatial patterns of limnological variables and long-term environmental change in a southern Canadian lake district. Freshwater Biology 48:1676–1697. https://doi.org/10.1046/j.1365-2427.2003.01105.x

R Core Team (2018) R: A language and environment for statistical computing. Vienna, R Foundation for Statistical Computing. https://www.R-project.org

Rodrigues G, Hartz SM (2001) Food dynamics of fish and the interaction with macroinvertebrates from a shallow lake in southern Brazil. Vereinigung für Theoretische und Angewandte Limnologie 27: 3309–3314. https://doi.org/10.1080/03680770.1998.11902439

Rodrigues-Filho C, Gurgel-Lourenço R, Bezerra L, Oliveira E, Leitão R, Garcez D, Sánchez-Botero J (2018) How are local fish communities structured in Brazilian semiarid headwater streams? Hydrobiologia 819: 93–108. https://doi.org/10.1007/s10750-018-3650-8

Schäfer A (1992) Ecological characteristics of the coastal lakes in southern Brazil: a synthesis. Acta Limnologica Brasiliensia 4: 111–122.

Schäfer A, Marchet C, Ahlert S, Schuh S (2013) Volume de água e tendência de eutrofização das Lagoas Costeiras. In: Lanzer R, Schäfer A, Jahn M (Eds) Gestão dos Recursos Hídricos, Municípios de Cidreira, Balneário Pinhal e Palmares do Sul: Recursos Hídricos e Toxicologia. Caxias do Sul, EDUCS, 22–35.

Scheffer M, van Geest G, Zimmer K, Jeppesen E, Sondergaard M, Butler M, Hanson M, Dederck S, De Meester L (2006) Small habitat size and isolation can promote species richness: second-order effects on biodiversity in shallow lakes and ponds. Oikos 112: 227–231. https://doi.org/10.1111/j.0030-1299.2006.14145.x

Schwarzbold A, Schäfer A (1984) Gênese e Morfologia das Lagos Costeiras do Rio Grande do Sul – Brasil. Amazoniana 9: 87–104.

Sibbing F, Nagelkerke L (2001) Resource partitioning by Lake Tana barbs predicted from fish morphometrics and prey characteristics. Reviews in Fish Biology and Fisheries 10: 393–437. https://doi.org/10.1023/A:1012270422092

Shipley, B (2000) Cause and correlation in biology: a user's guide to path analysis, structural equations and causal inference. Cambridge, Cambridge University Press, 317 pp.

Sokal R, Rohlf J (1994) Biometry: the principles and practices of statistics in biological research. New York, W.H. Freeman Publishers, 880 pp.

Sosa-López A, Mouillot D, Ramos-Miranda J, Flores-Hernandez D, Chi T (2007) Fish species richness decreases with sanility in tropical coastal lagoons. Journal of Biogeography 34: 52–61. https://doi.org/10.1111/j.1365-2699.2006.01588.x

Souza C, Batista V, Fabré N (2018) What are the main local drivers determining richness and fishery yields in tropical coastal fish assemblages? Zoologia 35: e12898. https://doi.org/10.3897/zoologia.35.e12898

Spasojevic MJ, Copeland S, Suding KN (2014) Using functional diversity patterns to explore metacommunity dynamics: a framework for understanding local and regional influences on community structure. Ecography 37: 939–949. https://doi.org/10.1111/ecog.00711

Stuart-Smith R, Bates A, Lefcheck J, Duffy JE, Baker S, Thomson R, Stuart-Smith J, Hill N, Kininmonth S, Airoldi L, Becerro M, Campbell S, Dawson T, Navarrete S, Soler G, Strain E, Willis T, Edgar G (2013) Integrating abundance and functional traits reveals new global hotspots of fish diversity. Nature 501: 539–541. https://doi.org/10.1038/nature12529

Tonn W, Magnuson J (1982) Patterns in the species composition and richness of fish assemblages in Northern Wisconsin lakes. Ecology 63: 1149–1166. https://doi.org/10.2307/1937251

Villéger S, Brosse S, Mouchet M, Mouillot D, Vanni M (2017) Functional ecology of fish: current approaches and future challenges. Aquatic Science: online published. https://doi.org/10.1007/s00027-017-0546-z

Villéger S, Mason N, Mouillot D (2008) New multidimensional functional diversity indices for a multifaceted framework in functional ecology. Ecology 89: 2290–2301. https://doi.org/10.1890/07-1206.1

Villéger S, Miranda J, Hernández D, Mouillot D (2010) Contrasting changes in taxonomic vs. functional diversity of tropical fish communities after habitat degradation. Ecological Applications 20: 1512–1522. https://doi.org/10.1890/09-1310.1

Villéger S, Miranda J, Hernández D, Mouillot D (2012) Low functional beta-diversity despite high taxonomic beta-diversity among tropical estuarine fish communities. Plos One 7: e40679. https://doi.org/10.1371/journal.pone.0040679

Webb C, Ackerly D, Kembel S (2008) Phylocom: software for the analysis of phylogenetic community structure and trait evolution. Bioinformatics 24: 2098–2100. https://doi.org/10.1093/bioinformatics/btn358

Winemiller K (1989) Patterns of variation in life history among South American fishes in seasonal environments. Oecologia 81: 225–241. https://doi.org/10.1007/BF00379810

Würdig N (1987) Alguns aspectos físicos e químicos do Sistema lagunar de Tramandaí, RS. Pesquisas 20: 49–74. https://seer.ufrgs.br/PesquisasemGeociencias/article/view/21672

Zuur A, Ieno EN, Walker N, Saveliev AA, Smith GM (2009) Mixed effects models and extensions in ecology with R. New York, Springer, 574 pp.

Author Contributions: SMH provided logistical support, guided and reviewed the research project and contributed to the writing of the paper; EAR conducted the fieldwork, obtained the traits fish data and contributed to the writing of the paper; ALL and FTB analyzed the data and contributed to the writing of the paper; TFRG and FGB conducted the field work and contributed to the logistical support.

Competing Interests: The authors have declared that no competing interests exist.

7

Richness, abundance and microhabitat use by Ardeidae (Aves: Pelecaniformes) during one seasonal cycle in the floodplain lakes of the lower Amazon River

Giulianne Sampaio Ferreira[1], **Danilo Augusto Almeida dos Santos**[1], **Edson Varga Lopes**[1,2]

[1] *Programa de Pós-Graduação em Recursos Aquáticos Continentais Amazônicos, Universidade Federal do Oeste do Pará. Rua Vera Paz, 68035-110 Santarém, PA, Brazil.*
[2] *Instituto de Biodiversidade e Florestas, Universidade Federal do Oeste do Pará. Rua Vera Paz, 68035-110 Santarém, PA, Brazil.*
*Corresponding author: Giulianne Sampaio Ferreira (*giulya_sampaio@hotmail.com*)*

ABSTRACT. The Amazon floodplains become periodically submerged as result of seasonal changes in the water levels throughout the year. These changes influence the availability of microhabitats and consequently the abundance of organisms in these ecosystems. In this study we investigated 1) how changes in the water level affect the richness and abundance of ardeid birds in the lowland floodplain lakes of the lower Amazon River, and 2) the microhabitats used by these birds throughout the seasonal cycle. Ten lakes were surveyed at each of the four phases of the seasonal cycle. In total, 3,280 individuals of 11 species were recorded. Of these, eight species occurred in the four phases, and three were observed in one or two phases. In the analysis including the entire family, there were more individuals in the phase with waters at lowest level and less in the phase that the water level was lowering. Many species were present throughout the seasonal cycle, suggesting that they might be resident species. However, their abundance varied throughut the cycle, suggesting that parts of their populations temporarily migrate elsewhere. The microhabitat that was most commonly used by most species at all phases of the seasonal cycle, with the excetions noted below, was "aquatic macrophytes", suggesting that ardeid birds have a strong preperence for this kind of habitat. Three species – *Egretta caerulea* (Linnaeus, 1758), *Nycticorax nycticorax* (Linnaeus, 1758) and *Bubulcus ibis* (Linnaeus, 1758) – preferred other microhabitats at some phase of their seasonal cycle. The present study shows that the floodplain lakes of the lower Amazon River are richer in ardeid bird species than other areas of the Amazon biome and other biomes in Brazil. The fact that we found rare species in our study and that they depend on aquatic macrophytes demonstrates the importance of conserving the floodplain lakes of the lower Amazon River.

KEY WORDS. Amazonian, aquatic macrophytes, ecological partitioning, waterbirds, wetlands.

INTRODUCTION

Since Huchtnson (1957) proposed the concept of ecological niche, our understanding of biodiversity and how species are distributed in an environment have evolved significantly. For instance, narrow species' niches are often associated with higher species richness locality. This is particularly noteworthy in the tropics and decreases North and Southwards, resulting in a latitudinal gradient in species richness (e.g., Rohde 1992, Fine 2015, Willig and Presley 2018).

The spacial distribution of species within habitats reflects their ecological partitioning, and the number of habitats used by each species reflects their degree of specialization in habitat use (Stotz et al. 1996). Most ardeid birds are associated with wetlands (Sick 1997, Accordi 2010). However, the use of habitats by these birds can vary considerably according to their degree of specialization (Dimalexis et al. 1997, Bancroft et al. 2002, Isacch and Martínez 2003). While some species use various types of habitats, others are strongly associated with a particular habitat or microhabitat (Accordi and Hartz 2006, Gimenes and Anjos 2011, Alves et al. 2012, Pinto et al. 2013).

Some wetlands change considerably throughout a year (Sioli 1985). For example, Amazon floodplains are strongly influenced by the flood pulse and become seasonally flooded as a consequence of river overflows (Sioli 1985, Piedade 1995, Junk et al. 1989, Fraxe et al. 2007, Ribeiro 2007). A seasonal cycle in these floodplains includes four distinct phases: 1) rising – elevation of water level, 2) peak-flood – sustained water level acme, 3) ebb – descent of water level, and 4) dry – waters at lowest level (Fraxe et al. 2007). This variation in water levels drastically alters the landscape (Piedade 1995), possibly influencing the availability of suitable habitats for ardeid birds. These birds, in turn, may vary in their response to changes in the floodplain landscape, depending on the breadth of their ecological niche and/or the availability of similar habitats at each stage of the seasonal cycle (structure, food availability, refuge conditions) (Figueira et al. 2006). If the habitat of a highly specialized species disappears at some stage of the seasonal cycle, such species is expected to move or migrate elsewhere (Thompson and Townsend 2006), while more generalist species are more likely to find available habitats locally year-round (Stotz et al. 1996, Petermann 1997).

There are numerous lakes in Amazon floodplains (Piedade 1995). These lakes are extremely important for organisms associated with wetlands, since they generally do not dry out completely during the dry season of the flood cycle, with waters at lowest level (Fraxe et al. 2007). Corroborating this idea, several studies have shown that lakes in periodically flooded areas concentrate food resources for waterbirds during the driest periods of the year (e.g., Antas 1994, Stotz et al. 1996, Silva et al. 2002, Figueira et al. 2006).

Even though both ardeid birds and floodplains are relatively common in the Amazon (Hancock and Elliott 1978, Cintra et al. 2007), information on how these birds use their habitat throughout the year is scarce. Is this study we investigate the variations in the abundance of Ardeid bird species and their habitat use in floodplain lakes throughout a seasonal cycle.

Considering that these birds are influenced by the dynamics of the flood pulses of the floodplains, we aimed to answer the following questions: 1) which species are present throughout the year? 2) In which phase of the seasonal cycle is each species more abundant? 3) which microhabitats are explored by ardeid birds in the floodplain lakes of the lower Amazon River? 4) Are microhabitats used by these birds the same at each stage of the seasonal cycle? Upon answering these questions, we hope to contribute to the conservation of the various species and their ecosystems in the Amazon floodplains.

MATERIAL AND METHODS

This study was carried out in a lowland area of the lower Amazon River, municipality of Santarém, western state of Pará, settlement of Santa Maria do Tapará (2°21'22.25"S; 54°34'15.79"W; Fig. 1). The climate is hot and humid, with average annual temperature ranging from 25 to 28°C and average annual rainfall 1,920 mm. The region is characterized by a period of more rain between December and May, and a drier period from June to November (INMET 2013).

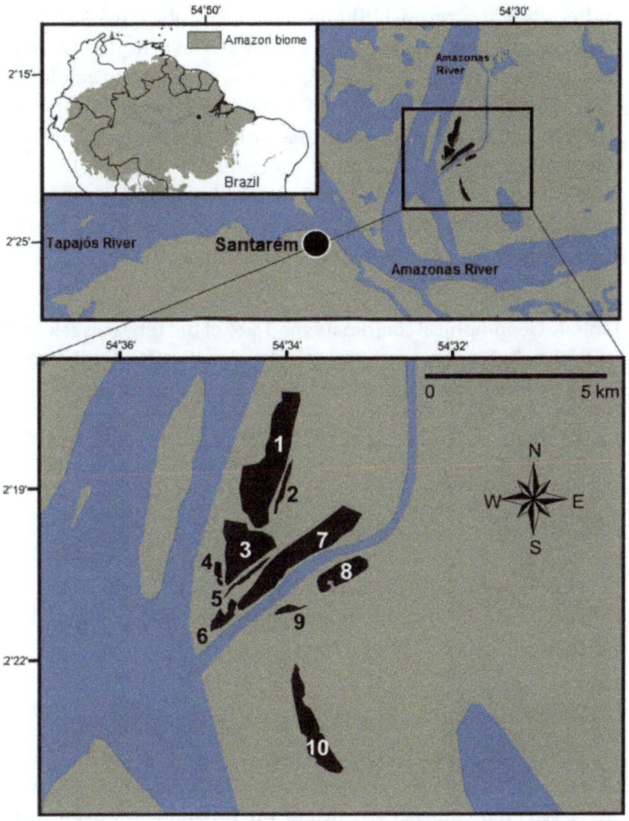

Figure 1. Map of the study area at the low Amazon River. It shows the location of the city of Santarém, at the confluence of the rivers Tapajós and Amazonas, and an enlargement of the study area, highlighting the 10 lakes selected for the survey.

We selected ten lakes for this study (Table 1). All of them are connected to the Amazon River when the water level is high, but only a few remain connected to rivers or canals when the water is at lowest level. The vegetation around the lakes is composed of 'várzea' forest (a type of forest typical of floodplains in the Amazon), agglomerations of plants of *Cecropia* (locally known as 'embaubal'), and natural fields. In the lakes there are aquatic macrophytes such as *Eichhornia crassipes* (Mart) Solms, 1883 (Pontederiaceae), *Montrichardia linifera* (Arruda) Schott, 1854 (Araceae) and *Victoria amazonica* (Poepp.) J.C. Sowerby (Nymphaeaceae).

Data collection followed the line transect method in Bibby et al. (1992). During rising, peak-flood, and ebb phases we used a small wooden boat to transect each lake, about 20 m from the margin. The boat's engine, known locally as 'rabeta', travels at an average speed of approximately 10 km/h. During the dry season, with lowest water levels, access to most lakes using boats became very difficult. Consequently, during this period, the transect was covered by foot, along the banks of the lake, at an average speed of 1 km/h. We sampled three lakes per day, starting at 07:00 am and ending at around 11:00 am. Between July 2013 and May 2014 (an entire seasonal cycle), each lake was sampled twice at each phase of the cycle, totaling eight samplings at each lake.

The species were identified through direct visual contact or with binoculars (8 x 42). We recorded all individuals of Ardeidae visualized during each sampling. Individuals flying over the lake were not counted, and we tried to avoid registering the same individual more than once. Thus, in addition to documenting each species, we also estimated their abundance.

Table 1. Geographical coordinates and size of the ten lakes selected for this study. The lakes were measured during the phase with waters at lowest level. Calculations done in the program GEPath 1.4.6.

Lake	Coordinates	Area (km2)
Aninga	2°19'15.42"S; 54°34'0.91"W	8.12
Botal	2°19'17.76"S; 54°33'41.30"W	2.72
Caiçara	2°20'3.17"S; 54°34'10.78"W	4.88
Espurú	2°20'27.26"S; 54°34'33.46"W	1.73
Figueiredo	2°21'0.47"S; 54°34'23.87"W	2.24
Pitomba	2°20'29.34"S; 54°34'11.19"W	3.36
Poço	2°20'54.73"S; 54°33'31.54"W	1.88
Pucu	2°20'27.89"S; 54°33'48.65"W	8.27
Purus	2°22'35.35"S; 54°33'18.18"W	7.06
São Francisco	2°20'29.75"S; 54°32'51.27"W	3.29

We characterized the use of microhabitat for each individual or flock (three or more individuals) registered at a lake or around it, up to 20 m from the bank. We considered the microhabitat exploited by a bird as being the predominant habitat within an estimated radius of five meters around that bird at the moment of the record (Lopes et al. 2006). We created microhabitat categories based on Alves et al. (2012), to standardize data collection, as follows: 1) Tree vegetation - woody plants three meters in height, or more, isolated or in clusters; 2) Shrub vegetation - woody plants less than three meters high; 3) Grass - herbaceous undergrowth, few centimeters in height; 4) Bare margin - area around lakes, without vegetation cover; 5) Lake perch - structure placed at the interior of the lake, such as wooden sticks to hold small boats; 6) Marginal perch - similar to the previous one, but the structure was fixed at the edge of the lake, out of the water; 7) Aquatic macrophytes - aquatic plants inside the lake; 8) Coastal zone - within the water, from the margin up to five meters; 9) Limnetic zone - in the water, beginning five meters from the shore. This last category was included because, during periods when the water level was at its lowest, the lakes became very shallow and the birds were able to enter the them.

To verify if there was variation in species' abundance among the four phases of the seasonal cycle, we used the non-parametric statistical test Chi-square (χ^2). Since two samplings were carried out in each phase, we analyzed, for each species, the sample encompassing the largest number of recorded individuals. This analysis was done in the STATISTICA 7.0 program (Statsoft 2004) when the frequencies recorded were higher than five (Fowler and Cohen 1995), at significance level of 5% ($\alpha = 0.05$).

In order to determine the preference of the species for a specific microhabitat at each stage of the seasonal cycle, we used a simple correspondence analysis (CA) according to Legendre and Legendre (1998). This graphical analysis was developed to represent the association between rows and columns of a contingency table. The proximity of the points referring to the row and the column indicates an association between these variables and their distancing represents a repulsion (Gonçalves and Santos 2009). The CA was carried out in the Paleontological Statistics-PAST program (Hammer et al. 2001). We performed ordinations between the species studied and the categories of microhabitats considered. For this analysis we constructed matrices with the percentage of occurrence of each species in each microhabitat. We adopted this procedure for the four phases of the seasonal cycle, and included only those species for which we had five or more records. Thus, a species may have been analyzed for one phase but not for another when records of it were too few or lacking.

RESULTS

Considering all phases of the seasonal cycle, we recorded 3,280 individuals from 11 species of Ardeidae (Table 2). *Ardea alba* Linnaeus, 1758 was the most abundant species, corresponding to 39.8% of the total records (Table 2). *Bubulcus ibis* (Linnaeus, 1758) and *Egretta thula* (Molina, 1782) were also relatively abundant (24.7% and 19.6% of the total records, respectively). Less than 200 individuals were recorded for each of the other species during the seasonal cycle (Table 2). *Cochlearius cochlearius* (Linnaeus, 1766), *Ixobrychus exilis* (Gmelin, 1789) and *Pilherodius pileatus* (Boddaert, 1783) were not included in the analyzes of abundance and microhabitat use because they were very low in numbers, and nocturnal habitat of the first species mentioned.

The highest and the lowest abundance of ardeid individuals occurred when the water was in the lowest levels and whem the water level was lowering (dry and ebb phases), respectively (Table 2). Eight of the 11 species recorded were present in the four phases of the seasonal cycle, six of which significantly varied in numbers throughout the seasonal phases. *Ardea alba*, *Butorides striata* (Linnaeus, 1758), *Egretta caerulea* (Linnaeus, 1758), *E. thula* and *Nycticorax nycticorax* (Linnaeus, 1758) were more abundant in the phase with lowest water levels. In contrast, *B. ibis* was more abundant in the peak-flood phase. The abundance of two species, *Tigrisoma lineatum* (Boddaert, 1783) and *Ardea cocoi* Linnaeus, 1766, did not vary among the four phases of the seasonal cycle. *Ixobrychus exilis* and *P. pileatus*, on the other hand, were recorded only in the phase with waters at lowest level and *C. cochlearius* was recorded in both this and rising phases. (Table 2).

Table 2. Species abundance registered during each phase of the seasonal cycle (see the text for details of the characteristics of each phase). The last row shows the p values of the χ^2 test, comparing the abundance of each species among the phases of the seasonal cycle. The taxonomy follows CBRO (2015).

Species	Phases				
	Ebb	Dry	Rising	Peak-flood	p
Tigrisoma lineatum (Boddaert, 1783)	32	49	40	40	0.3100
Cochlearius cochlearius (Linnaeus, 1766)	0	9	2	0	–
Ixobrychus exilis (Gmelin, 1789)	0	1	0	0	–
Nycticorax nycticorax (Linnaeus, 1758)	4	23	16	7	0.0040
Butorides striata (Linnaeus, 1758)	67	91	6	17	0.0001
Bubulcus ibis (Linnaeus, 1758)	135	192	114	361	0.0001
Ardea cocoi Linnaeus, 1766	18	30	26	17	0.1500
Ardea alba Linnaeus, 1758	80	817	259	137	0.0001
Pilherodius pileatus (Boddaert, 1783)	0	1	0	0	–
Egretta thula (Molina, 1782)	64	363	179	32	0.0001
Egretta caerulea (Linnaeus, 1758)	4	41	1	14	0.0001
Total	404	1617	643	625	0.0001

In the use of the microhabitat, in the phase that the water level was lowering, axes 1 and 2 of the CA ordination account for 64% and 19% of the data variation, respectively. In this phase, we found that the abundance of *T. lineatum* (100%), *A. cocoi* (80%), *E. thula* (73%), *A. alba* (69%), *B. striata* (67%) and *B. ibis* (64%) was influenced by aquatic macrophytes (Fig. 2, Table 3).

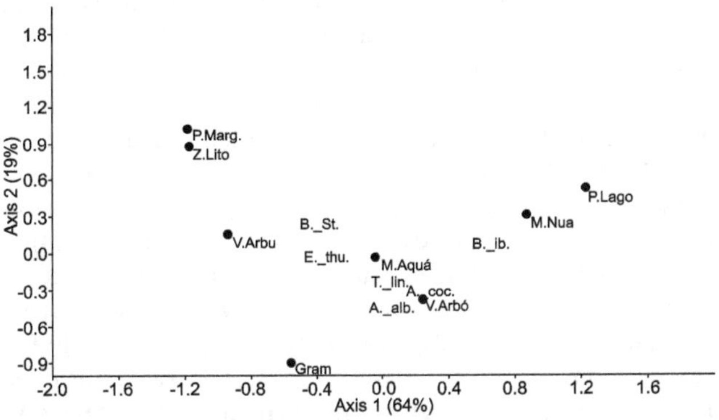

Figure 2. Correspondence analysis of the species of the family Ardeidae with the microhabitats explored by them during the ebb phase (descent of water level) in floodplain lakes of the low Amazon River. Acronyms for species' names and microhabitats are presented in Table 3.

In the phase with lowest water levels, the ordination of the CA data on axes 1 and 2 was responsible for 64% and 36% of the variation, respectively. The results of the CA suggest that the occurrence of *T. lineatum* (76%), *B. striata* (68%), *N. nycticorax* (57%), *E. thula* (45%) and *A. cocoi* (40%) might have been influenced by aquatic macrophytes. Records of *A. alba* were also associated with this microhabitat (37%) and with arboreal vegetation (26%). The occurrence of *E. caerulea* (66%) was strongly influenced by the presence of shrubs, whereas *B. ibis* occurred in association with bare margin (28%) and grass (27% Fig. 3, Table 3).

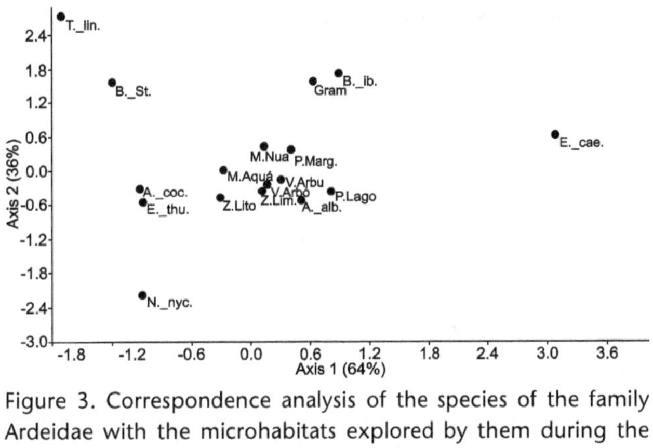

Figure 3. Correspondence analysis of the species of the family Ardeidae with the microhabitats explored by them during the dry phase (waters at lowest level) in floodplain lakes of the low Amazon River. Acronyms for species names and microhabitats are presented in Table 3.

In the water level rise phase, ordination of the CA data on axes 1 and 2 represented 90% and 6% of the data variation, respectively. In this phase *E. thula* (71%), *A. cocoi* (69%), *T. lineatum* (68%), *A. alba* (64%), and *B. striata* (50%) have been influenced by aquatic macrophytes. *Bubulcus ibis* (88%) was influenced by perch margin and *N. nycticorax* (81%) was associated with arboreal vegetation (Fig. 4, Table 3).

In the peak-flood phase, ordination of the CA data on axes 1 and 2 was responsible for 65% and 26% of the data variation, respectively. In this phase, the occurrence of *A. cocoi* (94%), *B. ibis* (84%), *A. alba* (73%), *T. lineatum* (72,5%), and *E. thula* (69%)

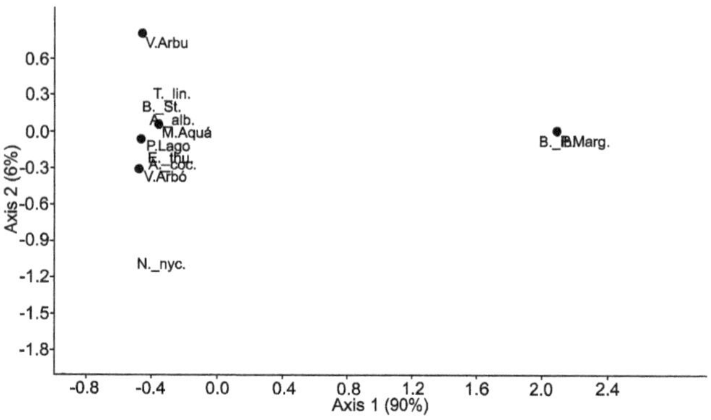

Figure 4. Correspondence analysis of the species of the family Ardeidae with the microhabitats explored by them during the rising phase(elevation of water level) in floodplain lakes of the low Amazon River. Acronyms for species names and microhabitats are presented in Table 3.

have been influenced by aquatic macrophytes. *Egretta caerulea* and *B. striata* were influenced by aquatic macrophytes (57%, 53%) and arboreal vegetation (43%, 42%), respectively. *Nycticorax nycticorax* occurred in association with two microhabitats at this stage, arboreal vegetation and shrubby vegetation (57% and 43%), respectively (Fig. 5, Table 3).

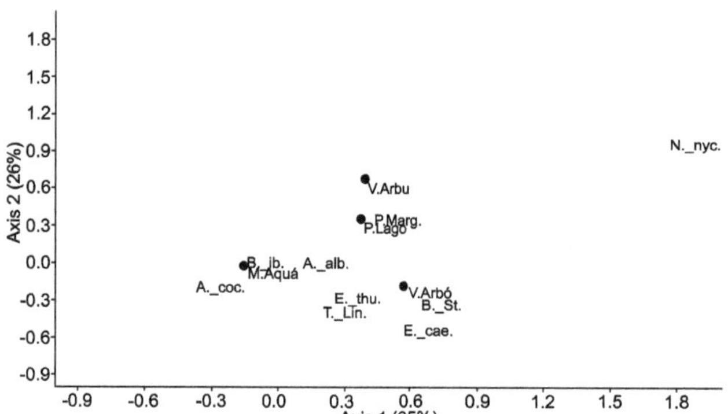

Figure 5. Correspondence analysis of the species of the family Ardeidae with the microhabitats explored by them during the peak-flood phase (sustained water level acme) in floodplain lakes of the low Amazon River. Acronyms for species names and microhabitats are presented in Table 3.

DISCUSSION

In our data, the abundance of most Ardeidae recorded varied throughout the seasonal cycle of the Amazon floodplains. Ardeid birds also seemed to have a preference for the aquatic macrophytes microhabitat (Hancock and Kushlan 1984, Gimenes and Anjos 2006, Pimenta et al. 2007, Kushlan, 2011). Some species were more abundant at a given phase and others were recorded at only some phases of the seasonal cycle, which suggests that the entire population or part of the population of some species move away at some phase of the seasonal cycle. However, it is important to point out that even though moving away is a plausible hypothesis for the observed variations in the numbers of individuals recorded at each phase, this hypothesis needs further testing. Other studies also had shown that part of the populations of some Ardeidae species migrate our move seasonally (Olmos and Silva 2001, Antas and Palo-Júnior 2004, Nunes and Tomas 2008). This displacement may be correlated with two factors: 1) the presence (or absence) of resources (i.e. food); 2) the degree of specialization of the species. The microhabitat explored by some, but not all, ardeid birds, varied throughout the year.

Among the six species that varied in abundance between the four phases of the seasonal cycle, five were more abundant during the dry season, with waters at lowest level. In this phase, there is frequently an increase in the abundance of waterbirds in lakes located in areas that get periodically flooded (e.g., Guadagnin et al. 2005, Accordi and Hartz 2006, Soares and Rodrigues 2009, Gimenes and Anjos 2011, Cintra 2012, Tavares and Siciliano 2014). The most common justification for this pattern is that food resources become very concentrated in these lakes during the dry period, which the waterbirds come to exploit (Gimenes and Anjos 2006, Cintra et al. 2007).

Bubulcus ibis was the only Ardeidae recorded in this study that was more abundant in the peak-flood phase, with waters at highest level. This species is native to Africa and Mediterranean Europe (Rice 1956) and is associated with large grazing mammals rather than with aquatic environments (Martínez-Vilalta et al. 2014a). This species was first recorded in Brazil in 1964 (Sick 1965), where its occurrence is facilited by the presence of livestock (Sick 1997, Bella and Azevedo-Júnior 2004, Seedikkoya et al. 2005, Nunes et al. 2010, Moralez-Silva and Lama 2014). It is commonly found foraging among cattle (Gasset et al. 2000, Martínez-Vilalta et al. 2014a) and capturing the insects that are associated with or get displaced by these large animals. In the study region, during the peak-flood phase, herds are transferred from the floodplain to slightly higher and non-flooded regions, and during this time *B. ibis* individuals concentrate at the marginal vegetation of the lakes. However, our results do not clarify why these birds did not accompany the cattle herds when they were transferred to drier areas, or even whether some birds did follow the cattle there.

In general, in the present study, the microhabitat preferred by most species of Ardeidae throughout the year was aquatic macrophytes. This preference is possibly related to the diet of these birds (Gimenes and Anjos 2006, Alves et al. 2012). Aquatic organisms, such as fish and invertebrates, on which the Ardeidae feed, use macrophytes for shelter and reproduction (Oliveira and Goulart 2000). Another possibility, suggested by Gimenes and Anjos (2011), is that this microhabitat allows some species, especially the smaller ones, to use areas located further away from

Table 3. Percentage relation of correspondence analysis among species and microhabitats during each phase of the season cycle. Between parentheses, the species and the microhabitats presented in the CCA of each phase.

Phases	Species	Aquatic macrophyte	Arboreal vegetation	Shrub vegetation	Grass	Naked margin	Lake perch	Lake margin	Limnetic zone	Coastal zone
Ebb	*Tigrisoma lineatum*	100	0	0	0	0	0	0	0	0
	Nycticorax nycticorax	0	0	0	0	0	0	0	0	0
	Butorides striata	67	6	15	0	3	0	6	0	3
	Bubulcus ibis	64	8	0	0	27	1	0	0	0
	Ardea cocoi	80	13	0	0	7	0	0	0	0
	Ardea alba	69	15	8	4	5	0	0	0	0
	Egretta thula	73	0	17	3	3	0	2	0	2
	Egretta caerulea	0	0	0	0	0	0	0	0	0
Dry	*Tigrisoma lineatum*	76	8	8	2	4	0	2	0	0
	Nycticorax nycticorax	57	35	9	0	0	0	0	0	0
	Butorides striata	68	14	9	2	2	0	2	2	0
	Bubulcus ibis	20	11	9	27	28	0	6	0	0
	Ardea cocoi	40	17	3	0	13	0	3	3	20
	Ardea alba	37	26	20	0	5	2	3	3	8
	Egretta thula	45	9	18	1	11	0	0	2	13
	Egretta caerulea	12	2	66	0	7	5	0	2	5
Rising	*Tigrisoma lineatum*	68	20	13	0	0	0	0	0	0
	Nycticorax nycticorax	19	81	0	0	0	0	0	0	0
	Butorides striata	50	33	17	0	0	0	0	0	0
	Bubulcus ibis	12	0	0	0	0	0	88	0	0
	Ardea cocoi	69	31	0	0	0	0	0	0	0
	Ardea alba	64	26	9	0	0	0	1	0	0
	Egretta thula	71	28	0	0	0	1	0	0	0
	Egretta caerulea	0	0	0	0	0	0	0	0	0
Peak-flood	*Tigrisoma lineatum*	73	28	0	0	0	0	0	0	0
	Nycticorax nycticorax	0	57	43	0	0	0	0	0	0
	Butorides striata	53	42	5	0	0	0	0	0	0
	Bubulcus ibis	84	10	6	0	0	0	0	0	0
	Ardea cocoi	94	6	0	0	0	0	0	0	0
	Ardea alba	73	18	7	0	0	1	1	0	0
	Egretta thula	69	28	3	0	0	0	0	0	0
	Egretta caerulea	57	43	0	0	0	0	0	0	0

the shore, allowing greater exploration of the area, regardless of the depth of the lake. Those authors observed this behavior in *B. striata* at the Paraná River.

According to Gimenes and Anjos (2011), A. cocoi, A. alba and T. lineatum did not demonstrate a clear preference for a particular habitat of the Paraná River, whereas Pinto et al. (2013) reported that *A. cocoi* and *A. alba* used similar substrates in the presence or absence of water. This results differ from ours, since in our data these species showed preference for aquatic macrophytes in the four phases of the seasonal cycle. These species have a wide geographical distribution and it is possible that their supply of resources and habitats vary throughout their distribution range. We suggest that these Ardeid birds are generalists, but generally prefer to use macrophytes when such plants are available. We are aware of one exception to this, in the results of Pimenta et al. (2007), who demonstrated that *E. thula* does not occur in association with aquatic macrophytes at the *Pampulha* Lake, even though these plants were abundant at the site. This difference in habitat use is not surprising since these species have a wide geographic distribution and possibly the supply of resources and habitats should change throughout their distributions.

Relatively few Ardeidae species used microhabitats other than macrophytes. *Nycticorax nycticorax* was recorded more often in arboreal vegetation and shrub vegetation in two phases. This species exhibits crepuscular and nocturnal habits (Martínez-Vilalta et al. 2014b), as well as *Cochlearius cochlearius*. It is possible that the microhabitat where *N. nycticorax* was more often recorded in this study corresponds to its resting habitat, and that it uses other microhabitats to forage in the twilight and nocturnal periods. *Egretta caerulea* showed a preference for shrub vegetation in the phase with waters at lowest level, and for tree vegetation in the peak-flood phase. According to Martínez-Vilalta et al. (2014c), *E. caerulea* is partly insectivorous, being able to look for insects in vegetation distant from the water. In addition, Sick (1997) mentioned that this species also feeds on the insects that are scared by cattle. This is also the case of *B. ibis*. This may be the reason why we recorded it, in this study, on shrub vegetation and arboreal vegetation.

In two phases of the cycle, the occurrence of *B. ibis* was more strongly correlated with microhabitats other than aquatic macrophytes: bare and grassy margin in the phase with waters at lowest level and perch margin in the rising phase. As mentioned above, this species displays an opportunistic feeding behavior (Gasset et al. 2000), foraging among bovine cattle. Pinto et al. (2013) observed that this species has a preference for dry microhabitats, which is possibly related to its association with large grazing mammals. In the present study, this species used macrophytes and perch margin only when the microhabitats such as bare margin and grasses were submerged.

The preference of the Ardeid birds for aquatic macrophytes throughout the seasonal cycle suggests that this may be a key microhabitat for the maintenance of this family (and probably of its preys) in Amazonian floodplains. This preference may be related to the diet of these birds, or to a greater exploitation of the environment during the different phases of the seasonal cycle. In addition, the occurrence of rare species (eg. *I. exilis* and *P. pileatus*) and the strong preference of ardeid species for specific microhabitats demonstrate the importance of preserving this region. It is important to emphasize that the floodplains are one of the most fragile Amazon ecosystems and are amongst the most affected by anthropic activities, which makes them even more relevant to conservation.

ACKNOWLEDGMENTS

We thank Universidade Federal do Oeste do Pará and Programa de Pós-Graduação em Recursos Aquáticos Continentais for the structure and logistic support; Coodenação de Aperfeiçoamento de Pessoal de Nível Superior (CAPES) for the fellowship granted to the first and second authors (prosseces 1224657 and 1231796, respectively); the residents of the community of Santa Maria do Tapará, especially Lauro Almeida and Rosenira Almeida, for logistical support during field work.

LITERATURE CITED

Accordi IA, Hartz SM (2006) Distribuição espacial e sazonal da avifauna em uma área úmida costeira do sul do Brasil. Revista Brasileira de Ornitologia 14(2): 117–135.

Accordi IA (2010) Pesquisa e conservação de aves em áreas úmidas. In: Matter SV, Straube FC, Accordi IA, Piacentini V, Cândido-JR JF (Eds) Ornitologia e Conservação: ciência aplicada, técnicas de pesquisa e levantamento. Rio de Janeiro, Technical Books, 191–216.

Alves MAS, Lagos AR, Vecchi MB (2012) Uso do habitat e táticas de forrageamento de aves aquáticas na Lagoa Rodrigo de Freitas, Rio de Janeiro, Brasil. Oecologia Australis 16(3): 525–539. https://doi.org/10.4257/oeco.2012.1603.12

Antas PTZ (1994) Migration and other movements among the lower Paraná River valley wetlands, Argentina, and the south Brazil/Pantanal wetlands. Bird Conservation International 4: 181–190.

Antas PTZ, Palo-Júnior H (2004) Guia de aves: espécies da reserva particular do patrimônio natural do SESC Pantanal. Rio de Janeiro, SESC Nacional, 249 pp.

Bancroft TG, Gawlick DE, Rutchey K (2002) Distribution of wading birds relative to vegetation and water depths in the northern Everglades of Florida, USA. Waterbirds 25: 265–277. https://doi.org/10.1675/1524-4695(2002)025[0265:DOWBRT]2.0.CO;2

Bella SD, Azevedo-Júnior SM (2004). Consideracões sobre a ocorrência da Garca-vaqueira, *Bubulcus ibis* (Linnaeus) (Aves, Ardeidae) em Pernambuco, Brasil. Revista Brasileira de Zoologia 21(1): 57–63. https://doi.org/10.1590/S0101-81752004000100011

Bibby CJ, Burguess NDE, Hill DA (1992) Bird census techniques. London, Academic Press, 257 pp.

Cintra R (2012) Ecological Gradients Influencing Waterbird Communities in BlackWater Lakes in the Anavilhanas Archipelago, Central Amazonia. International Journal of Ecology 2012: e801683. https://doi.org/10.1155/2012/801683

Cintra R, Santos PMRS, Leite CB (2007) Composition and structure of the lacustrine bird communities of seasonally flooded wetlands of western Brazilian Amazonia at high water. Waterbirds 30: 521–540.

Dimalexis A, Pyrovetsi M, Sgardelis S (1997) Foraging ecology of the Grey Heron (*Ardea cinerea*), Great Egret (*Ardea alba*) and Little Egret (*Egretta garzetta*) in response to habitat, at 2 Greek wetlands. Colonial Waterbirds 20: 261–272.

Figueira JEC, Cintra R, Viana LR, Yamashita C (2006) Spatial and temporal patterns of bird species diversity in the Pantanal of Mato Grosso, Brazil: implications for conservation. Brazilian Journal of Biology 66(2A): 393–404. https://doi.org/10.1590/S1519-69842006000300003

Fine VAP (2015) Ecological and evolutionary drivers of geographic variation in species diversity. Annual Review of Ecology, Evolution, and Systematics 46: 369–392. https://doi.org/10.1146/annurev-ecolsys-112414-054102

Fowler J, Cohen L (1995) Statistics for ornithologists. Norwich, British Trust for Ornithology, Guide 22, 2nd ed., 176 pp.

Fraxe TJP, Pereira HS, Witkoski AC (2007) Comunidades Ribeirinhas Amazônicas: modos de vida e uso dos recursos naturais. Manaus, Universidade Federal do Amazonas, Projeto Piatam, 223 pp.

Gasset JW, Folk TH, Alexy KJ, Miller KV, Chapman BR, Boyd FL, Hall DI (2000) Food habits of Cattle Egrets on St. Croix, U.S. Virgin Islands. Wilson Bulletin 112(2): 268–271.

Gimenes MR, Anjos L (2006) Influence of Lagoons Size and Prey Availability on the Wading Birds (Ciconiiformes) in the Upper Paraná River Floodplain, Brazil. Brazilian Archives of Biology and Technology 49(3): 463–473. https://doi.org/10.1590/S1516-89132006000400015

Gimenes MR, Anjos L (2011) Quantitative Analysis of Foraging Habitat Use by Ciconiiformes in the Upper Paraná River Floodplain, Brazil. Brazilian Archives of Biology and Technology 54 (2): 415–427. https://doi.org/10.1590/S1516-89132011000200025

Gonçalves MT, Santos RS (2009) Aplicação da Análise de Correspondência à Avaliação Institucional da Fecilcam. Revista Nupam 1(1): 1–14.

Guadagnin DL, Peter AS, Perello LFC, Maltchik L (2005) Spatial and temporal patterns of waterbird assemblages in fragmented wetlands of southern Brazil. Waterbirds 28(3): 261–272.

Hammer Ø, Harper DAT, Ryan PD (2001) PAST: Paleontological Statistics Software Package for Education and Data Analysis. Palaeontologia Electronica 4: 1–9. Available online at: http://palaeo-electronica.org/2001_1/past/issue1_01.htm [Accessed: 23/07/2014]

Hancock J, Elliott H (1978) The herons of the world. London, Editions, 304 pp.

Hancock J, Kushlan JA (1984) The herons handbook. London, Croom Helm, 288 pp.

Huchtnson GE (1957) Concluding Remarks. Cold Spring Harbor Symposia on Quantitative Biology 22: 415–427.

Inmet (2013) Instituto Nacional de Meteorologia. Available online at: http://www.inmet.gov.br [Acessed: 23/06/2013]

Isacch JP, Martínez MM (2003) Habitat use by nonbreeding shorebirds in flooding pampas grasslands of Argentina. Waterbirds 26: 494–500. https://doi.org/10.1675/1524-4695(2003)026[0494:HUBNSI]2.0.CO;2

Junk WJ, Bayley PB, Sparks RE (1989) The Flood Pulse Concept in River-Floodplain Systems. In: Dodge DP (Ed.) Proceedings of the International Large River Symposium. Champaign, Canadian Special Publication of Fisheries and Aquatic Sciences, 110–127.

Kushlan JA (2011) The Terminology of Courtship, Nesting, Feeding and Maintenance in Herons. Heron Conservation. Available online at: http://www.heronconservation.org [Accessed: 08/01/2019]

Legendre P, Legendre L (1998) Numerical Ecology. Amsterdam, Elsevier Science, 852 pp.

Lopes EV, Volpato GH, Mendonça LB, Fávaro FL, Anjos L (2006) Abundância, microhabitat e repartição ecológica de papa-formigas (Passeriformes, Thamnophilidae) na bacia hidrográfica do rio Tibagi, Paraná, Brasil. Revista Brasileira de Zoologia 23(2): 395–403. https://doi.org/10.1590/S0101-81752006000200013

Martínez-Vilalta A, Motis A, Kirwan GM (2014a) Cattle Egret (*Bubulcus ibis*). In: Del Hoyo J, Elliott A, Sargatal J, Christie DA, De Juana E (Eds) Handbook of the Birds of the World Alive. Barcelona, Lynx Edicions. Avaliable online at: http://www.hbw.com/node/52697 [Acessed: 14/05/2015]

Martínez-Vilalta A, Motis A, Kirwan GM (2014b) Black-crowned Night-heron (*Cochlearius cochlearius*). In: Del Hoyo J, Elliott A, Sargatal J, Christie DA, De Juana E (Eds) Handbook of the Birds of the World Alive. Barcelona, Lynx Edicions. Avaliable online at: http://www.hbw.com/node/5270. [Acessed: 14/05/2015]

Martínez-Vilalta A, Motis A, Kirwan GM (2014c) Little Blue Heron (*Egretta caerulea*). In: Del Hoyo J, Elliott A, Sargatal J, Christie DA, De Juana E (Eds) Handbook of the Birds of the World Alive. Barcelona, Lynx Edicions. Avaliable online at: http://www.hbw.com/node/52692. [Acessed: 14/05/2015]

Moralez-Silva E, Lama SND (2014) Colonization of Brazil by the cattle egret (*Bubulcus ibis*) revealed by mitochondrial DNA. NeoBiota 21: 49–63. https://doi.org/10.3897/neobiota.21.4966

Nunes AP, Tomas WM (2008) Aves migratórias e nômades ocorrentes no Pantanal. Corumbá, Embrapa Pantanal, 124 pp.

Nunes MFC, Barbosa-Filho RC, Roos AL, Mestre LAM (2010) The Cattle Egret (*Bubulcus ibis*) on Fernando de Noronha Archipelago: history and population trends. Revista Brasileira de Ornitologia 18(4): 315–327.

Oliveira EF, Goulart E (2000) Distribuição espacial de peixes em ambientes lênticos: interação de fatores. Acta Scientiarum 22(2): 445–453.

Olmos F & Silva R (2001) The avifauna of a southeastern Brazilian mangrove swamp. International Journal of Ornithology 4 (3/4): 135–205.

Petermann P (1997) The birds. In: Junk WJ (Ed.) The Central Amazon Floodplain: Ecology of a Pulsing System. Heidelberg, Springer-Verlag, Berlin, 419–451.

Piedade MTF (1995) Influência do Pulso de Cheias e Vazantes na Dinâmica Ecológica de Áreas Inundáveis. Projeto de Pesquisa Dirigida. Brasília, INPA, 446 pp.

Pimenta FE, Drummond JCP, Lima AC (2007) Aves aquáticas da Lagoa da Pampulha: seleção de habitats e atividade diurna. Lundiana 8(2): 89–96.

Pinto DP, Chivittz CC, Tozetti AM (2013) Microhabitat use by three species of egret (Pelecaniformes, Ardeidae) in southern Brazil. Brazilian Journal of biology 73(4): 791–796.

Ribeiro NV (2007) Atlas da várzea: Amazônia Brasil. Manaus, Ibama, 132 pp.

Rice DW (1956) Dynamics of range expansion of Catle Egrets in Florida. The Auk 73(2): 259–266.

Rohde K (1992) Latitudinal Gradients in Species Diversity: The Search for the Primary Cause. Oikos 65(3): 514–527.

Seedikkoya K, Azeez PA, Shukkur AA (2005) Cattle Egret *Bubulcus ibis* habitat use and association with cattle. Forktail 21: 174–175.

Sick H (1965) *Bubulcus ibis* (Linnaeus) na Ilha de Marajó, Pará: Garça ainda não registrada no Brasil. Anais da Academia

Brasileira de Ciências 37: 567–570. https://doi.org/10.1590/S0101-81752004000100011

Sick H (1997) Ornitologia brasileira. Rio de Janeiro, Nova Fronteira, 862 pp.

Silva JMC, Bates JM (2002) Biogeographic patterns and conservation in the South American Cerrado: a Tropical Savanna Hotspot. BioScience 52: 225–233.

Sioli H (1985) Amazônia: Fundamentos da ecologia da maior região de florestas tropicais. Petrópolis, Editora Vozes, 69 pp.

Soares RKP, Rodrigues AAF (2009) Distribuição espacial e temporal da avifauna aquática no Lago de Santo Amaro, Parque Nacional dos Lençóis Maranhenses, Maranhão, Brasil. Revista Brasileira de Ornitologia 17(3–4): 173–182.

Statsoft (2004) Statistica. Statsoft Inc, v. 7.0. Available online at: http://www.statsoft.com

Stotz DF, Fitzpatrick JW, Parker III TA, Moskovits DK (1996) Neotropical Birds: Ecology and Conservation. Chicago, The University of Chicago Press, 478 pp.

Tavares DC, Siciliano S (2014) Variação temporal na abundância de espécies de aves aquáticas em uma lagoa costeira do Norte Fluminense, sudeste do Brasil. Biotemas 27: 121–132.

Thompson R, Townsend C (2006) A truce with neutral theory: local deterministic factors, special traits and dispersal limitations determine patterns of diversity in streams invertebrates. Journal of Animal Ecology 75: 464–474.

Willig MR, Presley SJ (2018) Latitudinal Gradients of Biodiversity: Theory and Empirical Patterns. Encyclopedia of the Anthropocene 3: 13–19. https://doi.org/10.1016/B978-0-12-809665-9.09809-8

Author Contributions: GSF and DAAS had collected data in field work and the three authos had contributed equally to the construction of this article.
Competing Interests: The authors have declared that no competing interests exist.

8

The sexual dimorphic inguinal glands of the frog species *Ololygon centralis* (Anura: Hylidae) at light and transmission electron microscopy

Pedro Brito[1], **Cíntia P. Targueta**[2], **Walquíria Arruda**[1], **Fernanda Santos**[3], **Rogério Bastos**[4]

[1]*Laboratório de Estudos Morfológicos, Departamento de Histologia, Embriologia e Biologia Celular, Instituto de Ciências Biológicas, Universidade Federal de Goiás. Avenida Esperança, Campus Samambaia, Setor Itatiaia, 74690-900 Goiânia, GO, Brazil. pedrovalebrito@yahoo.com.br, walquiriaufg@gmail.com*
[2]*Laboratório de Genética e Biodiversidade, Departamento de Genética, Instituto de Ciências Biológicas, Universidade Federal de Goiás. Avenida Esperança, Campus Samambaia, Setor Itatiaia, 74690-900 Goiânia, GO, Brazil. cincintia@hotmail.com*
[3]*Laboratório de Histofisiologia, Departamento de Histologia, Embriologia e Biologia Celular, Instituto de Ciências Biológicas, Universidade Federal de Goiás. Avenida Esperança, Campus Samambaia, Setor Itatiaia, 74690-900 Goiânia, GO, Brazil. fer-alcantara@hotmail.com*
[4]*Laboratório de Herpetologia e Comportamento Animal, Departamento de Ecologia, Instituto de Ciências Biológicas, Universidade Federal de Goiás. Avenida Esperança, Campus Samambaia, Setor Itatiaia, 74690-900 Goiânia, GO, Brazil. rogerioiscinax@gmail.com*
*Corresponding author: Pedro Brito (*pedrovalebrito@yahoo.com.br*)*

ABSTRACT. The anuran skin characteristically has different types of glands, most of which are microscopic and are spread throughout the skin. Some species have specialized regions where glands agglomerate, forming macroglands. The description of the external morphology of *Ololygon centralis* (Pombal & Bastos, 1996) revealed the presence of an inguinal gland. *Ololygon centralis* is the only species of the genus that has a macrogland. The present study found these inguinal macroglands to be present only on male specimens, thus characterizing it as a sexually dimorphic skin gland. Microscopic analysis revealed that these glands are composed of many syncytial units involved by myoepithelial cells. The center of the syncytium is full of a proteinaceous secretion with a basic pH and the absence of sugar residues. Similar glands observed in other anuran species have been associated with pheromone production, suggesting that the inguinal glands described for *O. centralis* males may have a similar function.

KEY WORDS. Histochemistry, morphology, skin, serous gland.

INTRODUCTION

The skin of anurans characteristically has different types of glands, the secretions of which are responsible for a variety of different functions including poison production, antimicrobial protection and the maintenance of skin moisture (Duellman and Trueb 1994, Azevedo et al. 2005, Felsemburgh et al. 2009). The most common types of anuran glands are mucous glands, which mainly secrete glycoproteins, and granular (poison) glands (Alvarez et al. 2005, Lenzi-Matos et al. 2005, Felsemburgh et al. 2007, Rigolo et al. 2008, Jared et al. 2009, Felsemburgh et al. 2009, Brunetti et al. 2012). Granular glands are composed of a syncytial structure that secretes variable protein products and depending on the species, also produce amines, alkaloids, and bufadienolides (Duellman and Trueb 1994, Sciani et al. 2013). Some species, particularly some arboreal species, possess lipid glands, which seem to function in the prevention of water loss (Lacombe et al. 2000, Felsemburgh et al. 2007).

Most anuran skin glands are distributed throughout the body surface with some differences in distribution between dorsal, ventral and lateral regions. Some species have specialized regions with large glandular acini agglomerates that form skin protuberances. These macroglands are named according to their location on the body, such as parotoid, inguinal and lumbar glands. Macroglands usually produce toxins that are used in active or passive defense (Jared et al. 2009, Antoniazzi et al. 2013,

Mailho-Fontana et al. 2014). Some species have sexually dimorphic skin glands (SDSGs), which are involved in reproduction, such as in *Boana punctata* Schneider, 1799 (Brunetti et al. 2012).

The skin glands of at least 12 species of the related genera *Scinax* Wagler, 1830 and *Ololygon* Fitzinger, 1843 were investigated (Terreni et al. 2002, Silva et al. 2017). All these species have both mucous and granular glands distributed in their skin. Only four of these species have clusters of granular glands: *O. angrensis* Lutz, 1973, *O. flavoguttata* Lutz & Lutz, 1939, *O. albicans* Bokermann, 1967 and *S. hayii* Barbour, 1909 (Silva et al. 2017) – but with no skin protuberances typical of macroglands.

The description of the external morphology of *Ololygon centralis* (Pombal & Bastos, 1996) revealed a sizable inguinal gland (Pombal and Bastos 1996). This species was initially placed *Scinax*, and grouped in the *Scinax catharinae* Boulenguer, 1888, species group. Later, using a phylogeographic approach, Duellman et al. (2016) suggested that all the species included in the *S. catharinae* group should be placed in *Ololygon*.

Ololygon centralis has a small body with the inguinal region and parts of the thighs being yellowish and dark brownish, where it is possible to find the well distinct glands in males (Pombal and Bastos 1996). This species occurs mainly in or near gallery forest of the Brazilian Cerrado (Frost 2018) and is one of the few species of the genus to occur in this biome (Pombal and Bastos 1996, Frost 2018). The type locality for *O. centralis* is Floresta Nacional de Silvânia (FLONA) in the state of Goiás, Brazil.

Ololygon centralis is the only species of the genus known to possess a macrogland; however, there has been no research on the morphology or the products of this gland. Therefore, herein we describe the inguinal gland of *O. centralis* using light and electron microscopy.

MATERIAL AND METHODS

Adult individuals of *O. centralis* were collected at night at the reproductive site when they were in reproductive activity. The specimens of *O. centralis* were collected at the type-locality (Floresta Nacional de Silvânia, Silvânia, GO, 16°38′04.4″S, 48°39′31.2″W) and transported to the campus of the Universidade Federal de Goiás (UFG, Goiânia, GO, Brazil). The specimens were euthanized with a topical application of lidocaine and the inguinal skin dissected and processed for microscopy. The specimens were deposited at the Zoology Collection of the Universidade Federal de Goiás (ZUFG) (Appendix 1). This study was approved by UFG's ethics committee (#109/14), according to federal regulations (CEUA 2018).

Inguinal skin samples from the male and female specimens were fixed in methacarn (1:3:6, acetic acid, chloroform, and methanol), dehydrated and embedded in paraffin. Histological sections (5 µm) were stained with Harris hematoxylin for one minute and aqueous eosin for approximately four minutes (HE) for general morphological analysis (modified from Suvarna et al. 2018).

Some skin samples were fixed in buffered 4% paraformaldehyde, dehydrated in alcoholic series and embedded in glycol methacrylate resin (Historesin®, Leica). To determine the chemical nature of the secretion of the glands, three micrometer thick sections were cut on a Leica RM2245 microtome and submitted to the histochemical staining procedures. Photomicrographs were taken with an Olympus BX-41 photomicroscope and morphometric analysis done with IMAGE PRO PLUS software. At least ten aleatory regions of each structure were measured, and the mean and the standard error calculated.

Periodic acid-Schiff (PAS). This reaction was performed to investigate the presence of basic glycoprotein in the glandular secretion. Sections were treated with 1% periodic acid for 20 minutes, rinsed in distilled water and then covered with Schiff's reagent for five minutes. The sections were then rinsed in running water for 10 minutes and then counterstained with Harris hematoxylin for approximately one minute (modified from Pearse 1960).

Nile blue. This procedure was performed to investigate the presence of melanin in skin layers. Some sections were stained with sulfuric Nile blue [0.05% Nile blue in 1% aqueous sulfuric acid solution (w/v)] for 20 minutes at room temperature and then rinsed for 20 minutes in running water. Other sections were treated with 0.25% potassium permanganate for 45 minutes followed by 5% oxalic acid for eight minutes for melanin bleaching, followed by staining with sulfuric Nile blue (modified from Pearse 1960).

Toluidine Blue. This procedure was performed for metachromatic detection of acid components. Sections were stained in toluidine blue solution [1% toluidine blue PA; 1% sodium tetraborate (w/w) in distilled water] for one minute and then rinsed in running water.

Ponceau Xylidine. This method was performed to investigate the presence of basic protein in glandular secretion. Sections were stained in Ponceau Xylidine solution [0.1% Ponceau Xylidine in 2.5% acetic acid (w/v)] for 20 minutes at room temperature and then rinsed in distilled water (modified from Suvarna et al. 2018).

Bromophenol Blue. This procedure was performed to determine the approximated pH of glandular secretion. Sections were stained in Bromophenol blue solution [0.5% Bromophenol blue ACS, 10% mercuric chloride in 2% acetic acid (w/w/v)] for two hours at room temperature, rinsed in 0.5% acetic acid and then rinsed briefly in tertiary-butyl alcohol (modified from Pearse 1960).

Alcian Blue 8GS pH 2.5. This procedure was performed to detect the presence of acid mucopolysaccharides in glandular secretion. Sections were stained in Alcian Blue at pH 2.5 [1% alcian blue 8GS in 3% acetic acid (w/v)] for 30 minutes at room temperature and then rinsed for 10 minutes in running water (modified from Junqueira and Junqueira 1983).

Alcian Blue 8GS pH 1.0. This procedure was performed to detect the presence of strongly sulfated mucosubstances in glandular secretion. Sections were stained in Alcian Blue at pH 1.0 [1% Alcian Blue 8GS in Hydrochloric acid 0.1N (w/v)] for 30 minutes at room temperature and then dried with a filter paper (modified from Junqueira and Junqueira 1983).

For TEM, samples were fixed with 2.5% glutaraldehyde and 0.2% picric acid solution in 0.1M cacodylate buffer, pH 7.2, with 2% sucrose. The material was post-fixed in 1% osmium tetroxide in the same buffer, dehydrated in an acetone series and embedded in epoxy resin. Ultrathin sections were stained with 2% uranyl acetate and 0.2% lead citrate and analyzed using a JEOL, JEM 2100 Transmission Electron Microscope under 80kV.

RESULTS

Males of *O. centralis* possess a round gland in the inguinal region that varies in size from approximately two to three millimeters in diameter (n = 8). Females have no gland structure, despite them having the same pigmentation pattern in the inguinal region (Figs 1–2). Observation under magnification revealed that the gland is composed of many round yellowish-colored dots (Fig. 1).

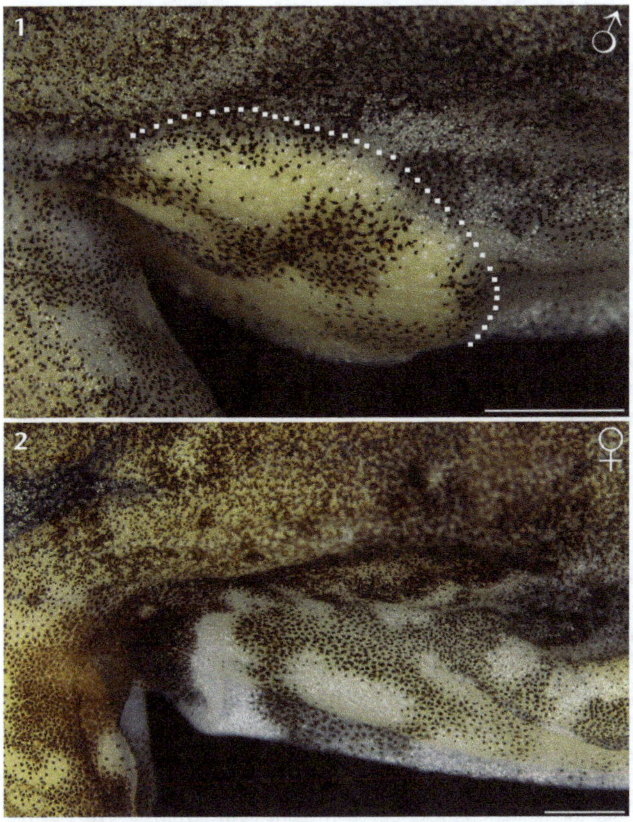

Figures 1–2. Photographs of the lateral sides of a *O. centralis* male (1) and female (2). The dashed line marks the limit of the inguinal gland in males that are absent in females. Scale bars: 1 mm.

Histological sections of the skin around four male inguinal glands revealed the presence of a stratified epithelial layer of approximately 14.5 (±1.8) µm thick supported by 30 (±12) µm of the dermis (Fig. 3). The dermis is divided into spongious dermis, where a layer of melanocytes is usually present, and dense dermis, with an abundance of collagen fibers. Roundish mucous glands of approximately 31 (±4) µm in diameter were observed in the spongious dermis (Fig. 3). In females, the entire region equivalent to the inguinal gland in males has the same histological characteristics of the skin described above, with no specialized structure (not shown).

Histological sections of male inguinal glands revealed many syncytial serous glands, of more than 500 (±7.5) µm in height, approximately 78 (±17) µm in width and filled with colloidal eosinophilic secretion (Fig. 4). The syncytial glands are in the enlarged spongious dermis. Among the syncytial glands are roundish mucous glands near the epidermis, some conjunctive cells, blood vessels and loose extracellular material (Figs 5–7, 10). There is a short duct connected to the secretory portion that passes through the epidermis (Figs 5, 6). The central portion of the syncytium is filled with colloidal eosinophilic secretion (Figs 5, 6). This syncytium has basal nuclei, with irregular outline (Figs 13, 18), and agglomerations of colloidal material in the cytoplasm (Figs 8, 11–15). The colloidal secretion is electron dense and is also present in the cortex of the syncytium interspaced with cytoplasmic organelles (Figs 11–15). The cortex region of the syncytium, near the nuclei, is rich in rough endoplasmic reticulum and free ribosomes (Figs 12, 13). A discontinuous layer of myoepithelial cells is present external to the syncytial glands (Figs 5, 6, 11, 16, 17). There is a layer of extracellular matrix with thin collagen fibrils separating the myoepithelial cells of neighbor syncytia (Figs 16, 17).

The roundish mucous glands comprise few cells (1–3) with cytoplasm that reacted positively to PAS (Fig. 7). No positive reaction was observed in the lumen of the mucous gland (Fig. 7), which appeared to be empty. The results of the histochemistry tests of the gland secretion of the serous syncytium are summarized in Table 1 and suggest that their secretion is composed mainly of basic proteins, with no, or very few, glycosylated portions. It is important to note that despite the pale blue color of the secretion due to Toluidine Blue, the cytoplasm

Table 1. Summary of the histochemistry tests used to study male serous inguinal glands of *Ololygon centralis*.

Technique	Results (gland secretion)	Significance
Xylidine Ponceau	Positive*	Presence of basic amino acid groups
Toluidine Blue	Pale	Absence of acid-negative groups
PAS	Negative	Absence of neutral sugar residues
Bromophenol Blue	Violet*	pH > 4.6
Alcian Blue pH 2.5	Negative*	Absence of mucopolysaccharides
Alcian Blue pH 1.0	Negative*	Absence of sulfated mucosubstances

*Data not shown.

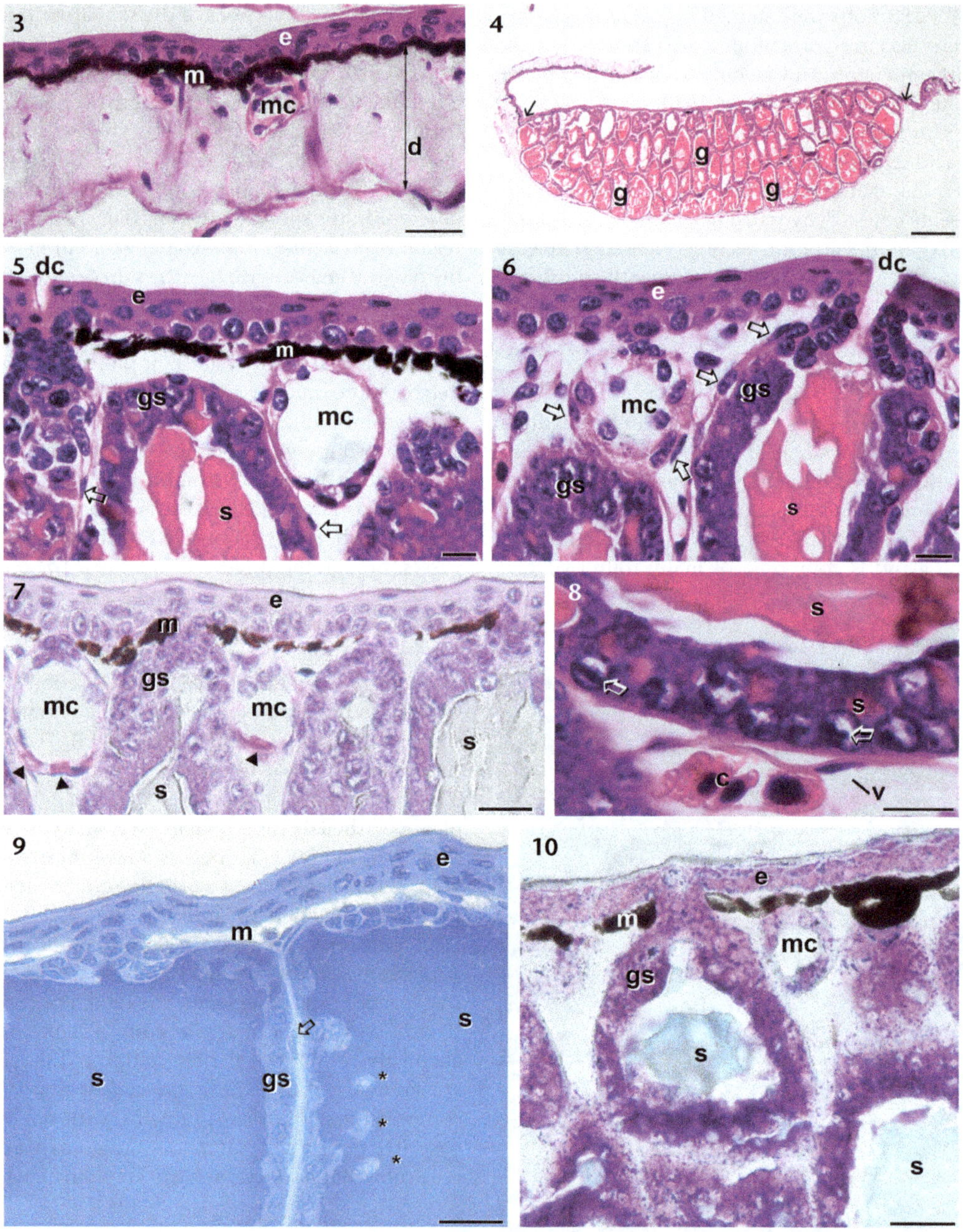

Figures 3–10. Photomicrographs of histological sections of the male inguinal gland region of *O. centralis*. (3–6, 8) Histological sections stained with HE. 3) Section of skin from the peripherical region of the inguinal gland. Notice that only mucous glands are present. 4) Low magnification micrograph showing the presence of many syncytial glands (g), with arrows indicating the lateral limits of the inguinal gland. (4–6) Major magnifications of the glandular apical portion, with many melanocytes (m), mucous glands (mc) and myoepithelial cells (open arrows). Note the glandular ducts (dc). 7) Histological section submitted to PAS reaction. Notice that only some cells of the mucous glands (mc) exhibit a positive reaction (arrowheads). (8) Major magnification of the lateral base portion of the syncytium, with colloidal secretion (s) in syncytium cytoplasm. Note also a blood vessel in the connective tissue. (9) Methacrylate section treated with potassium permanganate and oxalic acid and stained with Nile blue. Notice the bleached melanocytes (m) and some syncytial cytoplasmic projections (*) through the glandular secretion (s). (10) Methacrylate section stained with toluidine blue. Notice the pale blue color of the secretion suggesting it is alkaline, contrasting with the dark blue color of the glandular syncytium (gs). (e) epidermis; (d) dermis; (black open arrow) myoepithelial cells; (c) blood cells. Scale bars: 5, 6, 8 = 10 μm, 3, 7, 9, 10 = 20 μm; 4 = 200 μm.

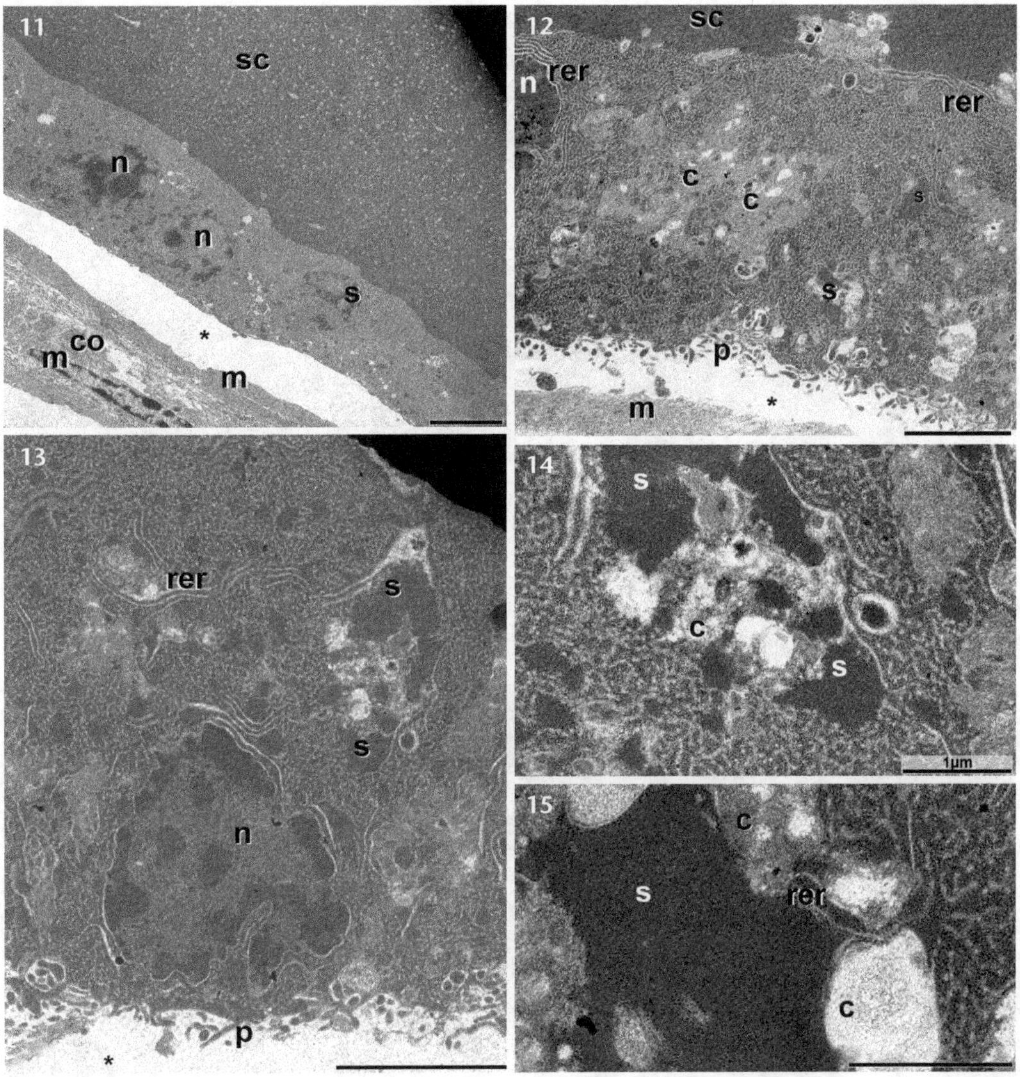

Figures 11– 15. Electron micrographs of the serous glands of the inguinal region. (11) Low magnification of the secretory syncytium with two visible nuclei (n) and also a sizeable cytoplasmic secretion aggregate (s). Notice the syncytium center (sc) filled with electron dense secretion and also the clear space (*) between syncytium basis and myoepithelial cells (m). Around the myoepithelial cells are some collagen fibrils (co). (12–13) Medium magnification of syncytium, where it is possible to notice some cytoplasmic secretion aggregate (s) and some regions of the cytoplasm with medium electron density (c). (14–15) Major magnifications of two large cytoplasmic secretion aggregate, with mixed portions of electron dense secretion (s) with medium electron density cytoplasm (c). (p) basal digitiform projections; (rer) rough endoplasmic reticulum. Sacale bars: 14, 15 = 1 μm, 12, 13 = 3 μm, 11 = 5 μm.

of the syncytium is dark blue/purple (Fig. 10). The potassium permanganate bleaching test confirmed the presence of melanin in the melanocyte layer below the epidermis (Fig. 9).

The base of the syncytium has many digitiform projections extending into the intercellular space among these cells and the myoepithelial cells (Figs 11–13, 16, 18). This space is probably an artifact of fixation. The delicate labyrinth formed by the projections does not exhibit many adhesive structures, which probably contributed to cell detachment (Fig. 18).

DISCUSSION

Males of *O. centralis* have quite unusual glands in the skin of the inguinal region. These macroglands are comprised mostly of large syncytial units. They differ from those of the more than 40 species that have been studied, in which serous skin glands comprise a syncytium with a central region constituted by a heterogeneous mixture of cell fragments (Delfino et al. 2001, Terreni et al. 2002, Alvarez et al. 2005, Lenzi-Matos et al. 2005,

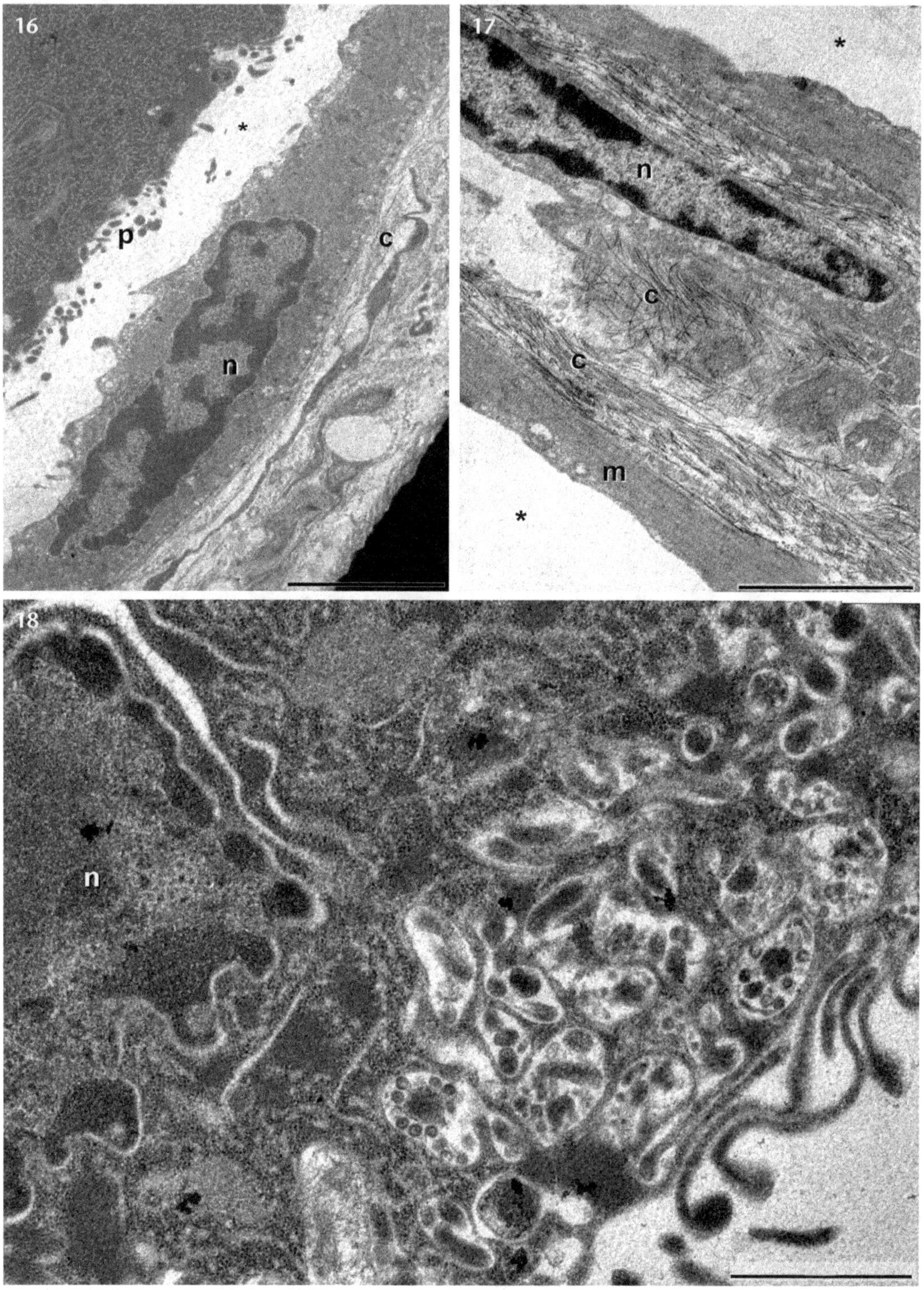

Figures 16–18. (16) The basal portion of the syncytium with digitiform projections (p) and the clear space (*) between them and the myoepithelial cells. Notice the myoepithelial cells nuclei (n) and the collagen fibrils. (17) Detail of the connective tissue between two neighbor alveoli, with myoepithelial cells (m) and collagen fibrils (c). (18) The basal portion of a syncytium with intricate projection labyrinth. Notice the syncytium nucleus with irregular outline (n). Scale bars: 18 = 1 μm, 16, 17 = 3 μm.

Felsemburgh et al. 2007, 2009, Rigolo et al. 2008, Gonçalves and Brito-Gitirana 2008, Jared et al. 2009, Silva et al. 2017). An exception is *Odontophrynus americanus* Duméril & Bibron, 1841, which has typical mucous and granular glands, but also a third tubule-alveolar gland with the same morphological and histochemical characteristics as observed in *O. centralis* (Felsemburgh et al. 2007). These glands, however, are spread throughout the body instead of being concentrated in a specific region (Felsemburgh et al. 2007).

One species of *Scinax* (*S. hayii*) and two species of *Ololygon* (*O. agrensis*, *O. flavoguttata*) have serous skin glands with cell fragments interspaced by colloidal material in the "lumen" (Silva et al. 2017). These glands, however, are also not organized as macroglands.

Few species of frogs have had sexual dimorphic glands histochemically analyzed. The sexually dimorphic glands of *B. punctata* differ from those of *O. centralis*, by not being organized as macroglands and their secretion is formed of cell fragments (Brunetti et al. 2012). Males of *Cycloramphus fuliginosus* Tschudi, 1838, have inguinal macroglands (Gonçalves and Brito-Gitirana 2008) that are similar to those observed in *O. centralis*, and which produce an acidophilic serous secretion. Gonçalves and Brito-Gitirana (2008) suggested two possible functions for these glands: i) since they are observed only in males of *C. fuliginosus*, they may be involved in the production of aquatic sexual pheromones; or ii) the production of antimicrobial substances to protect ova. Unlike *C. fuliginosus*, males of *O. centralis* males do not attract females in an aquatic environment, but instead, embrace females on dry ground and then enter the water puddles (personal observation).

Most known amphibian sexual pheromones are dispersed in water (Wabnitz et al. 1999, Kikuyama et al. 2002, Houck 2009, Belanger and Corkum 2009). However, Belanger and Corkum (2009) argued that the olfactory organs of anurans could detect pheromones spread via water and air. It has been shown that male frogs of the family Mantellidae from Madagascar can attract specific females by volatile pheromones (Poth et al. 2012). Another hypothesis is that *O. centralis* males have a novel way of applying pheromones that involves delivering it directly to the female during amplexus, as is done by some salamander species during courtship (Houck 2009). These observations reinforce the possibility of that the inguinal glands of *O. centralis* function in secreting pheromones. This hypothesis, however, has yet to be experimentally tested.

Regardless of the function(s) of the inguinal gland of *O. centralis*, the confirmation of a sexually dimorphic gland in this species raises several questions and possibilities to be tested in future research. Surveying for similar glands among other species of the same genus, chemical analysis of the glandular contents, and testing the effects of the male secretion on female behavior are all examples of research that would shed more light on the function of this gland.

The cytoplasm of the syncytium of the serous gland of *O. centralis* is electron dense and possesses secretion aggregates of various sizes. The larger aggregates seem to be formed by the function of smaller aggregates and have an irregular outline and regions of variable electron density, suggesting that portions of cytoplasm are inside the secretion. These two characteristics suggest that there is no surrounding membrane. We believe that the electron dense (colloidal) secretion is produced by free polyribosomes, which differs from the idea proposed by Delfino et al. (2001). These authors described the maturation process occurring in vesicles from Golgi apparatus in serous glands of many frog species. Nonetheless, the morphological characteristics of the secretion in that study differ from those described in the present study, reinforcing the unusual characteristics of glands of *O. centralis*.

The intense Toluidine Blue coloration in the syncytium cytoplasm of the serous glands of *O. centralis* suggests it is rich in ribosomes. Although Toluidine Blue is not a specific stain, it interacts with anionic or polyanionic components (i.e., nucleic acids and glycosaminoglycans). Since the PAS and Alcian Blue tests do not suggest the presence of glycosaminoglycans, and the Bromophenol Blue test suggests a basic pH for the colloid, the intense blue/purple coloration of the syncytium cytoplasm indicates ribosome richness.

The basal portion of the syncytium, with an intricate labyrinth of digitiform projections, suggests intense traffic of substances between the syncytium and the connective tissue. Furthermore, the presence of myoepithelial cells indicates that they may be involved with secretion ejection.

Located between the serous glands of the inguinal macroglands of *O. centralis* are round mucous glands. These glands are very similar to the mucous glands reported for other anuran species (Lenzi-Matos et al. 2005, Alvarez et al. 2005, Rigolo et al. 2008, Jared et al. 2009, Felsemburgh et al. 2009, Brunetti et al. 2012), with round alveoli, possessing by PAS-positive epithelial cells. These glands are probably responsible for maintaining skin moisture.

Future studies must be performed to confirm the hypothesis of pheromone production by the inguinal glands of *O. centralis*. Validating this hypothesis would help to understand the complex mating behavior of this species, and perhaps change conservation strategies directed at them.

ACKNOWLEDGEMENTS

We would like to acknowledge the Brazilian Agencies Fundação de Amparo à Pesquisa de Goiás (DOC-FIX/FAPEG 04/2014). RPB thanks the Conselho Nacional de Desenvolvimento Científico e Tecnológico for providing fellowship (CNPq 309894/2017-4). We would also like to thank Instituto Chico Mendes de Conservação da Biodiversidade-ICMBio (46522-4) and Floresta Nacional de Silvânia for authorizing the collection of specimens. Lastly, we thank LabMic (Laboratory of High-Resolution Microscopy) for technical support. The manuscript grammar review was carried out by Erik Wild and performed again through the website Grammarly®.

LITERATURE CITED

Alvarez BB, Delfino G, Nosi D, Terreni A (2005) Ultrastructure of poison glands of South American frogs: A comparison between *Physalaemus albonotatus* and *Leptodactylus chaquensis* (Anura: Leptodactylidae). Journal of Morphology 263: 247–258. https://doi.org/10.1002/jmor.10301

Antoniazzi MM, Neves PR, Mailho-Fontana PL, Rodrigues MT, Jared C (2013) Morphology of the parotoid macroglands in *Phyllomedusa* leaf frogs. Journal of Zoology 291: 42–50. https://doi.org/10.1111/jzo.12044

Azevedo RA, Pelli AA, Ferreira-Pereira A, De Jesus Santana AS, Felsemburgh F, Brito-Gitirana L (2005) Structural aspects of the Eberth-Katschenko layer of *Bufo ictericus* integument: Histochemical characterization and biochemical analysis of the cutaneous calcium (Amphibian, Bufonidae). Micron 36: 61–65. https://doi.org/10.1016/j.micron.2004.06.004

Belanger RM, Corkum LD (2009) Review of Aquatic Sex Pheromones and Chemical Communication in Anurans. Journal of Herpetology 43: 184–191. https://doi.org/10.1670/08-054R1.1

Brunetti AE, Hermida GN, Faivovich J (2012) New insights into sexually dimorphic skin glands of anurans: The structure and ultrastructure of the mental and lateral glands in *Hypsiboas punctatus* (Amphibia: Anura: Hylidae). Journal of Morphology 273: 1257–1271. https://doi.org/10.1002/jmor.20056

CEUA (2018) Comissão de Ética no Uso de Animais. Available online at: https://ceua.prpg.ufg.br/p/3313-comissao-de-etica-no-uso-de-animais [Accessed: 10/10/2018]

Delfino G, Nosi D, Giachi F (2001) Secretory granule-cytoplasm relationships in serous glands of anurans: ultrastructural evidence and possible functional role. Toxicon 39: 1161–1171. https://doi.org/10.1016/S0041-0101(00)00253-1

Duellman WE, Trueb L (1994) Biology of amphibians. Baltimore, Johns Hopkins University Press, 670 pp.

Duellman WE, Marion AB, Hedges SB (2016) Phylogenetics, classification and biogeography of the treefrogs (Amphibia: Anura: Arboranae). Zootaxa 4104(1): 001–109. https://doi.org/10.11646/zootaxa.4104.1.1

Felsemburgh FA, Carvalho-e-Silva SP, Brito-Gitirana L (2007) Morphological characterization of the anuran integument of the *Proceratophrys* and *Odontophrynus* genera (Amphibia, Anuran, Leptodactylidae). Micron 38: 439–445. https://doi.org/10.1016/j.micron.2006.06.015

Felsemburgh FA, de Almeida PG, Carvalho-e-Silva SP, Brito-Gitirana L (2009) Microscopical methods promote the understanding of the integument biology of *Rhinella ornata*. Micron 40: 198–205. https://doi.org/10.1016/j.micron.2008.09.003.

Frost DR (2018) Amphibian species of the world: an online reference. New York, American Museum of Natural History, v. 6.0 Available online at: http://research.amnh.org/herpetology/amphibia/index.html [Accessed: 10/10/2018]

Gonçalves VF, Brito-Gitirana L (2008) Structure of the sexually dimorphic gland of *Cycloramphus fuliginosus* (Amphibia, Anura, Cycloramphidae). Micron 39: 32–39. https://doi.org/10.1016/j.micron.2007.08.005

Houck LD (2009) Pheromone Communication in Amphibians and Reptiles. Annual Review of Physiology 71: 161–176. https://doi.org/10.1146/annurev.physiol.010908.163134

Jared C, Antoniazzi MM, Jordão AEC, Silva JRMC, Greven H, Rodrigues MT (2009) Parotoid macroglands in toad (*Rhinella jimi*): their structure and functioning in passive defence. Toxicon 54: 197–207. https://doi.org/10.1016/j.toxicon.2009.03.029

Junqueira LCU, Junqueira LMMS (1983) Técnicas básicas de citologia e histologia. São Paulo, Livraria Editora Santos, 123 pp.

Kikuyama S, Yamamoto K, Iwata T, Toyoda F (2002) Peptide and protein pheromones in amphibians. Comparative Biochemistry and Physiology 132: 69–74. https://doi.org/10.1016/S1096-4959(01)00534-6

Lacombe C, Cifuentes-Diaz C, Dunia I, Auber-Thomay M, Nicolas P, Amiche M (2000) Peptide secretion in the cutaneous glands of South America tree frog *Phyllomedusa bicolor*: an ultrastructural study. European Journal of Cell Biology 79: 631–641.

Lenzi-Mattos R, Antoniazzi MM, Haddad CFB, Tambourgi DV, Rodrigues MT, Jared C (2005) The inguinal macroglands of the frog *Physalaemus nattereri* (Leptodactylidae): structure, toxic secretion and relationship with deimatic behaviour. Journal of Zoology 266: 385–394. https://doi.org/10.1017/S095283690500703X

Mailho-Fontana P, Antoniazzi MM, Toledo LF, Verdade VK, Sciani JM, Bárbaro KC, Pimenta DC, Rodrigues MT, Jared C (2014) Passive and active defense in toads: The parotoid macroglands in *Rhinella marina* and *Rhaebo guttatus*. Journal of Experimental Zoology 321: 65–77. https://doi.org/10.1002/jez.1838

Pearse A (1960) Histochemistry: theoretical and applied. London, J&A Churchill, 2nd ed., 998 pp.

Pombal J, Bastos R (1996) Nova espécie de *Scinax* Wagler, 1830 do Brasil Central (Amphibia, Anura, Hylidae). Boletim do Museu Nacional, Nova Série, Zoologia 1: 371.

Poth D, Wollenberg KC, Vences M, Schulz S (2012) Volatile Amphibian Pheromones: Macrolides from *Mantellid* Frogs from Madagascar. Angewandte Chemie International Edition 51: 2254–2254. https://doi.org/10.1002/anie.201109230

Rigolo JR, Almeida JA, Ananias F (2008) Histochemistry of skin glands of *Trachycephalus aff. venulosus* Laurenti, 1768 (Anura, Hylidae). Micron 39: 56–60. https://doi.org/10.1016/j.micron.2007.08.006

Sciani JM, Angeli CB, Antoniazzi MM, Jared C, Pimenta DC (2013) Differences and similarities among paratoid macrogland secretions in south American toads: a preliminary biochemical delineation. The Scientific World Journal 2013: e937407. https://doi.org/10.1155/2013/937407

Silva HAM, Silva-Soares T, Brito-Gitirana L (2017) Comparative analysis of the integument of different tree frog species from *Ololygon* and *Scinax* genera (Anura: Hylidae). Zoologia 34: e20176. https://doi.org/10.3897/zoologia.34.e20176

Suvarna KS, Layton C, Bancroft JD (2018) Bancroft's Theory and Practice of Histological Techniques. Elsevier Health Sciences, 8th ed., 672 pp.

Terreni A, Nosi D, Brizzi R, Delfino G (2002) Cutaneous serous glands in South American anurans: An ultrastructural comparison between hylid and pseudid species. Italian Journal of Zoology 69: 115–123. https://doi.org/10.1080/11250000209356448

Wabnitz P, Bowie JH, Tyler MJ, Wallace JC, Smith BP (1999) Aquatic sex pheromone from a male tree frog. Nature 401: 444–445. https://doi.org/10.1038/46724

Author Contributions: PB and CPT wrote the paper; WA and FS conducted the experiments; RB designed the experiments.
Competing Interests: The authors have declared that no competing interests exist.

Appendix 1. Information about specimens used in the study.

Voucher	Sex	SVL[1] (mm)	Stereomicroscopy	Light Microscopy	Electron Microscopy
ZUFG9287	F	24.60		Hematoxylin-Eosin	
ZUFG9292	F	24.80		Hematoxylin-Eosin	
ZUFG10377	F	24.86	X		
ZUFG9288	M	21.00		Histochemistry	
ZUFG9289	M	21.40		Histochemistry	
ZUFG10374	M	20.32	X		
Lost specimen 1*	M	20.00		Hematoxylin-Eosin	X
Lost specimen 2*	M	18.45		Hematoxylin-Eosin	X

[1]Snout-vent length.
*These specimens were destroyed during dissection and could not be deposited in the zoological collection.

9

Gonadal development and reproductive period of the characin *Astyanax* aff. *bimaculatus* (Characiformes: Characidae) in a tropical reservoir in southeastern Brazil

Francisco G. Araújo[1], Aparecida A. Nascimento[2], Iracema D. Gomes[1], Armando Sales[2], Beatriz A. Chagas de Oliveira[2]

[1]*Laboratório de Ecologia de Peixes, Universidade Federal Rural do Rio de Janeiro. Rodovia BR 465, km 7, 23890-000 Seropédica, Rio de Janeiro, Brazil.*
[2]*Laboratório de Histologia e Embriologia Animal, Universidade Federal Rural do Rio de Janeiro. Rodovia BR 465, km 7, 23890-000 Seropédica, Rio de Janeiro, Brazil.*
Corresponding author: Francisco G. Araújo (gersonufrrj@gmail.com)

ABSTRACT. The gonadal development of the characin *Astyanax* aff. *bimaculatus* from Funil Reservoir, an impoundment in the middle reaches of the Paraiba do Sul River in southeastern Brazil was analysed through histological, histochemical and histometric techniques, and the spawning season was determined. Unbalanced sex ratio was found, with females outnumbering and reaching larger sizes than males. The following stages of oogenic cell development were identified for females: primary growth, pre-vitellogenic (cortical alveolar) and vitellogenic. In males, the following stages were determined: spermatogonia primary and secondary, spermatocyte primary and secondary, spermatid and spermatozoa. The spawning occurs most of the year, peaking in January-February, as indicated by the gonadosomatic index. In the peak of the spawning period, the condition factor and hepatosomatic indices decreased suggesting use of stored body energy reserves during the reproduction. Spawning in batches, small-sized oocytes and a long reproductive season suggest that *Astyanax* aff. *bimaculatus* fits to an opportunist strategy. Such strategy may facilitate the use of a wide range of lentic and lotic body waters in the Neotropical region.

KEY WORDS. Freshwater fish, histology, oocyte, spawning type, spermatocyte.

INTRODUCTION

The life cycle patterns of fish are the most varied among the vertebrates and different reproductive strategies have enabled them to obtain success in different environments (Stearns 1992, Winemiller and Rose 1992, Blanck et al. 2007, Belova 2008). The reproductive strategies used by different fish species vary markedly (Dala-Corte and Azevedo 2010, Lowerre-Barbieri et al. 2011) and the success achieved in different environments can have ecological and evolutionary implications (Rizzo et al. 2002, Jamieson 2009). There is a need of studies on different fish groups, which is crucial to understand their reproductive strategies and tactics (West 1990, Parenti and Grier 2004, Belova 2008).

The gonadal development is cyclical and seasonal in most Teleostei. Germ cell renewal, differentiation, development and the release of sperm/oocytes throughout each reproductive cycle result in gonadal alterations that characterize different reproductive phases. Variations in the gonadal fish morphology reflect important ecological and behavioural adaptations during reproduction (Coward et al. 2002, Fishelson and Gon 2008, Martins et al. 2012). In addition, the phenotypic plasticity of gonadal morphology reflects the adaptability to environmental changes (Defalco and Capel 2009, Galvão et al. 2016). The histological stages of oocyte and spermatogenesis development together with macroscopic aspects of the gonads are used to characterize reproductive phases (Brown-Peterson et al. 2011).

The migratory behaviour and reproductive period are among the major life-history traits in the reproductive process. Such traits generate trade-offs in the development of any reproductive strategy (Trujillo-Jiménez et al. 2013). The spawning season can be assessed by changes in ovaries/testes maturation along the annual cycle (Brown-Peterson et al. 2011). The biological indices related to reproduction can indicate the way in which fish use environmental and energetic resources. The gonadosomatic index (GSI) is a good indicator of reproductive activity that is also determined by the stages of gonadal maturation (Le Cren 1951,

Hojo et al. 2004, Hismayasari et al. 2015). Other indices such as the hepatosomatic index (HSI) and the condition factor (K) are also used to assess the reproductive period (Bolger and Connolly 1989, Jayasankar and Alagarswami 1994, Barbieri et al. 1996, Giosa et al. 2014). The HSI is usually used in fisheries science as an indicator of energy reserves in the liver (Cerda et al. 1996, Hismayasari et al. 2015). Variations in the HSI are related to the capacity of the liver to store glycogen, reflecting the physiological conditions and the availability of energy (mainly lipids) for the reproduction activity (Carvalho et al. 2009). The condition factor (K) can indicate the degree of fitness of the individuals and can be related to the reproductive process by reflecting the allocation of energy (mainly proteins) during the spawning period, in which the physiological state of the individual changes (Carvalho et al. 2009). Therefore, fish nutritional status and/or spending of reserves are a result of their relationship with the environment reflecting in these indices (Medeiros and Maltchik 2001, Lizama and Ambrosio 2002, Silva et al. 2018).

Environmental factors such as water temperature and rainfall are important drivers of the reproductive process for several fish species in the Neotropical Region (Hokanson 1977, Jonsson 1991, Bailly et al. 2008). Water temperature and rainfall can trigger the spawning process of many fish species and, therefore, their reproductive success may be related to increases in temperature and in water volume that is associated to increases in rainfall (Abrial et al. 2014). Seasonal variations in water level also have direct influence on the reproductive process because they cause changes in several water characteristics (e.g., transparency, temperature, and input of allochthonous materials), increasing habitat availability (mainly in the riverine zone) and food resources (Agostinho et al. 2004, Bailly et al. 2008, Espínola et al. 2016).

Fish of the Characiformes order exhibit a wide variety of life strategies, with an adaptive divergence that does not match any other animal order (Fink and Fink 1981, Nelson 2006). *Astyanax* (Baird & Girard, 1854) is the most diversified genus of the Characidae with more than a hundred species widely distributed in the Brazilian watersheds (Reis et al. 2003, Hirt et al. 2011, Silva et al. 2012). This genus probably has great ecological importance and great adaptive plasticity (Gurgel 2004, Orsi et al. 2004, Abelha et al. 2006). The characin *Astyanax* aff. *bimaculatus* is widely distributed in southeastern Brazil. The studied species belongs to the *Astyanax bimaculatus* complex but still there is no available literature clarifying its taxonomic status. The studied species has a large orbital diameter (35.0–38.3% head length), great height (43.5–45.7% standard length, only pentacuspidated teeth in the dental bone, and two circular foramens in the specialized neural process.

This species seems to perform small movements for reproduction in both lentic and lotic environments (Godinho et al. 2010, Weber et al. 2012). Species of *Astyanax* are pelagic spawners normally reproducing in schools during upstream movements (Breder and Rosen 1966, Mazzoni and Iglesias-Rios 2004, Súarez et al. 2017). Garutti (1989) and Gennari-Filho and Braga (1996) reported *Astyanax bimaculatus* (Linnaeus, 1758) as total or batch spawner, depending on the environmental conditions. Winemiller (1989) classified *A. bimaculatus* as having a seasonal reproductive strategy, based on duration of the breeding season, average female reproductive bouts per year, oocyte diameter and mean generation time, among other features. The objective of this study was to assess macro and microscopic characteristics of the gonadal development of *Astyanax* aff. *bimaculatus* from Funil Reservoir, an impoundment in the middle reaches of the Paraiba do Sul River in southeastern Brazil, and to describe the gametogenesis through histological, histochemical and histometric techniques. We assessed the reproductive strategy and analysed changes in the reproductive endpoints of GSI, K and HSI to determine the reproductive period.

MATERIAL AND METHODS

The Funil Reservoir (22°30′–22°40′S; 44°30′–44°45′W, 440 m) is located in the middle reaches of the Paraíba do Sul River basin, in southeastern Brazil. The reservoir has an extension of approximately 20 km, area of 40 km^2, and mean depth of 22 m. The retention time is short (10–50 days), with great variation in water levels and substantial erosion of the banks. According to Branco et al. (2002), the Funil Reservoir is developing increasingly eutrophic conditions due to anthropogenic influences. There is little vegetation cover around the reservoir because of previous agricultural use for coffee plantation and pasture.

The reservoir was built in 1969 to generate hydroelectric power and to reduce the floods in the area downstream the dam. In addition, domestic and public drinking water supply, irrigation and aquaculture are among other uses of the reservoir. The climate is subtropical with monthly mean water temperatures of 18–24 °C, with maximum in January-February and minimum in July-August. Rainfall is at highest levels in the summer months (December–January; 200–250 mm per month) and at the lowest in the winter months (June-August), with less than 50 mm per month (Marengo and Alves 2005).

The fish were captured bimonthly by gill nets from September 2006 to October 2007. Three gill nets (50 x 3 m; stretch mesh 25, 50 and 75 mm) were set up at sunset and retrieved in the following morning at four sites randomly chosen across the reservoir area. All individuals were killed by immersion in water at 4 °C, identified and measured for total length (TL, nearest 1 mm), and weighted for total mass (TW, nearest 0.01 g). A ventral incision was made to expose gonads for determination of the sex and the gonadal development phases. Gonads were removed and weighed wet (GW, nearest 0·01 g). A portion of each gonad was preserved in Bouin's solution during eight hours for histological analyses following Vazzoler (1996). Then, gonads were transferred to 70% ethanol for preservation.

The gonads were subject to histological techniques and embedded in paraffin. Transversal sections (5 μm of thickness) were cut, mounted on glass slides and stained in haematoxylin and eosin (HE). To determine the histochemical content of the oocyte structures, the following techniques were used: Alcian Blue (AB) pH 2.5 and Schiff's acid + reactive acid (PAS) for the detection of acid and neutral glycoproteins, respectively (Myers et al. 2008).

Gonad sections collected in different regions (proximal, medium and distal) were examined. Microphotographs were taken with a Sony Cyber Shot DSC-W 230 digital camera coupled to an Olympus B941 (Tokyo, Japan). To characterize the reproductive phases, we used the terminology proposed by Quagio-Grassiotto et al. (2013), which combines the macro and microscopic aspects of gonads throughout the reproductive cycle of teleosts. Voucher specimens were deposited in the Fish Collection of the Laboratory of Fish Ecology, Universidade Federal Rural do Rio de Janeiro, under number LEP-UFRRJ #1873.

Size structure was assessed by length-frequency distributions of the individuals grouped into 20 mm TL size classes. The sex ratio was compared for each size class and for the pooled individuals. A chi-square (χ^2) test was used to assess the significance of differences at confidence level of 95% ($p < 0.05$).

The gonads were assigned to developmental stages, based on form, size, mass, colour and vascularization. However, gonads were ultimately classified as either immature (juveniles and inactive stages) or mature (developing, spawning capable, regressing and regenerating phases) to reduce the chances of identification mistakes of gonadal stages. The gonad classification was adapted from Nuñez and Duponchelle (2009) and West (1990). Oocytes were classified according to their morphology, their affinity to the dyes, and the presence of specific inclusions (lipid droplets, yolk granules, cortical alveoli).

The criteria for identification of oocyte stages and postovulatory follicles (POFs) were adapted from Brown-Peterson et al. (2011). The diameters of the first 50 oocytes and their nuclei were measured to 0.0001 mm using an ocular micrometer. Measurements were taken only on oocytes sectioned through the nucleus in fishes randomly chosen from the monthly samples. Histological identification of the various maturity stages were determined according to development of the ovary and testes and by the presence/absence of different types of oocytes (i.e., whether organized by ovarian lamellae or not) and spermatocytes. Histological classification of ovaries was based on oocyte stage and the occurrence of different stages of postovulatory follicles (POFs). The diameters of the vitellogenic oocytes and the spermatogenic cell nuclei were determined using the computerized image analyser Image-J 1.48 (Schneider et al. 2012), in 20 non-overlapping random fields of the histological slide prepared for each specimen.

In males, the histological characterization of the reproductive phases were based on the morphological changes that occur in the epithelium of the seminiferous tubules throughout the reproductive cycle. This characterization considers the presence or absence of spermatocytes and the types of germinative cells contained in the spermatocytes besides eventual alternations of a continuous or discontinuous germinal epithelium (Brown-Peterson et al. 2011).

The spawning period was determined by variations in the gonadosomatic index, $GSI = 100 \times (GW \times TW^{-1})$. The Fulton's condition factor (K) and hepatosomatic index (HSI) were calculated as indirect indices of energy status. The Fulton's condition factor (K) was calculated following the equation $K = 100\times(TW\times TL^{-3})$. The hepatosomatic index (HSI) was calculated as, $HSI = 100 \times (LW \times TW^{-1})$, where LW is the weight of the liver.

RESULTS

Sex ratio

A total of 94 specimens (56 females, 38 males) were examined. The total length (TL) ranged from 70 mm to 170 mm (female) and from 74 mm to 150 mm (male). The total weight (TW) ranged from 10.7 to 64.3 g (female) and from 7.1 to 43.0 g (male). Females significantly outnumbered males in size larger than 130 mm TL ($p < 0.01$). Highly significant differences were found for the pooled fish ($\chi^2_{(4, 90)} = 29.38$; $p < 0.01$) (Table 1).

Table 1. Chi-square (χ^2) test for sex ratio comparisons of *Astyanax* aff. *bimaculatus* in Funil Reservoir. (EF) Expected frequency, (TL) total length (mm).

Size classes (TL)	Female	Male	Total	EF	χ^2	Significance
70–90	1	6	7	3.5	3.57	*
90–110	2	10	12	6.0	5.33	*
110–130	26	19	45	22.5	1.08	ns
130–150	22	3	25	12.5	14.4	**
150–170	5	0	5	2.5	5.0	*
Total	56	38	94	47	29.38	**

(ns) Non-significant, (*) significant at $p < 0.05$, (**) significant at $p < 0.01$.

Histological characteristics of cells of the oogenic lineage

Primary growth stage

Oogonia, chromatin nucleolar and perinucleolar stages were present in the ovary throughout the entire annual cycle, and are referred to as primary growth stages (PG). They were clearly observed with the Alcian-blue technique (Table 2).

Oogonia: In this stage there is a predominance of the smallest cells of the oogenic lineage. Oogonia stage were characterized by cells with a large (8–10 µm), spherical and basophilic nucleus, and basophilic cytoplasm. The cells can be found isolated or in nests in the ovuligerous lamellae (Fig. 1, Table 2). Chromatin nucleolar (Fig. 2): The cells are similar to oogonia, although somewhat larger (two-fold larger, 20 µm) (Table 2).

Perinucleolar: The cells showed strongly basophilic cytoplasm with irregular contour with large, spherical and well-defined nuclei (Figs 3, 4). There was an eccentric nucleolus. The advanced oocyte had a rounder shape. The cytoplasm was less basophilic and presented irregular contour. The nucleus was

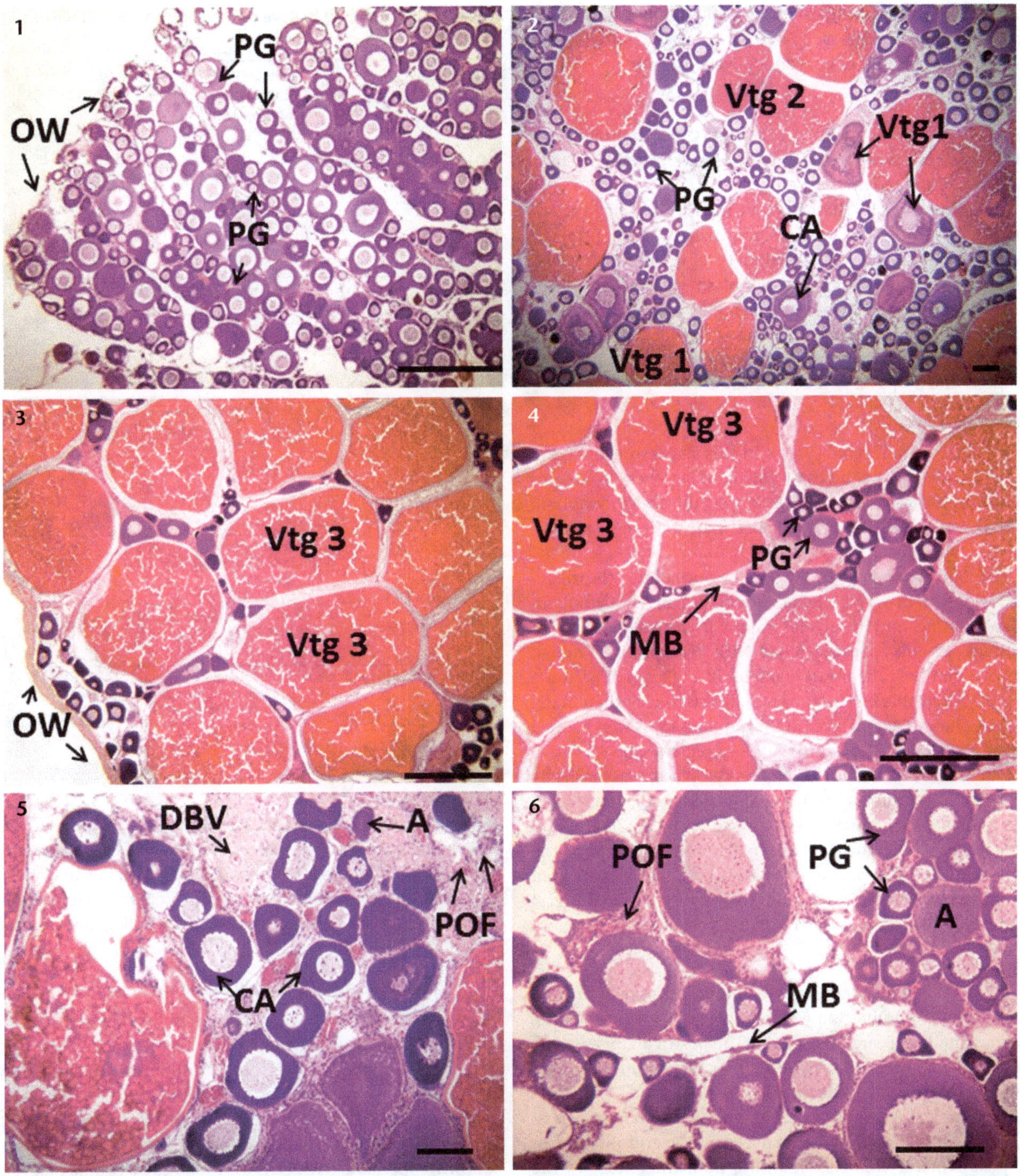

Figures 1–6. Photomicrographs of ovaries of *Astyanax* aff. *bimaculatus* in different phases of gonadal maturation. (1) Immature. Oocytes in primary growth (PG) ovuligerous lamellae (OL). (2) Developing. Presence of oocytes of different sizes (PG, CA, Vitg1 and Vtg2). (3–4) Spawning Capable. Predominance of vitellogenic oocytes (Vtg3). (5) Regressing. Presence of different types of oocytes (PG, CA, Vitg1, POF) and oocytes in atresia (A). (6) Regenerating. Oocytes in primary growth (PG) and POF. (OW) ovarian wall, (CA) cortical alveoli, (Vtg1) primary vitellogenic, (Vtg2) secondary vitellogenic, (Vtg3) tertiary vitellogenic, (MB) muscle bundle, (DBV) dilated blood vessels, (POF) postovulatory follicles, (A) atresia. Staining haematoxylin and eosin (HE). Scale bar: 100 μm.

large (28 μm) and acidophilic and some basophilic nucleoli were seen on the periphery (Table 2). The follicular envelope was composed of only one layer of pavement cells.

Secondary growth stage

This stage includes Cortical alveoli (CA) and Vitellogenic (Vtg), which is divided into three substages: primary (Vtg1), secondary (Vtg2), and tertiary (Vtg3).

Cortical alveoli: The cells had nuclei with an irregular contour (Figs 2, 5). The main characteristic was the presence of vesicles and alveoli in the periphery of the cytoplasm and the large diameter (42 μm). The cytoplasm was less basophilic than the previous stage. The nucleus had several nucleoli in its periphery (Table 2). The zona radiata was evident, rather thin and translucent, surrounded by a layer of cubic cells of the granulosa and by pavement cells of the theca.

Vitellogenic stage

Primary vitellogenic (Vtg1): In this stage, the yolk granules, also called yolk spheres or yolk globules, were numerous containing cortical alveoli and occupying the entire cytoplasm (Fig. 2). The nucleus became smaller than the previous stage (18 μm) (Table 2).

Secondary vitellogenic (Vtg2): The nucleus had the irregular contour with many peripheral nucleoli (Fig. 2). The cytoplasm was acidophilic, and the alveoli occupy its cortical portion. At this stage, alveoli appeared, filled with acidophilic material, and yolk granules. The mean diameter of Vtg2 was 378.7μm ± 315.3 μm s.d. (Table 2). The zona radiata was thicker than in the previous phase. The cell layer of the granulosa was well defined and the theca remained composed of pavement cells.

Tertiary vitellogenic (Vtg3): The cytoplasm was markedly acidophilic and completely filled with vitellogenic granules (Figs 3, 4). The lipid inclusions were dispersed in the cytoplasm and in the nucleus. The nucleus had smaller size compared to the previous stage. The mean diameter of Vtg3 was 585.4 ± 450.1 μm s.d. The granulosa cells were long, with a very irregular apical surface and the theca presented a high degree of vascularization.

Atresia

The cells of the granular layer migrated to the interior of the ooplasm, absorbing the yolk. At the end of this stage, the zona radiata disappeared (Figs 5, 6). The postovulatory follicle was recognizable by its disorganized structure, abundant vacuoles and a convoluted follicular wall (Table 2).

Atresia was frequently observed during the oocytes regressing, undergoing various phases of degeneration and absorption. The post-ovulatory follicles, resulting from the release of the mature oocyte are formed by hypertrophied granulosa cells. On the other hand, the theca cells do not undergo any changes with oocyte release.

During oocyte development, the follicle formation occurs, with the zona radiate (ZR) separating the oocyte from the follicular wall with a basal membrane between the follicular cell layer

Table 2. Stages of oocyte development in *Astyanax* aff. *bimaculatus*.

Stage	Characteristics	Image
Primary growth	Oogonias (nests) Chromatin nucleolar Perinucleolar	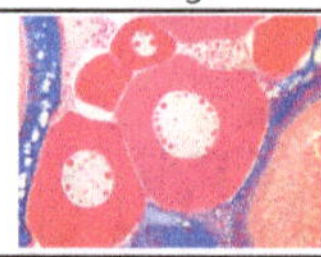
Pre-vitellogenic	Cortical alveoli	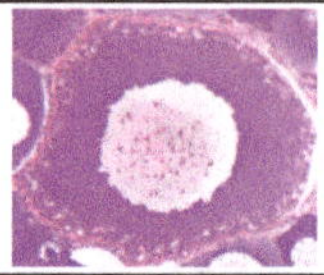
Vitellogenic	Primary vitellogenic -Vtg1	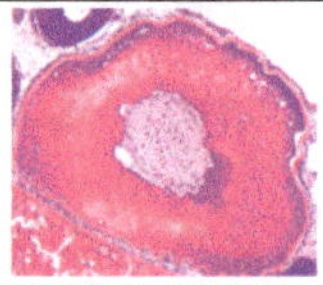
	Secondary vitellogenic – Vtg2 (378.7μm ± 315.3 μm s.d. diameter)	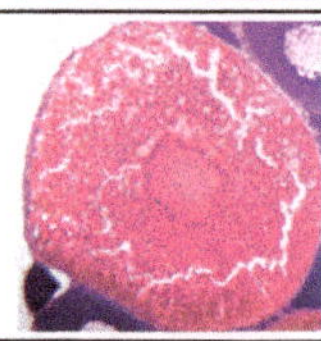
	Tertiary vitellogenic – Vtg3 (585.4 μm ± 450.1 μm s.d. diameter)	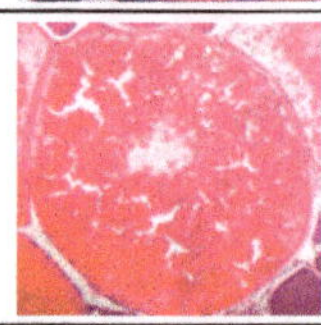
Postovulatory follicles and atresic	String formation of granulosa cells disorganised or absent. Disintegration of the nucleus	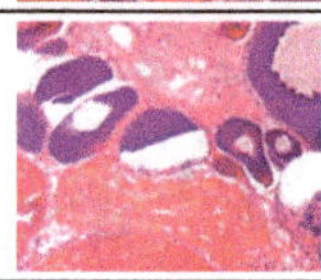

(granulosa cells) and the theca layer of connective tissue. Only in the vitellogenic oocytes, the presence of acid glycoproteins in the theca and granulosa cells was detected with positive reaction to AB. In addition, the neutral glycoproteins were found in the ZR, in the cortical alveoli and between the yolk globules because of the positive reaction to the PAS.

Histological characteristics of cells of the spermatogenic lineage

The testes were covered by a capsule of dense connective tissue, the tunica albuginea, which protrudes into the organ delimiting and supporting the seminiferous tubules. The determination of the types of male germ cells was performed according to the histological characteristics of the cytoplasm, nucleus and size of the cells. Based on these observations, the following spermatogenic cells were identified: primary (sg1) and secondary (sg2) spermatogonia; primary (sc1) and secondary (s2) spermatocytes; spermatids, (sd) and spermatozoa.

Primary spermatogonia (G1): They were the largest (10 μm) cells of the germ lineage with abundant eosinophilic cytoplasm (Figs 7–12), and large spherical nucleus with isolated nucleolus. The nucleus had a mean diameter of 5.76 ± 0.79 μm

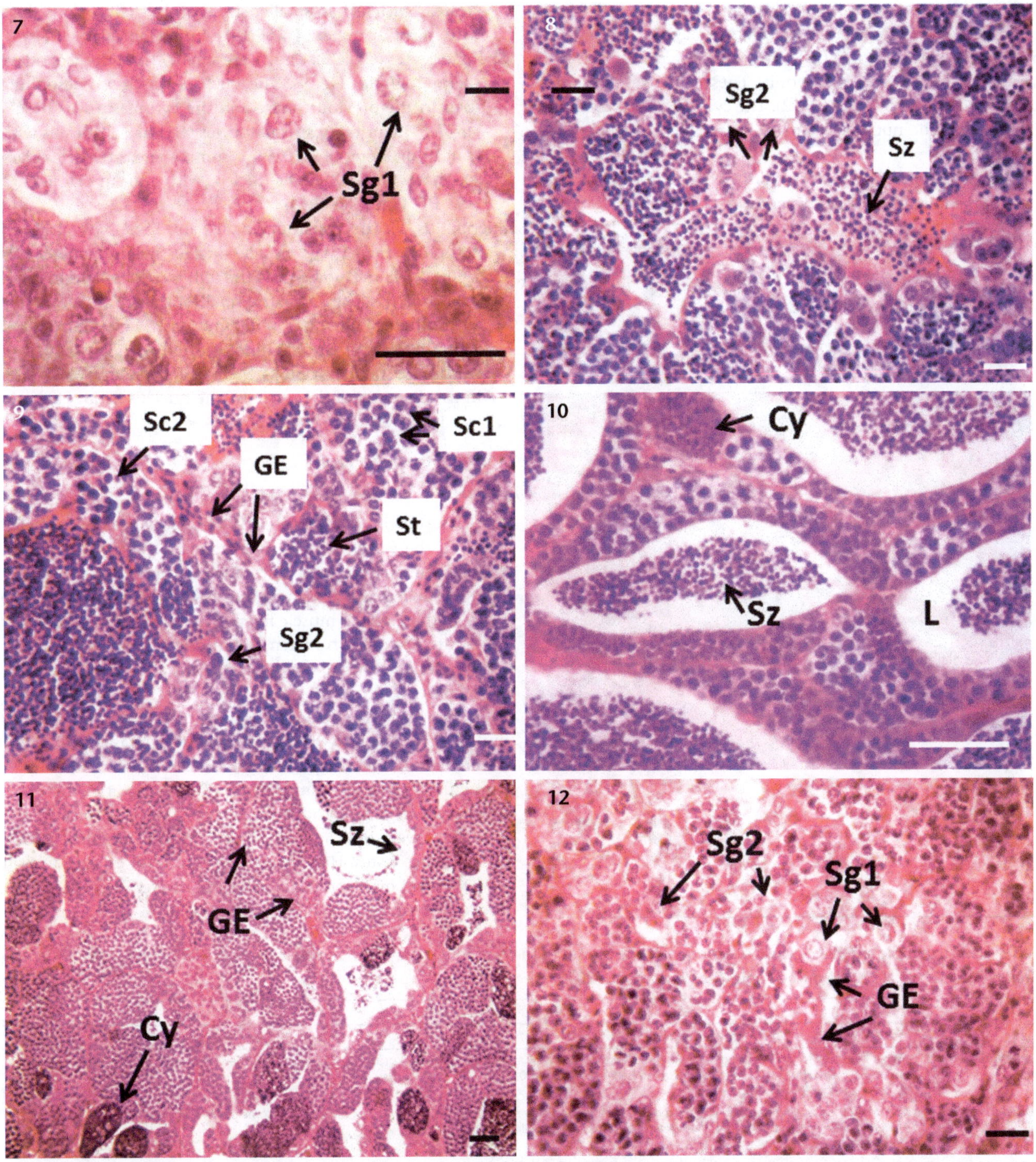

Figures 7–12. Photomicrographs of testes of *Astyanax* aff. *bimaculatus* in different phases of gonadal maturation. (7) Immature. Only primary spermatogonia (Sg1) without lumen. (8–9) Developing. Various types of spermatocytes evident along lobules (Sg2, Sc1, Sc2, St, Sz) and germinal epithelium (GE) continuous throughout. (10) Spawning Capable. Predominance of Sz in lumen seminiferous tubules. (11) Regressing. Presence of cysts (Cy), residual spermatozoa (Sz) and germinal epithelium (GE) in regeneration. (11) Regenerating. Proliferation of spermatogonia (Sg1, Sg2) and GE continuous throughout. Staining haematoxylin and eosin (HE). Scale bar: 20 µm.

(Table 3). They were observed isolated from the wall of the seminiferous tubules at all maturation stages.

Secondary spermatogonia (G2): They originated from the division of primary spermatogonia. These cells were observed grouped in cysts (Figs 8, 9, 12). The cytoplasm was clear and reduced. The nucleus was more basophilic; nucleolus were clearer. Nucleus had smaller size than in the previous stage with mean diameter of 4.50 ± 0.64 µm (Table 3).

Table 3. Stages of nuclear diameter (µm) of the spermatogenic cells development in *Astyanax* aff. *bimaculatus*.

Spermatogenic cells	Nuclear diameter ± s.d.	Image
G1 – primary spermatogonia	5.76 ± 0.79	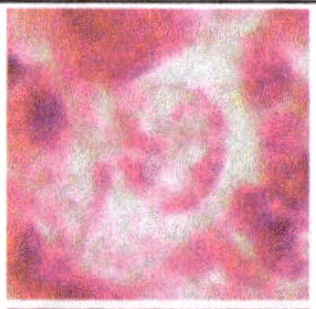
G2 – secondary spermatogonia	4.50 ± 0.64	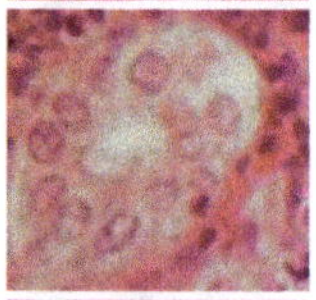
C1 – primary spermatocyte	3.70 ± 0.56	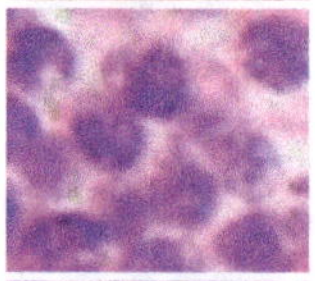
C2 – secondary spermatocyte	3.53 ± 0.48	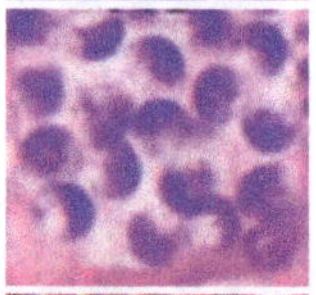
T – spermatid	1.80 ± 0.54	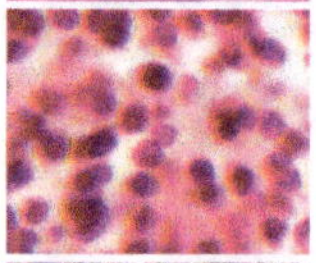
Z – spermatozoa	1.67 ± 0.42	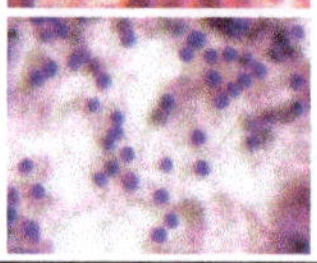

Spermatocytes (C1, C2): They were smaller than the secondary spermatogonia (Fig. 9). The nucleus was more basophilic than in the previous phase and the cytoplasm was hardly visible. The primary spermatocytes (C1) had a nucleus with a mean diameter of 3.70 ± 0.56 µm, whereas the secondary spermatocytes (C2) had a nucleus with a mean diameter of 3.53 ± 0.48 µm (Table 3).

Spermatids (T): They had indistinguishable limits, spherical and basophilic nuclei (Fig. 9). The nucleus mean diameter was 1.80 ± 0.54 µm (Table 3) and the cells were observed in cysts.

Spermatozoa (Z): They were the smallest cells of the germ line with spherical, very basophilic nucleus. They occupied the central region of the seminiferous tubules (Figs 8, 10, 11). During the spermatogenesis, there was a reduction of 80–90% in the diameter of the nucleus. The nucleus mean diameter was 1.67 ± 0.42 µm (Table 3).

Phases of the reproductive cycle

Females

The ovaries were enveloped by simple pavement epithelial tissue in the early stages of development changing to simple cubic in the final stages of maturation. Underneath the epithelium, we found the albuginea composed by dense connective tissue and regions of muscle fibers. The tunica albuginea emits septa into the ovarian lumen, delimiting the ovigerous lamellae where oogonia and oocytes are found at different stages of development.

The microscopic characteristics of the ovaries (Table 4) and testes (Table 5) were composed of five phases with slight differences in microscopic characteristics.

Immature: It was the primary stage of young ovaries that have not yet begun reproductive activity. Macroscopically, the ovaries were thin and translucent, not distinguished from the males testes. Histologically, a thin layer of ovarian wall (OW) and ovuligerous lamellae (OL) were observed, which were occupied by PG (Fig. 1, Table 4).

Table 4. Macroscopic and histological description of the phases of the reproductive cycle of female of *Astyanax* aff. *bimaculatus*. Adapted Brown-Peterson et al (2011) and Quagio-Grassiotto et al. (2013).

Phases	Features of ovaries	
	Macroscopic	Histological
Immature	Laminar form and small occupying less than one – third of the celomatic cavity, translucent, measuring 15–25 mm and weighing between 0.01 and 0.03 g; oocytes not visible to the naked eye	Primary growth (PG) present and the ovarian wall is thin
Developing	Wider ovaries, occupying less than one-third of the coelomatic cavity measuring 19–36 mm and weighing 0.7–2.3 g. Pale cream colour to whitish-yellow, visible blood vessels	PG, pre-vitellogenic (CA) and some in early vitellogenesis (Vtg1, Vtg2)
Spawning Capable	Oval and large shape occupying entire celomatic cavity, (24–43 mm; 2.5–4.8 g), yellowish-green colour, visible oocytes, blood vessels more evident	Prevalence of large vitellogenic oocytes (Vtg3), but oocytes in other stages of development is observed
Regressing	Flaccid, occupying less than one-third of the coelomic cavity (20–35 mm; 0.1–1.2 g) slightly brown and orange with haemorrhagic appearance; small oocytes visible to the naked eye	Disorganization of ovarian tissue with reabsorption of empty follicles (POF) and atresic oocytes (A); dilated blood vessels. Many different oocytes are still found
Regenerating	Small and broad ovaries, slightly brown colour (28–50 mm; 0.9–1.1 g), blood vessels less prominent	Presence of empty follicles (POF), oocytes in atresia and muscle bundle. Ovarian lamella partially occupied by oocytes at oocytes in development

Table 5. Macroscopic description of the phases of the reproductive cycle of male of *Astyanax* aff. *bimaculatus*. Adapted Brown-Peterson, Wyanski, Saborido-Rey, Macewicz and Lowerre-Barbieri et al. (2011) and Quagio-Grassiotto et al. (2013).

Phase	Features of the testes	
	Macroscopic	Histological
Immature	Filiform, occupying less than one-third of the coelomic cavity, measuring 14–30 mm and weighing between <0.01 and 0.01 g, translucent	Prevalence of primary spermatogonia (Sg1). Lumen of the tubules imperceptible
Developing	Flat shape, occupying less than one-third of the coelomic cavity, measuring 15–35 mm and weighing between 0.03 and 0.17 g, whitish colour.	Secondary spermatogonia (Sg2) and spermatocytes primary (Sc1) identified in spermatocytes along the germinal epithelium. Presence of primary, secondary spermatocytes (Sc1, Sc2), spermatids (St) and spermatozoa (Sz) in the lumen of the seminiferous tubules
Spawning Capable	Sinuous and flat shape, occupying one-third and nearly two-third of the coelomic cavity, measuring 24–45 mm and weighing between 0.12 and 0.47 g. Opaque white colour, unclear irrigation	Spermatozoa (Sz) present in the lumen of seminiferous tubules. Spermatogonia (Sg1, Sg2), spermatocytes (Sc1, Sc2) and spermatids (St) may be present in the spermatocytes
Regressing	Flaccid, haemorrhagic and occupy an average of 2/3 of the coelomic cavity, measuring 24–30 mm and weighing between 0.17 and 0.25 g. Reddish white colour	Spermatozoa present in the lumen of disorganized seminiferous tubules; germinal epithelium may be continuous or discontinuous. Spermatocytes containing non-released spermatids (St) dispersed by the seminiferous tubules
Regenerating	Small and bulky measuring 17–20 mm and weighing between 0.08 and 0.12 g	Lumen of the seminiferous tubules discrete or undetectable. Spermatogonia Sg1 and Sg2 in proliferation. Germinal epithelium continuous

Developing: The ovaries have began to mature; in this stage they presented whitish-yellow color, and increase in weight and length. Histologically, the alveolar cortical oocytes were also observed (Fig. 2, Table 4).

Spawning capable: The ovaries were distinctly large (24–43 mm) and occupy a large part of the coelomatic cavity. They were greenish yellow and vitellogenic oocytes, and a thick zona radiata was observed. (Figs 3, 4, Table 4).

Regressing: At this stage, there was a change in color and reduction in size (less than half of the anterior phase) and weight of the ovaries. The ovaries were flaccid and wrinkled. Some unreleased oocytes begun the process of atresia and postovulatory follicles were visible (Fig. 5, Table 4).

Regenerating: The ovaries begun the cycle of gonadal development by increasing weight and becoming more turgid (Fig. 6, Table 4).

Males

Immature: The testes were like two silvery or translucent threads, thinner and longer than immature ovaries (Fig. 7, Table 5).

Developing: The testes were translucent and thin. testes are longer, wider, often of triangular or circular section and whitish to pinkish (Figs 8, 9, Table 5).

Spawning capable: The testes were larger and whitish. Histologically, the tubules filled with spermatozoa were observed (Fig. 10, Table 5).

Regressing: The testes were shorter and lighter, but flaccid, empty-like (Fig. 11, Table 5).

Regenerating: In this phase, the testes begun the cycle of gonadal development. A gonad of a regenerating male is difficult to distinguish from a regressing phase, except at the end of the regeneration period, when the testes are relatively turgid. (Fig. 12, Table 5).

Spawning season

The highest percent of ovaries in the ripe stage (spawning capable) occurred in January-February (more than 80%), followed by September-October (30–75%), and the lowest between March and August (<10%) (Fig. 13). Similarly, the highest occurrence of testes in the ripe stage occurred in January-February (45%), followed by September to December (40–50%) and the lowest in March-April (18%).

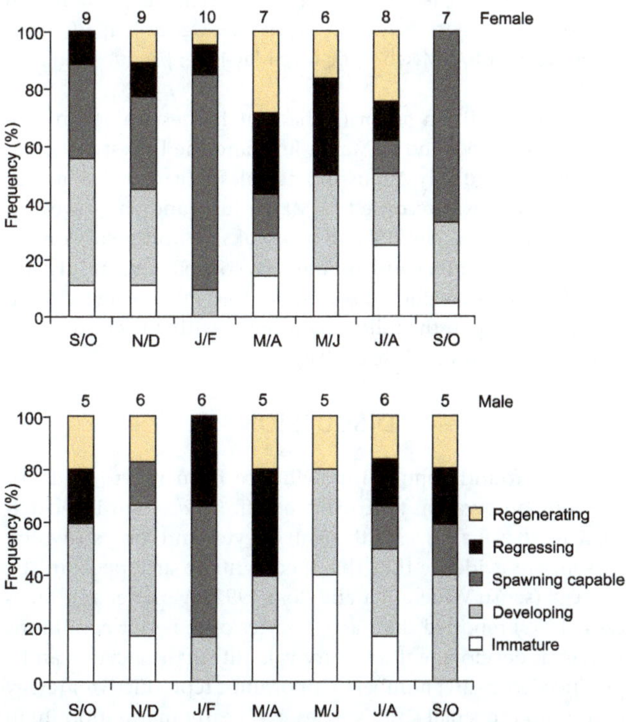

Figure 13. Bi-monthly changes in frequency distribution of the reproductive phases of *Astyanax* aff. *bimaculatus*. Number of individuals indicated above each bar. S/O, September-October; N/D, November-December; J/F, January-February; M/A, March-April; M/J, May-June; J/A, July-August.

The highest GSI values were recorded in January-February for both sexes and the lowest in May-August for females, and in March-June for males (Fig. 14). Values comparatively high of GSI (> 5) were also recorded for both sexes between September and December, and for females in March-April. The GSI values was directly related to the water temperature and reached the peak in January-February, when the rainfall was at the highest levels.

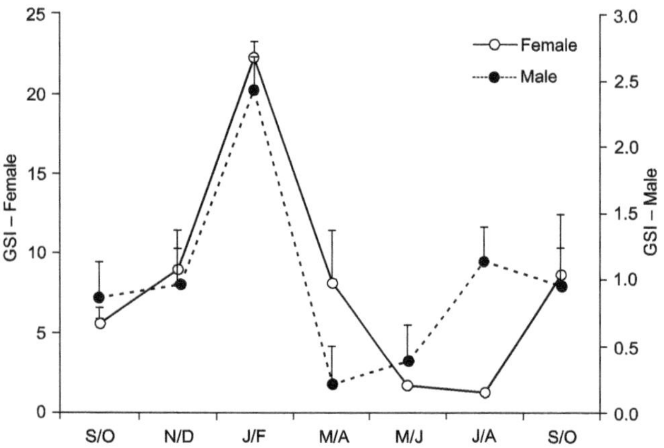

Figure 14. Bi-monthly changes in gonadosomatic index (means + standard error) for female and male of *Astyanax* aff. *bimaculatus* S/O, September-October; N/D, November-December; J/F, January-February; M/A, March-April; M/J, May-June; J/A, July-August.

The condition factor (K) had the highest values in September-December and in March-June and the lowest in January-February and July-August for females (Fig. 15). For males, the highest K was recorded in March-June and the lowest in January-February. The HSI had two peaks for both sexes, one in November-December and another less conspicuous in July-August. The highest values in November-December were followed by a decreasing in the following months with a slight increase in July-August for both sexes (Fig. 16).

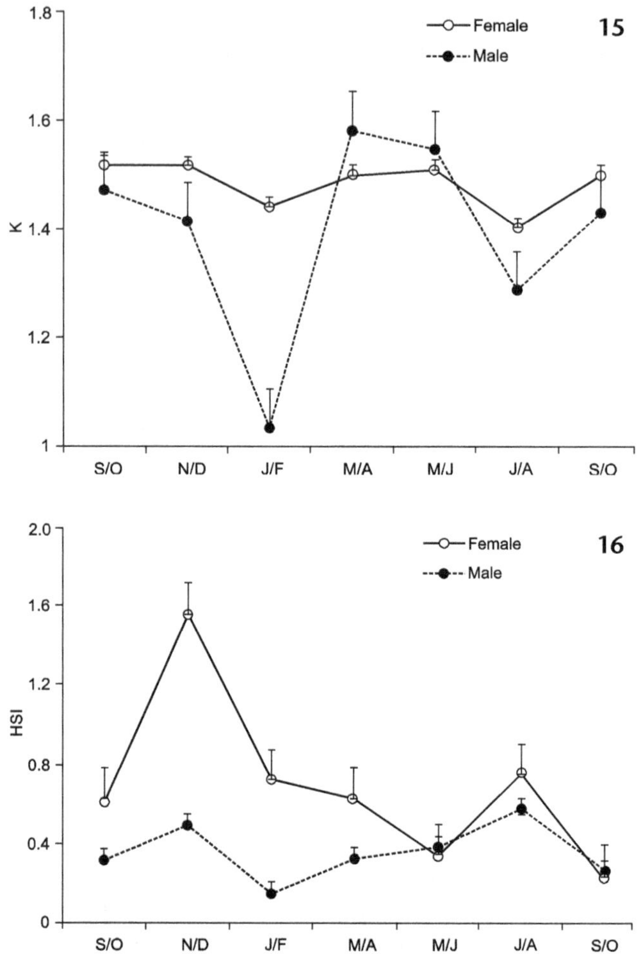

Figures 15–16. Bi-monthly changes in (15) condition factor (means + standard error) and (16) hepatosomatic index (means + standard error) for female and male of *Astyanax* aff. *bimaculatus*. S/O, September-October; N/D, November-December; J/F, January-February; M/A, March-April; M/J, May-June; J/A, July-August.

DISCUSSION

We found compelling evidence from macro and microscopy observation that *Astyanax* aff. *bimaculatus* is a batch spawner. Its small size, with small oocytes and long spawning season, are evidence that this species fits to an opportunistic strategy (sensu Winemiller and Rose 1992). Vazzoler and Menezes (1992) reported that large oocytes offer better conditions for larval development and survival, but small oocytes can be produced in greater numbers. Opportunist reproductive strategy is common in small-sized species, with early maturation, high mortality rates, multiple spawning, small eggs, low fecundity, rapid population turnover and capacity for rapid colonization (Lamouroux et al. 2002, Blanck et al. 2007).

The opportunist strategy is not universal among species of *Astyanax*. Winemiller (1989) classified *A. bimaculatus* and *Astyanax henseli* de Melo & Buckup, 2006 as having a seasonal reproductive strategy. Among the life history strategies proposed by Winemiller (1989), the reproductive characteristics presented by *A. henseli* are more similar to the seasonal strategy, i.e., late maturation, seasonal reproduction, intermediate or high fecundity and low investment in the offspring, with absence of parental care and reduced juvenile survival. In addition, synchronous oocyte development in two groups presented by *A. henseli* (Dala-Corte and Azevedo 2010) suggests that the species presents total spawning, that is, a single batch of oocytes is spawned within a reproductive period. On the other hand, Abelha and Goulart (2008) found a large reproductive period for *Astyanax paranae* Eigenmann, 1914 (October to April) in a small reservoir from state of Paraná, which is an indicative of opportunist strategy. Similarly, the reproductive period for *A.*

paranae in the stream population was characterized as long, with spawning extending from July to March (Veregue and Orsi 2003), which also did not fit to a seasonal reproductive strategy.

We observed ovaries in the regression phase in different months and this in an indication that *A.* aff. *bimaculatus* spawns in parcels (batch spawn). This type of spawning favours reduction of predation on offspring and competition among individuals for food and shelter (Suzuki et al. 2000, Hojo et al. 2004, Bailly et al. 2008). Spawning in parcel (batch spawning) has been described for *A. bimaculatus* (Agostinho et al. 1984), *A. scabripinnis* (Jenyns, 1842) (Barbieri 1992, Veloso-Júnior et al. 2009) and *A. fasciatus* (Cuvier, 1819) (Carvalho et al. 2009), whereas total spawning has been reported for *Astyanax schubarti* Britski, 1964 (Nomura 1975), A. fasciatus (Gurgel 2004) and *Astyanax lacustris* (Lütken, 1875) (Súarez et al. 2017). Batch spawning allows several spawning events during the same reproductive cycle. Consequently, different niches in space and time are occupied with different size classes in a variety of habitats. This leads to lower competition among adults for spawning sites and among larvae for available food sources (Ratton et al. 2003). Garutti (1989), studying populations of *A. lacustris* from the Paraná River Basin, suggested that the species has batch spawning and prolonged reproductive period in habitats such as streams and headwaters, and total spawning and short reproductive period in rivers with greater water volume, where individuals would be less exposed to abrupt changes. Fish can transition from total spawn to fractional spawn and vice-versa because of physiological or environmental alterations.

We observed that the spermatozoa of *A.* aff. *bimaculatus* had rounded heads, a characteristic of fish with external fertilisation (Grier 1981). The reduction in size of spermatogenic lineage cells throughout the development is a widely accepted general rule (Sprando and Russel 1988), and it was found in *A.* aff. *bimaculatus*. During the spermatogenesis, a reduction of 80–90% of the nuclear diameter of the spermatogenic cells was observed.

Astyanax aff. *bimaculatus* has an unbalanced sex ratio, with females outnumbering and reaching larger sizes than males. These results are in accordance with the majority of freshwater fish in the tropics (e.g., Duarte and Alcaraz 1989, Duarte et al. 2007, Gomes et al. 2011, 2015) with predominance of females in the largest size classes. Sex ratio is an important trait to estimate the reproductive biomass and total population fecundity, being also one of the most important drivers of the reproductive potential (Marshall et al. 2006). An imbalance in the sex ratio, particularly in adults, is relatively common in fish and is related to sex differences in growth, mortality and/or the energy costs of reproduction (Potts and Wootton 1984). Predominance of females in larger sizes favours higher fecundity (Nikolsky 1963, Shine 1990, Gross 2005) since large females have a larger peritoneal cavity and, thus, can lay a larger number of eggs. Although fecundity and the number of fertilized eggs increase with female body size, a general pattern in teleosts (Gross and Sargent 1985, Duarte and Alcaraz 1989), the ecological importance of the sex ratio is still not fully explained (Gross 2005). The smaller-sized male may be a consequence of selection for early male maturation and reproductive effort, which reduce male growth compared to that of females (Parker 1982, Endler 1983, Andersson 1994). Smaller-sized males were also observed for the congeneric *Astyanax fasciatus* in another reservoir in southeastern Brazil (Carvalho et al. 2009), which are in accordance with our findings.

The peak of reproductive activity indicated by GSI was January-February, followed by a less conspicuous period of reproductive activity in September-October. Studies on the reproductive biology of other species of *Astyanax* have shown that there are variations in the reproductive period, ranging from only two months (e.g., Godinho et al. 2010) to nine months (e.g., Veregue and Orsi 2003, Mazzoni et al. 2005). Overall, species of *Astyanax* have seasonal reproductive strategy, peaking in the rainy season between spring and summer (Dala-Corte and Azevedo 2010). Few studies with species of *Astyanax* reported individuals sexually active during the autumn or winter (Garutti 1989, Veregue and Orsi 2003). Gurgel (2004) found that the congeneric *A. fasciatus* has a long reproductive period, with the highest GSI in mid-summer, coinciding with the peaks in rainfall. These results are in accordance with our findings for *A.* aff. *bimaculatus*.

During the peak of GSI, we also observed decreases in the condition factor (K) and in the hepatosomatic index (HSI). The inverse relationship between both the HSI and K with the GSI suggests the mobilization of hepatic energy and body reserves to gonadal development during the spawning season, which is likely to be associated concomitantly with a decrease of feeding activity. Moreover, the highest K values between March and June for both sexes, after the peak of GSI, indicate a recover in body mass after the reproductive effort. The K factor has been used as a proxy of the spawning period, because in this period, the food intake may cease and K should reach the lowest values (Barbieri et al. 1996).

The HSI has been reported as a more accurate condition index to measure the energy reserves of fish compared with other indices (Dominguez-Petit and Saborido-Rey 2010, Alonso-Fernandez and Saborido-Rey 2012). In the present study, HSI had two peaks for both sexes, one in November-December and another less conspicuous in July-August. During the reproductive period individuals allocate less effort to food search, and consume reserves stored in the liver (capital breeders *sensu* McBride et al. 2015), resulting in a reduction in HSI values, particularly in females (Nikolsky 1963). These highest HSI values before the peaks of GSI are indications of the use of stored energy reserves in liver for gonadal maturation. These findings corroborate the hypothesis that changes in HSI are associated with the role of the liver in the reproductive activity (Alonso-Fernandez and Saborido-Rey 2012) and ovarian maturation. Other species of *Astyanax* had similar patterns of inverse relationship between GSI and HSI. In *Astyanax aeneus* (Günther, 1860), the highest HSI values were found during periods of reproductive inactivity and this was interpreted as an increase in the reserve materials stored in the liver for subsequent use in gamete production (Trujillo-Jiménez et al. 2013). *Astyanax henseli* had higher HSI before the breeding period and lower at the reproductive peak, suggesting greater use of liver reserves for vitellogenesis and gonadal maturation (Dala-Corte and Azevedo 2010, Trujillo-Jiménez et al. 2013), thus confirming the pattern of transference of energy from the liver to the gonads during the reproductive process

Opportunistic species have a higher resistance to environmental and anthropogenic alterations (Winemiller 1989). However, *A. bimaculatus* showed disruptions in their reproductive activity when present in environments with adverse physical-chemical conditions (Bailly et al. 2008, Vasconcelos et al. 2014). Disturbances such are dam construction, chemical pollution and habitat degradation that are frequent in rivers and reservoirs in developing countries can impair the reproductive success of opportunist species. Considering that environmental pressures vary according to the characteristics of each system, variations in the reproductive tactics of closely related species are expected (Dala-Corte and Azevedo 2010). The influence of damming tended to be with lower intensity on fish with opportunist reproductive strategy (Vasconcelos et al. 2014). Conversely, long-distance migratory species respond more markedly to spatiotemporal variations, indicating that the ecosystem dynamics exert greater effects on populations of these species (Bailly et al. 2008). Short-lived species with fast growth or reproductive compensation are expected to have survival advantages, and this may explain, at least in part the great success of *A.* aff. *bimaculatus* and other congeneric species to colonize successfully a large number of lotic and lentic aquatic environments in the Neotropical region.

ACKNOWLEDGEMENTS

The authors thank several undergraduate students from the Laboratório de Ecologia de Peixes, Universidade Federal Rural do Rio de Janeiro for helping with the fieldwork. This research was partially funded by Conselho Nacional de Desenvolvimento Científico e Tecnológico (Process 304954/2011-0) and by Fundação Carlos Chagas de Amparo à Pesquisas do Estado do Rio de Janeiro, CNE Process: E-26/201.248/2014, through the Grant "Cientista do Nosso Estado" for the first author. This research was conducted under SISBIO Collection of Species Permit number 10707 issued by ICMBio, Instituto Chico Mendes de Conservação da Biodiversidade. All animal handling procedures have been approved and are in accordance with practices approved by the institutional ethics committee (CEUA/Universidade Federal Rural do Rio de Janeiro/ICBS, Protocol 12179).

LITERATURE CITED

Abelha MCF, Goulart E (2008) Population structure, condition factor and reproductive period of *Astyanax paranae* (Eigenmann, 1914) (Osteichthyes: Characidae) in a small and old Brazilian reservoir. Brazilian Archives of Biology and Technology 51(3): 503–512.

Abelha MCF, Goulart E, Kashiwaqui EAL, Silva MR (2006) *Astyanax paranae* Eigenmann, 1914 (Characiformes: Characidae) in the Alagados Reservoir, Paraná, Brazil: diet composition and variation. Neotropical Ichthyology 4(3): 349–356.

Abrial E, Rabuffetti AP, Espínola LA, Amsler ML, Blettler MCM, Paira AR (2014) Influence of hydrological changes on the fish community in two lotic environments of the Middle Paraná floodplain, Argentina. Aquatic Ecology 48: 337–349.

Agostinho AA, Gomes LC, Veríssimo SV, Okada EK (2004) Flood regime, dam regulation and fish in the Upper Paraná River: effects on assemblage attributes, reproduction and recruitment. Reviews in Fish Biology and Fisheries 14: 11–19.

Agostinho CA, Molinari SL, Agostinho AA, Verani JR (1984) Reproductive cycle and size at first maturation of females of *Astyanax bimaculatus* in the Ivaí River, Paraná State. Revista Brasileira de Biologia 44(1): 31–36.

Alonso-Fernandez A, Saborido-Rey F (2012) Relationship between energy allocation and reproductive strategy in *Trisopterus luscus*. Journal of Experimental Marine Biology and Ecology 416: 8–16. https://doi.org/10.1016/j.jembe.2012.02.001

Andersson M (1994) Sexual selection. Monographs in behavior and ecology. Princeton, New Jersey, Princeton University Press, 624 pp.

Bailly D, Agostinho AA, Suzuki HI (2008) Influence of the flood regime on the reproduction of fish species with different reproductive strategies in the Cuiabá River, Upper Pantanal, Brazil. River Research and Applications 24(9): 1218–1229. https://doi.org/10.1002/rra.1147

Barbieri GM (1992) Biology of *Astyanax scabripinnis paranae* (Characiformes, Characidade) in the Fazzari Creek, São Carlos. São Paulo State. II. Quantitative aspects of the reproduction. Revista Brasileira de Biologia 52(1): 589–596.

Barbieri G, Hartz SM, Verani RJ (1996) The condition factor and gonadosomatic index as indicator of the spawning period for *Astyanax fasciatus* (Cuvier, 1819) in the Lobo Reservoir, São Paulo (Osteichthyes, Characidae). Iheringia, Serie Zoologia 81: 97–100.

Belova GV (2008) Oocyte morphology of several mesopelagic fishes in connection with their taxonomic status and habitat conditions. Russian Journal of Marine Biology 34(2): 110–117. https://doi.org/10.1134/S1063074008020041

Blanck A, Tedesco PA, Lamouroux N (2007) Relationships between life-history strategies of European freshwater fish species and their habitat preferences. Freshwater Biology 52: 843–859.

Bolger T, Connolly PL (1989) The selection of suitable índices for the measurement and analysis of fish condition. Journal of Fish Biology 34: 171–182. https://www.researchgate.net/publication/227832970

Branco CWC, Rocha MIA, Pinto GFS, Gômara GA, De Filippo R (2002) Limnological features of Funil Reservoir (RJ, Brazil) and indicator properties of rotifers and cladocerans of the zooplankton community. Lakes Reservoir: Research and Management 7: 87–92.

Breder CM, Rosen DE (1966) Modes of reproduction in fishes. Neptune City, T.F.H. Publications, 941 pp.

Brown-Peterson NJ, Wyanski DM, Saborido-Rey F, Macewicz BJ, Lowerre-Barbieri SK (2011) A standardized terminology for describing reproductive development in fishes. Marine and Coastal Fisheries: Dynamics, Management, and Ecosystem Science 3: 52–70. https://doi.org/10.1080/19425120.2011.555724

Carvalho PA, Paschoalini AL, Santos GB, Rizzo E, Bazzoli N (2009) Reproductive biology of *Astyanax fasciatus* (Pisces: Characiformes) in a reservoir in southeastern Brazil. Journal Application Ichthyology 25: 306–313.

Cerda JB, Calman G, Lafleur JRGJ, Limesand S (1996) Patterns of vitellogenesis and follicle maturational competence during the ovarian follicular cycle of *Fundulus heteroclitus*. General and Comparative Endocrinology 103(1): 24–35. https://doi.org/10.1006/gcen.1996.0090

Coward K, Bromage NR, Hibbitt O, Parrington J (2002) Gametogenesis, fertilization and egg activation in teleost fish. Reviews in Fish Biology and Fisheries 121: 33–58. https://doi.org/10.1023/A:1022613404123

Dala-Corte R, Azevedo MA (2010) Reproductive biology of *Astyanax henseli* (Teleostei, Characidae) in the upper reaches of the Sinos River, RS, Brasil. Iheringia, Série Zoologia 100: 259–266. https://doi.org/10.1590/S0073-47212010000300012

Defalco T, Capel B (2009) Gonad morphogenesis in vertebrates: divergent means to a convergent end. Annual Review of Cell and Developmental Biology 25: 457–482. https://doi.org/10.1146/annurev.cellbio.042308.13350

Dominguez-Petit R, Saborido-Rey F (2010) New bioenergetic perspective of European hake (*Merluccius merluccius* L.) reproductive ecology. Fisheries Research 104: 83–88. https://doi.org/10.1016/j.fishres.2009.09.002

Duarte CM, Alcaraz M (1989) To produce many small or few large eggs: a size-independent reproductive tactic of fish. Oecologia 80(3): 401–404.

Duarte S, Araújo FG, Sales A, Bazzoli N (2007) Morphology of gonads, maturity and spawning season of *Loricariichthys spixii* (Siluriformes, Loricariidae) in a subtropical reservoir. Brazilian Archives of Biology and Technology 50: 1019–1032.

Endler JA (1983) Natural and sexual selection on color patterns in poeciliid fishes. Environmental Biology of Fishes 9: 173–190.

Espínola L, Rabuffetti A, Abrial E, Amsler M, Blettler M, Paira A, Simes N, Santos LN (2016) Response of fish assemblage structure to changing flood and flow pulses in a large subtropical river. Marine and Freshwater Research 68(2): 319–330. https://doi.org/10.1071/MF15141.

Fink SV, Fink WL (1981) Interrelationships of the ostariophysan fishes (Teleostei). Zoological Journal of the Linnaean Society 72: 297–353.

Fishelson L, Gon O (2008) Comparative oogenesis in cardinalfishes (Apogonidae, Perciformes), with special focus on the adaptative structures of the egg envelopes. Environmental Biology of Fishes 81: 397–414.

Galvão GA, Silva ALB, Cardoso AS, Santos HS, Pereira PAN, Ribeiro LB (2016) Comparative gonadal histomorphometry of *Astyanax lacustris* (Lütken, 1875) and *Psellogrammus kennedyi* (Eigenmann, 1903) (Characiformes, Characidae) from a reservoir in Brazilian semiarid. Boletim do Instituto de Pesca 42: 734–748.

Garutti V (1989) Contribution to the reproductive knowledge of *Astyanax bimaculatus* (Ostariophysi, Characidae), in waters from the Paraná watershed. Revista Brasileira de Biologia 49(2): 489–495.

Gennari-Filho O, Braga FMS (1996) Fecundity and spawning of *Astyanax bimaculatus* and *A. schubarti* (Characidae, Tetragonopterinae) in the Barra Bonita Reservoir, Piracaba River (SP). Revista UNIMAR 18(2): 241–254.

Giosa M, Czerniejewski P, Rybczyk A (2014) Seasonal changes in condition factor and weight-length relationship of invasive *Carassius gibelio* (Bloch, 1782) from Leszczynskie Lakeland, Poland. Advances in Zoology Article ID 678763, 7 pages. https://doi.org/10.1155/2014/678763

Godinho AL, Lamas IR, Godinho HP (2010) Reproductive ecology of Brazilian freshwater fishes. Environmental Biology of Fishes 87: 143–62.

Gomes ID, Araújo FG, Uehara W, Sales A (2011) Reproductive biology of the armoured catfish *Loricariichthys castaneus* (Castelnau, 1855) in Lajes reservoir, southeastern Brazil. Journal of Applied Ichthyology 27: 1322–1331.

Gomes ID, Araújo FG, Nascimento AA, Sales A (2015) Equilibrium reproductive strategy of the armored catfish *Hypostomus auroguttatus* (Siluriformes, Loricariidae) in a tropical river in Southeastern Brazil. Environmental Biology of Fishes 98: 249–260.

Grier HJ (1981) Cellular organization of the testis and spermatogenesis in fishes. Integrative and Comparative Biology 21: 345–357. https://doi.org/10.1093/icb/21.2.345

Gross MR (2005) The evolution of parental care. The Quarterly Review of Biology 80: 37–45.

Gross MR, Sargent RC (1985) The evolution of male and female parental care in fishes. American Zoologist 25(3): 807–822. https://www.jstor.org/stable/3883036

Gurgel HCB (2004) Populational structure and reproduction season of *Astyanax fasciatus* (Cuvier) (Characidae:Tetragonopterinae) in the Ceará Mirim River, Poço Branco, Rio Grande do Norte, Brazil. Revista Brasileira de Zoologia 21(1): 131–135. https://doi.org/10.1590/S0101-81752004000100022

Hirt LM, Araya PR, Flores SA (2011) Population structure, reproductive biology and feeding of *Astyanax fasciatus* (Cuvier, 1819) in an Upper Paraná River tributary, Misiones, Argentina. Acta Limnologica Brasiliensia 23(1): 1–12.

Hismayasari IB, Marhendra APW, Saidin SR, Supriyadi SD (2015) Gonadosomatic index (GSI), hepatosomatic index (HSI) and proportion of oocytes stadia as an indicator of rainbowfish *Melanotaenia boesemani* spawning season. International Journal of Fisheries and Aquatic Studies 2(5): 359–362.

Hojo RES, Santos GB, Bazzoli N (2004) Reproductive biology of *Moenkhausia intermeida* (Eigenmann) (Pisces, Characiformes) in Itumbiara Reservoir, Goiás, Brazil. Revista Brasileira de Zoologia 21(3): 519–524. https://doi.org/10.1590/S0101-81752004000300015

Hokanson KEF (1977) Temperature requirements of some percids and adaptations to the seasonal temperature cycle. Journal of the Fisheries Research Board of Canada 34(10): 1524–1550. https://doi.org/10.1139/f77-217

Jamieson BGM (2009) Reproductive biology and phylogeny of fishes (Agnathans and Bony Fishes). Enfield, Science Publishers, 788 pp.

Jayasankar P, Alagarswami K (1994) Histological and histochemical observations on the oocytes in the sand whiting *Sillago sihama* (Forsskal). Proceedings of the Indian Natural Science Academy, B60(2): 173–182.

Jonsson N (1991) Influence of water flow, water temperature and light on fish migration in rivers. Nordic Journal of Freshwater Research 66: 20–35.

Lamouroux N, Poff NL, Angermeier PL (2002) Inter-continental convergence of stream fish community traits along geomorphic and hydraulic gradients. Ecology 83: 1792–1807.

Le Cren ED (1951) The length-weight relationship and seasonal cycle gonad weight and condition in the perch, *Perca fluviatilis*. Journal of Animal Ecology 2092: 201–219.

Lizama MAP, Ambrosio AM (2002) Condition factor in nine species of fish of the Characidae family in the upper Paraná River floodplain, Brazil. Brazilian Journal of Biology. 62: 113–124. https://doi.org/10.1590/S1519-69842002000100014

Lowerre-Barbieri SK, Brown-Peterson NJ, Murua H, Tomkiewicz J, Wyanski D, Saborido-Rey F (2011) Emerging issues and methodological advances in fisheries reproductive biology. Marine and Coastal Fisheries 3: 32–51.

Marengo JÁ, Alves LM (2005) Hydrological trends in the Paraíba do Sul River basin. Revista Brasileira de Meteorologia 20: 215–226.

Marshall CT, Needle CL, Thorsen A, Kjesbu OS, Yaragina NA (2006) Systematic bias in estimates of reproductive potential of an Atlantic cod (*Gadus morhua*) stock: implications for stock-recruit theory and management. Canadian Journal of Fisheries and Aquatic Sciences 63: 980–994.

Martins YS, Arantes FP, Sato Y, Santos JE, Rizzo E, Bazzoli N (2012) Comparative analysis of gonadal morphology in six fish species of the Incertae Sedis genera in Characidae of occurrence in the São Francisco River Basin, Brazil. Acta Zoologica 93: 48–56.

Mazzoni R, Iglesias-Rios R (2004) Longitudinal segregation of *Astyanax janeiroensis* in the Rio Ubatiba: A neotropical stream of south-east Brazil. Ecology of Freshwater Fish 13: 231–234.

Mazzoni R, Mendonça RS, Caramaschi EP (2005) Reproductive biology of *Astyanax janeiroensis* (Osteichthyes, Characidae) from the Ubatiba River, Maricá, RJ, Brazil. Brazilian Journal of Biology 65(4): 643–649.

McBride RS, Somarakis S, Fitzhugh GR, Albert A, Yaragina NA, Wuenschel MJ, Basilone G (2015) Energy acquisition and allocation to egg production in relation to fish reproductive strategies. Fish and Fisheries 16: 23–57. https://doi.org/10.1111/faf.12043

Medeiros ESF, Maltchik L (2001) Fish assemblage stability in an intermittently flowing stream from the Brazilian semiarid region. Austral Ecology 26(2):156–164.

Myers BM, Fredenburgh JL, Grizzle WE (2008) Carbohydrates. In: Bancroft JD, Gamble M (Eds) Theory and practice of histological techniques. Philadelphia, Elsevier, 6th ed., 161–187.

Nelson JS (2006) Fishes of the world. New Jersey, John Wiley & Sons, 4th ed., 601 pp.

Nikolsky GV (1963) The ecology of fishes. Academic Press, London, 352 pp.

Nomura H (1975) Fecundidade, maturação sexual e índice gônado-somático de lambaris do gênero *Astyanax* Baird e Girard, 1854 (Osteichthyes, Characidae), relacionados com fatores ambientais. Revista Brasileira de Biologia 35(4): 775–798.

Nuñez J, Duponchelle F (2009) Towards a universal scale to assess sexual maturation and related life history traits in oviparous teleost fishes. Fish Physiology and Biochemistry 35:167–180.

Orsi ML, Carvalho ED, Foresti F (2004) Biologia populacional de *Astyanax altiparanae* Garruti & Britski (Teleostei: Characidae) do médio rio Paranapanema, Paraná, Brasil. Revista Brasileira de Zoologia 21(2): 207–218. https://doi.org/10.1590/S0101-81752004000200008

Parenti LR, Grier HJ (2004) Evolution and phylogeny of gonad morphology in bony fishes. Integrate Comparative Biology 44: 333–348.

Parker GA (1982) Why are there so many tiny sperm? Sperm competition and the maintenance of two sexes. Journal of Theoretical Biology 96: 281–294.

Potts GW, Wootton RJ (1984) Fish reproduction strategies and tactics. London, Academic Press, 410 pp.

Quagio-Grassiotto I, Wildner DD, Ishiba R (2013) Gametogênese de peixes: aspectos relevantes para o manejo reprodutivo. Revista Brasileira de Reprodução Animal 37(2): 181–191.

Ratton TF, Bazzoli N, Santos GB (2003) Reproductive biology of *Apareiodon affinis* (Pisces: Parodontidae) in the Furnas Reservoir, Minas Gerais, Brazil. Journal of Applied Ichthyology 19(6): 387–390.

Reis RE, Kullander SO, Ferraris-Jr CJ (2003) Check list of the freshwater fishes of South and Central America. Porto Alegre, Edipucrs, 742 pp.

Rizzo E, Sato Y, Barreto BP, Godinho HP (2002) Adhesiveness and surface patterns of eggs in neotropical freshwater teleosts. Journal of Fish Biology 61(3): 615–632.

Schneider CA, Rasband WS, Eliceir KW (2012) NIH Image to ImageJ: 25 years of image analysis. Nature Methods 9(7): 671–675. https://doi.org/10.1038/nmeth.2089

Shine R (1990) Proximate determinants of sexual differences in body size. The American Naturalist 135: 278–283.

Silva DA, Pessoa EKR, Costa SAGL, Chellappa NT, Chellappa S (2012) Reproductive ecology of *Astyanax lacustris* (Osteichthyes: Characidae) in the Piató Lagoon, Assú, Rio Grande do Norte, Brazil. Biota Amazonica 2(2): 54–61.

Silva EP, Duarte MRN, Medeiros ESF (2018) Length weight relationship of two fish species from a dryland intermittent river in northeastern Brazil. Neotropical Biological Conservation 13(1): 90–93.

Sprando RL, Russel LD (1988) Spermiogenesis in the bluegill (*Lepomis macrochirus*): a study of cytoplasmic events including cell volume changes and cytoplasmic elimination. Journal of Morphology 198: 165–177.

Stearns SC (1992) The evolution of life histories. Oxford, University Press, 264 pp.

Súarez YR, Silva EA, Viana LF (2017) Reproductive biology of *Astyanax lacustris* (Characiformes: Characidae) in the southern Pantanal floodplain, upper Paraguay River basin, Brazil. Environmental Biology of Fishes 100: 775–783.

Suzuki HI, Agostinho AA, Winemiller KO (2000) Relationship between oocyte morphology and reproductive strategy in loricariid catfishes of the Paraná River, Brazil. Journal of Fish Biology 57: 791–807.

Trujillo-Jiménez P, Sedeño-Díaz JE, Camargo JA, López-López E (2013) Reproductive traits and population structure of *Astyanax aeneus* (Characiformes: Characidae) from a subtropical river in Mexico. Revista de Biologia Tropical 61(2): 769–786.

Vasconcelos LP, Alves DC, Gomes LC (2014) Fish reproductive guilds downstream of dams. Journal of Fish Biology 85(5): 1489–1506. https://doi.org/10.1111/jfb.12501

Vazzoler AEAM (1996) Biologia da reprodução de peixes teleósteos: Teoria e prática. São Paulo, SBI, 169 pp.

Vazzoler AEAM, Menezes NA (1992) Synthesis of knowledge on the behavior of South American Characiformes (Teleostei, Ostariophysi). Revista Brasileira de Biologia 52: 627–640.

Veloso-Júnior VC, Guimarães-Cruz RJ, Barros MDM, Barata RSL, Santos JE (2009) Reproduction of the lambari *Astyanax scabripinnis* (Jenyns, 1842) (Pisces: Characidae) in a small stream in Southeastern Brazil. Journal of Applied Ichthyology 25(3): 314–320.

Veregue AML, Orsi ML (2003). Reproductive biology of *Astyanax scabripinnis paranae* (Eigenmann) (Ostheichthyes, Characidae), in the Marrecas Creek, Tibagi River, Paraná. Revista Brasileira de Zoologia 20(1): 97–105. https://doi.org/10.1590/S0101-81752003000100012

Weber AA, Arantes FB, Sato Y, Rizzo E, Bazzoli N (2012) Oocyte adhesiveness and embryonic development of *Astyanax bimaculatus* (Linnaeus, 1758) (Pisces: Characidae). Zygote 21: 198–202.

West G (1990) Methods of assessing ovarian development in fishes: a review. Australian Journal of Marine and Freshwater Research 41: 199–222.

Winemiller KO (1989) Patterns of variation in life history among South American fishes in seasonal environments. Oecologia 81: 225–241.

Winemiller KO, Rose KA (1992) Patterns of life-history diversification in North American fishes: implications for population regulation. Canadian Journal of Fisheries and Aquatic Sciences 49: 2196–2218.

Author Contributions: FGA and IDG conceived and designed the study, AAN and IDG carried out the fieldwork, AS, AAN and BACO performed the laboratorial techniques, and all authors wrote the paper.

Competing Interests: The authors declare that they have no conflict of interest.

10

Composition and structure of the frugivorous butterfly community (Lepidoptera: Nymphalidae) at the Serra Azul State Park (PESA), Mato Grosso, Brazil

Wesley O. de Sousa[1], **Lincey E. Sousa**[2], **Fátima R. J. da Silva**[1], **Wildio I. da Graça Santos**[3], **Rodrigo Aranda**[1]

[1]*Departamento de Biologia, Universidade Federal de Mato Grosso, Campus Universitário de Rondonópolis. Avenida dos Estudantes 5055, Cidade Universitária, 78736-900 Rondonópolis, MT, Brazil.*
[2]*Instituto Federal de Mato Grosso. Rodovia BR 158, Radial José Maurício Zampa, Industrial, 78600-000 Barra do Garças, MT, Brazil.*
[3]*Departamento de Zoologia, Setor de Ciências Biológicas, Universidade Federal do Paraná. Jardim das Américas, Caixa Postal 19020, 81531-980 Curitiba, PR, Brazil.*
Corresponding author. Wesley Oliveira de Sousa (entomoi@hotmail.com)

ABSTRACT. Due to the important ecological role of Nymphalidae as consumers of fermented fruits, excrement, exudates of decomposed plants and animals, as well as bioindicators of environmental quality that attract the interest of the general public, this study aimed to characterize the fauna of frugivorous butterflies at the Serra Azul State Park – PESA (Mato Grosso, Brazil) with regards to composition, taxonomic richness and spatial distribution. Collections were carried out in 2014 from six 250 m plots in a RAPELD module (5x5 km). We defined five sampling points in each plot and placed a Van Someren-Rydon trap (VSR), with a bait made from banana fermented in cane juice, positioned 1 m above the ground for 24 hours. We collected 204 specimens of Nymphalidae from five subfamilies, 22 genera, and 40 species. Satyrinae was the most abundant and species-rich subfamily. The community structure varied spatially, with the gallery forest presenting the greatest richness, diversity, and equitability, while 'cerrado ralo' presented higher abundance and less diversity and equitability. Therefore, according to the pattern of abundance in the PESA, species were best adapted to the log series model. *Yphthimoides renata* (Stoll, 1780) (Satyrinae), *Hamadryas feronia* (Linnaeus, 1758), and *Callicore sorana* (Godart, 1824) (Biblidinae) were the most abundant species, 19 species were singletons, and eight were doubletons. Richness estimators showed that 63% of all species were sampled. Our results indicate that the frugivorous butterfly community is structured according to phytophysiognomies, and the dominance of species that are tolerant to disturbances indicates landscape fragmentation within the PESA, which interferes with species substitution pattern and their abundances in different environments of the park.

KEY WORDS. Cerrado, conservation, inventory, Nymphalidae, Phytophysiognomy.

INTRODUCTION

Research about conservation monitoring and planning is most effective when supported by prior knowledge about the distribution of organisms and patterns of occurrence within an area. Such information is collected through biodiversity inventories, which provide important information for restoration, composition maintenance, and structure and function of biological communities (Bellaver et al. 2012, Martins et al. 2017a, Pérez et al. 2017) in different regions. In the Brazilian Cerrado, documenting and understanding the fauna is an urgent task, since habitat loss and fragmentation has rapidly increased during the past few decades in central Brazil, resulting in more endangered species and a loss of species richness at the local and regional scales (Brannstrom et al. 2008, Klink and Machado 2005, Santos et al. 2017).

Arthropods are useful ecological indicators due to their great diversity and abundance, ease of collection, and rapid responses to environmental changes (Leivas and Fischer 2008, Lewinsohn et al. 2005, Marques et al. 2010, 2016, Soares et al. 2012). Among the arthropods, butterflies are the most taxonomically well known in the Neotropical region (Bonebrake et al. 2010).

Furthermore, frugivorous Nymphalidae butterflies stand out in environmental monitoring and evaluation studies due to their biological and ecological characteristics, since they depend on specific micro-habitats and adequate resources to survive (Martins et al. 2017b). Thus, changes in habitat quality caused by deforestation, fire, and global climate change affect butterfly populations. The short life cycles of frugivorous butterflies also make them useful organisms for biomonitoring (Uehara-Prado and Ribeiro 2012, Pérez et al. 2017, Pinheiro et al. 2008).

Using frugivorous butterflies in ecological studies has methodological and sampling advantages, since simple traps, which are easily made and maintained, are used in field collections (Pinheiro et al. 2008). Furthermore, these butterflies are considered flagships by the scientific community, representing the environmental cause and capturing the public interest, as well as demonstrating the importance of species conservation in the ecosystems where humans live (Bonebrake et al. 2010, DeVries and Walla 2001). However, relevant information, such as local and regional inventories, is still needed to generate knowledge about the richness, abundance, and distribution, as well as data about ecological relationships in communities, such as host associations or responses to abiotic components in a region (DeVries and Walla 2001).

The state of Mato Grosso is potentially highly biodiverse because three of the main Brazilian biomes are present within its borders: Amazon tropical rainforest, Cerrado, and Pantanal. However, the frugivorous butterfly fauna in Brazil is still poorly contemplated in most inventories (Santos et al. 2008, Silva et al. 2015), especially in the central western and eastern regions of Mato Grosso (Queiroz-Santos et al. 2016).

Furthermore, the ecological importance and status of frugivorous butterflies as bioindicators were so far poorly considered in studies carried out in the region (Uehara-Prado et al. 2007, 2009). Knowledge about Mato Grosso butterflies is fragmentary and information is scattered across various sources (Santos et al. 2008, Queiroz-Santos et al. 2016).

In this context, this study aims to increase knowledge and fuel the discussion about the importance of services provided by Serra Azul State Park (PESA) in preservation and public use by maintaining the ecological functions and high biological diversity of the Cerrado. In addition, there is a lack of knowledge about the Cerrado biodiversity, which highlights the importance of understanding the biology and taxonomy of terrestrial invertebrates, especially for groups of poorly studied arthropods, helping future studies that aim to select criteria and biological models for conservation of this biome. Herein, we (1) inventoried the frugivorous butterfly fauna at PESA, (2) described the ecological metrics of the communities, and (3) analyzed the community composition in relation to the phytophysiognomies in the study area.

MATERIAL AND METHODS

The study area is located in the Serra Azul State Park (PESA) in the municipality of Barra do Garças, eastern region of the state of Mato Grosso (15°51'S, 51°16'W), with an area of approximately 11,000 ha and an altitude between 350 m and 750 m above sea level (Santos et al. 2017) (Fig. 1). The collections were carried out during January 2014 from a RAPELD module (Magnusson et al. 2005) located inside the PESA. The module comprises two five-kilometer lines, each one divided into five plots consisting of transects of 250 m in length, separated by about 1 km, totalizing 10 plots. Six of these plots were used in this study (Table 1).

Table 1. Geographic coordinates and phytophysiognomies of the plots in the RAPELD module installed in the Serra Azul State Park (PESA), Barra do Garças, Mato Grosso, Brazil.

Plots	Coordinates	Phytophysiognomy
1	15°50'58,7"S; 52°15'47"W	Cerrado sensu stricto
3	15°50'22,2"S; 52°14'49"W	Gallery forest
5	15°49'42,5"S; 52°13'49"W	Cerrado ralo – open savanna
6	15°51'07,6"S; 52°14'49"W	Deciduous forest
8	15°50'33,5"S; 52°13'56"W	Cerrado sensu stricto
10	15°49'53,8"S; 52°13'05"W	Cerrado ralo – open savanna

The PESA plots were characterized according to their vegetation type (phytophysiognomy) following the protocol from the Research Program on Biodiversity (PPBio) with standardized adaptations by the ComCerrado Network. Of our six plots, four presented Cerrado phytophysiognomies, Cerrado sensu stricto, and 'cerrado ralo' (open savannah with sparse trees), with different tree densities (356 to 1,463 individuals (Δ) Ha^{-1}). The other two plots were characterized as gallery forest (684 Δ Ha^{-1}) and deciduous forest (586 Δ Ha^{-1}) (Table 1). In the Cerrado plots, vegetation height varied from 1.0 to 13 m, while in the gallery and deciduous forests vegetation height varied between 2.2 and 25 m. The most common phytophysiognomies in the PESA module were low density woody elements and the species distribution pattern was probably related to the substrata where they develop (Maryland S. Lacerda, unpublished data).

For the collections, we defined five sampling points (100 m^2) in each plot spaced 50 m apart from each other. At each sampling point we installed a Van Someren-Rydon (VSR) trap (Rydon 1964), made up of a cylindrical thin-film cone (black tulle) (110 cm high by 35 cm diameter) closed at the top with thin white fabric and a 20 cm bottom opening, where butterflies can enter. A total of 30 traps were installed in the vegetation, approximately one meter above the ground. Each trap contained a bait made of banana fermented in cane juice, which remained in the field for 24 hours. Total sampling effort was 720 hours.

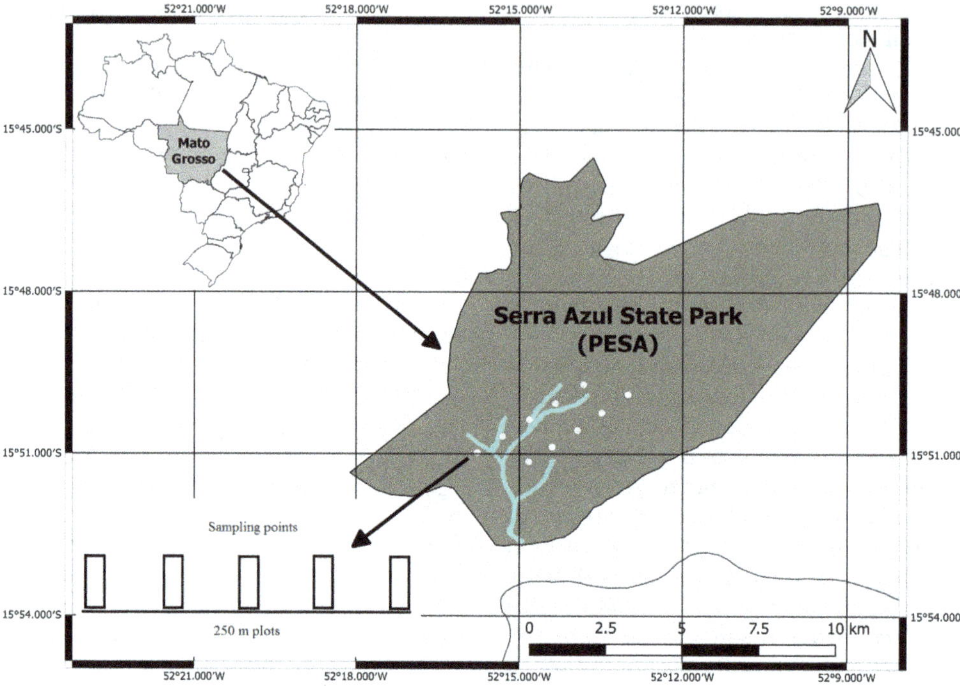

Figure 1. Location of the Serra Azul State Park (PESA), Barra do Garças, Mato Grosso, Brazil. Study area showing the RAPELD module transects with delimitation of RAPELD module with five sampling points (100 m^2) spaced 50 m apart from each other.

All specimens were individually wrapped in silk paper envelopes, labeled, and mounted on a wooden stretcher with entomological pins. Specimens were then dried in an oven at 40°C for three days, labeled, and deposited at the Entomological Reference Collection of the Departmento de Biologia, Universidade Federal de Mato Grosso, Campus Universitário de Rondonópolis (DBFR). Identifications were made based on specific literature (Lamas 2004, Uehara-Prado et al. 2004, Freitas et al. 2012, Seraphim et al. 2014) and confirmed by the fifth author and Dr Amabílio José A. de Camargo (EMBRAPA Cerrado, Brasília, DF), specialists in Lepidoptera.

We calculated the absolute richness and relative frequency of each species in each phytophysiognomy. To describe the community of frugivorous butterflies in the sampled areas, we used the diversity indices of Shannon-Wiener (H'), Pielou equitability (J), and Chao species richness estimator, as well as a rarefaction curve. Shannon-Wiener index was compared using t-test. The pattern of species abundance was tested in relation to the four predictive models (broken-stick, log normal, log series, and geometric). To verify if species composition differed with the phytophysiognomies, we performed a non-metrical dimensional scaling (NMDS – Bray Curtis similarity index). Similarity percentage (SIMPER) was used to assess which taxa were most responsible for the similarity between samples. Seriation test was used to order the species in the localities, presuming the faunal gradient for reorganizing the data matrix when the presences of the species are concentrated along the diagonal gradient, with the 'Monte Carlo' simulation. All analyses were performed in the Past® program and Sigmaplot® was used for graphic construction.

RESULTS

We obtained a total of 204 individuals of Nymphalidae, distributed in five subfamilies, 22 genera, and 40 species. Satyrinae presented the highest abundance, number of genera, and number of species, and Biblidinae presented the second largest abundance. The most dominant species were *Yphthimoides renata* (Stoll, 1780) (N = 77, 37.75%), *Hamadryas feronia* (Linnaeus, 1758) (N = 30, 14.71%), and *Callicore sorana* (Godart, 1824) (N = 10, 4.9%) (Table 2).

The species abundance pattern was best fit to the log series model (alpha = 14.9, $\chi^2 = 111$, $p < 0.001$; Broken stick: $\chi^2 = 185.4$, p =0.003; Geometric: $\chi^2 = 541.8$, p =0.009; log-normal distribution: $\chi^2 = 2.25$, p =0.32) (Fig. 2). The log-series model explains the presence of a few dominant species and a large number of rare species, influenced by the high abundance of *Y. renata* and *H. feronia*. Of the least abundant species, 19 species were singletons and eight doubletons, representing about 67% of the total observed richness. The cerrado ralo presented the highest abundance of Nymphalidae (N = 92), followed by the Cerrado sensu stricto (N = 61), gallery forest (N = 30), and deciduous forest (N = 21) (Table 2).

Table 2. Abundance (N) and relative frequency (%) of the subfamilies and species of Nymphalidae sampled from the four phytophysiognomies in the Serra Azul State Park (PESA), Barra do Garças, Mato Grosso, Brazil.

Nymphalidae	Cerrado sensu stricto		Gallery forest		Deciduous forest		Cerrado ralo		Total	
	(N)	(%)	(N)	(%)	(N)	(%)	(N)	(%)	(N)	(%)
Biblidinae	33	16.17	4	1.96	2	0.98	13	6.37	52	25.49
Callicore sorana (Godart, 1824)	8	3.92	0	0.00	2	0.98	0	0.00	10	4.90
Catonephele acontius (Linnaeus, 1771)	1	0.49	1	0.49	0	0.00	0	0.00	2	0.98
Hamadryas amphinome (Linnaeus,1767)	0	0.00	0	0.00	0	0.00	1	0.49	1	0.49
Hamadryas februa (Hubner, 1823)	3	1.47	0	0.00	0	0.00	2	0.98	5	2.45
Hamadryas feronia (Linnaeus, 1758)	21	10.29	0	0.00	0	0.00	9	4.41	30	14.71
Nica flavilla (Godart, 1824)	0	0.00	3	1.47	0	0.00	0	0.00	3	1.47
Temenis laothoe (Cramer, 1777)	0	0.00	0	0.00	0	0.00	1	0.49	1	0.49
Charaxinae	5	2.45	6	2.94	2	0.98	2	0.98	15	7.35
Archaeoprepona demophon (Linnaeus, 1758)	0	0.00	1	0.49	0	0.00	0	0.00	1	0.49
Archaeoprepona sp.	0	0.00	1	0.49	0	0.00	0	0.00	1	0.49
Archeoprepona amphimachus (Fabricius, 1775)	0	0.00	1	0.49	1	0.49	0	0.00	2	0.98
Memphis moruus (Fabricius, 1775)	1	0.49	1	0.49	0	0.00	0	0.00	2	0.98
Prepona laertes (Hübner, 1811)	0	0.00	1	0.49	0	0.00	1	0.49	2	0.98
Prepona pheridamas (Cramer, 1777)	0	0.00	1	0.49	1	0.49	0	0.00	2	0.98
Siderone galanthis (Cramer, 1775)	4	1.96	0	0.00	0	0.00	1	0.49	5	2.45
Heliconiinae	0	0.00	1	0.49	0	0.00	0	0.00	1	0.49
Heliconiinae sp.	0	0.00	1	0.49	0	0.00	0	0.00	1	0.49
Nymphalinae	0	0.00	2	0.98	0	0.00	0	0.00	2	0.98
Colobura dirce (Linnaeus, 1758)	0	0.00	2	0.98	0	0.00	0	0.00	2	0.98
Satyrinae	23	11.27	17	8.33	17	8.33	77	37.75	134	65.68
Amphidecta calliomma (Felder &Felder, 1862)	0	0.00	0	0.00	1	0.49	0	0.00	1	0.49
Amphidecta reynoldsi (Sharpe, 1890)	0	0.00	0	0.00	1	0.49	0	0.00	1	0.49
Caligo brasiliensis (Felder 1862)	1	0.49	0	0.00	0	0.00	0	0.00	1	0.49
Chloreuptychia arnaca (Fabricius, 1776)	0	0.00	3	1.47	0	0.00	0	0.00	3	1.47
Cissia myncea (Cramer, 1780)	0	0.00	4	1.96	2	0.98	1	0.49	7	3.43
Cissia sp.	0	0.00	2	0.98	0	0.00	0	0.00	2	0.98
Hermeuptychia hermes (Fabricius, 1775)	3	1.47	0	0.00	1	0.49	0	0.00	4	1.96
Morpho helenor achillides Felder & Felder 1867	0	0.00	1	0.49	0	0.00	0	0.00	1	0.49
Opsiphanes invirae (Hübner, 1808)	4	1.96	0	0.00	0	0.00	2	0.98	6	2.94
Paryphthimoides undulata (Butler, 1867)	1	0.49	0	0.00	0	0.00	0	0.00	1	0.49
Posttaygetis penelea (Cramer, 1777)	0	0.00	0	0.00	1	0.49	0	0.00	1	0.49
Satyrini sp. 1	0	0.00	1	0.49	0	0.00	0	0.00	1	0.49
Satyrini sp. 2	0	0.00	1	0.49	0	0.00	0	0.00	1	0.49
Satyrini sp. 3	0	0.00	0	0.00	0	0.00	1	0.49	1	0.49
Taygetina oreba (Butler, 1870)	0	0.00	0	0.00	2	0.98	2	0.98	4	1.96
Taygetis acuta Wetmer, 1910	0	0.00	1	0.49	0	0.00	1	0.49	2	0.98
Taygetis cleopatra Felder & Felder, 1867	0	0.00	1	0.49	0	0.00	0	0.00	1	0.49
Taygetis mermeria (Cremer, 1776)	0	0.00	0	0.00	4	1.96	1	0.49	5	2.45
Taygetis rufomarginata Staudinger, 1888	0	0.00	1	0.49	0	0.00	0	0.00	1	0.49
Taygetis sp.	0	0.00	1	0.49	0	0.00	0	0.00	1	0.49
Taygetis virgilia (Cramer, 1776)	0	0.00	0	0.00	0	0.00	1	0.49	1	0.49
Yphthimoides pacta (Weymer, 1911)	1	0.49	0	0.00	0	0.00	6	2.94	7	3.43
Yphthimoides renata (Stoll, 1780)	12	5.88	1	0.49	5	2.45	59	28.92	77	37.75
Yphthimoides sp.	0	0.00	0	0.00	0	0.00	1	0.49	1	0.49
Total	61	30.00	30	15.00	21	10.00	92	45.00	204	100.00

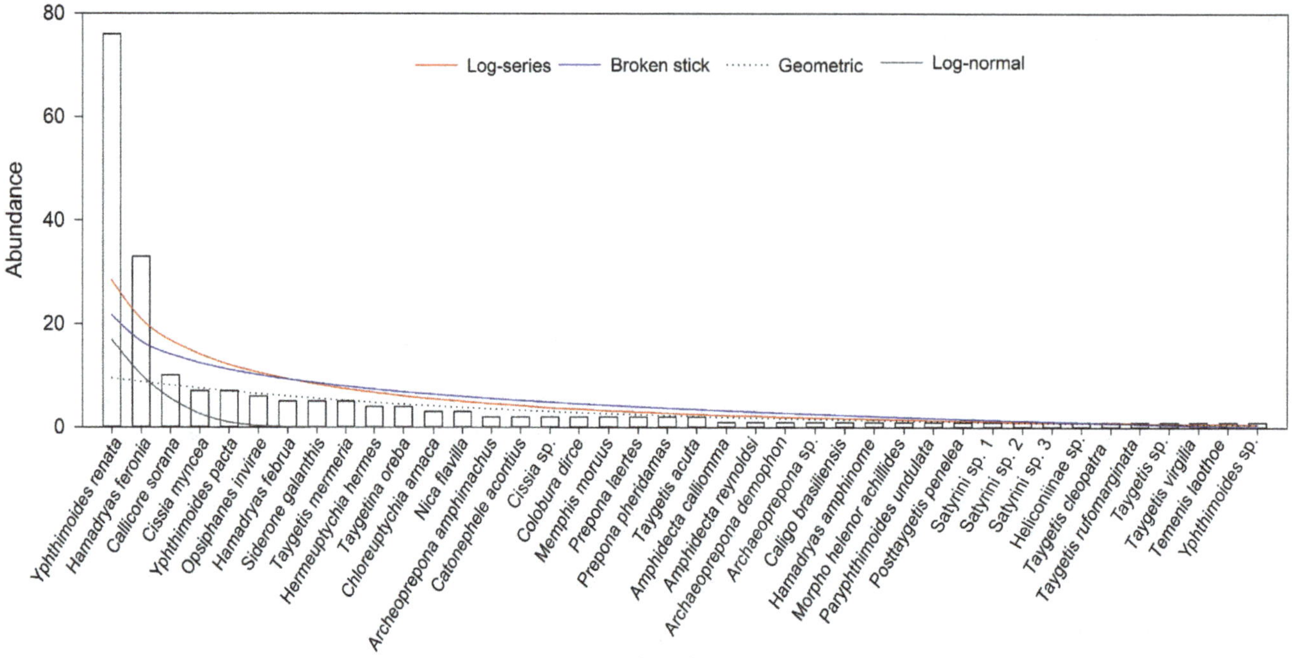

Figure 2. Abundance of Nymphalidae species sampled at the Serra Azul State Park (PESA). Barra do Garças, Mato Grosso, Brazil. Abundance model fits to log serial distribution (alpha = 14.9, $\chi^2 = 111$, $p < 0.05$).

According to the Chao2 richness estimator we expected to find 65 (± 15) species in the sampled area. However, only ~63% of the estimated species richness was sampled (82% to 51% considering the SD of the estimator) (Fig. 3). The richest phytophysiognomy was the gallery forest, followed by cerrado ralo, Cerrado sensu stricto, and deciduous forest (Table 3). The diversity and equitability index showed that the gallery forest was the most diverse and equitable, followed by the deciduous forest, Cerrado sensu stricto, and cerrado ralo, being statistically significant between Cerrado sensu stricto and gallery forest and gallery forest and deciduous forest (Table 3). Among the species sampled, 16 of them were responsible for more than 80% of the similarity between the areas, and four of these were responsible for more than 50% of the similarity, with *Y. renata* accounting for

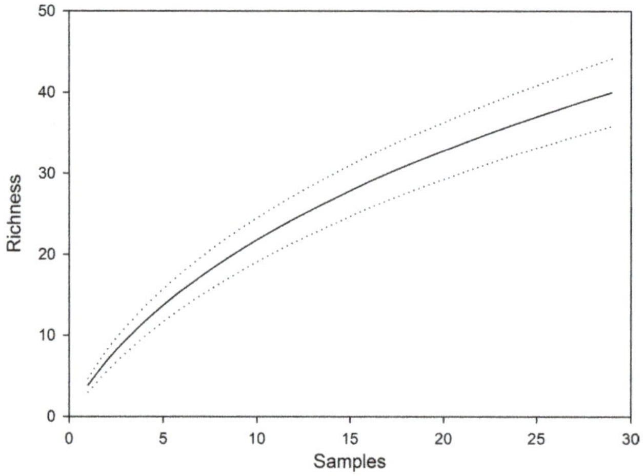

Figure 3. Rarefaction curve (± SD) of Nymphalidae species richness sampled at the Serra Azul State Park (PESA), Barra do Garças, Mato Grosso, Brazil.

Table 3. Ecological metrics of Nymphalidae in phytophysiognomies of the Serra Azul State Park (PESA), Barra do Garças, Mato Grosso, Brazil. Richness (S), Abundance (N), Shannon index (H'), Pielou equitability (J), and species richness estimator (Chao-2). Shannon diversity t test index (H') comparison among phytophysiognomies. Blank cell: t-test result. Bold: p values.

Phytophysiognomies	S	N	H'	J	Chao-2
Cerrado sensu stricto	12	61	1.945	0.7826	22
Cerrado ralo	16	91	1.469	0.5299	27
Deciduous forest	11	22	2.199	0.9171	15
Gallery forest	21	30	2.904	0.9539	61
Shannon diversity t	t/p	CR	C	GF	DF
	CR	–	**0.29**	**0.250**	**0.18**
	C	1.06	–	**0.03***	**0.77**
	GF	-1.16	2.19	–	**0.02***
	DF	1.35	-0.29	2.47	–

CR: Cerrado ralo, C: Cerrado sensu strict, GF: Gallery forest, DF: Deciduous forest. *Statistical significance.

28.94%, *H. feronia* 14.17%, *C. sorana* 5.99%, and *Cissia myncea* (Cramer, 1780) 5.1% (Table 4).

Table 4. Individual and cumulative contribution of species to the similarity of Nymphalidae community composition in the Serra Azul State Park (PESA), Barra do Garças, Mato Grosso, Brazil.

Species	Contribution (%)	
	Individual	Cumulative
Yphthimoides renata	28.94	28.94
Hamadryas feronia	14.17	43.11
Callicore sorana	5.99	49.10
Cissia myncea	5.21	54.31
Taygetis mermeria	3.40	57.71
Chloreuptychia arnaca	3.04	60.75
Opsiphanes invirae	2.87	63.62
Yphthimoides pacta	2.64	66.26
Siderone galanthis	2.46	68.72
Taygetina oreba	2.24	70.96
Hermeuptychia hermes	2.08	73.04
Hamadryas februa	1.82	74.86
Archeoprepona amphimachus	1.72	76.58
Prepona pheridamas	1.69	78.28
Cissia sp.	1.57	79.85
Catonephele acontius	1.51	81.36

The frugivorous butterfly community presented different species compositions among the phytophysiognomies (Stress = 0.17, R^2 = Axis 1 = 0.42 and Axis 2 = 0.22). The largest difference occurred between the Cerrado phytophysiognomies (Cerrado sensu stricto and cerrado ralo) and the gallery forest. The deciduous forest presented species from both the Cerrado and the gallery forest (Fig. 4). There was a substitution of species occurrence in relation to the phytophysiognomies (Criterion 0.38, Z score: -2.034, $p < 0.05$) (Fig. 5).

DISCUSSION

Regarding richness and abundance, the most dominant subfamilies of frugivorous butterflies in the PESA were Satyrinae and Biblidinae. Furthermore, these taxa showed higher abundance and species richness in other studies from the Neotropical region (Lamas 2004, Martins et al. 2017b, Silva et al. 2012, Santos et al. 2011, Uehara-Prado et al. 2007).

The diversity of Satyrinae is related to its biology. They have diurnal habits, feed on fruits in different stages of decomposition and with fungi, have low dispersal capacity, fly near the ground, and prefer shaded areas. They are dominant members of the butterfly communities in different Neotropical (Marín et al. 2011, Paluch et al. 2016) and Cerrado phytophysiognomies in the PESA.

The complexity of interspecific or biotic interactions is considered the main reason for diversity in tropical environments, and the intimate relationships between butterflies and their host plants could help clarify the origin of butterfly megadiversity in the tropics (Bonebrake et al. 2010). In this sense, the abundance of Satyrinae was dependent on the distribution pattern of its host plants, since the larval stage of these butterflies generally develop on monocotyledons, whose growth, abundance, and nutritional content are strongly seasonal. Our samples were obtained in the rainy season, which increases food resource for butterflies. Therefore, plants are adequate resources for Satyrinae larvae, and due to such food availability for the larval stage, we observed peak recruitment of adults in these areas (Silva et al. 2007).

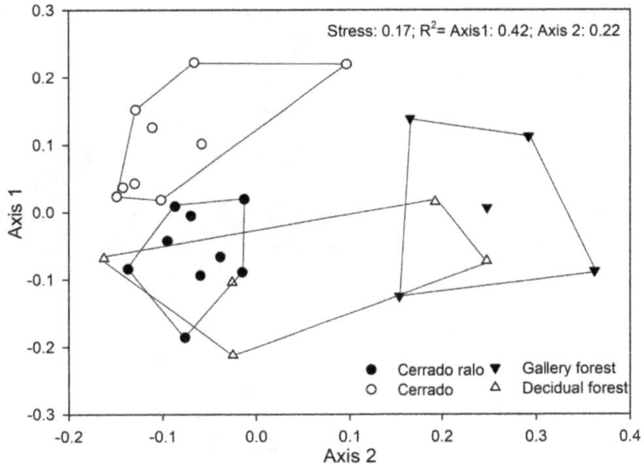

Figure 4. Non-metric dimensional scaling of Nymphalidae community sampled at the Serra Azul State Park (PESA), Barra do Garças, Mato Grosso, Brazil.

Our results indicate that the association of Satyrinae with monocotyledons may explain the higher abundance and dominance of certain species in the Cerrado areas, as the cerrado ralo, which is typically more open and has more monocotyledons as compared to forested areas. This was observed for *Yphthimoides* Forst, 1964, as *Y. renata* was the most abundant of the three species of *Yphthimoides* recorded in PESA. *Yphthimoides* is a Neotropical genus associated with Poaceae (Freitas 2004, Marín et al. 2011, Freitas et al. 2012, Barbosa et al. 2016). Individuals of the genus are distributed in different habitats, including open environments such as forest edges, savannas, and secondary environments (e.g., open pasture areas in urban environments), and are abundant in areas with different degrees of disturbance and anthropogenic impacts (Freitas et al. 2012, Barbosa et al. 2015, 2016).

The abundance of Biblidinae, explained by the dominance of *H. feronia* in the Cerrado sensu stricto, also reinforces the previous hypothesis since this species is associated with anthropogenically impacted areas (e.g., Martins et al. 2017b, Uehara-Prado et al. 2007).

In this way, the log-series distribution found may be indicative of the disturbed environment in the PESA, which

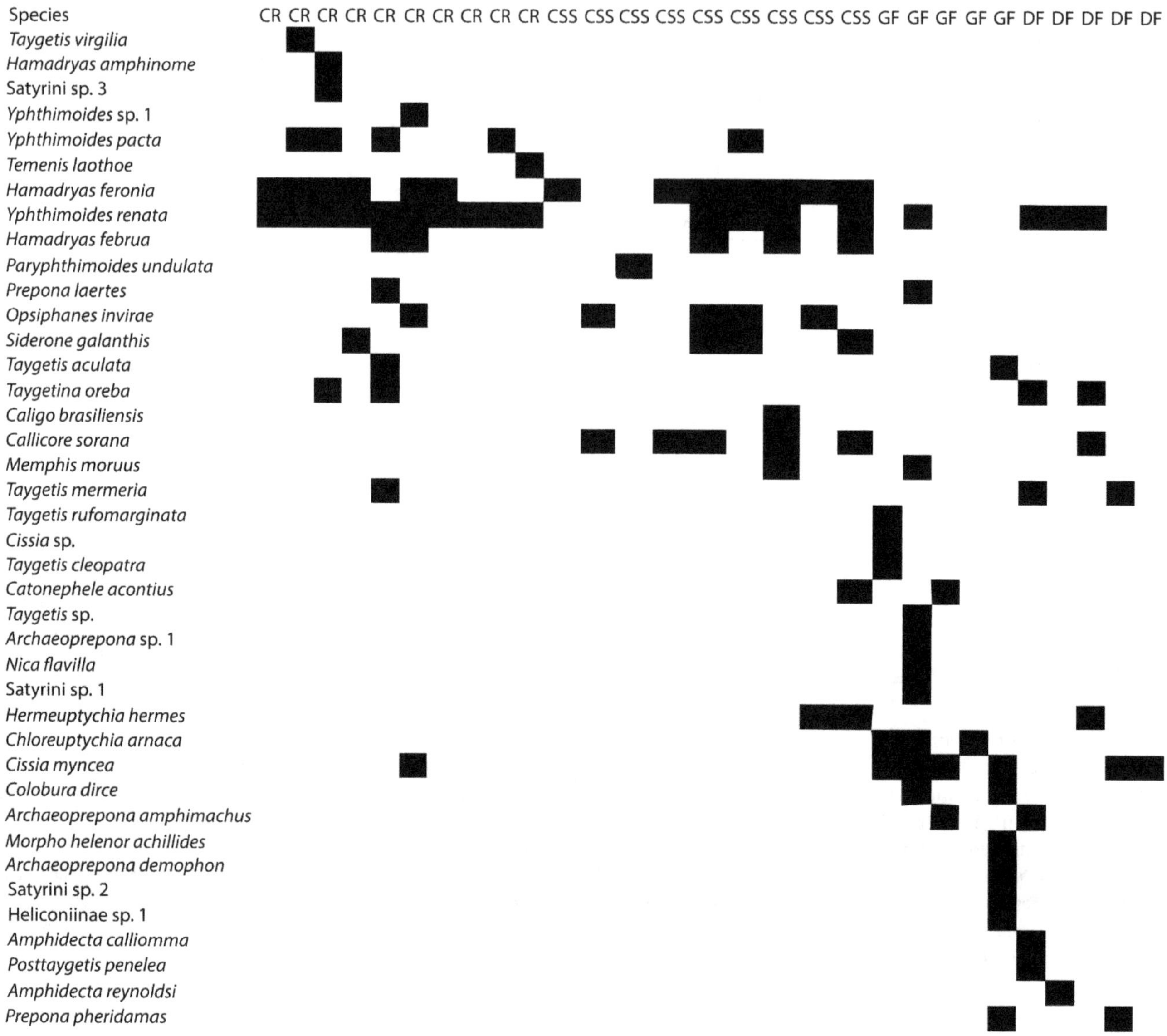

Figure 5. Nymphalidae community ordination in relation to sampled sites (CR: Cerrado ralo, C: Cerrado sensu strict, GF: Gallery forest, DF: Deciduous forest) in Serra Azul State Park (PESA), Barra do Garças, Mato Grosso, Brazil.

has a large amount of grasses, due to the region's fire history. The two most abundant species in the frugivorous butterfly community were the ones considered tolerant to anthropogenic effects, which contributed to the abundance model and to the low diversity and equitability indexes registered for the Cerrado physiognomies of the PESA. In this sense, both *Y. renata* and *H. feronia* can be considered indicators of disturbances and/ or landscape fragmentation within the PESA, as observed by Martins et al. (2017b).

The occurrence of species with low abundance and the 27 rare species recorded in the PESA can be considered high when compared to inventories carried out in the areas with marginal influences of savannah, Chaco, and Pantanal (e.g., De Souza and Guillhermo-Ferreira 2015). Rare species are those that are difficult to sample because their populations are apparently very small. Besides that, these species may be considered rare in one spot, but not in others, due to differences in the availability of food resources, host plants, or microclimatic factors (Brown Jr and Freitas 2002, De Souza and Guillhermo-Ferreira 2015).

The observed richness was well below the estimated richness for the frugivorous butterfly community distributed in the PESA, demonstrating that other species could still be collected with a larger sampling effort. When compared to other studies developed in the Cerrado (Silva et al. 2012, 2015) or predominantly forested areas (Uehara-Prado et al. 2007), the richness and/or diversity recorded for the PESA was low, but was higher than the richness/diversity recorded in fragmented areas from other regions in Brazil (Silva et al. 2007, Martins et al. 2017b).

While it is difficult to compare richness and diversity of frugivorous butterflies from different studies due to the peculiarities of each study, especially regarding differences in collection methods, sample effort and size, and disturbed and undisturbed areas (Martins et al. 2017b). A careful evaluation of species abundance in different areas can reveal, for example, changes in the composition and structure of the hostplants. Such information should be investigated in future studies for species such as *C. sorana*, *Hamadryas amphinome* (Linnaeus, 1767), *H. februa* (Hübner, [1823]), *H. feronia*, *Colobura dirce* (Linnaeus, 1758), and *Opsiphanes invirae* (Hübner, 1808), which are considered common and abundant on regional scales (Silva et al. 2015), but registered low abundances in the PESA.

Our results demonstrate that the community of frugivorous butterflies at the PESA is spatially structured, with the highest richness, diversity, and equitability in the gallery forest. However, the second largest richness and lower diversity and equitability were found in the cerrado ralo. Cerrado sensu stricto and deciduous forest showed similar ecological metrics. Such pattern may be related to periodic disturbances, such as the frequency and intensity of fires (e.g., Santos et al. 2017), which alter the floristic structure of the PESA physiognomies (e.g., Peixoto et al. 2012, Santos et al. 2017, Ribeiro et al. 2012) and, consequently, the organisms that depend on such physiognomies.

In this study, we found a significantly higher number of species in the forest with higher frequency of burning (Peixoto et al. 2012). This reinforces the differences in richness, diversity, and equitability values for the frugivorous butterfly community in the PESA phytophysiognomies, with higher values in forested areas when compared to typical Cerrado areas.

The frugivorous butterfly community in the cerrado ralo and gallery forest indicates that these phytophysiognomies are subsets of the butterfly communities in the deciduous forest, since, in terms of composition and dominance, they share the same species. Furthermore, authors such as Pinheiro and Ortiz (1992) and Emery et al. (2006) proposed that a variety of phytophysiognomies promotes a higher environmental and spatial heterogeneity for butterflies, since variations in microenvironments favor a wider range of species.

Our results confirm that frugivorous butterfly guilds are promising biological models for developing future environmental monitoring studies, and that patterns of species richness, abundance, and distribution can be used to make conservation and management decisions in the PESA and other areas. The spatial and temporal monitoring of the butterflies can be used in the selection of species for indicating the degree of degradation of the habitat.

This initial inventory can serve as a starting point for long-term monitoring of the PESA butterfly community, which could provide data to support decisions about area management. In addition, our data highlights the need for more studies in areas that have not yet been inventoried. Finally, as more species are recorded and/or discovered, information is increased and can be used for Neotropical biodiversity conservation.

ACKNOWLEDGEMENTS

We thank the Rede ComCerrado, the SISBIOTA Project, Conselho Nacional de Desenvolvimento Científico e Tecnológico (CNPq process 563134/2010-0) and the PPBIO Project (CNPq process 457497/2012-2) for financial support. We thank also all teams linked to PPBio – Rede ComCerrado of the Federal University of Mato Grosso, Campus do Araguaia, who contributed directly and indirectly to the logistic support, collections, and/or mounting of the butterflies. We also thank Amabílio José A. de Camargo (EMBRAPA Cerrado, Brasília, DF) for identifying the frugivorous butterflies.

LITERATURE CITED

Barbosa EP, Silva AK, Paluch M, Azeredo-Espin, AML, Freitas AVL (2015) Uncovering the hidden diversity of the Neotropical butterfly genus *Yphthimoides* Forster (Nymphalidae: Satyrinae): description of three new species based on morphological and molecular data. Organisms Diversity & Evolution 15: 577–589. https://doi.org/10.1007/s13127-015-0221-y

Barbosa E, Marín M, Giraldo CE, Uribe S, Freitas A (2016) Description of two new species of the Neotropical genus *Yphthimoides* Forster, 1964 (Lepidoptera: Nymphalidae: Satyrinae) from the '*renata* clade'. Neotropical Biodiversity 2(1): 87–98. https://doi.org/10.1080/23766808.2016.1217582

Bellaver J, Iserhard CA, Santos JP, Silva AK, Torres M, Siewert RR, Moser A, Romanowski HP (2012) Borboletas (Lepidoptera: Papilionoidea e Hesperioidea) de Matas Paludosas e Matas de Restinga da Planície Costeira da região Sul do Brasil. Biota Neotropica 12(4): 181–190. http://dx.doi.org/10.1590/S1676-06032012000400019

Bonebrake TC, Ponisio LC, Boggs CL, Ehrlich PR (2010) More than just indicators: A review of tropical butterfly ecology and conservation. Biological Conservation 143: 1831–1841. https://doi.org/10.1016/j.biocon.2010.04.044

Brannstrom C, Jepson W, Filippi AM, Redo D, Xu Z, Ganesh S (2008) Land change in the Brazilian Savanna (Cerrado), 1986–2002: comparative analysis and implication for land-use policy. Land Use Policy 25(4): 579–595. https://doi.org/10.1016/j.landusepol.2007.11.008

Brown Jr KS, Freitas AVL (2002) Butterfly communities of urban forest fragments in Campinas, São Paulo, Brazil: Structure, instability, environmental correlates, and Conservation.

Journal of Insect Conservation 6(4): 217–231. https://doi.org/10.1023/A:1024462523826

De Souza PRB, Guillermo-Ferreira R (2015) Butterflies of the Bodoquena Plateau in Brazil (Lepidoptera, Papilionoidea). ZooKeys 546: 105–124. https://doi.org/10.3897/zookeys.546.6138

Devries PJ, Walla TR (2001) Species diversity and community structure in neotropical fruit-feeding butterflies. Biological Journal of the Linnean Society 74(1): 1–15. https://doi.org/10.1006/bijl.2001.0571

Emery EO, Brown Jr KS, Pinheiro CEG (2006) As borboletas (Lepidoptera, Papilionoidea) do Distrito Federal, Brasil. Revista Brasileira de Entomologia 50(1): 85–92. http://dx.doi.org/10.1590/S0085-56262006000100013

Freitas AVL (2004) A new species of *Yphthimoides* (Nymphalidae, Satyrinae) from Southeastern Brazil. Journal of the Lepidopterists' Society 58(1): 7–12.

Freitas AVL, Kaminski LA, Mielke OHH, Barbosa EP, Silva-Brandão KL (2012) A new species of *Yphthimoides* (Lepidoptera: Nymphalidae: Satyrinae) from the southern Atlantic forest region. Zootaxa 3526: 31–44.

Klink CA, Machado RB (2005) A conservação do Cerrado brasileiro. Megadiversidade 1(1): 147–155.

Lamas G (2004) Checklist: Part 4A. Hesperioidea - Papilionoidea. In: Heppner JB (Ed.). Atlas of Neotropical Lepidoptera. Gainesville, Association for Tropical Lepidoptera. Scientific Publishers, 439 pp.

Leivas FWT, Fischer ML (2008) Avaliação da composição de invertebrados terrestres em uma área rural localizada no município de Campina Grande do Sul, Paraná, Brasil. Biotemas 21(1): 65–73. https://doi.org/10.5007/2175-7925.2008v21n1p65

Lewinsohn TM, Freitas AVL, Prado PI (2005) Conservação de invertebrados terrestres e seus habitats no Brasil. Megadiversidade 1(1): 62–69.

Magnusson WE, Lima AP, Luizão R, Luizão F, Costa FRC, Castilho CV, Kinupp VF (2005) RAPELD: A modification of the Gentry method for biodiversity surveys in long-term ecological research sites. Biota Neotropica 2: 1–6.

Marín MA, Peña C, Freitas AVL, Wahlberg N, Uribe SI (2011) From the Phylogeny of the Satyrinae Butterflies to the Systematics of Euptychiina (Lepidoptera: Nymphalidae): History, Progress and Prospects. Neotropical Entomology 40(1): 1–13.http://dx.doi.org/10.1590/S1519-566X2011000100001

Marques MI, De Sousa WO, Santos GB, Battirola LD, Anjos KC (2010) Fauna de artrópodes de solo. In: Fernandes IM, Signor CA, Penha J (Eds) Biodiversidade no Pantanal de Poconé. Cuiabá, Centro de Pesquisa do Pantanal, 73–102.

Marques MI, Figueiredo AM, Santos GB, De Sousa WO, Ide S, Battirola LD (2016) Diversity of Soil Beetles (Hexapoda, Coleoptera) in an Area at the Pantanal of Poconé, Mato Grosso, Brazil. Entomobrasilis 9(2): 89–96. https://doi.org/10.12741/ebrasilis.v9i2.539

Martins LP, Araujo Junior EC, Martins ARP, Colins MS, Almeida GCF, Azevedo GG (2017a) Butterflies of Amazon and Cerrado remnants of Maranhão, Northeast Brazil. Biota Neotropica 17(3): 1–11. http://dx.doi.org/10.1590/1676-0611-bn-2017-0335

Martins LP, Araujo Junior EC, Martins ARP, Duarte M, Azevedo GG (2017b) Species diversity and community structure of fruit-feeding butterflies (Lepidoptera: Nymphalidae) in an eastern Amazonian forest. Papéis Avulsos de Zoologia 57(38): 481–489. https://doi.org/10.11606/0031-1049.2017.57.38

Paluch M, Mielke OHH, Linharess LM, Silva DC (2016) Butterflies (Lepidoptera: Papilionoidea and Hesperioidea) of the Private Reserve of Natural Heritage Fazenda Lontra/Saudade, Itanagra, Northern Coast of Bahia, Brazil. Biota Neotropica 16(1): 1–7. http://dx.doi.org/10.1590/1676-0611-BN-2014-0085

Peixoto KS, Sanchez M, Pedroni F, Ribeiro MN, Facure KG, Gomes-Klein VL, Guilherme FAG (2012) Dinâmica da comunidade arbórea em uma floresta estacional semidecidual sob queimadas recorrentes. Acta Botanica Brasilica 26(3): 697–708. http://dx.doi.org/10.1590/S0102-33062012000300020

Pérez JH, Gaviria-Ortiz FG, Santos WIG, Carneiro E, Mielke OHH, Casagrande MM (2017) Long term survey of the butterfly fauna of Curitiba, Paraná, Brazil: How does a scientific collection gather local biodiversity information? (Lepidoptera: Papilionoidea). SHILAP Revista de Lepidopterologia 45(179): 433–446.

Pinheiro CEG, Ortiz JVC (1992) Communities of fruit-feeding butterflies along a vegetation gradient in Central Brazil. Journal of Biogeography 19(5): 505–511.

Pinheiro CEG, Malinov IC, Andrade TO, Maravalhas JB, Andrade MBM, Deus LPA, Pedrosa LGP, Zanatta GV (2008) As borboletas (Lepidoptera, Papilionoidea) do Campus Universitário Darcy Ribeiro (Distrito Federal, Brasil). Biota Neotropica 8(4): 139–144. http://dx.doi.org/10.1590/S1676-06032008000400013

Queiroz-Santos L, Dias FMS, Dell'Erba R, Casagrande MM, Mielke OHH (2016) Assessment of the current state of biodiversity data for butterflies and skippers in the state of Mato Grosso, Brazil (Lepidoptera, Papilionoidea and Hesperioidea) ZooKeys 595: 147–161. http://dx.doi.org/10.3897/zookeys.595.7856

Ribeiro MN, Sanchez M, Pedroni F, Peixoto KS (2012) Fogo e dinâmica da comunidade lenhosa em cerrado sentido restrito, Barra do Garças, Mato Grosso. Acta Botanica Brasilica 26(1): 203–217. http://dx.doi.org/10.1590/S0102-33062012000100020

Rydon AHB (1964) Notes on the use of butterfly traps in East Africa. Journal of the Lepidopterists' Society 18: 51–58.

Santos EC, Mielke OHH, Casagrande MM (2008) Butterfly inventories in Brazil: the state of the art and the priority-areas model for research aiming at conservation. Natureza & Conservação 6(2): 176–198.

Santos JO, Lenza E, Maracahipes-Santos L, Forsthofer M, Roberto RL, Eisenlohr PV, Sanchez M, Pedroni F (2017) Assessment of differences between two vegetation physiognomies, including the variation among strata, provides insights on biodiversity conservation of "Cerrado". Brazilian Journal of Botany 40: 923–932. https://doi.org/10.1007/s40415-017-0405-0

Santos JP, Iserhard CA, Teixeira MO, Romanowski HP (2011) Fruit-feeding butterflies guide of subtropical Atlantic Forest and Araucaria Moist Forest in State of Rio Grande do Sul, Brazil. Biota Neotropica 11(3): 253–274. http://dx.doi.org/10.1590/S1676-06032011000300022

Seraphim N, Marín MA, Freitas AVL, Silva-Brandão KL (2014) Morphological and molecular marker contributions to disentangling the cryptic *Hermeuptychia hermes* species complex (Nymphalidae: Satyrinae: Euptychiina). Molecular Ecology Resources 14: 39–49. http://dx.doi.org/doi:10.1111/1755-0998.12161

Silva ARM, Landa GG, Vitalino RF (2007) Borboletas (Lepidoptera) de um fragmento de mata urbano em Minas Gerais, Brasil. Lundiana 8(2): 137–142.

Silva ARM, Castro CO, Mafia PO, Mendonça MOC, Alves TCC, Beirão MV (2012) Fruit-feeding butterflies (Lepidoptera: Nymphalidae) in an urban area (Area of Special Protection Manancial Cercadinho) in Belo Horizonte, Minas Gerais State, Brazil. Biota Neotropica 12(3): 292–297. http://dx.doi.org/10.1590/S1676-06032012000300028

Silva ARM, Pontes DV, Guimarães MP, Oliveira MV, Assis LTF, Uehara-Prado M (2015) Fruit-feeding butterflies (Lepidoptera: Nymphalidae) of the Área de Proteção Especial Manancial Mutuca, Nova Lima and Species list for the Region of Belo Horizonte, Minas Gerais, Brazil. Biota Neotropica 15: 1–9. http://dx.doi.org/10.1590/1676-06032015011814

Soares GR, Oliveira AAP, Silva ARM (2012) Butterflies (Lepidoptera: Papilionoidea and Hesperioidea) from an urban park in Belo Horizonte, Minas Gerais State, Brazil. Biota Neotropica 12(4): 209–217. http://dx.doi.org/10.1590/S1676-06032012000400022

Uehara-Prado M, Freitas AVL, Francini RB, Brown Jr KS (2004) Guia das borboletas frugívoras da Reserva Estadual do Morro Grande e região de Caucaia do Alto, Cotia (São Paulo). Biota Neotropica 4(1): 1–25. http://dx.doi.org/10.1590/S1676-06032004000100007

Uehara-Prado M, Brown Jr KS, Freitas AVL (2007) Species richness, composition and abundance of fruit-feeding butterflies in the Brazilian Atlantic Forest: comparison between a fragmented and a continuous landscape. Global Ecology and Biogeography 16: 43–54. https://doi.org/10.1111/j.1466-8238.2006.00267

Uehara-Prado M, Fernandes JO, Bello AM, Machado G, Santos AJ, Vaz-de-Mello FZ, Freitas AVL (2009) Selecting terrestrial arthropods as indicators of small-scale disturbance: first approach in the Brazilian Atlantic forest. Biological Conservation 142(6): 1220–1228. http://dx.doi.org/10.1016/j.biocon.2009.01.008

Uehara-Prado M, Ribeiro DB (2012) Borboletas em Floresta Atlântica: métodos de amostragem e inventário de espécies na Serra do Itapeti. In: Morini MSC, Miranda VFO (Eds) Serra do Itapety: Aspectos Históricos, Sociais e Naturalísticos. Bauru, Canal6 Editora, 167–186.

Author Contributions: WOS and LES designed and conducted the experiments; WOS LES and WIGS processed and identified the insects; WOS FRJS and RA analyzed the data; WOS LES FRJS WIGS and RA wrote the paper.

Competing Interests: The authors have declared that no competing interests exist.

11

Two new species and new records of Neanuridae (Hexapoda: Collembola) from Brazilian central Amazonia

Ana Carolina da Rocha Neves[1], **Maria Cleide de Mendonça**[1], **Gabriel Costa Queiroz**[1]

[1]*Departamento de Entomologia, Museu Nacional, Universidade Federal do Rio de Janeiro. Quinta da Boa Vista, São Cristóvão, 20940-040 Rio de Janeiro, RJ, Brazil.*
Corresponding author: Ana Carolina da Rocha Neves (ana_rneves@yahoo.com.br)

ABSTRACT. Two new species of Neanuridae from the Amazon Rainforest of Northern Brazil are described and illustrated. The new species *Friesea multiclavata* **sp. nov.** (holotype male in MNRJ: Brazil, Amazonas State) belongs to the reducta-group and represents the first record of Frieseinae for Amazonas State in Brazil. Within Pseudachorutinae, the new species *Furculanurida boiuna* **sp. nov.** (holotype male in MNRJ: Brazil, Amazonas State) represents the third species of this genus without eyes and body pigment, together with *Furculanurida africana* (Massoud, 1963) – type species of the genus – and *Furculanurida emucronata* Zon et al., 2014, both from Ivory Coast, in Africa. Lastly, an indefinite species of *Ectonura* sp. (Neanurinae) is the second record of the genus for the Neotropical Region.

KEY WORDS. Frieseinae, Neanurinae, Neotropics, Pseudachorutinae, taxonomy.

INTRODUCTION

Collembola has a worldwide distribution with a large number of species (Bellinger et al. 2018). Despite the several taxonomic contributions about springtails that have accumulated over the last years, there are few studies on the Brazilian fauna, which has resulted in vast areas lacking any springtail record (Abrantes et al. 2010, 2012, Zeppelini et al. 2018). In the state of Amazonas, for example, there are still several completely unexplored ecosystems for springtails (Abrantes et al. 2010, 2012). As for now, the Collembola fauna from the Amazon is second in terms of recorded species to the Atlantic Forest (Zeppelini et al. 2018).

Small river branches in the Amazon, known as 'Igarapés', are considered an important component of the forest. They provide an essential structural heterogeneity and sustain a highly diverse fauna. Their primary source of energy comes from organic matter from the riparian vegetation of adjacent forests (Lima and Gascon 1999, Nessimian et al. 1998).

Poduromorpha is the second most numerous among Collembola orders, with 11 families, 331 genera and about 3300 species (Bellinger et al. 2018). Brazil has representatives of seven families – Neanuridae, Brachystomellidae, Hypogastruridae, Onychiuridae, Tullbergiidae, Odontellidae and Isotogastruridae (Zeppelini et al. 2018).

Neanuridae comprises six subfamilies, three of which occur in Brazil: Frieseinae, Neanurinae and Pseudachorutinae. Pseudachorutinae and Frieseinae, with 466 and 111 species, respectively and Neanurinae with only six recorded species (Zeppelini et al. 2018).

Among the 178 species of *Friesea* Dalla Torre, 1895, 11 are recorded for Brazil. Nine of those were recorded only for the state of Rio de Janeiro: *Fr. boitata* Queiroz & Mendonça, 2015, *Fr. claviseta* Axelson, 1900, *Fr. cubensis* Potapov & Banasko, 1985, *Fr. curupira* Queiroz & Mendonça, 2015, *Fr. josei* Palacios-Vargas, 1986, *Fr. jurubatiba* Silveira & Mendonça, 2018, *Fr. magnicornis* Denis, 1931, *Fr. mirabilis* (Tullberg, 1871) and *Fr. reducta* Denis, 1931. The other two species, *Fr. sublimis* Macnamara, 1921 and *Fr. arlei* Massoud & Bellinger, 1963, were recorded for Espírito Santo and Mato Grosso States, respectively (Bellinger et al. 2018, Zeppelini et al. 2018).

Furculanurida Massoud, 1967 includes 14 described species from Africa and Americas, while in Brazil there are four recorded species: *Fu. nessimiani* Fernandes & Mendonça, 2002 and *Fu. tropicalia* Queiroz & Fernandes, 2011 from the Atlantic Forest and *Fu. belemensis* Arlé & Rufino, 1976 and *Fu. goeldiana* Arlé & Rufino, 1976 from the Amazon Rainforest (Bellinger et al. 2018, Zeppelini et al. 2018).

Ectonura has 17 species described in the world and only one, *E. snowdeni* Queiroz & Deharveng, 2015, recorded for Brazil (Bellinger et al. 2018, Zeppelini et al. 2018).

In this paper we describe, illustrate and provide diagnosis for two new species of Neanuridae of the genera *Friesea* and *Furcu-*

lanurida. In addition, *Ectonura*, poorly known in the neotropics, is recorded for the first time in the Amazon, representing the second record for the Neotropical Region. The new species, *Furculanurida boiuna* sp. nov. is the third of the genus without eyes and pigment and the first with these characteristics found outside Ivory Coast, in Africa (type locality of the genus). *Friesea multiclavata* sp. nov., represents the first record of this genus for the state of Amazonas.

MATERIAL AND METHODS

Specimens were collected using trays, also known as yellow traps, placed upon soil litter found near Igarapés in five municipalities from central Amazon, Amazonas State. These trays, containing a mixture of water and detergent, were kept for 3–5 days and the captured fauna was sorted and mounted on glass slides, using Mark-André II as medium, and studied under optic microscope.

The terminology for antennal chaetotaxy is after D'Haese (2003); dorsal head chaetotaxy after Cassagnau (1974); tibiotarsal chaetotaxy after Deharveng (1983); labial chaetotaxy according to Massoud (1967); thorax and abdomen chaetotaxy according to Potapov and Banasko (1985).

Abbreviations as follows: (AM) Amazonas State; (CM) Cleide Mendonça code collection; (ES) Espírito Santo State; (MNRJ) Coleção Entomológica do Museu Nacional, Universidade Federal do Rio de Janeiro, Rio de Janeiro, RJ, Brazil; (NC) North Carolina State; (PA) Pará State; (RJ) Rio de Janeiro State; (Ant) antennal segment; (Sgd) dorsal guard S-chaeta of antennal III organ; (Abd) abdominal segment; (Scx) subcoxa; (Th) thoracic segment; (Tita) tibiotarsus(i).

TAXONOMY

Friesea Dalla Torre, 1895

Type species: *Triaena mirabilis* Tullberg, 1871

Friesea multiclavata sp. nov.

Figs 1–9, Tab. 1

Description. Body length: 0.51 mm (Holotype). Habitus cylindrical and robust, typical of *Friesea*. Color bluish-gray with ventral of head, legs and sternites white. Secondary granules moderately developed. Antennae shorter than cephalic diagonal. Ratio antenna: cephalic diagonal = 1: 1.4. Ant IV with apical bulb simple and subapically displaced on ventral side; subapical organite, dorsolateral S-microchaeta and 6 S-chaetae present dorsally (Fig. 1). Sensory organ of Ant III formed by five S-chaetae: two internal S-microchaetae bent externally and freely exposed; two subcylindrical guard S-chaetae and one smaller ventral S-microchaeta (Fig. 2). Ant I with 7 simple chaetae and Ant II with 11 chaetae, of which 10 are simple and one is strongly spatulate dorsoexternally (see detail in Fig. 7).

Head with 8+8 eyes in heavily pigmented eye-patch. Mandible strong with 5 unequal teeth; basal tooth measuring approximately twice the size of the others; maxilla head typical of the genus, with three lamellae; internal lamella with about 6 denticles (Fig. 3). Buccal cone short; Pre-labral/labral chaetae arranged according to the formula: 2/5,5,4. Labium typical of the genus, with papillated chaeta L, chaeta F about twice the length of E (Fig. 4).

Chaetotaxy of legs I, II, III. Subcoxae I 1,3,3; Subcoxae II 0,2,2; Coxae 3,6,7; Trochanters 5.4,4; Femora 11,10,10; Tibiotarsi 18,18,17, without M chaeta. Tibiotarsi I-III with 4,5,5 clavate tenent hairs, respectively (2 dorsal and 2 ventral on Tita I; 3 dorsal and 2 ventral on Tita II and III) (Figs 5, 6). Subcoxa I of legs I-III and lateral region of Abd V with hemispheric tegumentary protuberance constituted of primary granules (see detail of Fig. 7). Ungues toothless.

Dorsal body. Composed of short ordinary chaetae, slender S-chaetae, slightly longer than ordinary, and strongly clavate chaetae laterally on head and abdominal tergites IV-VI (Fig. 7). Head chaetotaxy composed of a0, d0, d2-5, sd1-5, oc1-3, c1-2, p1-2. Th I with 2+2 simple chaetae. Abd IV with 2+2 small clavate dorsolateral chaetae; Abd V with 2+2 long clavate dorsal chaetae; Abd VI with 10 strongly clavate chaetae arranged in three linear rows, as 4,4,2 (Fig. 7 and detail). Formula of S-chaetae by half tergite: 022/11111.

Ventral tube with 3+3 chaetae. Abdominal sternites II-V with 4+4, 5+5, 12+12, 5+5 chaetae, respectively (Fig. 8). Tenaculum and furca absent; furcal area with 2+2 microchaetae. Anal valves with 11-12 chaetae and one hr chaeta. Male genital plate with 2+2 pregenital chaetae, 4+4 eugenital and 9 circungenital chaetae (Figs 8, 9).

Material examined. Holotype male. BRAZIL, Amazonas State: Presidente Figueiredo municipality, forest leaf litter of Amazon Rainforest, coordinates 02°02′56″S, 60°06′08″W, 23.IV.2008, Hamada, Azevedo, Neiss, Silva & Meneses leg. (CM/MNRJ slide number 2538).

Etymology. The specific name multiclavata is derived from Latin, meaning bearing nails, and is an allusion to the numerous clavate chaetae arranged along the tergites.

Remarks. *Friesea multiclavata* sp. nov. can be included in the reducta-group, which is composed of 14 species, according to Queiroz and Mendonça (2015), based on the following characters: 8+8 eyes and absence of furca and anal spines (Table 1). Despite the absence of anal spines, most species in this group, with the exception of *Friesea africana* Delamare Deboutteville, 1953, have some degree of chaetae modification on Abd VI. In this sense, the new species is unique within this group since it also presents modified chaetae on head and antennae.

Friesea multiclavata sp. nov. is very similar to *Fr. albithorax* Massoud & Thibaud, 1980 (Antilles), *Fr. lobulata* Palacios-Vargas & Díaz, 1986 and *Fr. mucumontana* Palacios-Vargas & Díaz, 1986 (Venezuela) mainly for sharing the 2+2 chaetae on Th I, clavate chaetae on body and 18,18,17 chaetae on tibiotarsi I-III.

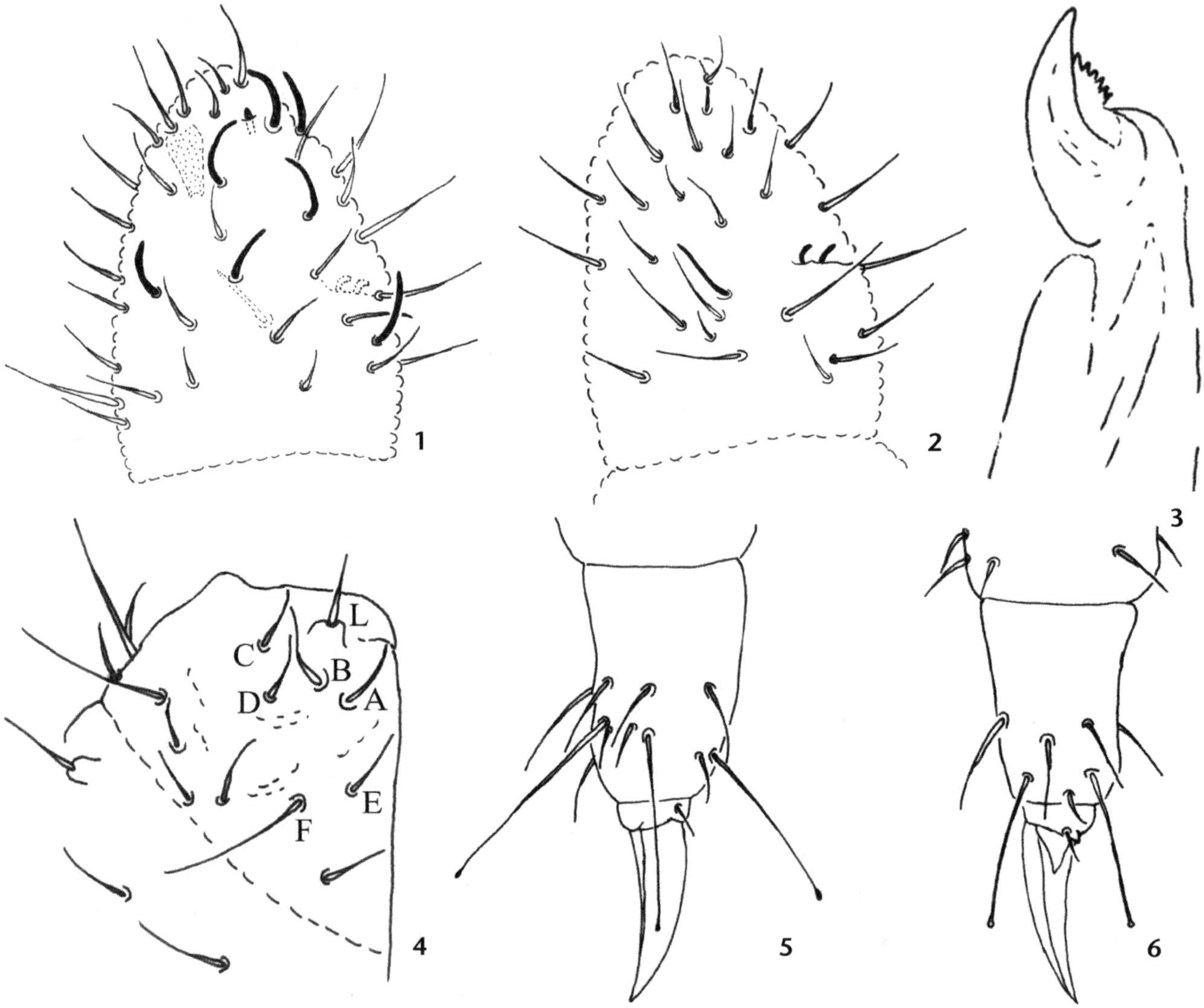

Figures 1–6. *Friesea multiclavata* sp. nov. holotype male: (1) Ant III-IV in dorsal view; (2) Ant III-IV in ventral view; (3) maxillae in dorsal view; (4) half labium in ventral view; (5) Tita II in ventrolateral view; (6) Tita II in dorsolateral view.

However, *Fr. multiclavata* sp. nov. has a simple apical bulb, 4,5,5 tenent hairs on tibiotarsi and ungues without teeth while *Fr. lobulata* has a trilobed apical bulb, only two clavate tenent hairs on legs I-III and toothed unguis. *Friesea multiclavata* sp. nov. is uniformly bluish-gray colored, while *Fr. albithorax* presents white thoracic tergites and the rest of the body is grey. Moreover, *Fr. multiclavata* sp. nov. is unique among the four species here referred, by the 10 strongly clavate chaetae on Abd VI, while the others have six (*Fr. lobulata* and *Fr. mucumontana*) or eight (*Fr. albithorax*) clavate chaetae.

Friesea multiclavata sp. nov. has an inconspicuous tegumentary protuberance on subcoxa I of legs I-III and ventrolaterally on Abd V (Fig. 7), also found in *Fr. reducta* and *Fr. boitata*. This structure was first mentioned by Massoud and Thibaud (1980) in specimens of *Fr. reducta* from Lesser Antilles. Later on, *Fr. boitata* was also found to possess the same structures on subcoxae I and Abd V. According to Queiroz and Mendonça (2015), it is possible that these structures were not mentioned in other descriptions as they may have not been properly visualized.

Although there are no phylogenetic studies for Neotropical species of *Friesea*, the proposition of a group relying on morphological similarities, such as reducta-group, represents the first step towards a better understanding of the genus' morphological diversity in the biogeographical region. As proposed by Queiroz and Mendonça (2015), it is possible to distinguish "subgroups" of species within this group based on the number of chaetae on Th I, being 2, 3 or 4 by half tergite. Despite that, we believe that the striking resemblance of these species, especially those

Table 1. Main characters of species of *Friesea*, belonging to reducta-group. Modified from Queiroz and Mendonça (2015).

Species	Apical bulb	Th I chaetae	Chaetae on Tita I-III	Clavate tenent hairs Legs I-III	Furcal area chaetae	Modified chaetae on Abd VI [number (arrangement)]	Type Locality
Fr. africana Delamare Deboutteville, 1953	?	?	?	–	?	–	Tanzania
Fr. albithorax Massoud & Thibaud, 1980	simple	2+2	(18,18,17)?	4,5,5	?	clavate = 8 (4,4)	Antilles
Fr. boitata Queiroz & Mendonça. 2015	simple	4+4	18,18,17	4,5,5	4	slightly clavate = 6 (4,2)	Brazil (RJ)
Fr. bonariensis Izarra, 1965	simple	4+4	?	2,2,2	?	blunt = 6 (4,2)	Argentina
Fr. josei Palacios-Vargas, 1986	simple	3+3	17,17,16	3,3,3	?	spiniform = 10	Puerto Rico
Fr. jurubatiba Silveira & Mendonça, 2018	simple	4+4	18,18,17	4,5,5	4	clavate = 8 (4,4)	Brazil (RJ)
Fr. lobulata Palacios-Vargas & Díaz, 1986	trilobed	2+2	18,18,17	2,2,2	?	slightly clavate = 6?	Venezuela
Fr. marianoius Palacios-Vargas, 2005	simple	?	17,17,16	–	6	strongly clavate = 10 (4,4,2)	México
Fr. mucumontana Palacios-Vargas & Díaz, 1986	simple	2+2	18,18,17	4,5,5	?	clavate = 6 (4,2)	Venezuela
Fr. reducta Denis, 1931*	simple	3+3	(18,18,17)?	4,5,5	?	strongly clavate = 10 (4,4,2)	Costa Rica
Fr. sensillata Palacios-Vargas & Díaz, 1986	simple	4+4	18,18,17	8,9,9	?	strongly clavate = 10 (4,4,2)	Venezuela
Fr. steineri Simón, 1975	simple	3+3	17,17,16	–	6	slightly clavate = 9 (4,4,1)	Spain
Fr. tepetlana Palacios-Vargas, 1986	simple	4+4	17,17,17	4,5,5	4	spiniform = 6 (4,2)	México
Fr. tzontli Palacios-Vargas & Acosta, 1994	simple	4+4	17,17,16	4,5,5	?	clavate = 10 (4,4,2)	México
Fr. xitlensis Palacios-Vargas & Acosta, 1994	simple	4+4	18,18,17	4,4,5	?	spiniform = 6 (4,2)	México
Fr. multiclavata sp. nov.	simple	2+2	18,18,17	4,5,5	4	strongly clavate = 10 (4,4,2)	Brazil (AM)

*In part after illustrations and characterization of *Fr. reducta* by Massoud and Thibaud (1980).

previously mentioned with 2+2 chaetae on Th. I, suggests close relationship between them.

While further studies may clarify phylogenetic relationships, comparative morphological studies are still the mainframe for species recognition. In this sense, all described species in the mentioned subgroup can be distinguished mainly based on body chaetotaxy, especially chaetae morphology, such as clavate chaetae. This character is considered to be consistent and, therefore, it is widely used, especially for recently described species.

Furculanurida Massoud, 1967

Type species: *Micranurida africana* Massoud, 1963

Furculanurida boiuna sp. nov.

Figs 10–22, Tab. 2

Description. Body length 0.9 mm (Holotype). Habitus elongated and cylindrical, paratergites not developed. Secondary granules moderately developed. Color white, in ethanol. Antennae shorter than cephalic diagonal. Ratio antenna: cephalic diagonal = 1: 1.6. Ant IV with trilobed apical bulb, subapical organite apically displaced, next to apical bulb; 6 long S-chaetae (S1-S4, S8, S9?); dorsolateral S-microchaeta absent. Ant III and Ant IV dorsally fused. Antennal organ III with two inner small and curved S-microchaetae, two subcylindrical guard S-chaetae and one S-microchaeta ventrally; dorsal guard S-chaeta apically displaced, towards Ant IV and aligned to S2 and S3, and as long as S-chaetae of Ant IV (Figs 10, 11). Ant II with 11 chaetae and Ant I with 6 chaetae.

Head. Eyes absent, PAO circular with 8–9 vesicles, rosette-like (Fig. 12). Central head chaetotaxy with d1–5, sd3–5 (chaetae sd1,2 absent – probable homology), oc1–3, p1–3; c row of chaetae is absent. Maxilla styliform; Mandible with 7 teeth: 2 large basal, 3 subequal intermediate and 2 apical (Fig. 13). Pre-labral/labral chaetae arranged according to the formula: 4/2,3,5,2 (Fig. 14). Labium truncate with C and D chaeta apically displaced (Fig. 15).

Chaetotaxy of legs I, II, III. Subcoxa I 1,3,3; Subcoxa II 0,2,2; Coxa 3,6,8; Trochanter 6,6,6; Femur 12,12,11; Tibiotarsus 19,19,18; M chaetae basally displaced (Figs 16–18). Unguis of leg I with a strong basal tooth on inner edge (Fig. 16); unguis of legs II and III with small basal tooth on inner edge (Fig. 18).

Dorsal chaetotaxy of tergites consisting of simple, short and subequal chaetae and thin and long S-chaetae. Formula of S-chaetae by half tergite: 022/11111. Ratio ordinary chaeta: S-chaeta= 1: 4.5. Th I with 2+2 chaetae. Th II with dorsolateral S-microchaetae; Th II and III with one lateral ordinary chaeta posteriorly displaced. Abd VI with 8 chaetae, which 4 arranged in a row (Fig. 19).

Ventral tube with 3+3 chaetae. Abdominal sternites II-V with 2+2, 2+2, 5+5 and 3+3 chaetae, respectively (Fig. 20). Tenaculum with 3+3 teeth. Furca well developed: manubrium with 20 chaetae, dens with 5–6 chaetae; mucro with two lamellae and slightly curved apex (Fig. 21). Ratio mucro: dens = 1: 2.3. Each anal valve with 13 chaetae and 3 hr. Genital plate of male with 5+5 eugenital chaetae and 10 circungenital chaetae (Fig. 22).

Material examined. Holotype male. BRAZIL, Amazonas State: Manacapuru municipality, forest leaf litter of Amazon Rainforest, coordinates 03°12'23"S, 60°40'21,0"W, 29.III.2008, Nessimian, Querino, Pepinelli, Azevedo & Neiss leg. (CM/MNRJ slide number 2523).

Etymology. The name of this species is a reference to the legendary Amazonian giant snake called Boiuna, in Amerindian

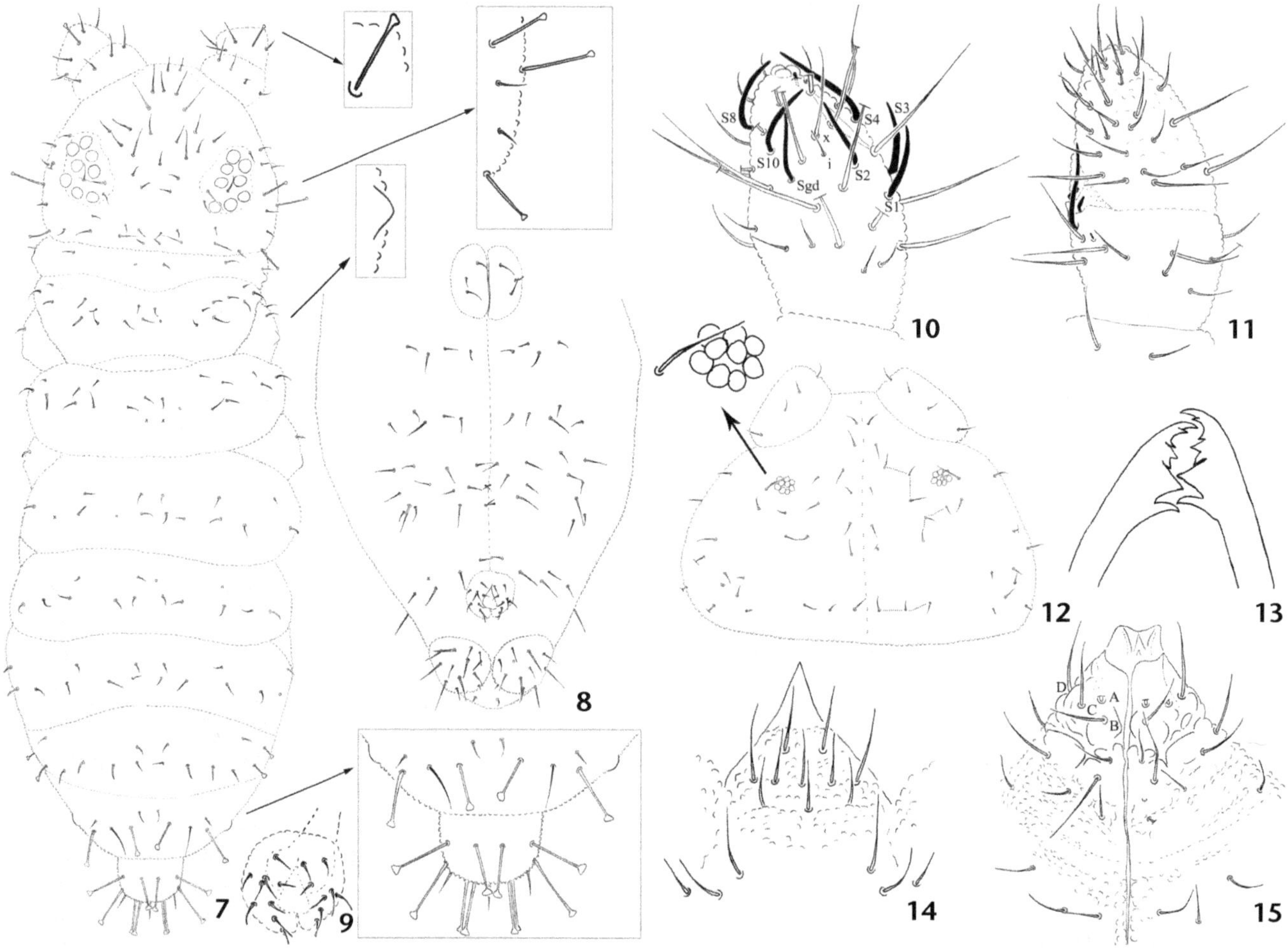

Figures 7–15. (7–9) *Friesea multiclavata* sp. nov. holotype male: (7) dorsal body chaetotaxy with details of clavate chaetae of Ant II, lateral head and Abd V-VI and tegumentary protuberances of Scx I of legs; (8) ventral chaetotaxy of Abd I-VI; (9) genital plate of male. (10–15) *Furculanurida boiuna* sp. nov. holotype male: (10) dorsal view of Ant III-IV; (11) ventral view of Ant III-IV; (12) dorsal head chaetotaxy with detail of PAO; (13) mandibles; (14) labrum; (15) labium.

Tupi Language. According to the legend, a huge snake grows to unrealistic proportions and, as it crawls through dry land, it leaves behind the grooves that later on become the Igarapés.

Remarks. The broadness and vagueness of the current diagnosis for *Furculanurida* was discussed in some studies involving species of this genus and its similarity to other genera (Thibaud and Palacios-Vargas 2000, Queiroz and Fernandes 2011, Zon et al. 2014). Recently, Zon et al. (2014) drew attention to the fact that *Furculanurida* has a poor definition, with many morphological features overlapping with the diagnosis of *Pseudachorutes* Tullberg, 1871 and *Stachorutes* Dallai, 1973 (see Table 2). The most conflicting aspects are related to number of eyes, PAO shape and number of vesicles and furcal development, as well as antennal chaetotaxy, more specifically the presence/absence of dorsolateral S-microchaeta on Ant IV (Queiroz and Fernandes 2011, Zon et al. 2014). Therefore, a revision of the genus is needed, along with a redescription of the type species, *Fu. africana* (Massoud, 1963).

According to Massoud (1967), *Fu. africana* has no eyes and no body pigments but has a well-developed furca. Regarding these characteristics, until now, *Fu. emucronata* Zon et al., 2014 was the only species similar to *Fu. africana*, except for the absence of mucro. In this sense, the new species *Fu. boiuna* sp. nov. represents the third in the genus without eyes and body pigment.

In relation to these three species, the most remarkable similarity among them is: the presence of a somewhat swollen branch of mandible and two basal strong teeth, although some minor differences can be observed in the total number of apical teeth. *Furculanurida boiuna* sp. nov., together with *Fu. emucronata*, has seven teeth on the mandible and the ungues exhibit a tooth on their inner edge, while *Fu. africana* mandible has nine teeth and the ungues are devoid of teeth.

Table 2. Main characters of all known species of *Furculanurida*. Modified from Queiroz and Fernandes (2011).

Species	Color	Ant IV S-chaetae	Ant IV S-microc.	Eyes per side	PAO vesicles	PAO form	Mandible teeth	VT chaetae	Chaetae on each dens	Locality
Fu. africana (Massoud, 1963)	white	6	?	–	8-10	circular	9	?	6	Ivory Coast
Fu. arawakensis Thibaud & Massoud, 1983	blue	6	–	4	5-9	circular	7	3+3	6	Lesser Antilles
Fu. belemensis Arlé & Rufino, 1976	blue-gray	6	?	5	8-9	circular	4-6	?	6	Brazil (PA)
Fu. duodecimoculata Thibaud & Massoud, 1980*	light blue	7	?	6	11	circular	4-5	3+3	6	Morocco
Fu. emucronata Zon, Tano & Deharveng, 2014	white	7	–	–	13-16	elliptical	7	3+3	6	Ivory Coast
Fu. furculata (Salmon, 1956)	deep blue	7	?	2	4	circular	7	?	6	Rwanda
Fu. goeldiana Arlé & Rufino, 1976	blue-gray	6	?	7	7-10	circular	4	?	6	Brazil (PA)
Fu. grandcolasorum Weiner & Najt, 1998	blue-gray	6	–	5	9-10	elliptical	6	3+3	5	Tanzania
Fu. guatemalensis Palacios-Vargas & Gao, 2009	gray	6	+	5	15	elliptical	3-4	4+4	6	Guatemala
Fu. langdoni Bernard, 2007	violet-blue	?	+	5	20-22	elliptical	6	4+4	5	USA (NC)
Fu. longisensillata Najt, Thibaud & Weiner, 1990	deep blue	6	–	6	6-7	circular	10	3+3	6	French Guiana
Fu. nessimiani Fernandes & Mendonça, 2002	deep blue	7	–	6	6	circular	5	3+3	6	Brazil (SP)
Fu. septemoculata Palacios-Vargas & Gao, 2009	gray-violet	6	+	7	14	elliptical	2	4+4	6	Guatemala
Fu. tropicalia Queiroz & Fernandes, 2011	blue-gray	6	–	8	8-10	circular	4	3+3	6	Brazil (ES)
Fu. boiuna sp. nov.	white	6	–	–	8-9	circular	7	3+3	6	Brazil (AM)

**Furculanurida duodecimoculata* is the only species with 2+2 teeth on tenaculum, while the others show 3+3 teeth. The descriptions of *Fu. africana*, *Fu. belemensis* and *Fu. goeldiana* makes no reference to number of teeth on tenaculum.

Despite the mentioned similarities, *Fu. boiuna* sp. nov. shows a complete furca, while in *Fu. emucronata* it is incomplete, without mucro. Moreover, *Fu. emucronata* has 7 S-chaetae on Ant IV and between 13–16 vesicles on PAO of an elliptical shape, while *Fu. boiuna* sp. nov. shows 6 S-chaetae on Ant IV and 8–9 vesicles arranged in circular shape.

From a biogeographical point of view, the fact that the Neotropical *Fu. boiuna* sp. nov. is the first species outside of Africa with these set of characters is of considerable relevance. This indicates that a more widespread distribution, i.e. holotropical, is possible and raises questions regarding Pseudachorutinae distribution throughout the tropics. In this sense, despite their rareness, since few specimens are known from these three species, e.g., *Fu. africana* is known only by the holotype, these eyeless species are important for Pseudachorutinae taxonomy.

It must be highlighted that the new species clearly fits the recently proposed Arlesia-group of genera (Queiroz and Zeppelini 2017). Except for an elongated Sgd, almost subequal to Ant IV S-chaetae, Ant III-IV chaetotaxy is doubtlessly similar to the mentioned group of genera. For example *Fu. boiuna* sp. nov. has trilobed apical bulb; absence of ms; presence of S1–4, S8 and S10; x chaeta between a1 and i chaeta; and apically displaced Sgd. Regarding head and thorax chaetotaxy, the pattern similarity is also evident. On head: the reduced sd chaetae, absence of c row of chaetae, as well as only p1–3 chaeta on head. On thorax: Th. I with only 2+2 chaetae; Th. II and III with one dorsolateral chaeta posteriorly displaced.

In the same sense, according to original illustrations provided by Zon et al. (2014) for *Fu. emucronata*, it is possible to recognize one main chaetotaxal difference from the proposed *Arlesia* group: the absence of one dorsolateral chaeta on Th. II and III, probably the posteriorly displaced one. However, regarding all other chaetotaxy features of Ant III-IV (S10 is interpreted as S9 by Zon et al. 2014), head (sd with fewer chaetae – sd2–5? – and c row absent) and tibiotarsi (although not drawn, M chaetae position is clearly basally displaced), it can also be placed inside the *Arlesia* group.

In short, as advocated by Queiroz and Fernandes (2011) and Zon et al. (2014), it should be reinforced that *Furculanurida*, now with 15 species (see Table 2), needs to be revised. We estimate that after such review, only these three species, *Furculanurida boiuna* sp. nov., *Fu. emucronata* and *Fu. africana*, would remain within the genus. However, further analyses must be made to stablish species relationships as well as their character evolution.

Ectonura Cassagnau, 1980

Ectonura sp.

Material examined. One on slide. BRAZIL, Amazonas State: Presidente Figueiredo municipality, forest leaf litter of Amazon Rainforest, coordinates 02°01′13″S, 59°49′29″W, 20.IV.2008, Hamada, Azevedo, Neiss, Pes & Meneses leg. (CM/MNRJ slide number 2535).

Remarks. Neanurinae includes seven tribes worldwide (Lobellini, Morulodini, Neanurini, Paranurini, Paleonurini and Sensillanurini) and is poorly known in the Neotropical region, with 75 recorded species – around 10% of a total of approximately 750 species. Only Paleonurini occurs in Brazil, with five genera and six species, *Australonura gili* Queiroz & Deharveng, 2014, *Australonura neotropica* Queiroz & Deharveng, 2014, *Ectonura snowdeni*, *Itanura brasiliensis* Arlé, 1959, *Paleonura nuda* Cassagnau & Oliveira, 1990 and *Pronura amazonica* Cassagnau & Oliveira, 1990, the last two found in the Amazon Region (Zeppelini et al. 2018). *Ectonura* includes 17 species around the world, 11 of them from New Caledonia, five from South Africa and only one, *Ectonura snowdeni*, from mountains in the state

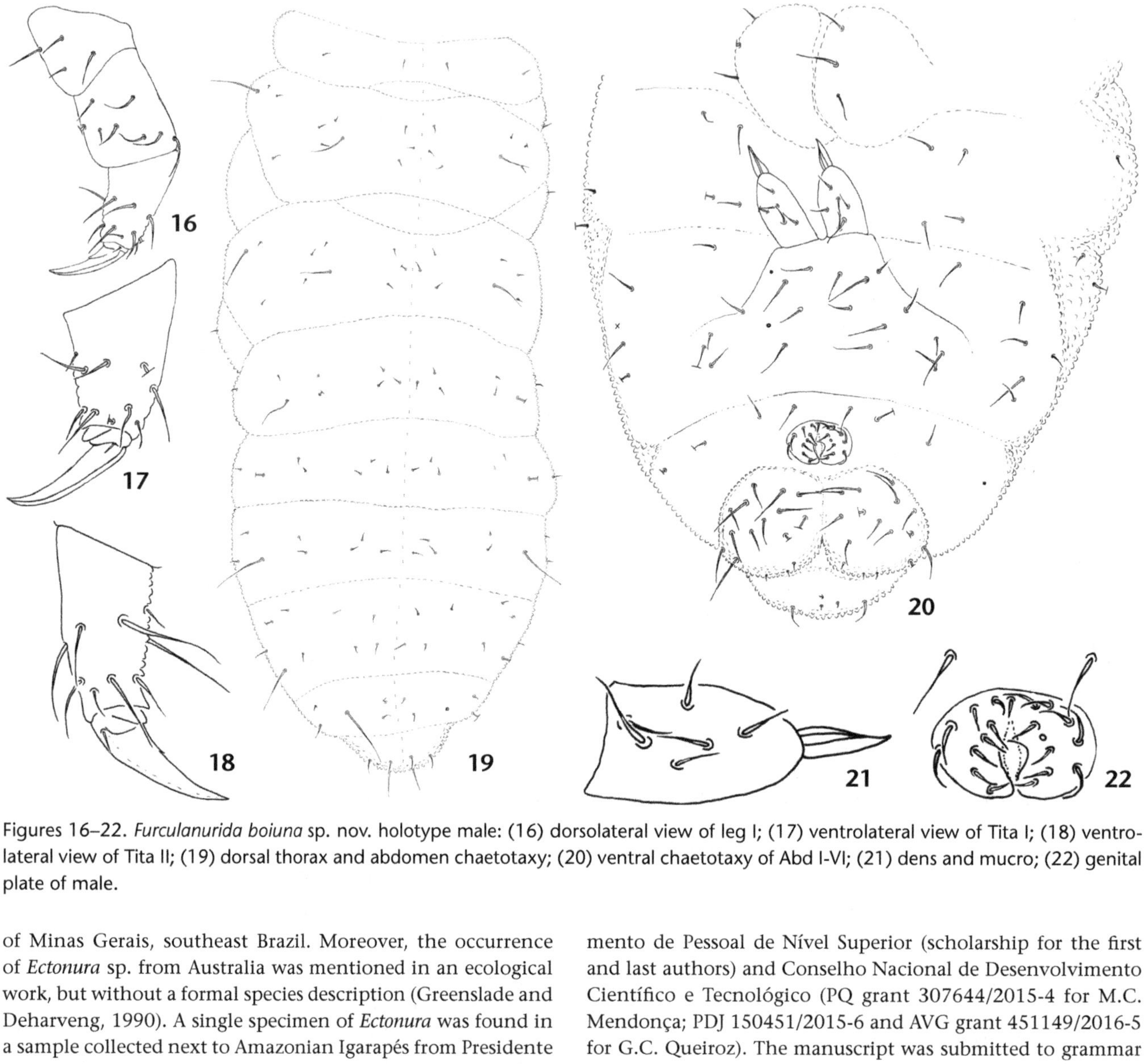

Figures 16–22. *Furculanurida boiuna* sp. nov. holotype male: (16) dorsolateral view of leg I; (17) ventrolateral view of Tita I; (18) ventrolateral view of Tita II; (19) dorsal thorax and abdomen chaetotaxy; (20) ventral chaetotaxy of Abd I-VI; (21) dens and mucro; (22) genital plate of male.

of Minas Gerais, southeast Brazil. Moreover, the occurrence of *Ectonura* sp. from Australia was mentioned in an ecological work, but without a formal species description (Greenslade and Deharveng, 1990). A single specimen of *Ectonura* was found in a sample collected next to Amazonian Igarapés from Presidente Figueiredo municipality. Due to its poor state of conservation, only diagnostic characteristics for the genus level were identifiable, especially the central cephalic region with few tubercles and chaetal group arrangement. Despite being impossible to identify at species level, it still accounts for the second record of the genus for Neotropical Region and the first for the State of Amazonas, which implies a wider distribution within the tropics of the new world.

ACKNOWLEDGEMENTS

Special thanks to Jorge Luiz Nessimian (UFRJ) for sample donation. Financial support from Coordenação de Aperfeiçoamento de Pessoal de Nível Superior (scholarship for the first and last authors) and Conselho Nacional de Desenvolvimento Científico e Tecnológico (PQ grant 307644/2015-4 for M.C. Mendonça; PDJ 150451/2015-6 and AVG grant 451149/2016-5 for G.C. Queiroz). The manuscript was submitted to grammar revision by reviewer André Hoffmann.

LITERATURE CITED

Abrantes EA, Bellini BC, Bernardo AN, Fernandes LH, Mendonça MC, Oliveira EP, Queiroz GC, Sautter KD, Silveira TC, Zeppelini D (2010) Synthesis of Brazilian Collembola: an update to the species list. Zootaxa 2388: 1–22.

Abrantes EA, Bellini BC, Bernardo AN, Fernandes LH, Mendonça MC, Oliveira EP, Queiroz GC, Sautter KD, Silveira TC, Zeppelini D (2012) Errata Corrigenda and update for the 'Synthesis of Brazilian Collembola: an update to the species list'. Abrantes et al., 2010. Zootaxa 2388: 1–22. Zootaxa 3168: 1–21.

Bellinger PF, Christiansen KA, Janssens F (2018) Checklist of the Collembola of the World. http://www.collembola.org [Accessed: 22/06/18]

Cassagnau P (1974) Chétotaxie et phylogénie chez les Collemboles Poduromorphes. Pedobiologia 14: 300–312.

Deharveng L (1983) Morphologie évolutive des Collemboles Neanurinae en particulier de la lignée néanurienne. Travaux du Laboratoire d'Écobiologie des Arthropodes Édaphiques, volume 4, Toulouse, 63 pp.

D'Haese CA (2003) Homology and morphology in Poduromorpha (Hexapoda, Collembola). European Journal of Entomology 101: 385–407. https://doi.org/10.14411/eje.2003.060

Greenslade P, Deharveng L (1990). Australian Species of the Genus *Australonura* Collembola: Neanuridae). Invertebrate Taxonomy 3: 565–93. https://doi.org/10.1071/IT9890565

Lima MG, Gascon C (1999) The conservation value of linear forest remnants in central Amazonia. Biological Conservation 91: 241–247. https://doi.org/10.1016/S0006-3207(99)00084-1

Massoud Z (1967) Monographie des Neanuridae, Collemboles Podurmorphes à pièces buccales modifièes. Biologie de L'Amerique Australe 3: 1–399.

Massoud Z, Thibaud J-M (1980) Les Collemboles des Petites Antilles II. – Neanuridae. Révue d'Écologie et de Biologie du Sol 17(4): 591–605.

Nessimian JL, Dorvillé LFM, Sanseverino AM, Baptista DF (1998) Relation between flood pulse and functional composition of the macroinvertebrate benthic fauna in the lower Rio Negro, Amazonas, Brazil. Amazoniana 15: 35–50.

Potapov MB, Banasko JA (1985) A new species of springtail from Cuba with comments on the role of chaetotaxy in diagnostics of the *Friesea* (Collembola, Neanuridae) species. Zoologicheskii Zhurnal 64(8): 1162–1167.

Queiroz GC, Fernandes LH (2011) New Brazilian species of *Furculanurida* Massoud, 1967 (Collembola: Neanuridae). Zootaxa 2805: 57–64.

Queiroz GC, Mendonça MC (2015) Two new Brazilian species of *Friesea* Dalla Torre, 1895 (Collembola, Neanuridae). Austral Entomology 54: 110–116. https://doi.org/10.1111/aen.12101

Queiroz GC, Zeppelini D (2017) Neotropical Pseudachorutinae (Hexapoda: Collembola: Neanuridae): A comparative morphological study with emphasis on endemic taxa. Zoologischer Anzeiger 269: 127–154. http://dx.doi.org/10.1016/j.jcz.2017.08.005

Thibaud J-M, Palacios-Vargas JG (2000) Remarks of *Stachorutes* (Collembola: Pseudachorutinae) with a new mexican species. Folia Entomologica Mexicana 109: 107–112.

Zeppelini D, Queiroz GC, Bellini BC (2018) Collembola. In: Boeger WA, Zaher H, Rafael JA, Valim MP. Taxonomic Catalog of the Brazilian Fauna. PNUD. http://fauna.jbrj.gov.br/fauna/faunadobrasil/379 [Accessed: 22/06/18]

Zon SD, Tano Y, Deharveng L (2014) A new species of *Furculanurida* (Collembola: Neanuridae: Pseudachorutinae) from Ivory Coast, with comments on related genera. Zootaxa 3878(5): 491–497. http://dx.doi.org/10.11646/zootaxa.3878.5.8

Author Contributions: ACRN applied the methodology; ACRN and MCM conducted the experiments; ACRN, MCM and GCQ analyzed the data; ACRN, MCM and GCQ wrote the paper.
Competing Interests: The authors have declared that no competing interests exist.

12

Taxonomic revision of *Megalothoraca* (Diptera: Richardiidae) with description of a new species, synonyms and new combination

Lisiane Dilli Wendt[1], **Luciane Marinoni**[1]

[1] *Departamento de Zoologia, Universidade Federal do Paraná. Caixa Postal 19020, 81531-980 Curitiba, PR, Brazil.*
*Corresponding author: Luciane Marinoni (*luciane.marinoni@gmail.com*)*

ABSTRACT. A taxonomic revision of *Megalothoracha* Hendel, 1911 is provided, including a description of *Megalothoraca rosalyae* Wendt, **sp. nov.** from Colombia. The genus encompasses large species from Colombia, Ecuador, Peru and Bolivia (new record), and each species was described based either on males or females. These species can be diagnosed by long and slender mid and hind legs and wing longer than body, vein R_{2+3} with two strong curves on apical third, and short spurious vein medially. And males have costal vein conspicuously bowed anteriorly and thickened on medial portion. The following nomenclatural changes are made: *Megalothoraca hendeli* Enderlein, 1912, **syn. nov.** is a junior synonym of *M. pterodontida* Hendel, 1911; *Batrachophthalmum teleopsis* (Hennig, 1938) **comb. nov.** is transferred from *Megalothoraca*, and *Batrachophthalmum quimbaya* Carvalho, Wolff & Wendt, 2011, **syn. nov.** is a junior synonym of *B. teleopsis*. A key to species and illustrations of the female and male terminalia are provided.

KEY WORDS. Identification key, Neotropical, new record, Richardiinae, taxonomy, Tephritoidea.

INTRODUCTION

With about 210 valid species (Steyskal 1968, Wendt and Ale-Rocha 2014a, 2015) classified in 32 genera (Hancock 2010), richardiid flies have been insufficiently sampled in most regions of the world, and many of the included genera need to be revised. The biology of richardiid species has been poorly investigated, but it is known that many are potential pests of some types of fruits (Peñaranda and Ospina 1999, Montilla et al. 2007, 2008, Arellano et al. 2015), while others are saprophagous species associated with animal decomposition (Lopes de Carvalho and Linhares 2001, Moretti et al. 2008, Cavallari et al. 2015).

Megalothoraca Hendel, 1911 includes six large species (8.0–12.4 mm) with long and slender mid and hind legs. The genus has been recorded from Peru, Ecuador (Steyskal 1968), and Colombia (Wendt and Ale-Rocha 2016). Hendel (1911b) first proposed the genus for a species from Peru – *M. pterodontida*, described based on males. Later, Enderlein (1912) described two additional species, also based on males – *M. flava* and *M. hendeli*. In the same paper, the author proposed *Phlebacrocyma*, which he described two new species based on female specimens. Hennig (1938b) described one more species of *Megalothoraca*, an enigmatic fly with long eye stalks – *M. teleopsis*. Three decades later, Steyskal (1968), in his catalogue, synonymized *Phlebacrocyma* with *Megalothoraca*.

Herein, we provide a new diagnosis and redescription of *Megalothoracha* and each of the six previously included species, based on the revision of all types. Additionally, one species from Colombia is described, a new combination and synonymies are proposed, and an identification key and illustrations of the male and female terminalia are provided.

MATERIAL AND METHODS

The external morphological terminology follows Cumming and Wood (2009), and the terminology for the male and female terminalia follows Norrbom and Kim (1988) and White et al. (1999), respectively. Measurements of the head are presented as proportions, which were calculated with the head in anterior view for (1) head width-height: the greatest width of the head, including the eyes, divided by the greatest height of the head, measured from the vertex to the ventral limit of the face; (2) frons-head: the greatest width of the frons divided by the greatest width of the head; (3) frons width-height ratio: the greatest width of the frons divided by the greatest length of the frons, measured from the vertex to the dorsal margin of the lunule; and the head in anterofrontal view for (4) ocelli-distance: distance from the dorsal margin of anterior ocellus to the ventral margin posterior ocelli divided by the distance between the posterior ocelli. Some length measurements differed a little from the measurements given in the original descriptions, which are presented between parentheses after the measurements.

Label data of types are given exactly as written and presented as follow: individual lines are separated with a backslash (\); individual labels are separated with quotes (""), square brackets ([]) indicate information not present on the specimen labels.

All type material of *Megalotharaca* was studied during visits to the following collections (abbreviation and curators are between parenthesis): Hungarian Natural History Museum, Budapest, Hungary (HNHM, Z. Soltész), Polish Academy of Science, Museum of the Institute of Zoology, Warsaw, Poland (MZPW, T. Huflejt), Senckenberg Deutsches Entomologisches Institut, Müncheberg, Germany (SDEI, Frank Menzel), and Museum für Naturkunde, Leibniz Institute for Evolution and Biodiversity Science, Berlin, Germany (ZMHB, J. Ziegler).

Figures 1, 3, 20, 21, 54 to 56 were taken in 2010 by Dominika Mierzwa (MZPW). Figures 2, 6, 12, 14, 16–18, 59, 61–65 were photographed with a camera Canon Powershot SX50HS. The photographs (Figs 4, 5, 7–11, 13, 15, 19, 22, 57, 58, 60) were taken at MZPW and ZMHB, using local equipment, a stereomicroscope Leica with camera attached and were assembled with software Auto-montage® (MZPW) and Helicon Focus (ZMHB).

The holotype of the new species described here, and one non-type specimen of *M. pterodontida* Hendel were borrowed from Colección Entomológica, Universidad de Antioquia, Medellín, Colombia (CEUA, Marta Wolff) and from ZMHB, respectively. This material was studied with a stereoscopic microscope Nikon SMZ800 with camera lucida attached (for confection of drawings). Images of the relevant external structures (Figs 30–40, 48) and measurements of the holotype were obtained through a stereoscopic microscope Leica® M205C, equipped with a digital camera and the software Auto-Montage®. In order to study and to illustrate the male and female terminalia the abdomens were detached and treated according to the protocol of Cumming (1992) and stored in plastic microvials containing glycerine. The photographs of sternites and terminalia of holotypes (Figs 42–47, 49–53) were taken with digital camera (Moticam 2300) attached to a compound microscope with the aid of the software Motic Images Plus 3.0ML (Motic BA210). The wing of the new species was also detached and mounted in Entelan between cover slips, and then glued to a cardboard. Both the vial and the cardboard were attached to the pin of the respective specimen.

TAXONOMY

Megalothoraca Hendel, 1911

Megalothoraca Hendel, 1911a: 395 (type species: *Megalothoraca pterodontida* Hendel, 1911, by monotypy); Hendel 1911a: 182 (key); 1911b: 5, 50–52 (key, redescription, note, distribution, spp. list); Hennig 1938a: 121 (citation); 1838b: 9 (citation); Aczél 1950: 43 (catalogue); Steyskal 1968: 5 (catalogue); Hancock 2010: 877 (key); Carvalho et al. 2011: 43 (key); Alencar et al. 2013: 202 (citation); Wendt and Ale-Rocha 2014b: 306, 308 (citation, discussion); 2016: 592 (catalogue of Colombia, note).

Phlebacrocyma Enderlein, 1912: 105 (type species: *Phlebacrocyma undulosum* Enderlein, 1912, original designation); Hennig 1938a: 121 (citation); Aczél 1950: 44 (catalogue); Steyskal 1968: 5 (catalogue; synonymy).

Diagnosis and comments. *Megalothoraca* species are large (8.0–12.4 mm), with thorax well developed and wing generally longer than body, mid and hind legs long and slender, fore leg generally conspicuously more robust than the others (at least in males), and abdomen oval. Males and females of the included species are unique in lacking spine-like seta on the posteroventral surface of the mid and hind femora; mid femur with only one anteroventral spine-like seta and hind femur with one or two such setae (e.g. Fig. 40); vein R_{2+3} with two conspicuous curves on apical third (Figs 16, 48, 57), and medial spurious vein short (Fig. 48) (absent in *M. flava* Enderlein, 1912). Additionally, *Megalothoraca* males have costal vein conspicuously bowed anteriorly and thickened on medial portion (unmodified in females) (Figs 5, 18, 22), thorax distinctly well developed, setae on the fore and mid coxae stout (Fig. 7) and hind coxa setulose on posterior surface (males of other Richardiinae genera have unmodified costal vein, thorax proportional to the rest of body, unmodified setae on the coxae and hind coxa bare posteriorly – also setulose in *Setellia* Robineau-Desvoidy, 1830).

The following features of *Megalothoraca* are shared with *Euolena egregia* (Gerstaecker, 1860) and species of *Bathachophtalmum* Hendel, 1911: antero- and posteroventral spine-like setae on fore femur distinctly more numerous and robust than on mid and hind femora; palpus straight. Moreover, species of *Megalothoraca* are similar to species of *Batrachophthalmum*, *Euolena* Loew, 1873 (in part) and *Hemixantha* Loew, 1873 (in part) in the male terminalia with medial surstylus bilobed at apex, and prensiseta long and inserted preapically on inner margin. Species of *Megalothoraca* can be easily differentiated by the venation of their wing, thorax well developed, number and thickness of the spine-like setae on mid and hind femora and the presence of setulae on hind coxa, as described above.

The general morphology of *Megalothoraca* species is quite regular, and the chaetotaxy of the head and especially the thorax is very similar among species. The main characteristics used to distinguish them are the coloration of the body and the wing spots. Since all species were described based on either male or female specimens and no species is known from both sexes, it is not possible to ascertain whether there is sexual dimorphism. *Megalothoraca undulosa* (Enderlein, 1912), *M. ferruginea* (Enderlein, 1912), and *M. rosalyae* sp. nov., (known only from females) lack a lobe on the costal vein; have narrower medial band on the wing, lack anterior spine-like setae on the fore and mid coxae, and their thorax is not as well developed as the thorax of males (*M. pterodontida* Hendel, 1911 and *M. flava*).

Included species. *Megalothoraca ferruginea*, *M. flava*, *M. pterodontida*, *M. rosalyae* sp. nov., *M. undulosa*.

Redescription. Large species, length 8.2–12.4 mm, commonly with wing longer than body (except in *M. ferruginea*).

Robust body, with general colour yellow (Fig. 5) or reddish brown (Fig. 1) to dark brown (Figs 15, 31, 55), generally non-metallic (except abdomen of *M. rosalyae* sp. nov., Fig. 30). **Head.** In dorsal view, slightly narrower than thorax; in frontal view, wider than high. Eye oval, higher than wide. Vertex, in frontal view, straight (Figs 10, 58) or slightly concave (in *M. rosalyae* sp. nov., Fig. 32). Frons not protruded anteriorly (Figs 9, 19, 34). Ocellar triangle slightly elongate, anterior ocellus inserted on dorsal third or fourth of frons; bright, non-pollinose; convex, contoured by a weak (or stronger in *M. rosalyae* sp. nov., Figs 32, 33) cavity. Lunule developed. Antennal bases narrowly separated by less than width of scape; scape slightly elongate, almost as long as pedicel. First flagellomere short, at most 2.5 times longer than high. Arista long pubescent. Clypeus well developed. Palpus elongate and linear, base, mid region and apex similar in width (Figs 9, 19). Postcranium, in lateral view, distinctly more developed on ventral two-thirds (Figs 9, 19, 34). Chaetotaxy: setae long and robust; ocellar seta present or absent, orbital and genal setae present, postocellar and postgenal setae absent, outer vertical seta distinctly longer and thicker than inner vertical seta, with alveolus robust and elevated (e.g. Figs 10, 32, 33). **Thorax.** Especially in males, thorax distinctly well developed and robust (Figs 5, 15). Scutum elongated, gradually widened posteriorly (Figs 3, 13, 37, 54), with two longitudinal and parallel excavated rows, distinctly visible (e.g. Figs 37, 38); setulae sparsely and randomly distributed; weakly pollinose. Scutellum bare. Subscutellum developed. Prosternum, proepimeron, metasternum, and metapleuron bare. Katatergite not setulose, sparsely pollinose. Metathoracic spiracle conspicuously setulose at posterior margin. Postmetacoxal bridge completely sclerotized. Chaetotaxy: very reduced in number, scapular seta, postpronotal seta (except in *M. rosalyae* sp. nov.), presutural supra-alar, postsutural surpa-alar, intra-alar, proepimeral and katepisternal setae absent; postalar seta present, robust; one dorsocentral and one scutellar setae stout; posterior notopleural seta distinctly stouter and longer than anterior notopleural seta; postalar seta developed, robust. **Legs.** Elongate. Males with fore (Fig. 7) and mid coxae with one robust anterior apical spine- like seta; females without these spines, only with long setae. Fore femur distinctly (in males, e.g. Figs 5, 15) or slightly (in female, e.g. Figs 39, 40) thicker than mid and hind femora (except in *M. ferruginea*, similar in thickness), with stout antero- and posteroventral spine- like setae (generally number in left and right femora varies in the same specimen), and more numerous than on mid and hind femora (e.g. Figs 15, 21, 39, 40). Mid and hind legs slender and long (Figs 1, 5, 15, 31). Mid and hind femora sparsely setulose, without posteroventral spine-like seta, mid femur with only one anteroventral spine-like seta (e.g. Fig. 1), and hind femur with one (e.g. Figs 1, 15) or two spine-like setae (e.g. Fig. 40). Hind coxa setulose on posterior surface in males, bare in females. Hind femur without anterodorsal preapical seta. **Wing** (Figs 16, 18, 22, 48, 57). Long, generally longer than body length, 8.0–13.8 mm. Pattern conspicuous, including basal, medial and apical marks (Fig. 22, each mark indicated). Males with costal vein conspicuously bowed anteriorly and thickened at their medial portion in males, unmodified in females. (Figs 18, 48). Veins bare. Vein R_{2+3} with two conspicuous or inconspicuous (in *M. rosalyae* sp. nov., Fig. 48) curves in apical third; with short postmedial spurious vein (absent in *M. flava*). Veins R_{4+5} and M converging apically. Vein r-m at distal third to fifth of cell dm. Vein A_1+CuA_2 long, becoming weaker towards margin. Alula and anal lobe well developed. **Abdomen.** Dark brown (e.g. Figs 20, 54), sometimes with blue sheen (Fig. 30); metallic or non-metallic; smooth, non-punctate, and sparsely pollinose or shining. Shape oval (Fig. 41), syntergite 1+2 with base about one third as wide as apex, and with one or two pairs of preapical laterodorsal setae. **Female sternites and terminalia** (only female of *M. rosalyae* sp. nov. was dissected and examined, Figs 42–47, 49–53). Sternite 1 present, setulose (Fig. 42). Sternites 3–5 with short basal apodemes (Figs 44–46). Sternite 6 with basal round hole (Fig. 47). Oviscape short, almost as wide as long (Fig. 53). Eversible membrane poorly sclerotized, with distal half covered with oblique rows of multidentate scales. Tergite and sternite 8 with pair of rows of moderately long setulae in distal two-thirds. Tip of cerci rounded (Figs 49, 50). Two subglobose spermathecae strongly sclerotized. **Male sternites and terminalia** (only male of *M. pterodontida* was dissected and observed, Figs 23–29). Sternite 1 present and setulose (Fig. 23). Sternites without basal apodemes (Figs 23–27). Spiracles 6 and 7 present. Epandrium densely setulose, globose (Fig. 28), in lateral view, higher than wide (Fig. 29). Cerci, in lateral view, developed, and globose, densely setulose (Fig. 29). Medial and lateral surstyli similar in length. Lateral surstylus without preapical projection. Medial surstylus with a submedial anterior rounded projection; bilobate apically; with long and acuminate preapical prensiseta.

Distribution. Colombia, Ecuador, Peru, Bolivia (new record).

Identification key to the species of *Megalothoraca* Hendel

1 Thorax and legs entirely yellow (Figs 5, 9). Vein R_{2+3} without spurious vein. [Ecuador]....*M. flava* Enderlein, 1912 [♂]

1′ Thorax and at least some parts of legs reddish brown, or pale brown to dark brown (Figs 1, 13–15, 21, 31, 55). Vein R_{2+3} with short spurious vein (Figs 22, 48, 57; spurious veins indicated by arrows) 2

2 Head and thorax entirely reddish brown (Figs 1, 3). Fore and mid coxae and femora pale brown (Fig. 1). [Ecuador] ....*M. ferruginea* (Enderlein, 1912) [♀]

2′ Head and thorax entirely dark brown (Figs 13, 15, 19, 20, 55), except in *M. rosalyae* sp. nov. with part of frons reddish brown (Figs 32–36). Fore and mid coxae and femora brown to dark brown. 3

3 Tarsi yellowish white, contrasting with dark brown leg (Figs 55, 56). Female with sternite 6 and oviscape yellow, and rest of abdomen dark brown, almost black (Figs 56, 60). [Ecuador] .... *M. undulosa* (Enderlein, 1912) [♀]

3′ Tarsi not contrasting with remaining other portions of leg, brownish (Fig. 31), or if yellow to brownish yellow, tibiae and part of femora also brownish yellow (Figs 14, 15). Female with sternite 6 and oviscape dark brown, almost black (Fig. 53) (female of *M. pterodontida* unknown)........ 4

4 Ocellar seta absent (Fig. 33) and postpronotal seta present. Postpronotal lobe paler than scutum (Figs 37, 38). Abdomen metallic dark blue (Fig. 30). [Colombia] .... *M. rosalyae* Wendt, sp. nov.

4′ Ocellar seta present and postpronotal seta absent. Postpronotal lobe concolorous with scutum (Fig. 15). Abdomen dark brown (Figs 13, 15). [Ecuador, Peru, Bolivia]..... .... *M. pterodontida* Hendel, 1911 [♂]

Megalothoraca ferruginea (Enderlein, 1912)

Figs 1–4

Phlebacrocyma ferruginea Enderlein, 1912: 106. Type locality: Santa Inez, Ecuador. Hennig 1938a: 121 (discussion); Aczél 1950: 44 (catalogue).

Megalothoraca ferruginea Steyskal, 1968: 5 (catalogue).

Diagnosis and comments. *Megalothoraca ferruginea*, known from a female specimen, is the only species within genus with head and thorax entirely reddish brown, fore and mid coxae, fore and mid femora pale brown (Figs 1, 3) (black in *M. undulosa* and most other species, except yellow in *M. flava*).

Redescription. Holotype female, MZPW: Measurements (in mm): body length 8.20 (8.50 according to Enderlein 1912: 106); wing length 8.00 (8.25 according to Enderlein 1912: 106). **Head** (Figs 1, 3). Extremely damaged, especially oral region and eyes, ratios not measured, and some parts not observed. Anterior ocellus inserted on posterior third of frons. Reddish brown, except ocellar triangle and vertex brown. Vertex in frontal view straight, not concave. Frons and postcranium sparsely whitish and brownish pollinose, respectively. Antenna brownish. Clypeus, palpus, gena and prementum missing. Ocellar seta slightly longer than orbital seta; inserted at same level. **Thorax** (Figs 1, 3). Entirely reddish brown, with scutellum, subscutellum and mediotergite slightly darker. Scutum 1.33 times longer than wide. Postpronotal seta absent. **Legs** (Fig. 1). Fore coxa, trochanter, and femur yellowish orange; mid and hind coxae, trochanters, and femora reddish brown; all tibiae brown; all tarsomeres yellowish. Fore and mid coxae without anterior spine. Femora similar in thickness. Left fore femur with two anteroventral and three posteroventral spine-like setae, and right fore femur with two antero- and posteroventral spine-like setae; setae different in size, occupying apical half of femur. Fore and mid coxae without spine-like seta. Mid and hind femora with one stout preapical anteroventral spine-like seta; stouter than those on fore femur. Hind coxa bare on posterior surface. **Wing** (Fig. 1). Generally hyaline except for the following marks: cell bc yellowish; faint and diffuse brownish basal band from vein R_1 at level of costal break, passing on base of bifurcation between R_{2+3} and R_{4+5}, and on vein bm-cu; oblique yellow area in cell r_1, from costal break reaching medial band; incomplete and narrow medial brown band, from costal margin to vein CuA_1, passing on spurious vein and on vein r-m, with irregular margins; and wider and conspicuous apical brown spot, faint in cell m, with proximal margin irregular, almost straight. Vein R_{2+3} from apical third equally twice sinuous. Spurious vein faint and short, at same level of vein r-m. Vein M, distal to vein r-m, sinuous and curved towards R_{4+5}. Vein r-m at distal third of cell dm length. Halter with stem brownish and knob brown. **Abdomen** (Figs 1, 3). Entirely dark brown, except contrasting tergite 5 yellowish orange. Tergites brown setulose, very sparsely whitish pollinose. Syntergite 1+2 with two pairs of weak preapical laterodorsal setae. Sternites 1–4 brownish, sternite 5 yellowish orange. **Female terminalia** (Fig. 4). Not dissected. Oviscape reddish yellow. Tip of cerci rounded.

Male. Unknown.

Type-material examined. Holotype female, MZPW: "Santa Inéz\(Ecuad. [Ecuador])\R.Haensch S."; "Type" [red label]; "Phlebacrocyma\ferrugineum\Type Enderl. ♀ [handwritten]\ Dr. Enderlein det. 1912" (Fig. 2). The holotype is pinned, in poor condition (Figs 1, 3): head damaged, with eyes, oral and genal regions, and postcranium ventrally missing, right anterior region of anepisternum and katepisternum with a hole; oviscape, apex of segment 8, and cerci exposed (Fig. 4).

Distribution. Ecuador.

Megalothoraca flava Enderlein, 1912

Figs 5–11

Megalothoraca flava Enderlein, 1912: 104. Holotype male (MZPW). Type locality: Santa Inez, Ecuador. Hennig 1938a: 121 (citation); Steyskal 1968: 5 (catalogue).

Diagnosis and comments. *Megalothoraca flava*, known only from a male, is distinct by having thorax and legs entirely yellow (Fig. 5) (other species are brownish, brown or black), spurious vein in R_{2+3} absent, and apical brown spot not completely homogenous, with faint hyaline areas on cells r_{2+3} and r_{4+5}.

Redescription. Holotype male, MZPW: Measurements (in mm): body length 12.4 (11.75 according to Enderlein 1912: 105); wing length 13.8 (13.5 according to Enderlein 1912: 105). **Head** (Figs 9–11). Head width-to-height ratio 1.36. Frons-to-head ratio 0.33. Frons width-to-height ratio 0.72. Ocelli-distance ratio 2.14; anterior ocellus inserted at dorsal third of frons (Figs 10, 11). Brown, except prementum, palpus, and posterior half of postcranium yellow (Fig. 9). Vertex, in frontal view, straight. Frons sparsely whitish pollinose. Parafacial, gena and ventral two-thirds of postcranium, especially at contour eye and extending to occipital foramen, densely whitish pollinose. Vertex and dorsal third of postcranium bright. Postcranium sparsely brown setulose. Orbital and outer vertical areas, extended to vertex, slightly tumescent and bright (Fig. 11). Ocellar seta present, longer than orbital seta, and inserted at same level. **Thorax** (Figs 5, 9). Entirely yellow. Scutum 1.54 times longer than wide. Postpronotal seta absent. **Legs** (Figs 5, 7–9). Entirely yellow. Fore (Figs 7, 8) and mid coxae with strong anterior spine-like

Figures 1–4. *Megalothoraca ferruginea*, holotype female, MZPW: (1) habitus lateral; (2) labels; (3) habitus dorsal; (4) oviscape, part of segment 8, and cerci. Photographs 1 and 3 were provided by Dominika Mierzwa, MZPW. Scale bars: 1, 3, 4 = 0.2 mm, 2 = 0.5 mm.

seta. Fore femur thicker than mid and hind femora. Left fore femur with five anteroventral and three posteroventral spine-like setae, and right fore femur with six anteroventral and three posteroventral spine-like setae; setae different in size, occupying apical two-thirds of anteroventral surface and apical third of posteroventral surface of femur. Mid and hind femora with one preapical anteroventral spine-like seta, thinner than those on fore femur. Hind coxa setulose on posterior surface. **Wing.** Generally hyaline except for the following marks: conspicuous brown basal band from vein R_1, from base of bifurcation between R_{2+3} and R_{4+5} to vein bm-cu, distinctly wider in cell r_1; wide medial brown band, from costal vein to vein M, wider at distal margin in cell r_1, but not reach apical spot, and more diffuse in cell dm, with proximal and distal margins irregular; apical brown spot, as wide as medial band, with diffuses hyalines areas in cells r_{2+3} and r_{4+5}, very faint in cell m, with proximal margin irregular. Vein R_{2+3} from apical third equally twice conspicuous and shortly sinuous. Spurious vein absent. Vein M, distal to vein r-m, sinuous and curved towards R_{4+5}. Vein r-m at distal fifth of cell dm. Halter pale brown. Abdomen. Brown with metallic reflexes, except basal half of syntergite 1+2 and tergite 5 reddish brown. Tergites sparsely and shortly brownish setulose, non pollinose. Syntergite 1+2 with one pair of robust and one pair of weak preapical laterodorsal setae. Sternites brown. **Male terminalia.** Not dissected. Epandrium and cerci brownish yellow.

Female. Unknown.

Type-material examined. Holotype male, MZPW: "Santa Inéz\(Ecuad. [Ecuador])\R.Haensch S."; "Type" [red label]; "Megalothoraca\flava\Type Enderl. ♂ [handwritten]\Dr. Enderlein det. 1912" (Fig. 6). The holotype is pinned and relatively well preserved, with no part missing; scutum, scutellum, and median region of costal vein of left wing damaged (Fig. 5).

Distribution. Ecuador.

Megalothoraca pterodontida Hendel, 1911

Figs 12–19, 23–29

Megalothoraca pterodontida Hendel, 1911a: 395. Type locality: Peru, Cagon (1000m). Hendel 1911b: 52 (distribution, plate 3, figs 74, 75: habitus, head); Hennig 1938b: 9 (discussion); Aczél 1950: 43 (catalogue); Steyskal 1968: 5 (catalogue).
Megalothoraca hendeli Enderlein, 1912: 104. Type locality: Ecuador, Santa Jnéz. Hennig 1938a: 121 (material list); Steyskal 1968: 5 (catalogue from Neotropical and United States). **New synonymy.**

Diagnosis and comments. *Megalothoraca pterodontida*, known from male specimens, is recognized by body mostly dark brown (Figs 13–15); medial band of wing wide, complete from costal to posterior margins, and with a rounded hyaline area anterior to vein M in cell dm, band continuous in cell r_1 reaching apical spot (Figs 13, 16). Here, *M. hendeli* is recognized as a junior synonym of *M. pterodontida* since there are no significant differences between the three syntype of *M. hendeli* (Figs 20–22) from Ecuador and the holotype of *M. pterodontida* (Figs 12, 14, 16) from Peru.

Redescription. Holotype male (Figs 12, 14, 16, 18), HNHM: Measurements (in mm): body length 11.5 (12.0 according to Hendel 1911a: 396); abdomen length 4.1 (4.0 according to Hendel 1911a: 396); wing length 13.3 (13.0 according to Hendel 1911a: 396). **Head** (Fig. 14). Head width-to-height ratio 1.31. Frons-to-head ratio 0.39. Frons width-to-height ratio 0.98. Ocelli-distance ratio 2.8; anterior ocellus inserted at dorsal third of frons. Dark brown, with ocellar triangle and vertex almost black; clypeus and prementum dark brown, palpus brown. Vertex, in frontal view, straight. Frons and face sparsely brownish pollinose. Parafacial, gena, and ventral two-thirds of postcranium, especially at contour eye and extending to occipital foramen, densely whitish pollinose. Vertex and dorsal third of postcranium bright. Postcranium sparsely black setulose. Scape and

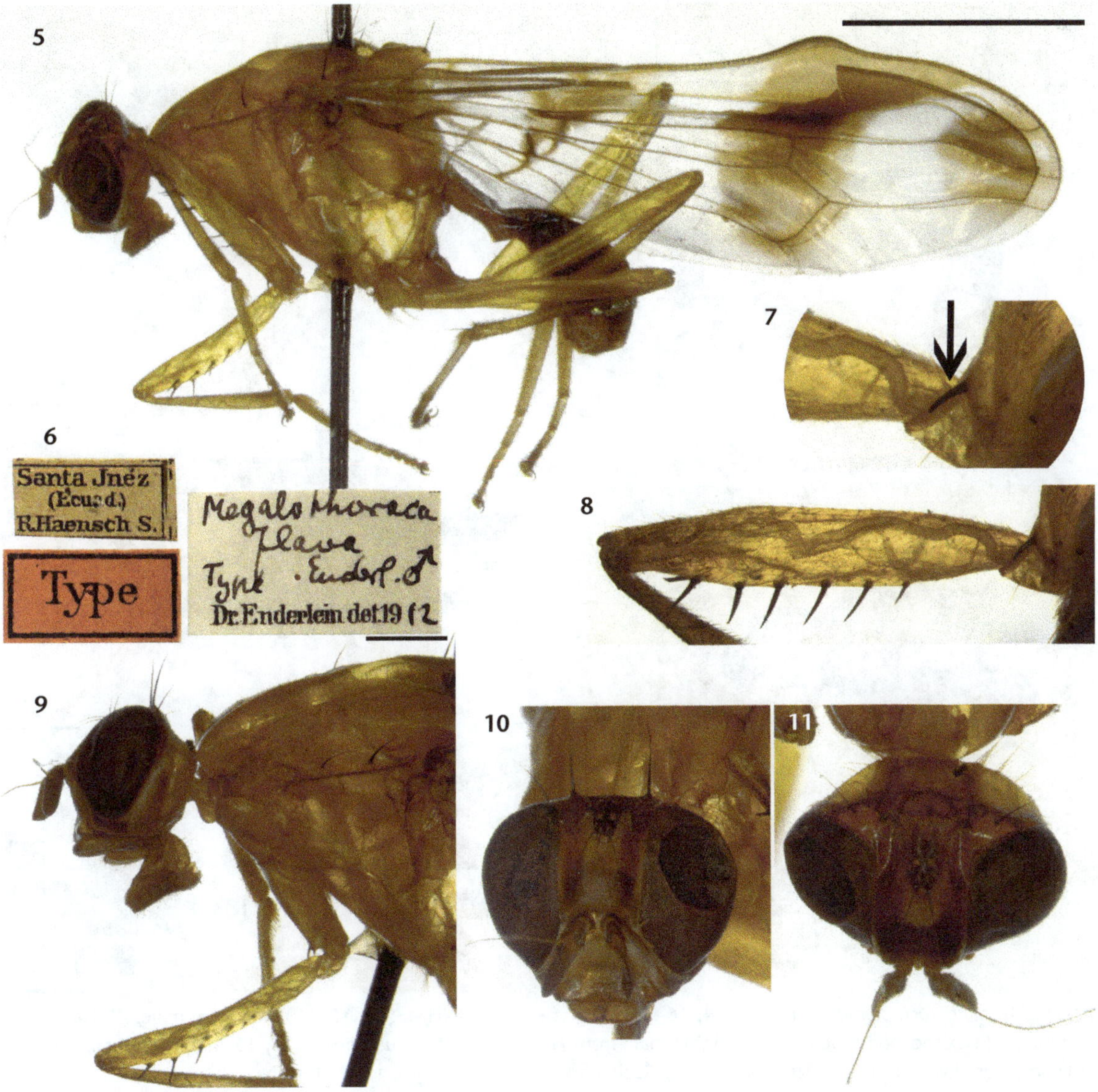

Figures 5–11. *Megalothoraca flava*, holotype male, MZPW: (5) habitus lateral; (6) label; (7) spine-like seta on fore coxa, in detail; (8) fore femur, anterior view; (9) head and thorax, lateral; (10) head, frontal; (11) head, dorsal. Scale bars = 0.5 mm.

pedicel dark brown, flagellomere and arista slightly paler. Orbital and outer vertical areas, extended to vertex, slightly tumescent and bright. Ocellar seta present, longer than orbital seta, and inserted at same level. **Thorax** (Fig. 14). Entirely reddish dark brown, with postpronotal lobe and notopleuron densely whitish pollinose. Scutum 1.41 times longer than wide. Postpronotal seta absent. **Legs** (Fig. 14). All coxae, trochanters, and fore femur reddish dark brown; all tarsomeres yellow; fore tibia, mid and hind femora at basal fourth and apical fifth, mid and hind tibiae at basal third brownish, remainder of mid and hind femora and of tibiae yellowish orange. Fore and mid coxae with strong anterior spine-like seta. Fore femur distinctly thicker than mid and hind femora. Left fore femur missing, right fore femur with seven anteroventral and four posteroventral spine-like setae; setae distinctly different in size, occupying apical three-fourths of anteroventral surface, and apical two-thirds of posteroventral surface of femur. Mid and hind femora with one preapical spine-like seta, thinner than those on fore femur. Hind coxa setulose on posterior surface. **Wing** (Figs 16, 18). Generally hyaline except for the following marks: cell bc yellow and cell c slightly yellowish; faint and diffuse brownish basal band from vein R_1, from base of bifurcation between R_{2+3} and R_{4+5} to vein bm-cu; yellow area in cell r_1 from costal break to proximal margin of medial spot; complete wide brown medial band, from costal vein to

Figures 12–19. *Megalothoraca pterodontida*: (12, 14, 16–18) Holotype male, HNHM; (13, 15, 19) Non-type male, ZMHB. (12) Photo of box with holotype (left) and non-type material (right) male from HNHM; (13) habitus dorsal; (14, 15) habitus lateral; (16) wing; (17) labels; (18) wing, costal vein in detail; (19) head, lateral. Scale bars: 14–17 = 2.0 mm, 19 = 1.0 mm.

posterior margin and continuous in cell r_1, reaching apical spot, with a rounded hyaline area anterior to vein M in cell dm, with proximal and distal margins irregular; and apical brown spot, as wide as medial band, very faint in cell m, with proximal margin rounded. Vein R_{2+3} from apical third equally twice conspicuous and shortly sinuous. Spurious vein faint, slightly distal to vein r-m. Vein M, distal to vein r-m, sinuous and curved towards R_{4+5}. Vein r-m at distal fifth of cell dm. Halter pale brown. **Abdomen.** Tergites brown (Fig. 14), sparsely and shortly brown setulose, sparsely whitish pollinose, denser at basal half of syntergite 1+2. Syntergite 1+2 with one pair of strong preapical laterodorsal setae. Sternites dark brown. Holotype not dissected, sternites visualized from one non-type specimen from Peru (ZMHB) (Figs 23–27): sternites without basal apodeme. Sternite 1 (Fig. 23) relatively well developed, one-third of sternite 2 length, convex at apex and almost straight at apex, densely short setulose, setulae similar in length. Sternite 2 (Fig. 24) divided into two plates, twice longer than wide, densely short setulose, setulae similar in length. Sternite 3 (Fig. 25) sub rectangular, with base straight and narrower, twice longer than wide, setulae different in size. Sternites 4 (Fig. 26) sub quadrate, as long as wide, with long and short setulae. Sternite 5 (Fig. 27) sub rectangular, slightly wider than long, with short setulae medially, and long setulae at margin. **Male terminalia.** Holotype not dissected, one non-type specimen from Peru (ZMHB) dissected (Figs 28, 29). Epandrium and cerci brown. Spiracles 6 and 7 presents, similarly developed. Epandrium long setulose. Cerci, in lateral view, developed, and globose, densely setulose. Medial and lateral surstyli similar

Figures 20–22. *Megalothoraca hendeli*, syntype male, MZPW, new synonymy of *M. pterodontida*: (20) habitus dorsal; (21) habitus ventral; (22) wing. Photographs 20 and 21 were provided by Dominika Mierzwa, MZPW. Abbreviations: A. Apical spot; B. Basal band; M. Medial band. Scale bars = 2.0 mm.

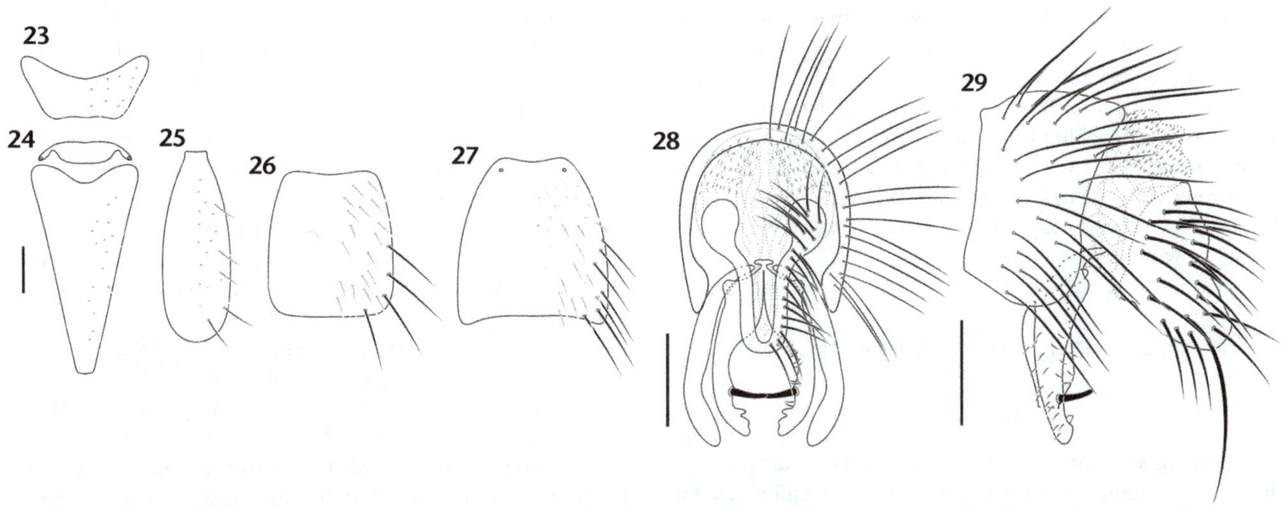

Figures 23–29. *Megalothoraca pterodontida*, non type male, ZMHB, drawing lines: (23–27) sternites, ventral: (23) sternite 1; (24) sternite 2; (25) sternite 3; (26) sternite 4; (27) sternite 5; (28, 29) epandrium, medial and lateral surstyli, and cerci; (28) posterior view; (29) lateral view. Scale bars = 0.2 mm.

in length. Lateral surstylus lateral flattened laterally, without preapical projection. Medial surstylus distinctly wider at basal half; bilobate apically, preapical lobe smaller than apical lobe. Distiphalus with ribbons weakly setulose, apex without set of setae, and with plate poorly sclerotized.

Female. Unknown.

Type-material examined. *Megalothoraca pterodontida* Hendel, holotype male, HNHM: "Cagon\Peru (1000 m.)"; "Megalothoraca\pterodontida\typus" [handwritten]; "TYPUS" [red label] (Fig. 17) (HNHM). Specimen pinned, and in excellent condition, with no part missing, except by the left fore leg from femur; just a little dust attached at body; wing well stretched (Figs 12 (left side), 14). *Megalothoraca hendeli* Enderlein, syntypes male, MZPW: "Santa Inéz\(Ecuad. [Ecuador])\R. Haensch S."; "Type" [red label]; "Megalothoraca\hendeli\Type Enderl. ♂ [handwritten]\Dr. Enderlein det. 1912" (1 male). Specimen well preserved, with no damage, with outstretched wings (Figs 20–22). **Comments:** In the original description, Enderlein (1912) did not designate any specimen as the holotype, nevertheless this specimen has a red label attached to it, where "Type" is written, and the other two specimens have yellow labels written "co-type". Revision of this species have never been provided before, then all specimens must be considered syntypes. "Santa Inéz\(Ecuad. [Ecuador])\R. Haensch S S."; "Co = Typus" [yellow label]; "Megalothoraca\hendeli\Type Enderl. ♂ [handwritten]\Dr. Enderlein det. 1912" (2 males). Both specimens are well preserved, with no damage.

Aditional material. Peru. Cagón, 1000 m [without date and collector], det. Kertész (1 male, HNHM); Department Cuzco, Callanga river, i. Thale des Piñipini [valley], 1900, 1500m, Garlepp col., det. Enderlein (2 males, ZMHB); Bolivia. [La Paz] Mapiri [without date and collector], det. Enderlein, 1919[?] (1 male, ZMHB).

Variation. **Legs.** The specimen from Bolivia has the fore coxa with anterior seta stout, but not as spine-like seta. The number of antero- and posteroventral spine-like on fore femur can moderately vary in greater or lesser number than on the holotype. **Wing.** Non-type material from Bolivia (Fig. 13) and syntypes of *M. hendeli* (Fig. 22, indicated with black arrow) have the extension between medial band and apical spot, in cell r_1, with a hyaline area anterior to first sinuosity of vein R_{2+3}. Besides, the material from Bolivia has the medial band in cell dm incomplete, not reaching the vein dm-cu, and hyaline area of cell dm, anteriorly to vein M, opened (Fig. 13, indicated with black arrow). **Abdomen.** Paler (e.g. non-type from Peru, ZMBH) or darker (e.g. syntypes of *M. hendeli*) than holotype, with tergites reddish brown to dark brown.

Distribution. Ecuador, Peru, and Bolivia (new record).

Megalothoraca rosalyae Wendt, sp. nov.

Figs 30–53

Diagnosis and comments. *Megalothoraca rosalyae* sp. nov., known from a female specimen, can be recognized by general dark brown (almost black) color of the thorax (Figs 31, 37), except postpronotal lobe paler, reddish brown (Fig. 38), and abdomen dark blue, distinctly metallic (Fig. 30), ocellar seta absent and postpronotal seta present; two anterodorsal spine-like setae on mid femur (Fig. 40), and sinuosity of vein R_{2+3} smother, asymmetrically twice sinuous (Fig. 48).

Description. Holotype female, CEUA: Measurements (in mm): body length 8.57; wing length 8.80; abdomen length 3.47. **Head** (Figs 32–36). Head width-to-height ratio 1.33. Frons-to-head ratio 0.35. Frons width-to-height ratio 0.90. Ocelli-distance ratio 1.81; anterior ocellus inserted at dorsal fourth of frons (Figs 33, 36). Frons dark brown on dorsal third and with polished appearance, reddish brown on ventral two-thirds and velvety appearance (Figs 32, 22); densely whitish pollinose. Lunule brown (Fig. 32). Face, gena, and postcranium dark brown, almost black (Figs 32, 34, 36). Parafacial reddish brown (Fig. 32). Vertex, in frontal view, concave (Fig. 32). Face sparsely whitish pollinose. Gena and postcranium on ventral two-thirds, especially at contour eye and extending to occipital foramen, densely whitish pollinose (Figs 34, 36). Vertex and postcranium on dorsal third bright. Postcranium sparsely black setulose. Antenna dark brown, except base of flagellomere slightly paler. Clypeus, palpus and prementum dark brown. Orbital and outer vertical area extended to vertex, distinctly tumescent and bright (Figs 32, 33). Orbital and outer vertical setae missing, visible only alveoli, distinctly marked (Figs 33, 35). Ocellar seta absent. **Thorax** (Figs 37, 38). Dark brown, almost black, except postpronotal lobe brown, with anatergite, mediotergite, and postmetacoxal bridge densely whitish pollinose. Scutum 1.33 times longer than wide. Postpronotal seta present (missing, visible only alveolus). **Legs** (Figs 39, 40). Entirely dark brown, with hind tarsomeres slightly paler. Fore and mid coxae without anterior spine-like seta. Fore femur slightly thicker than mid and hind femora. Left fore femur with one preapical antero- and posteroventral spine-like seta, and right fore femur with one preapical anteroventral and two posteroventral, on apical third, spine-like setae; setae similar in size. Mid and hind femora with one and two anteroventral preapical spine-like setae, respectively; similar in thickness with those on fore femur. Hind coxa bare on posterior surface. **Wing** (Fig. 48). Generally hyaline except for the following marks: cell bc brownish; basal band discontinuous in an irregular brown spot in cell r_1 between costal break and base of bifurcation between R_{2+3} and R_{4+5}, and faint brown spot on vein bm-cu; complete narrow brown medial band, from costal to posterior margin of wing, wider in cell r_1, with irregular margins; wider brown apical spot, faint in cell m, with proximal margin irregular. Vein R_{2+3} twice asymmetric and inconspicuously sinuous, proximal curve beginning from medial region, and distal curve shorter and more conspicuous. Spurious vein conspicuous, at same level of vein r-m. Vein M, distally to vein r-m, almost straight and inclined towards R_{4+5}. Vein r-m at distal two-fifths of cell dm. Halter whitish. **Abdomen** (Figs 30, 41–47). Tergites dark blue, with metallic blue reflexes (Fig. 30); sparsely and shortly whitish setulose, not pollinose, except basal third of syntergite 1+2 distinctly whitish pollinose (Fig. 41). Sternites dark brown (brown when dissected), densely setulose (Figs 42–47), sternites 3–5 with basal apodeme (Figs 44–46). Sternite 1 (Fig. 42) relatively well developed, one-third of sternite 2 size, semicircular, densely short setulose, setulae similar in length. Sternite 2 (Fig. 43) divided into two plates, twice longer than wide, densely short setulose, setulae similar in length. Sternite

Figures 30–40. *Megalothoraca rosalyae* sp. nov., holotype female, CEUA: (30) habitus dorsal; (31) habitus lateral; (32–36) head: (32) frontal; (33) laterofrontal, dorsal region, in detail; (34) lateral; (35) dorsal, ocellar triangle, in detail; (36) dorsal, general view; (37) thorax, dorsal; (38) thorax, lateral; (39) fore femur, posterior view; (40) mid and hind femur, anterior view. Scale bars: 30, 31 = 2.0 mm, 32, 34, 36–40 = 1.0 mm.

3 (Fig. 44) subrectangular, twice longer than wide, with long and short setulae. Sternites 4 (Fig. 45) and 5 (Fig. 46) spatulate, as long as wide (at largest width) (Fig. 45), and wider than long (Fig. 46), respectively, with long and short setulae. **Female terminalia** (Figs 49–53). Oviscape dark brown, with paler central apical mark (Fig. 53). Taenia brownish, slightly less sclerotized than oviscape; two-fifths oviscape length (Fig. 53). Eversible membrane as long as taenia, with denticles on basal half. Sternite and tergite 8 ciliated, except basally, similar in length (Figs 49, 51). Sternite 8 0.84 mm long. Tip of cerci rounded and base slightly concave centrally, vaguely resembling a heart shape (Figs 49, 50). Setae on cercus (Fig. 50): two long basal setae, one ventral and one more lateral dorsal; ventrally two very short postmedial setae; two long preapical setae, more apical one shorter; and very short and thin setulae distributed on cerci (Fig. 50, visible only alveoli of these setulae). Two spherical and strongly sclerotized spermathecae, one slightly larger than other (Fig. 52).

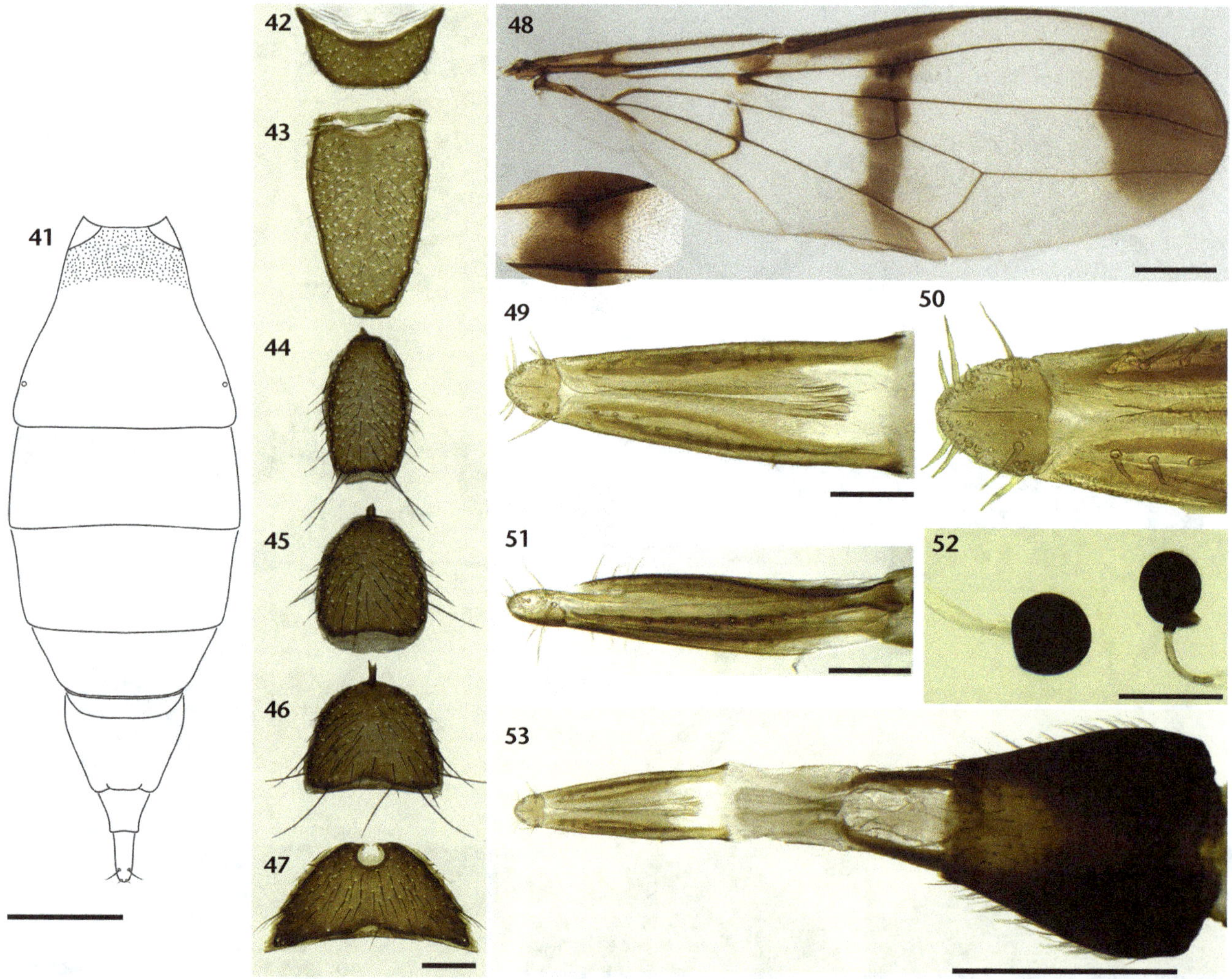

Figures 41–53. *Megalothoraca rosalyae* sp. nov., holotype female, CEUA: (41) Line drawing of abdomen, dorsal; (42–47) sternites, ventral: (42) sternite 1; (43) sternite 2; (44) sternite 3; (45) sternite 4; (46) sternite 5; (47) sternite 6; (48) wing (spurious vein in detail, left side); (49–51) cerci and segment 8: (49) general view, ventral; (50) in detail, ventral; (51) general view, lateral; (52) spermatechae; (53) oviscape, taenia, eversible membrane, segment 8, and cerci. Scale bars: 41, 48, 53 = 1.0 mm, 42–47, 49, 51, 52 = 0.2 mm.

Male. Unknown.

Type-material. Holotype female, CEUA: "Colombia, Antioquia, Bolivar\Vereda la Mina, Bosques\cercanos a la mina, 1900m.\5.80358°N 76.0691°W, Manual\24.iv.2011. E. Garcias col. CEUA"; "Holotype (♀) CEUA*Megalothoraca rosalyae*\Wendt & Marinoni, 201_" [red label] (CEUA). Specimen pinned, in excellent condition, with no damage (Figs 30, 31); almost setae on head and thorax missing (visible only alveoli); left wing detached and mounted between cover slips; abdomen dissected and stored in plastic microvial with glycerine.

Distribution. Colombia.

Etymology. The epithet honors the Brazilian dipterist Rosaly Ale-Rocha, in recognition of her extensive contributions to the taxonomy of Diptera.

Megalothoraca undulosa (Enderlein, 1912)

Figs 54–60

Phlebacrocyma undulosa Enderlein, 1912: 105 (fig. 6: wing). Type locality: Ecuador, Santa Inéz. Hennig 1938a: 121 (citation); Aczél 1950: 44 (catalogue).

Megalothoraca undulosa Steyskal, 1968: 5 (catalogue, new combination).

Diagnosis and comments. *Megalothoraca undulosa*, known from a female specimen, can be distinguished from all other species within the genus by having all tarsi yellowish white, sharply contrasted with the rest of leg, dark brown (Figs 55, 56). The other species have the legs unicolor. The female sternite 6 and oviscape are yellow, highlighted from remainder of abdomen dark brown, almost black (Figs 56, 60).

Figures 54–60. *Megalothoraca undulosa*, holotype female, MZPW: (54) habitus dorsal; (55) habitus lateral; (56) habitus ventral; (57) wing; (58) head, frontal; (59) label; (60) oviscape. Photographs 54 to 56 provided by Dominika Mierzwa, MZPW. Scales bar: 54–57 = 2.0 mm, 59 = 5.0 mm.

Redescription. Holotype female, MZPW: Measurements (in mm). Body length 10.0; abdomen length 4.6; wing length 9.5 (9.0 according to Enderlein 1912: 106). **Head** (Figs 54–56, 58). Head width-to-height ratio 1.34. Frons-to-head ratio 0.35. Frons width-to-height ratio 0.95. Ocelli-distance ratio not measured; anterior ocellus inserted at dorsal third of frons (Fig. 58). Entirely dark brown, with ocellar triangle and vertex darker. Vertex, in frontal view, straight (Fig. 58). Frons and face sparsely whitish pollinose. Gena and ventral two-thirds of postcranium, at contour eye and extended to occipital foramen, densely greyish pollinose. Vertex and dorsal third of postcranium with polished appearance. Postcranium sparsely black setulose. Antenna brown. Clypeus, prementum, and palpus dark brown. Orbital and outer vertical area, extended to vertex, slightly tumescent and bright. Ocellar and orbital similar in length and thickness, inserted at same level. **Thorax** (Figs 54, 55). Entirely reddish dark brown, with anatergite, mediotergite, and postmetacoxal bridge densely whitish pollinose. Scutum 1.33 times longer than wide. Postpronotal seta absent. **Legs** (Figs 55, 56). Reddish dark brown, with tarsomeres yellowish white, sharply contrasting with the rest of leg (Fig. 56). Fore and mid coxae without anterior spine. Fore femur slightly thicker than mid and hind femora. Left fore femur with two antero- and posteroventral spine-like setae, and right fore femur with two anteroventral and three posteroventral spine-like setae; different in size; occupying apical half of femur. Mid and hind femora with one preapical anteroventral spine like seta; stouter than those on fore femur. Hind coxa bare on posterior surface. **Wing** (Fig. 57). Generally hyaline except for the following marks: cell bc yellow, and cell c slightly yellowish; faint and diffuse brownish basal band from vein R_1 at level of costal break, from base of bifurcation between R_{2+3} and R_{4+5} to vein bm-cu; oblique brownish yellow area in cell r_1 from costal break to medial band; incomplete and very narrow brown medial band, from costal margin to vein CuA_1, passing on spurious vein and vein r-m, with irregular margins; and wider and conspicuous apical brown spot, faint in cell m, with proximal margin irregular. Vein R_{2+3} from apical third equally twice conspicuous and shortly sinuous. Spurious vein conspicuous, at same level of vein r-m. Vein M, distally to vein r-m, curved or inclined towards R_{4+5}. Vein r-m at distal third of cell dm. Halter brown. **Abdomen** (Figs 54–56). Entirely reddish dark brown, almost black, except contrasting tergite 6 yellow orange. Tergites brownish setulose, not pollinose. Syntergite 1+2 with two pairs of preapical laterodorsal setae. Sternites 1–4 black, sternite 6 yellow orange. **Female terminalia** (Fig. 60). Not dissected. Oviscape yellow, contrasting with remainder of abdomen, dark brown. Tip of cerci rounded.

Male. Unknown.

Figures 61–65. (61–64) *Batrachophthalmum teleopsis*, holotype male, SDEI: (61) habitus lateral; (62) labels; (63) habitus frontal; (64) wing; (65) *Batrachophthalmum quimbaya* (= *B. teleopsis*), paratype male, DZUP, hind leg, anterior view. Scale bars: 61, 63–65 = 1.0 mm, 62 = 0.5 mm.

Type-material examined. Holotype female, MZPW: "Santa Inéz\(Ecuad. [Ecuador])\R. Haensch S."; "Type" [red label]; "Megalothoraca\undulosum\Type Enderl. ♀ [handwritten]\Dr. Enderlein det. 1912" (Fig. 59). The holotype is well preserved, with no damage, except left wing missing (Figs 54–55); oviscape and cerci exposed (Fig. 60).

Distribution. Ecuador.

Batrachophthalmum teleopsis (Hennig, 1938), comb. nov.

Figs 61–64

Megalothoraca teleopsis Hennig, 1938b: 9 (fig. 3: head). Type locality: Colombia, San Antonio. Aczél 1950: 44 (catalogue); Steyskal 1968: 5 (catalogue); Carvalho et al. 2011: 42, 43 (citation, key).

Batrachophthamum quimbaya Carvalho, Wolff & Wendt, 2011: 44 (figs 6–10: male head, habitus, wing, and thorax). Type locality: Colombia, Pereira. Distribution: Colombia. References: Wendt and Ale-Rocha 2014b: 306–310 (female and terminalia descriptions; figs 1A-F, 2A-K; 3A-I: female habitus, head, thorax, terminalia, sternites, spermathecae; male terminalia and sternites); 2016: 587, 592 (catalogue of Colombia, material list). **New synonymy.**

Type-material examined. *Megalothoraca teleopsis* Hennig, holotype male, SDEI: "S. Antonio\Columbia"; "coll. [collector] Fassl" [handwritten]; "Holotypus" [red label]; "~~HOLOTYPUS~~" [red label] (Fig. 62). The holotype is pinned, in poor condition, very damaged: eyes, flagellomeres, left fore and mid legs, right hind leg, and left wing missing; head and abdomen glued to the thorax (Figs 61, 63); apex of abdomen damaged (Fig. 61). *Bratrachophtalmum quimbaya* Carvalho, Wolff & Wendt, holotype male, CEUA: ♂ "COLOMBIA. Risaralda. Pereira./Santuario de Flora y Fauna Otún-/Quimbaya Trampa VSR bosque/Mayo 2001 Grupo de Entomologia/CEUA 12796" (CEUA). The holotype is pinned, and in excellent condition, with no damage. Paratype male, DZUP: ♂ labelled identical to the holotype (DZUP).

Distribution. Colombia.

Synonyms and new combinations

Megalothoraca teleopsis was described by Hennig (1938b) as the first and only species within the genus with stalked eyes, and Carvalho et al. (2011) described an enigmatic stalk-eyed fly from Colombia – *Batrachophthalmum quimbaya*. Upon comparing the types of both species, we did not find any morphological differences to distinguish them (Figs 1–3 and see Carvalho et al. 2011: figs 6–10). Therefore, *B. quimbaya* is herein considered a junior synonymy of *M. teleopsis*. Additionally, although *M. teleopsis* has features similar to other species of *Megalothoraca* (thorax developed, setae on fore femur distinctly stouter than on mid and hind femora, mid and hind legs longer and more slender than fore leg), those features are also common among species of *Batrachophthalmum* and *Euolena* (in part).

Megalothoraca teleopsis and *Batrachophthalmum rufiventre* Hendel, 1911 type species of the genus, share several characteristics, then here we transfer *M. teleopsis* to *Batrachophthalmum*, based on: as well *B. rufiventre*, *M. teleopsis* has the mid and hind femora with rows of weak spine-like setae on antero- and posterodorsal surfaces, and contrasting with the mid and hind femora of the other species of *Megalothoraca*, which have one (sometimes two) stout spines only on anteroventral surface. Also, *M. teleopsis* has vein R_{2+3} without spurious vein and sinuous just before margin (Fig. 64), as in *B. rufiventre* (see Carvalho et al. 2011: figs 1, 2), and differing from species of *Megalothoraca* which have vein R_{2+3} with two conspicuous preapical curves and a small postmedial spurious vein (e.g. Figs 16, 22, 57).

Batrachophthalmum species are recognized by the following combination of characteristics: head wider than high, especially in males; arista densely pubescent; all femora with spine-like setae on antero- and posteroventral; setae on fore femur distinctly more robust than those on mid and hind femora; hind coxa bare on posterior surface; costal vein of male with no modification, R_{2+3} without spurious vein, with only one tender sinuosity on apical third sinuous just before margin; r-m at distal three-fourths of cell dm.

ACKNOWLEDGEMENTS

We are very grateful to the following curators for the loan of material and/or for access to the collections: M Wolff (CEUA), Z. Soltész (HNHM), T. Huflejt (MZPW), F. Menzel (SDEI) and J. Ziegler (ZMHB). We thank Dominika Mierzwa from MZPW for kindly taking the photos of some types from the collection. This research, as well as the travels to visit the collection were funded by Conselho Nacional de Desenvolvimento Científico e Tecnológico (CNPq) through the support of the project "Ampliação do conhecimento em taxonomia, sistemática e aspectos biológicos de Tephritoidea (Diptera) na Região Neotropical, em especial das famílias Tephritidae e Richardiidae" (process 400829/2014-2). L. Wendt and L. Marinoni also thank CNPq for the fellowships provided (process 152777/2016-4 and PQ307732/2015-0, respectively). Part of the photos were taken with Automontage equipment available to us through the project "Taxonline: Rede Paranaense de Coleções Biológicas" headed by L. Marinoni. Also, our thanks to Wayne Mathis from Smithsonian Institution, USA, for proofreading the manuscript.

LITERATURE CITED

Aczél M (1950) Catalogo de la sufamilia americana "Richardiinae" (Dipt. Acalypt. Otitidae). Acta Zoologica Lilloana 9: 5–47.

Alencar LB de, Wendt LD, Ale-Rocha R (2013) *Richardia* Robineau-Desvoidy (Diptera, Tephritoidea, Richardiidae) from Central Amazon, Brazil. Revista Brasileira de Entomologia 57(2): 202–208. https://doi.org/10.1590/S0085-56262013005000015

Arellano G, Vergara C, Bello S (2015). Plagas entomológicas y otros artrópodos en el cultivo de la piña (*Ananas comosus* var. *comosus* (L.) Merr., Coppens & Leal) en Chanchamayo y Satipo, Departamento de Junín, Perú. Ecología Aplicada 14(2): 175–189.

Carvalho CJB de, Wolff M, Wendt LD (2011) A new stalk-eyed species of *Batrachophthalmum* Hendel (Diptera, Richardiidae) from Colombia. Zootaxa 2932: 41–46. https://doi.org/10.5281/zenodo.278067

Cavallari ML, Baltazar FN, de Carvalho EC, Nihei SS, Muñoz DR, Tolezano JE (2015) Diptero fauna associated with *Sus scrofa* Linné, 1758, carcasses in urban and coastal regions of São Paulo State, Brazil. Psyche 2015: 1–12. https://doi.org/10.1155/2015/983980

Cumming JM (1992) Lactic acid as an agent for macerating Diptera specimens. Fly Times 8: 7.

Cumming JM, Wood DM (2009) Morphology and terminology. In: Brown BV, Borkent A, Cumming JM, Wood DM, Woodley NE, Zumbado MA (Eds) Manual of Central American Diptera. NRC Research Press, Ottawa, Canada, vol. 1, 9–50.

Enderlein G (1912) Die Richardiiden des Stettiner Museums. Sitzungsberichte der Gesellschaft Naturforschender Freunde zu Berlin 1912: 99–113.

Hancock EG (2010) Richardiidae (Richardiid flies). In: Brown BV, Borkent A, Cumming JM, Wood DM, Woodley NE, Zumbado MA (Eds) Manual of Central American Diptera). NRC Research Press, Ottawa, Canada, vol. 2, 871–879.

Hendel F (1911a) Die arten der dipteren-subfamilie Richardiinae. Deutsche Entomologische Zeitschrift 1911: 181–212; 239–270; 367–396.

Hendel F (1911b) Diptera. Fam. Muscaridae, Subfam. Richardiinae. Genera Insectorum 113: 1–56.

Hennig W (1938a) Beitrage zur Kenntnis der Richardiiden (Dipt.). Revista de Entomologia 8: 111–122.

Hennig W (1938b) Neue Beiträge zur Systematic der Richardiiden und Tyliden. Arbeiten über Morphologische und Taxonomische Entomologie aus Berlin-Dahlem 5: 8–15.

Lopes de Carvalho LM, Linhares AX (2001) Seasonality of Insect Succession and Pig Carcass Decomposition in a Natural Forest Area in Southeastern Brazil. Journal of Forensic Science 46(3): 604–608. https://doi.org/10.1520/JFS15011J

Montilla R, García JL, Lacruz L, Durán D (2007) *Spalangia drosophilae* Ashmead (Hymenoptera: Pteromalidae) parasitoide de pupas de la mosca de la piña *Melanoloma viatrix* Hendel (Diptera: Richardiidae) em Trujillo, Venezuela. Agronomía Tropical 57(2): 107–112.

Montilla R, Lacruz L, Durán D (2008) Distribución geográfica de *Melanoloma viatrix* Hendel (Diptera; Richardiidae) en Trujillo, Venezuela. Agronomía Tropical 58(4): 403–407.

Moretti T de C, Ribeiro OB, Thyssen PJ, Solis DR (2008) Insects on decomposing carcasses of small rodents in a secondary forest in Southeastern Brazil. European Journal of Entomology 105: 691–696. https://doi.org/10.14411/eje.2008.094

Norrbom AL, Kim KC (1988) Revision of the *schausi* group of *Anastrepha* Schiner (Diptera: Tephritidae), with a discussion of the terminology of the female terminalia in the Tephritoidea. Annals of the Entomological Society of America 81: 164–173. https://doi.org/10.1093/aesa/81.2.164

Peñaranda EA, Ospina MAO (1999) La mosca de la piña, nueva plaga en cultivos del Departamento de Antioquia. In: Castro RS, Reyes AG (Eds) Control de enfermedades y plagas en la piña. Subgerencia de Prevención y Control, Instituto Colombiano Agroapecuario, ICA, Santafé de Bogotá, 63–71.

Steyskal GC (1968) Famly Richardiidae. In: Papavero P (Ed.) A catalogue of the Diptera of the Americas South of the United States. Departamento de Zoologia, Secretaria da Agricultura, São Paulo, 1–20.

Wendt LD, Ale-Rocha R (2014a) Taxonomic revision of wasp-like richardiids, *Neoidiotypa* Osten-Sacken (Diptera, Tephritoidea, Richardiidae). Zootaxa 3857(4): 527–550. https://doi.org/10.11646/zootaxa.3857.4.4

Wendt LD, Ale-Rocha R (2014b) Description of the female of the enigmatic *Batrachophthalmum quimbaya* (Diptera: Richardiidae). Revista Colombiana de Entomologia 40(2): 306–310.

Wendt LD, Ale-Rocha R (2015) Antlered richardiid flies: new species of *Richardia* (Tephritoidea: Richardiidae) with antler-like genal processes. Entomological Science 18: 153–166. https://doi.org/10.1111/ens.12122

Wendt LD, Ale-Rocha R (2016) Family Richardiidae. Zootaxa 4122 (1): 585–595. http://dx.doi.org/10.11646/zootaxa.4122.1.49

White IM, Headrick DH, Norrbom AL, Carroll LE (1999) Glossary. In: Aluja M, Norrbom AL (Eds) Fruit flies (Tephritidae): Phylogeny and evolution of behavior. CRC Press, Boca Raton, 881–924.

Author Contributions: LDW revised, described, and illustrated the material. LDW and LM wrote the paper.
Competing Interests: The authors have declared that no competing interests exist.

13

Body mass index and glucose variations during the night in free-ranging *Artibeus planirostris* (Chiroptera: Phyllostomidae)

Luane Maria Melo Azeredo[1], Monique Silva Ximenes[2], Kleytone Alves Pereira[3], Maria Paula Aguiar Fracasso[4], Luiz Carlos Serramo Lopez[1]

[1]*Laboratório de Ecologia Comportamental e Psicobiologia, Departamento de Sistemática e Ecologia, Centro de Ciências Exatas e da Natureza, Universidade Federal da Paraíba. Campus I, Cidade Universitária, Castelo Branco, 58051-900 João Pessoa, PB, Brazil.*
[2]*Programa de Pós-Graduação em Ecologia e Monitoramento Ambiental, Universidade Federal da Paraíba. Campus IV, Rio Tinto, 58297-000 PB, Brazil.*
[3]*Programa de Pós-Graduação em Psicobiologia, Universidade Federal do Rio Grande do Norte. Campus Lagoa Nova, 59078-970 Natal, RN, Brazil.*
[4]*Centro Universitário de João Pessoa. Rodovia BR 230, km 22, Água Fria, 58053-000 João Pessoa, PB, Brazil.*
*Corresponding author: Luane Maria Melo Azeredo (*luaneazeredo@gmail.com*)*

ABSTRACT. Body condition is an important measure to estimate the energy reserve of an organism. Scientists frequently use body condition indices (BCIs) with morphometric measures but direct measurements, such as blood glucose, seem to be more reliable. We observed oscillations in the body condition and glucose indexes of individuals of *Artibeus planirostris* (Spix, 1823) during 13 nights in the field. We assume that if glucose levels are proportional to feeding state and body condition is a measure of energy reserve, blood glucose and BCI should be positively correlated and both are expected to increase during the night as the bats leave their diurnal roost to feed. To test this, we examined the relationship between blood glucose levels, BCI and reproductive phase of free flying male bats (n = 70) for 12 hours after sunset for 13 nights. Bats were captured in Reserva Biológica de Guaribas (Paraíba, Brazil) using mist nets. Blood glucose was analyzed with a portable glucometer. Supporting our assumptions, the number of hours after sunset and BCI presented significant positive correlations with glucose levels in *A. planirostris*. Reproductive phase did not present a significant correlation with any other variables. As we predicted, glucose level can be used as proxy for morphometric BCI and it can be measured with a simple portable glucometer. The increase both in glucose and BCI around the night can be explained by the efficient assimilation of nutrients present in fruits ingested by bats and the quick metabolism that increases the levels of glucose (an other nutrients) in blood, increasing the body mass.

KEY WORDS. Body condition, blood metabolites, foraging.

INTRODUCTION

Body condition is an important ecological attribute that provides an estimate of the energy reserve of an organism (Speakman 1997, Stevenson and Woods 2006, Labocha and Hayes 2012). Moreover, this measure may have an important role in estimating the reproductive success and survival rate of animals (Moya-Larano et al. 2008). Body condition can be estimated using morphometric body condition indices (BCIs). These indexes are based on the relationship between body mass and linear body measurements. The basic premise of this index is that if individuals in a population have similar biometric measurements, for instance length of forearm, the difference between their body masses would represent different measures of their energy reserves (Speakman and Racey 1986, Speakman 2001).

However, these BCI's need to be carefully analyzed and associated with other methods of estimating body condition, to avoid misinterpretations (Waye and Mason 2008). Besides that, morphometric BCIs may be not sensitive enough to detect variations in the body condition of animals in habitats that have a constant food supply, since there is no need to store fat for periods when resources are scarce (Jenni-Eiermann 1998). In addition to morphometry, measurements of internal nutrient levels, like the ones estimated from blood samples analysis, are more direct and can be used to determine the physical condition of individuals (Stevenson and Woods 2006, Wilder et al. 2016). These physiological data include hematocrit, triglycerides, glucose and ketones estimations, which results can be a proxy for health and feeding state (Mcguire et al. 2009, Morais et al. 2014, Azeredo et al. 2016). Besides being more direct, blood chemistry measures can be used for all individuals in a population (Stevenson and Woods 2006). A disadvantage of using only BCI is that it may not reliably infer body condition of some individuals, such as pregnant females for instance, due to the influence of the fetal body mass (Speakman 2001, Stevenson and Woods 2006).

Blood glucose is a metabolite that is directly correlated to feeding state (Jenni-Eiermann and Jenni 1997). Bats, can be good biological models to study the relationship between glucose and body condition because they expend a considerable amount of energy by moving through active flight. Therefore, they need specific physiological adaptations that facilitate locomotion fueled by a limited reserve like lipids and glycogen (Amitai et al. 2010).

Frugivorous bats consume carbohydrate and lipid-rich fruits and the digestion occurs immediately in an efficient way (Protzek et al. 2010, Batista et al. 2016). Therefore we assume that blood glucose will be positively correlated with body condition, since glucose is the first primary source of energy for bats and they can use this resource effectively. Many studies have evaluated the daily and seasonal body mass of mammals and also have associated the body weight variation with feeding state. Handley et al. (1991) stated that the frugivore *Artibeus jamaicensis* Leach, 1821 can assimilate approximately 93% of soluble carbohydrates present in fruits of its diet. The glucose ingested is mostly oxidized and used to convert carbohydrate in fat (McNab 1976). Therefore, the ideal body condition of a bat must be a balance between its short and long-term energy requirements with energy expenditure during the flight searching for food source (Laska 1990, Altringham 1996). As the vast majority of the bats do not feed during the day, we expected that our study species would consume their energy reserves during the day and restore them during the night hours while they are active.

The studied species was the flat-faced fruit-eating bat, *Artibeus planirostris* (Spix, 1823). This species is relatively large, compared with other neotropical bats such as *Carollia perspicillata* (Linnaeus, 1758) and *Sturnira lilium* (Geoffroy, 1810) with forearm length ranging from 62 to 73 mm and body mass ranging from 40 to 69 g (Hollis 2005). It is quite generalist in terms of habitat, occupying urban spaces, mainly tropical forests, savannas, open shrub areas with grasslands, and dry forests (Ballesteros and Racero-Casarrubia 2012). Artibeus planirostris is a frugivore, and thus acts as seed disperser (Oliveira and Lemes 2010). Due to the homoeothermic condition of bats and their relatively small body size when compared to other mammals, *A. planirostris* has a high metabolic rate, so it is possible to observe oscillations in its body condition and glucose based on its feeding state along the night. There are several studies on the species of *Artibeus*, but this is the first one to investigate the relationship between body condition and blood glucose during the night under natural conditions.

MATERIAL AND METHODS

This study was conducted at Reserva Biológica de Guaribas (Rebio Guaribas), a protected area spanning part of the municipalities of Mamanguape and Rio Tinto, in state of Paraíba, Brazil (6°44′02″S, 35°10′32″W and 6°40′53″S, 35°09′59″W) The climate in Rebio Guaribas is warm and humid with an average annual temperature of 26 °C (Alvares et al. 2013). The rainy season occurs from March to August and the dry season from September to February. The seasonality was estimated by the annual precipitation data of the city of Mamanguape from the last five years obtained from the website Proclima (CPTEC/INPE 2018).

Bats were caught with nine mist nets of 12 x 3 m in two week interval expeditions of three days each from June to November of 2012, totaling 13 days. Adult animals were identified based on the absence of the epiphyseal cartilage of the 4th finger, at the metacarpal – phalangeal junction (Kunz and Anthony 1982). Individuals were caught and released on the same night. The mist nets remained open for 12 hours straight (from 5PM to 5AM) and we checked the nets every 30 minutes. Walking to the laboratory to transport the bats in containment bags took approximately 15 minutes. All individuals were marked with plastic necklaces with color cylinders (Esbérard and Daemon 1999). Our sampling effort was calculated according to Straube and Bianconi (2002).

We collected blood samples from the femoral vein using disposable calibrated lancets GTech® inserted into a lancing device GTech®. The concentrations of blood glucose were measured using reagent test strips (Freestyle®) and a glucometer Optium™ Xceed® medsense in a similar procedure of Azeredo et al. (2016). Every sample collected belonged to different bats, since we had no recaptures. Permits to capture and handle the animals were duly provided by ICMBio (license #25891–3).

We calculated the body condition index (BCI) to assess the body condition of all individuals of *A. planirostris*. We evaluated the index using Le Cren's relative condition (Kn) (Le Cren 1951). The expected body mass was assessed using residuals of an ordinary least square (OLS) regression between body mass and the forearm length of *A. planirostris*. Due to the difficulty to identify pregnant females and the potential noise in the BCI caused by the fetal body mass we used only male bats in our analysis. We classified males as non-reproductive (with abdominal testis) or reproductive (scrotal testis) according to Zortéa (2003). Data were normalized by square root transformation before being analyzed. We performed a stepwise model selection using the Akaike's information criterion to choose the model that best fitted the relationship between blood glucose and the other variables (hours after sunset, BCI, rainy and dry seasons and reproductive phase). The best model was employed in a general linear model. We also performed a correlation matrix to observe the correlation among all continuous variables. We adopted confidence interval of 95% ($p < 0.05$) and the analyses were performed with R (R Core Team 2017).

RESULTS

The sampling effort was 37,908h*m². A total of 70 male bats were captured. The model that best fitted the relationship between our collected variables was that one that included glucose, BCI and hours after sunset (AIC = 205.51). Reproductive phase and the seasons (rainy and dry) were excluded by stepwise model selection and were considered non-significant. Glucose levels ranged from 1.11mmol/L to 20.36 mmol/L, (mean 7.50 ± SD 4.85), body mass ranged from 31.88 to 52.51 g (mean 43.28 ± 4.52) and forearm length ranged from 55.4 to 63.0 mm (mean 59.41 ± 1.81). The linear model showed that BCI and the hours after sunset influenced positively and significantly the levels of blood glucose (F-statistic: 16.03 on 2 and 67 DF, p-value: 2.051e-06). (see Table 1 for details). Blood glucose increased along the night so did the BCI. The correlation matrix with the selected variables is described in Table 2 and Figs 1 and 2 show the relationship between glucose and hours after sunset and BCI and hours after sunset. Table 3 shows in details how the data were organized before the analysis were performed.

Table 1. Summary of linear model results, showing the relationship between glucose concentration and the variables selected by AIC stepwise model selection, for *Artibeus planirostris* in Rebio Guaribas, PB.

Coefficients	Estimate	Std Error	t-value	p-value
Intercept	-4.08	4.89	-1.98	0.03
Hours after sunset	0.57	0.16	3.86	<0.001
BCI	5.29	1.98	2.67	<0.01

Table 2. Correlation matrix between glucose, BCI and hours after sunset for *Artibeus planirostris* in Rebio Guaribas, PB.

Variables	Glucose	Hours after sunset
Hours after sunset	0.51***	
BCI	0.41**	0.33*

*p < 0.01, **p < 0.001, ***p < 0.0001.

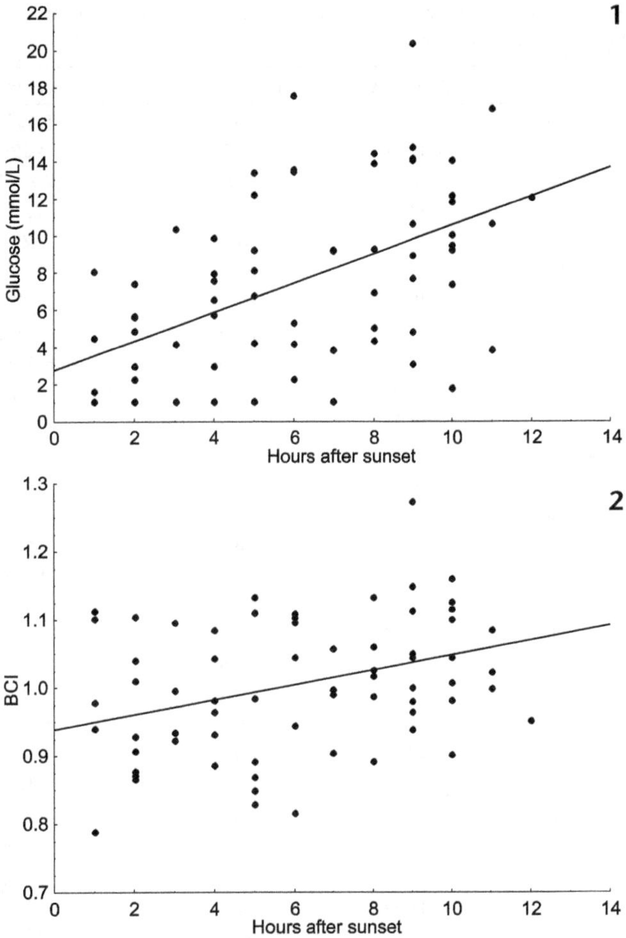

Figures 1–2. Correlation between (1) Glucose (mmol/L) and (2) BCI and the hours after sunset for individuals of *Artibeus planirostris* (N = 70) in Reserva Biológica de Guaribas, PB.

DISCUSSION

As predicted, glucose level was positively correlated with BCI and both variables increased their values along the night. The levels of blood glucose increased steadily after sunset, even with the bats flying and consequently expending energy. Our results were similar to Peng et al. (2017) who simulated natural conditions in captive frugivorous bats in Southwestern China and observed that although bats expended lots of energy flying, their glucose levels increased over time (a result that could be explained because they had been fed every 15 minutes during the experiment, thus keeping their homeostasis). Therefore, we assume that the positive correlation between blood glucose and the hours after sunset for *A. planirostris* is the result of the foraging habits of bats and the rapid assimilation of carbohydrates ingested as a way to keep their body condition (Handley et al. 1991). Consequently, in the first hours the levels of glucose were lower probably because bats were still leaving their roosts after a period of fasting during the day. During this fasting period the main fuel source probably was the glycogen stored in their livers that is used to regulated the homeostasis like was described for the congeneric species *A. jamaicensis* (McNab 1976).

Some studies pointed that high levels of glucose can have detrimental effects on the health of mammals (Brownlee 2001,

Table 3. Data collected to evaluate the body condition of individuals of *Artibeus planirostris* from Jun to November of 2012 in Reserva Biológica de Guaribas, PB.

Rep. Phase	Body mass	Forearm length	Glucose (mmol/L)	Month	Season	BCI	Hours after sunset
NR	42.98	60.5	1.6095	Jun	rainy	0.979179	1
R	38.47	61.2	4.884	Jun	rainy	0.865873	2
NR	46.36	58	2.2755	Jun	rainy	1.104203	2
NR	43.13	57.2	1.11	Jun	rainy	1.042414	4
R	50.31	61.1	6.771	Jun	rainy	1.134317	5
NR	45.71	56.8	14.763	Jun	rainy	1.112969	9
R	49.96	61.1	7.3815	Jun	rainy	1.126426	10
R	45.7	57.4	4.4955	Jun	rainy	1.100474	1
NR	44.53	55.4	8.0475	Jun	rainy	1.113125	1
NR	45.01	61.4	5.661	Jun	rainy	1.009597	2
NR	39.54	58.4	1.11	Jun	rainy	0.934969	3
NR	43.38	60.9	7.9365	Jun	rainy	0.981454	4
NR	48.01	59.7	13.431	Jun	rainy	1.109222	5
NR	43.99	58.2	5.328	Jun	rainy	1.043961	6
R	42.02	58.2	3.8295	Jun	rainy	0.997209	7
NR	44.27	60	4.329	Jun	rainy	1.017425	8
NR	47.09	58.3	12.1545	Jun	rainy	1.11551	10
NR	37.06	55.6	1.11	Jun	rainy	0.922886	3
NR	41.85	61.8	2.997	Jun	rainy	0.932316	4
NR	38.12	59	1.11	Jun	rainy	0.891736	5
R	43.5	60.2	4.1625	Jun	rainy	0.99623	3
R	47.6	60.5	7.548	Jun	rainy	1.084433	4
NR	42.8	61.1	5.7165	Jun	rainy	0.964993	4
R	38.87	56.9	4.1625	Jun	rainy	0.944673	6
NR	49.75	60.5	9.324	Jun	rainy	1.133415	8
R	50.61	60.1	9.4905	Jun	rainy	1.161094	10
NR	42.69	59	16.8165	Jun	rainy	0.998642	11
NR	39	59.3	1.11	Jul	rainy	0.90746	2
NR	40.98	57.2	1.11	Jul	rainy	0.99045	7
NR	44.18	58.4	3.108	Jul	rainy	1.044687	9
R	41.23	58	9.879	Aug	rainy	0.982017	4
R	43.56	61	9.2685	Aug	rainy	0.983824	5
R	49.82	61.9	17.538	Aug	rainy	1.107979	6
NR	47.99	62.3	14.43	Aug	rainy	1.060062	8
R	41.75	58.7	11.8215	Aug	rainy	0.981912	10
NR	41.3	61.2	5.6055	Aug	rainy	0.92957	2
R	48.7	61.2	10.3785	Aug	rainy	1.096127	3
NR	37.68	59.8	12.21	Aug	rainy	0.869024	5
R	49.15	61.4	2.2755	Aug	rainy	1.102459	6
NR	45.42	57.3	13.4865	Aug	rainy	1.095743	6
R	41.69	58.3	6.9375	Aug	rainy	0.98759	8
R	40.62	57.3	10.656	Aug	rainy	0.979944	9
R	48.86	58.7	20.3685	Aug	rainy	1.149131	9
R	45.05	59.2	4.8285	Aug	rainy	1.050098	9
NR	33.46	56.7	13.5975	Aug	rainy	0.816214	6
NR	40.05	58.1	12.0435	Aug	rainy	0.952182	12
NR	39.73	61.3	13.875	Aug	rainy	0.892696	8
NR	41.89	59.9	8.9355	Aug	rainy	0.964421	9

Continues

Table 3. Continued.

Rep. Phase	Body mass	Forearm length	Glucose (mmol/L)	Month	Season	BCI	Hours after sunset
NR	44.94	59.4	10.0455	Aug	rainy	1.043818	10
R	37.7	59.3	2.997	Sep	dry	0.877211	2
R	38.86	61.4	1.11	Sep	dry	0.871649	2
R	35.45	57.7	4.218	Sep	dry	0.848974	5
R	35.25	58.7	8.103	Sep	dry	0.82904	5
R	38.4	58.6	1.11	Sep	dry	0.904748	7
R	47.14	61.4	9.213	Sep	dry	1.057374	7
R	40.71	59.8	7.659	Sep	dry	0.938906	9
NR	38.31	58.6	1.776	Sep	dry	0.902627	10
NR	42.97	58.9	9.2685	Sep	dry	1.00699	10
R	44.74	60.3	10.656	Sep	dry	1.022838	11
R	48.26	61.3	3.8295	Sep	dry	1.084357	11
R	31.88	56	1.11	Sep	dry	0.787917	1
R	39.41	57.9	1.11	Oct	dry	0.940376	1
R	45.38	60.2	7.437	Oct	dry	1.039285	2
R	45.71	61.4	5.0505	Oct	dry	1.025298	8
R	44.08	60.7	14.1525	Oct	dry	1.000754	9
NR	49.09	61.5	14.0415	Oct	dry	1.099227	10
R	40.62	63	6.549	Nov	dry	0.886764	4
R	40.62	63	6.549	Nov	dry	0.886764	4
NR	52.51	57	14.0415	Nov	dry	1.273812	9
NR	52.51	57	14.0415	Nov	dry	1.273812	9

Ceriello 2003). However, just like birds, bats may present high levels of glucose with no adverse effects to their body condition (Braun and Sweazea 2008). Perhaps bats have evolved mechanisms to mitigate the negative effects of high levels of glucose, like high physical activity to regulate the homeostasis, such as what was described for the nectarivore *Glossophaga soricina* (Pallas, 1766) (Kelm et al. 2008) and the frugivores *Eonycteris spelaea* (Dobson, 1871) and *Cynopterus sphinx* (Vahl, 1797) (Peng et al. 2017). However, some studies have found negative correlations between blood glucose and BCIs for frugivorous bats and positive correlations for insectivorous bats in China (Meng et al. 2016). Indeed, bats from Pteropodidae may have different regulations of glucose to compare with phyllostomids bats. Recently, Amaral et al. (2018) described that the nectarivorous bat *Glossophaga soricina* (Pallas, 1766) does not have the hability to store energy becoming more susceptible to death after 18 hours of fasting. Therefore, the increasing of body mass in *A. planirostris* is more likely due to the water present in the fruits than a proper fat storage. However, we could not find other studies that have evaluated the relationship between daily body mass and blood glucose in bats. Another explanation for the different results between this study and Meng et al. (2016) is that the BCI used by the authors was the same for all species, without any kind of validation to check whether the BCI was adequate, as suggested in the study of Labocha et al. (2014). Sometimes, researchers make arbitrary choices based only on the most used indices in literature. However, it is important to consider that a good body condition index is one that can remove the effect of the body size measure on the value of body mass (Green 2001). Le Cren's index has the advantage of being adjusted to each individual population allowing to better comparisons within populations like it was the case in our study.

Although more direct measures are more reliable to predict body condition, many scientists still use morphometric BCIs, because the other measures can be expensive or difficult to perform under field conditions (Wilder et al. 2016). Here we used blood glucose assessed by tests strips and a portable glucometer that gave us satisfactory results. The portable glucometer, and inexpensive and practical tool, has been used by many studies, including ones involving free-ranging wild animals like birds (Lieske et al. 2002, Downs et al. 2010, Lobban et al. 2010). Our results were similar to Azeredo et al. (2016) who demonstrated a positive relation between glucose and the BCI of a neotropical passerine. We also could demonstrate that the body condition of *A. planirostris* can be evaluated by using blood glucose levels as the levels increased continuously along the night and accompanied the increase in the BCI.

In Brazil, some studies have used morphometric BCIs for bats as an attempt to assess the seasonality of their body condition, or used it to compare traits among populations of different habitats (Pereira et al. 2010, Morais et al. 2014, De Oliveira et al. 2017, Nunes et al. 2017). However, this is the first study to demonstrate the hourly variation and the relationship between glucose and a morphometric BCI. Our findings also highlight the importance of noting the hour the individual was collected. In future studies about corporal condition in bats, as these measures in can change along the night this potential confounding effect must be considered.

It is important to emphasize that we did not test the accuracy of our portable glucometer by comparing its reading with other laboratory methods. Some studies demonstrated that portable glucometers do not have the same accuracy of laboratory methods when used for animals (Hollis et al. 2008, Acierno et al. 2012, Bennett et al. 2017). However, other studies pointed that glucometers can be reliable tools for veterinary purposes (Hackett and Mccue 2010, Katsoulos et al. 2011). We used the portable glucometer Optium Xceed® medsense that has already been validated to measure ketoacidosis in dairy cows (Voyvoda and Erdogan 2010) and glucose in horses (Peugnet et al. 2014). Yet, according to Stoot et al. (2014) review, the glucometer must be calibrated for species-specific blood physiology, because the values can vary among taxa, although it is still a suitable alternative tool to traditional laboratory devices. Despite these limitations, the positive correlation between the glucose levels and traditional morphometric BCI indicate that human portable glucometers can be used to complement morphometric measurements in order to evaluate the nutritional status of bats in the wild.

The present study showed a positive relationship between blood glucose level and body mass index that indicates that body mass is probably associated with the nutritional state of *A. planirostris*. Increase in glucose levels along the night can be explained by the efficient assimilation of nutrients presents in fruits ingested by the bats and the quick metabolism that increases the levels of glucose in blood. Some studies pointed that high active flight is the key to regulate homeostasis of bats. We cannot confirm that the gain of weight for *A. planirostris* is due to storing energy reserves or just the water present in fruits. Thus, we suggest that in further studies, it would be interesting to observe the action of other metabolites besides glucose, in order to better explore the body condition of bats

ACKNOWLEDGEMENTS

We thank SISBIO – ICMBio for the license that allowed this study to be conducted. We also thank Jorge Nascimento (Julião) and Getulio Freitas managers of the Reserva Biológica de Guaribas, PB, for their logistical support. In addition we are thankful for Genildo, our forest guide, and all people who participated in the data collection: Paloma, Rumenigg, Karlla, Hannah, Mayara, Caio, Anderson, Eduardo, Júlia, Mariana, Letícia and Derick. This research was undertaken during the master's dissertation of MSX and postdoctoral research of MPAF. Lastly, we thank the Programa de Pós-graduação em Ecologia e Monitoramento ambiental (PPGEMA) from Universidade Federal da Paraíba, and Capes and CNPq for scholarships.

LITERATURE CITED

Acierno MJ, Schnellbacher R, Tully TN (2012) Measuring the level of agreement between a veterinary and a human point-of-care glucometer and a laboratory blood analyzer in Hispaniolan Amazon parrots (*Amazona ventralis*). Journal of Avian Medicine and Surgery 26(4): 221–224. https://doi.org/10.1647/2011-038R1.1

Altringham JD (1996) Bats: biology and behaviour. New York, Oxford University Press, 272 pp.

Alvares CA, Stape JL, Sentelhas PC, Goncalves JLD, Sparovek G (2013) Koppen's climate classification map for Brazil. Meteorologische Zeitschrift 22: 711–728. https://doi.org/10.1127/0941-2948/2013/0507

Amaral TS, Pinheiro EC, Freitas MB, Aguiar LMS (2018) Low energy reserves are associated with fasting susceptibility in Neotropical nectar bats Glossophaga soricina. Brazilian Journal of Biology. https://doi.org/10.1590/1519-6984.169674

Amitai O, Holtze S, Barkan S, Amichai E, Korine C, Pinshow B, Voigt CC (2010) Fruit bats (Pteropodidae) fuel their metabolism rapidly and directly with exogenous sugars. Journal of Experimental Biology 213: 2693–2699. https://doi.org/10.1242/jeb.043505

Azeredo LMM, Oliveira TC, Lopez LCS (2016). Blood metabolites as predictors to evaluate the body condition of Neopelma pallescens (Passeriformes: Pipridae) in northeastern Brazil. Zoologia (Curitiba) 33(6): e20160043. https://doi.org/10.1590/s1984-4689zool-20160043

Ballesteros J, Racero-Casarrubia J (2012) Murciélagos del área urbana en la ciudad de Montería, Córdoba – Colombia Urban Bats from the City of Monteria, Cordoba-Colombia. Revista MVZ Córdoba 17(3): 3193–3199.

Batista CB, Reis NR, Rezende MI (2016) Nutritional content of bat-consumed fruits in a forest fragment in Southern Brazil. Brazilian Journal of Biology 77(2): 244–250. https://doi.org/10.1590/1519-6984.10115

Bennett KA, Turner LM, Millward S, Moss SEW, Hall AJ (2017) Obtaining accurate glucose measurements from wild animals under field conditions: comparing a hand held glucometer with a standard laboratory technique in grey seals. Conservation Physiology 5(1): cox013. https://doi.org/10.1093/conphys/cox013

Braun EJ, Sweazea KL (2008) Glucose regulation in birds. Comparative Biochemistry and Physiology Part B: Biochemistry & Molecular Biology 151(1): 1–9. https://doi.org/10.1016/j.cbpb.2008.05.007

Brownlee M (2001) Biochemistry and molecular cell biology of diabetic complications. Nature 2001 414(6865): 813–820.

Ceriello A (2003). The possible role of postprandial hyperglycaemia in the pathogenesis of diabetic complications. Diabetologia 46(Suppl. 1): M9–M16. https://doi.org/10.1007/s00125-002-0931-5

CPTEC/INPE (2018) Programa de monitoramento climático em tempo real do nordeste. Available at: http://proclima.cptec.

inpe.br/balanco_hidrico/balancohidrico.shtml [Acessed: 09/10/2018]

De Oliveira HFM, De Camargo NF, Gager Y, Aguiar LMS (2017). The Response of Bats (Mammalia: Chiroptera) to Habitat Modification in a Neotropical Savannah. Tropical Conservation Science 10: 1–14. https://doi.org/10.1177/1940082917697263

Downs CT, Wellmann AE, Brown M (2010) Diel variations in plasma glucose concentrations of Malachite Sunbirds Nectarinia famosa. Journal of Ornithology 151: 235–239. https://doi.org/10.1007/s10336-009-0439-6

Esbérard CEL, Daemon C (1999) Um novo método para marcação de morcegos. Chiroptera Neotropical 5(1–2): 116–117.

Green AJ (2001) Mass/length residuals: Measures of body condition or generators of spurious results? Ecology 82(5): 1473–1483. https://doi.org/10.1890/0012-9658(2001)082[1473:MLRMOB]2.0.CO;2

Hackett ES, Mccue PM (2010) Evaluation of a Veterinary Glucometer for Use in Horses. Journal of Veterinary Internal Medicine 24(3): 617–621. https://doi.org/10.1111/j.1939-1676.2010.0481.x

Handley CO, Wilson DE, Gardner AL (1991) Demography and natural history of the common fruit bat, *Artibeus jamaicensis*, on Barro Colorado Island, Panamá. Smithsonian Contributions to Zoology: 1–173. https://doi.org/10.5479/si.00810282.511

Hollis AR, Schaer BLD, Boston RC, Wilkins PA (2008) Comparison of the accu-chek aviva point-of-care glucometer with blood gas and laboratory methods of analysis of glucose measurement in equine emergency patients. Journal of Veterinary Internal Medicine 22(5): 1189–1195. https://doi.org/10.1111/j.1939-1676.2008.0148.x

Hollis L (2005) *Artibeus planirostris*. Mammalian Species (775): 1–6. https://doi.org/10.1644/1545-1410(2005)775[0001:AP]2.0.CO;2

Jenniei-Ermann S, Jenni L (1997) Diurnal variation of metabolic responses to short-term fasting in passerine birds during the postbreeding, molting and migratory period. The Condor 99(1): 113–122. https://doi.org/10.2307/1370229

Jenni-Eiermann S, Jenni L (1998) What can plasma metabolites tell us about the metabolism, physiological state and condition of individual birds? An overview. Biologia e Conservazione della Fauna 102: 312–319.

Katsoulos PD, Minas A, Karatzia MA, Pourliotis K, Christodoulopoulos G (2011) Evaluation of a portable glucose meter for use in cattle and sheep. Veterinary Clinical Pathology 40(2): 245–247. https://doi.org/10.1111/j.1939-165X.2011.00309.x

Kelm DH, Schaer J, Ortmann S, Wibbelt G, Speakman JR, Voigt CC (2008) Efficiency of facultative frugivory in the nectar-feeding bat *Glossophaga commissarisi*: the quality of fruits as an alternative food source. Journal of Comparative Physiology B 178: 985–996. https://doi.org/10.1007/s00360-008-0287-3

Kunz TH, Anthony ELP (1982) Age Estimation and Post-Natal Growth in the Bat *Myotis lucifugus*. Journal of Mammalogy 63(1): 23–32. https://doi.org/10.2307/1380667

Labocha MK, Hayes JP (2012) Morphometric indices of body condition in birds: a review. Journal of Ornithology 153: 1–22. https://doi.org/10.1007/s10336-011-0706-1

Labocha MK, Schutz H, Hayes JP (2014) Which body condition index is best? Oikos 123(1): 111–119. https://doi.org/10.1111/j.1600-0706.2013.00755.x

Laska M (1990) Food transit times and carbohydrate use in three phyllostomid bat species. Zeitschrift für Säugetierkunde 55: 49–54.

Le Cren ED (1951) The length-weight relationship and seasonal cycle in gonad weight and condition in the perch (*Perca fluviatilis*). Journal of Animal Ecology 20: 201–219. https://doi.org/10.2307/1540

Lieske CL, Ziccardi MH, Mazet JaK, Newman SH, Gardner IA (2002) Evaluation of 4 handheld blood glucose monitors for use in seabird rehabilitation. Journal of Avian Medicine and Surgery 16(4): 277–285.https://doi.org/10.1647/1082-6742(2002)016[0277:EOHBGM]2.0.CO;210.1647/1082-6742(2002)016[0277:eohbgm]2.0.co;2

Lobban K, Downs C, Brown M (2010) Diel variations in plasma glucose concentration in some South African avian frugivores. Emu 110(1): 66–70. https://doi.org/10.1071/mu09088

Mcguire LP, Fenton MB, Faure PA, Guglielmo CG (2009) Determining feeding state and rate of mass change in insectivorous bats using plasma metabolite analysis. Physiological and Biochemical Zoology 82(6): 812–818.https://doi.org/10.1086/605951

Mcnab BK (1976) Seasonal fat reserves of bats in two tropical environments. Ecology 57(2): 332–338. https://doi.org/10.2307/1934821

Meng FX, Zhu L, Huang WJ, Irwin DM, Zhang SY (2016) Bats: body mass index, forearm mass index, blood glucose levels and SLC2A2 genes for diabetes. Scientific Reports 6: 29960. https://doi.org/10.1038/srep29960

Morais DB, Barros MS, Freitas MBD, Pinto Da Matta SL (2014) Seasonal assessment of the reproductive cycle and energy reserves of male bats *Sturnira lilium* (Chiroptera: Phyllostomidae). Journal of Mammalogy 95(5): 1018–1024. https://doi.org/10.1644/14-mamm-a-080

Moya-Larano J, Macias-Ordonez R, Blanckenhorn WU, Fernandez-Montraveta C (2008) Analysing body condition: mass, volume or density? Journal of Animal Ecology 77(6): 1099–1108. https://doi.org/10.1111/j.1365-2656.2008.01433.x

Nunes H, Lopes Rocha F, Cordeiro-Estrela P (2017) Bats in urban areas of Brazil: roosts, food resources and parasites in disturbed environments. Urban Ecosystems 20(4): 953–969. https://doi.org/10.1007/s11252-016-0632-3

Oliveira AKM, Lemes FTF (2010). *Artibeus planirostris* como dispersor e indutor de germinação em uma área do Pantanal do Negro, Mato Grosso do Sul, Brasil. Revista Brasileira de Biociências 8(1):49–52.

Peng X, He X, Liu Q, Sun Y, Liu H, Zhang Q, Liang J, Peng Z, Liu Z, Zhang L (2017). Flight is the key to postprandial blood glucose balance in the fruit bats Eonycteris spelaea and Cynopterus sphinx. Ecology and Evolution 7: 8804–8811. doi: 10.1002/ece3.3416

Pereira MJR, Marques JT, Palmeirim JM (2010) Ecological responses of frugivorous bats to seasonal fluctuation in fruit availability in Amazonian forests. Biotropica 42(6): 680–687. https://doi.org/10.1111/j.1744-7429.2010.00635.x

Peugnet P, Wimel L, Duchamp G, Sandersen C, Camous S, Guillaume D, Dahirel M, Dubois C, Jouneau L, Reigner F, Berthelot V, Chaffaux S, Tarrade A, Serteyn D, Chavatte-Palmer P (2014) Enhanced or reduced fetal growth induced by embryo transfer into smaller or larger breeds alters post-natal growth and metabolism in pre-weaning horses. Plos One 9: e102044. https://doi.org/10.1371/journal.pone.0102044

Protzek AOP, Rafacho A, Viscelli BA, Bosqueiro JR, Cappelli AP, Paula FMM, Boschero AC, Pinheiro EC (2010) Insulin and glucose sensitivity, insulin secretion and -cell distribution in endocrine pancreas of the fruit bat *Artibeus lituratus*. Comparative Biochemistry and Physiology Part A: Molecular & Integrative Physiology 157(2): 142–148. https://doi.org/10.1016/j.cbpa.2010.05.016

R Core Team (2017) R: A language and environment for statistical computing. Vienna, R Foundation for Statistical Computing, version 3.4.1. Available online at: https://www.R-project.org

Speakman J (1997) Factors influencing the daily energy expenditure of small mammals. Proceedings of the Nutrition Society 56(3): 1119–1136. https://doi.org/10.1079/pns19970115

Speakman J (2001) Body composition analysis of animals: a handbook of non-destructive methods. Cambridge, Cambridge University Press, 252 pp.

Speakman JR, Racey PA (1986) The influence of body condition on sexual development of male brown long-eared bats (*Plecotus auritus*) in the wild. Journal of Zoology 210(4): 515–525. https://doi.org/10.1111/j.1469-7998.1986.tb03653.x

Stevenson RD, Woods WA (2006) Condition indices for conservation: new uses for evolving tools. Integrative and Comparative Biology 46(6): 1169–1190. https://doi.org/10.1093/icb/icl052

Stoot LJ, Cairns NA, Cull F, Taylor JJ, Jeffrey JD, Morin F, Mandelman JW, Clark TD, Cooke SJ (2014) Use of portable blood physiology point-of-care devices for basic and applied research on vertebrates: a review. Conservation Physiology 2(1): cou011. https://doi.org/10.1093/conphys/cou011

Straube FC, Bianconi GV (2002). Sobre a grandeza e a unidade utilizada para estimar esforço de captura com utilização de redes-de-neblina. Chiroptera Neotropical 8(1–2): 150–152.

Voyvoda H, Erdogan H (2010) Use of a hand-held meter for detecting subclinical ketosis in dairy cows. Research in Veterinary Science 89(3): 344–351. https://doi.org/10.1016/j.rvsc.2010.04.007

Waye HL, Mason RT (2008) A combination of body condition measurements is more informative than conventional condition indices: Temporal variation in body condition and corticosterone in brown tree snakes (*Boiga irregularis*). General and Comparative Endocrinology 155(3): 607–612. https://doi.org/10.1016/j.ygcen.2007.08.005

Wilder SM, Raubenheimer D, Simpson SJ (2016) Moving beyond body condition indices as an estimate of fitness in ecological and evolutionary studies. Functional Ecology 30(1): 108–115. https://doi.org/10.1111/1365-2435.12460

Zortéa M (2003) Reproductive patterns and feeding habits of three nectarivorous bats (Phyllostomidae: Glossophaginae) from the Brazilian Cerrado. Brazilian Journal of Biology 63(1): 159–168. https://doi.org/10.1590/S1519-69842003000100020

Author's contribution: LCSL and MPAF delimited the experiments; MSX and KAP conducted the experiments; LMMA and LCSL analyzed the dataset; LMMA and MSX wrote the paper.
Competing Interests: The authors have declared that no competing interests exist.

14

Water acidification causes death of marine ornamental fish (Perciformes: Pomacentridae) during transport: contributing to the conservation of wild populations

Flavia D.F. Sampaio[1], Helena C. Silva-de-Assis[2], Franciele L. Bettim[2], Luís F. Fávaro[3], Carolina A. Freire[4]

[1]*Instituto Federal do Paraná. Rua João Negrão 1285, 80230-150 Curitiba, PR, Brazil.*
[2]*Departamento de Farmacologia, Setor de Ciências Biológicas, Universidade Federal do Paraná. 81531-980 Curitiba, PR, Brazil.*
[3]*Departamento de Biologia Celular, Setor de Ciências Biológicas, Universidade Federal do Paraná. 81531-980 Curitiba, PR, Brazil.*
[4]*Departamento de Fisiologia, Setor de Ciências Biológicas, Universidade Federal do Paraná. 81531-980 Curitiba, PR, Brazil.*
*Corresponding Author: Carolina A. Freire (*osmolab98@gmail.com*)*

ABSTRACT. Pomacentridae is a common family in the aquarium fish trade. Most species are harvested from nature. Here we evaluate the following water parameters in the pomacentrid sergeant major, *Abudefduf saxatilis* (Linnaeus, 1758), to assess their stress level during a 24, 48, and 72 hours transport: dissolved oxygen (DO), total ammonia, and pH. In addition, we evaluated the following physiological parameters: plasma osmolality, muscle water content, blood glucose, and the enzyme activities of the branchial carbonic anhydrase (CA), the hepatic glutathione S-transferase (GST), catalase (CAT), and superoxide dismutase (SOD). The mortality of fish measuring >6 cm total length was 22%, while no mortality was observed for fish measuring <6 cm. The pH of the water was significantly correlated with fish mortality, especially for the initial 24 hours of transport. Hypoxia after 24–48 hours also led to fish mortality, but build up ammonia was not a problem even after 72 hours. We suggest that a minimum water volume of 125 ml/g fish is necessary for safe and cost-effective transport of the sergeant major, preferably with <6 cm in total length.

KEY WORDS. Ammonia, aquarium trade, carbon dioxide, fish conservation, pH.

INTRODUCTION

The trade of marine ornamental fish (MOF) is essentially based on harvesting fish from nature. Fish are mostly caught from tropical countries and exported to developed nations (Wood 2001b, Wabnitz et al. 2003, Shuman et al. 2004, Alencastro et al. 2005). Intense and repeated harvesting from nature, which is magnified by unsuccessful transport events, may cause a problem to the conservation of the target species. If most fish survived their journeys to their final destinations, there would be less specimens captured, but in order to guarantee higher survival rates during transport, it is necessary to minimize the stress on fish, from capture to their target environment (Wood 2001b, Wabnitz et al. 2003). Despite all technology available for transport, however, the variables associated with fish survival during transport are not fully understood.

The conservation of MOF, or the assurance of sustainability in the MOF trade, has been extensively discussed in the literature: management strategies (Wood 2001a, b, Wabnitz et al. 2003, Alencastro et al. 2005), improvement of law enforcement (Gasparini et al. 2005, Sampaio and Ostrensky 2013, Sampaio et al. 2015), invasion by non-native species (Padilla and Williams 2004, Semmens et al. 2004, Weigle et al. 2005), or the need for some species to be reared in aquaculture (Job 2005). However, research describing the association between the stress of capture and transport, and its relationship with the conservation of MOFs, has not been found.

Transport of live fish is a crucial process for the aquarium business. Successful transport is an important step to minimize fish mortality (Wood 2001b, Wabnitz et al. 2003, Alencastro et al. 2005). Ornamental fish caught in the wild are frequently transported over very long distances (Brinn et al. 2012). For example, fish collected in the Philippines exported to the United States travel for about 20–25 hours until they arrive in Los Angeles, and more than 40 hours until they are in New York. Depending on the conditions of collection and transport, high mortality (90%) may occur (Rubec et al. 2001).

Live fish need specific handling protocols for effective transport, and protocols vary according to how fish tolerate the changes that occur in the water during transport (Treasurer 2012). Since data on ornamental fish tolerance to stress is very scarce, these protocols are usually unavailable. These protocols need to take into account the loss of water quality and the arousal of physiological disturbance to fish during transport.

Physiological tolerance to a decrease in water quality and temperature stress during transport are not well documented for MOF (Chow et al. 1994, Paterson et al. 2003, Hur et al. 2007). Among the physiological data available in the literature, the following research has been conducted and is relevant to our paper: the physiological response (oxygen consumption rate and excretion of metabolic wastes) of the clownfish *Amphiprion ocellaris* Cuvier, 1830 (Pomacentridae) to transport (Chow et al. 1994), the evaluation of transport and handling of the seahorse *Hippocampus abdominalis* Lesson, 1827 on blood glucose, cortisol and lactate (Wright et al. 2007), and the use of the essential oil of *Lippia alba* (Mill.) N.E. Br. ex Britton & P. Wilson (Plantae: Verbenaceae) as an anesthetic for the seahorse *Hippocampus reidi* Ginsburg, 1933, evaluating blood glucose and leucocyte count (Cunha et al. 2011).

It is generally understood that fish health and survival during post-transport is associated with the maintenance of water quality, therefore transport protocols that ensure good water quality should be part of the MOF industry goals (Bruckner 2005). Many studies focus on the use of anesthetics (Tondolo et al. 2013, Parodi et al. 2014, Zeppenfeld et al. 2014), or on ideal fish density inside the transport containers (Carneiro and Urbinati 2002, Urbinati et al. 2004, Tenningen et al. 2012). The evaluation of stress-associated responses of fish to transport have mostly focused on physiological stress markers, such as cortisol and glucose (Schreck et al. 1995, Sampaio and Freire 2016).

The ongoing/growing interest in the MOF trade has led to increased capture pressures on wild fish. This study evaluated physiological and biochemical responses (plasma osmolality, glucose, muscle hydration, and oxidative stress enzymes) of the sergeant major fish *Abudefduf saxatilis* (Linnaeus, 1758), and changes in the water of transport (dissolved oxygen, pH, and ammonia) in a situation simulating long transport. *Abudefduf saxatilis* belongs to Pomacentridae, one of the most relevant families in the worldwide MOF trade (Wood 2001b, Wabnitz et al. 2003). Fish of the genus *Abudefduf* are among the 10 most-exported species worldwide and the most imported in the United States (Wabnitz et al. 2003). The species chosen is widespread along the Brazilian coast (Atlantic sergeant) with its aesthetically attractive appearance as potential commercial relevance as an ornamental fish. The ultimate goal of this study is to foster studies on the successful transport of marine ornamental fish either from the wild or from cultured specimens, thus contributing to the conservation of their stocks in nature.

MATERIAL AND METHODS

Fish were caught in the state of Santa Catarina, city of Bombinhas, Sepultura Beach (27°08′28″S, 48°28′42″W), Brazil, between October 28th, 2013 and April 9th, 2014, using a handmade net. The total length of fish varied between 3.4 and 18.2 cm, mean ± SEM of 6.52 ± 0.31 cm. This size variability reflected the fact that they were sampled in nature. After being captured, the animals were individually placed in plastic bags of low-density polyethylene (LDPE), containing ~1/3 (volume) of seawater. The salinity of the seawater used to fill the bags was always of 35 psu, and temperature was of 21.2–25.6 °C. Two sizes of plastic bags were used (small: 12 cm X 33 cm, and large: 24 cm X 34 cm, both 0.2 mm-thick), and were chosen according to the size of the fish, in order to attenuate variability in the proportion of water volume to the weight of the fish. Achieving a precise proportion between fish weight and water volume inside the bags was not attempted here. The goal was to simulate normal catch conditions of wild fish. Each plastic bag was filled up with air of 100% oxygen to ~2/3 of its volume, allowing space only for closing the bag. The mouth of the bag was then twisted and tightly closed with several rounds of rubber bands. Bags were placed inside Styrofoam boxes, which were transported by car (4–5 h trip) to the Laboratory of Comparative Physiology of Osmoregulation in Curitiba. Upon arrival in the laboratory, the boxes were placed on the lab floor and kept closed. Fish were periodically (~every 12 hours) checked for mortality. Dead fish were sampled immediately, and this occurred in <24 h and between 24 and 48 h. After 48 h there was no further mortality. The final "n" for each group was: 16 live and 9 dead (≤24 h); 13 live and 5 dead (24–48 h); and 13 live fish (72 h).

Plastic bags were opened (total n = 56, 42 live and 14 dead) at the set times, and dissolved oxygen (DO), pH and temperature (21.4–26.0 °C) were immediately measured. Dissolved oxygen (DO), ammonia (N-NH_3), and pH were measured in the water of each sampled bag, including those that had dead fish, immediately after removal of the fish. Ammonia was measured using commercial kits (AlphaKit, Brazil) with absorbance read at 630 nm (spectrophotometer ULTROSPEC 2100 pro – Amersham Pharmacia Biotech, Sweden). DO and pH were respectively measured using a dissolved oxygen meter (Lutron, model DO-5519, Brazil) and a portable pH meter (pHtek, model PH100, Brazil).

After 24, 48 or 72 h of transport confinement, live fish were anaesthetized with benzocaine (80 mg/L seawater) in 2-liter plastic vials containing seawater. Within 2–3 min, fish were in complete anesthesia. Fish were measured for total length, weighed, and had a blood sample taken from the caudal vein using heparinized syringes. They were then euthanized by spinal cord section. A drop of blood was used to measure blood glucose (human glucotest Accu-Check®-Roche-Performa Nano model), and the remaining volume was centrifuged for 5 min at 2,100 xg to allow for plasma separation. Plasma samples were kept in the freezer at -20 °C for the osmolality assay. Tissues (gills, liver and axial muscle fragments) were removed and immediately frozen at -80 °C for posterior assays of enzyme activities. The experimental procedure described here was approved by the Committee on Ethics and Animal Welfare of the Federal University of Paraná (certificate #761/2014, issued on February 13, 2014).

An additional group of fish (n = 11) was collected and brought to the laboratory exactly as described above and transferred to a stock aquarium (20 L) in the lab. These specimens were kept for seven days in this aquarium and served as a reference (control) for the experimental groups that underwent 24, 48, and 72 h transport. This was performed in duplicate, with three fish in the first trial, and eight in the second trial. We decided to do this after considering that capturing/handling fish on the beach cause a certain level of stress to these animals, and since they were already stressed by then, they would could not be used for reliable physiological reference. Within a few hours (2–4 hours) in the stock tank control, fish recovered from the stress of handling, resuming a regular swimming pattern, feeding, and displaying their usual bright colors. During the maintenance period in the lab, fish were fed commercial fish flakes once a day. After this period, they were processed exactly as described for the experimental animals.

Plasma osmolality was read in undiluted samples using a vapor pressure osmometer (Wescor VAPRO 5520, USA). The percentage of water in muscle fragments was determined as the 24 h weight loss at 60 °C. Muscle slices were thawed and weighed (wet weight precision 0.1 mg, BioPar S22ST, Brazil), dried and weighed again (dry weight). The loss in weight was expressed as a percentage of the initial wet weight of the muscle slice.

Gill samples were weighed and homogenized in 10% w/v (weight/volume) with 10 mM phosphate buffer, pH 7.4 for the carbonic anhydrase (CA) assay. The homogenate was centrifuged at ~2,000 xg (5 min, room temperature) to precipitate cellular debris. The supernatant was separated. CA activity was quantified by the addition of the supernatant and distilled water saturated with CO_2 into a reaction medium containing mannitol (225 mM), sucrose (75 mM), and Tris-phosphate (10 mM), at pH 7.4. Immediately after the addition of the sample, the pH of the reaction vial was noted, every 4 s, for a total of 20 s (pH meter inoLAB benchtop pH Level 1 WTW®, Germany). Assay temperature was maintained at ~2.5 °C through an ice bath. The slope of the linear regression of pH versus time was the catalyzed reaction rate (CR) for the sample. The non-catalyzed rate of reaction (NCR, blank) was the slope of pH drop (same procedure used for the sample CR determinations) versus time using the buffer without the addition of tissue homogenate. The NCR was determined as the mean of three replicates, one taken at the beginning of the assay section, one every five sample readings, and another at the end of the session.

Glutathione S-transferase (GST), catalase (CAT) and superoxide dismutase (SOD) activities were assayed in liver samples, homogenized in phosphate buffer (0.1 M, pH 7.0), and centrifuged at 15,000 xg for 30 min, at 4 °C. Aliquots of the supernatant were separated for each enzyme assay. GST activity was quantified using 30 µl of the supernatant and 180 µl of the reaction medium containing 10 mM reduced glutathione (GSH), 60 mM 1-chloro-2,4-dinitrobenzene (CDNB) in 0.1M potassium phosphate buffer, at pH 6.5. Activities were read at 340 nm for 180 sec at 15 sec intervals (Keen et al. 1976). CAT activity was quantified according to the method in Aebi (Aebi 1984). A volume of 30 µL of supernatant was mixed to 990 µL of a reaction medium containing 30% H_2O_2, 0.5m Methylene-diamine-tetraacetic acid (EDTA), and 1 M Tris-base at pH 8.0. Activity was proportional to the reduction in H_2O_2 absorbance after 60 s at 240 nm. SOD activity was determined as the rate of inhibition of auto-oxidation of pyrogallic acid (Gao et al. 1998) at 440 nm. One unit of SOD was defined as the enzymatic activity necessary to inhibit the reduction of NBT by 50%. Total protein concentration for all enzyme assays were determined by the Bradford method (Bradford 1976) with bovine serum albumin as a standard, and expressed in $mg.mL^{-1}$, in triplicates. GST and CAT were expressed as µmol $(min.mg\ protein)^{-1}$ and SOD as $U.mg\ protein^{-1}$.

One-way ANOVA was used to compare the reference group with the experimental time groups (24, 48 and 72 h). Normality and homogeneity of variances among groups were tested using the Shapiro-Wilk test. The Kruskal-Wallis test was used for data that did not meet the requirements for normality. The results are expressed as the mean ± SEM for parametric data, and median (25–75% quartiles) for non-parametric data (Table 1). Student's t-tests (or Mann-Whitney Rank Sum Test, when normality failed) were employed for comparisons between live and dead fish (Figs 1–3). Pearson correlation was used to verify the association between variables. Analyses were performed using the SigmaPlot® 11.0 software, and the significance level was always set at 0.05.

RESULTS

Fish mortality during the transport experiments

The total number of fish used in this study was of 67, but 14 individuals (21%) died before being processed (nine before 24 h, and five between 24 and 48 h), inside the individual plastic bags. The length of live fish ranged between 3.4 and 8.1cm (0.75–8.52 g), and there was one exceptionally large fish measuring 14.2 cm (54 g). The mean size of live fish was 5.49 ± 0.14 cm, and 3.01 ± 0.25 g (mean ± SEM, without the outlier of 14.2 cm). Dead fish ranged between 6.4 and 18.2 cm long (4.5–125 g), with a mean size of 9.48 ± 0.93 cm, and weight of 23.28 ± 6.92 g. Observations of raw mortality data versus size promptly indicated that weight increased much faster than body length (Fig. 4). Furthermore, larger fish died in greater numbers. The weight x length graph clearly showed a steep rise in weight

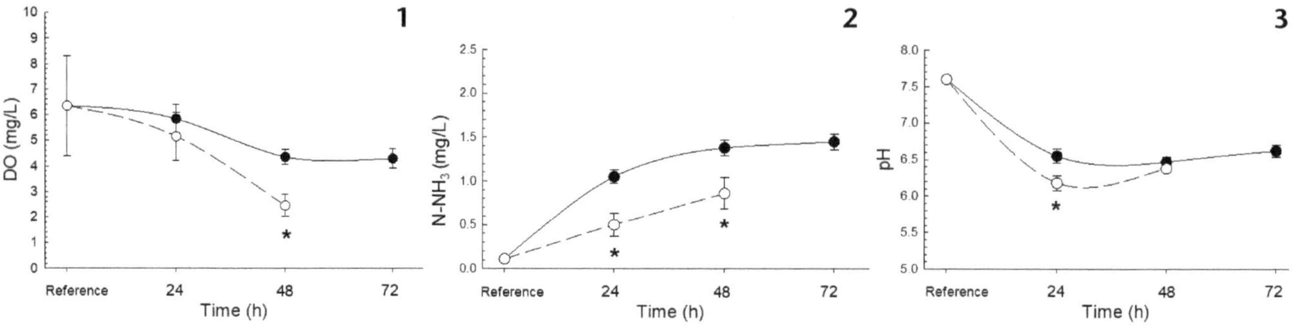

Figures 1–3. Time course of variation of water parameters for living and dead *A. saxatilis*: (1) water dissolved oxygen (DO, mg/L), (2) total ammonia-N (NH_3-N, mg/L) and (3) pH versus time (h). Values shown are means ± SEMs. When not apparent, SEM is smaller than the symbol. This happened for the water of the reference fish, ammonia and pH. Black circles: fish that were alive at the end of the experiment; white circles: fish that died during the experiment. *: water parameter for dead fish is different from that of live fish.

Table 1. Physiological parameters of live *Abudefduf saxatilis* of the reference and experimental groups maintained for 24, 48, or 72 hours in the plastic bags. Data are mean ± SEM or median* [25–75% quartiles](n). No significant differences were detected among groups, except for SOD (lower case letters).

Parameter	Reference	24 hours	48 hours	72 hours
Plasma osmolality	232±16	271±9	243±16	244±1
(mOsm/kg H_2O)	(7)	(7)	(10)	(6)
Glycemia*	51.0 [47.5–64.5]	76.0 [51.0–291.0]	53.5 [38.5–182.5]	41.0 [37.5–79.5]
(mg/dL)	(7)	(7)	(10)	(6)
Muscle water content*	79.0 [78.3–79.4]	78.3 [77.5–79.7]	78.2 [77.5–78.8]	78.4 [78.0–78.9]
(%)	(11)	(13)	(13)	(12)
Gill Carbonic Anhydrase*	25.7 [20.2–36.1]	21.7 [18.4–30.4]	20.2 [15.7–22.8]	19.7 [16.1–22.6]
(/mg protein)	(8)	(12)	(10)	(10)
Liver GST	199±20.2	223±12.8	221±14.4	222±13.8
(µmol/min.mg protein)	(5)	(10)	(8)	(9)
Liver CAT*	53.4 [39.8–89.6]	50.1 [45.9–75.2]	55.5 [50.7–59.7]	45.7 [36.3–64.9]
(µmol/min.mg protein)	(5)	(10)	(7)	(8)
Liver SOD	127.3±16.9^a	66.2±6.7^b	81.5±3.7^b	67.1±4.6^b
(U/mg protein)	(6)	(9)	(7)	(8)

for fish longer than 6–8 cm, which is associated with a high mortality rate – with the sole exception of the large 14.2 cm specimen, which survived the 24 h transport. When separate (live or dead fish) regression lines are fitted to the weight/length versus length data, the slope of the lines change from 0.27 (live) to 0.50 (dead fish, Fig. 5).

The ratio of bag water volume to the weight of *A. saxatilis* was variable especially for small fish, following a hyperbolic shape against fish length (Fig. 6). It was possible to see that the large dead fish had a much lower ratio of bag water volume to fish weight (Mann-Whitney $p < 0.001$) than the small live fish.

Water parameters changing upon transport

Dissolved oxygen (DO) did not decrease in the water containing live fish but decreased in the packages with dead fish between 24–48 h, when compared to the water containing fish that survived the 48 h transport (Fig. 1). Total ammonia in the water was greater after 24 and 48 h for live fish, when compared to the water containing dead fish (Fig. 2). The pH of the water in the bags with dead fish was lower after 24 h when compared to the pH of the water with live fish after the same amount of time, 24 h (Fig. 3).

Physiological parameters

The physiological parameters remained unchanged in live fish, in all transport groups (reference, 24, 48, 72 h, Table 1). The only significant difference was noted in the SOD activity of the liver, with transported fish showing lower values than the fish from the reference group (Table 1).

DISCUSSION

Fish mortality during the transportation experiments

The results show a very conspicuous change in the pattern of increase in mass/weight, as fish grew in length. This can be easily observed when separate regression lines (one for live, an-

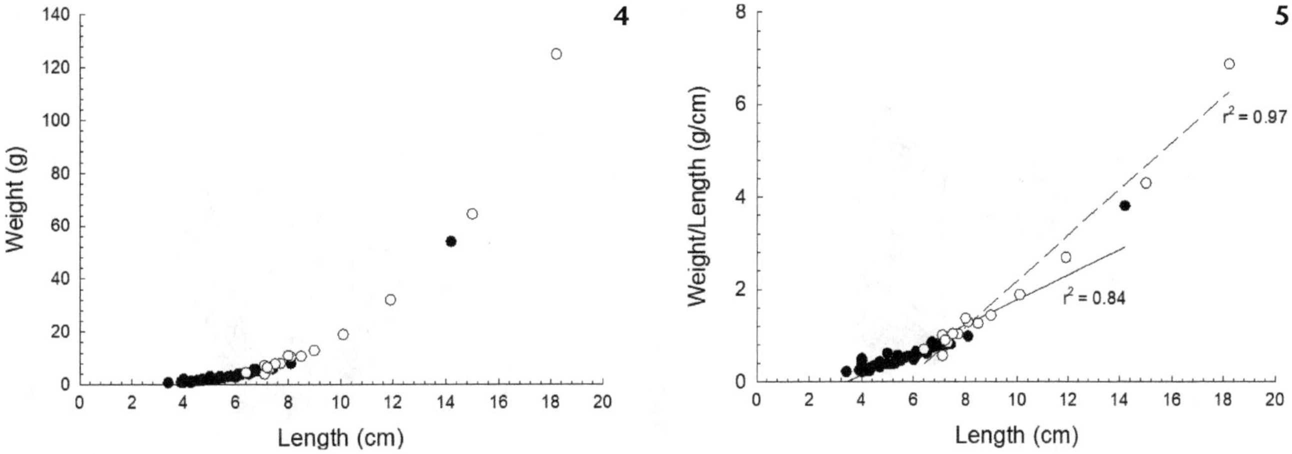

Figures 4–5. Relationship between weight (g) and total length (cm) of *A. saxatilis*: (4) weight (g) as a function of the total length (cm) of the fish. Black circles, fish that were alive at the end of the experiment (n = 53); white circles, fish that died during the experiment (n = 14). (5) Weight/length ratio versus fish length, with separate linear regressions for living (solid line) and dead (dashed line) fish, and respective r^2 values.

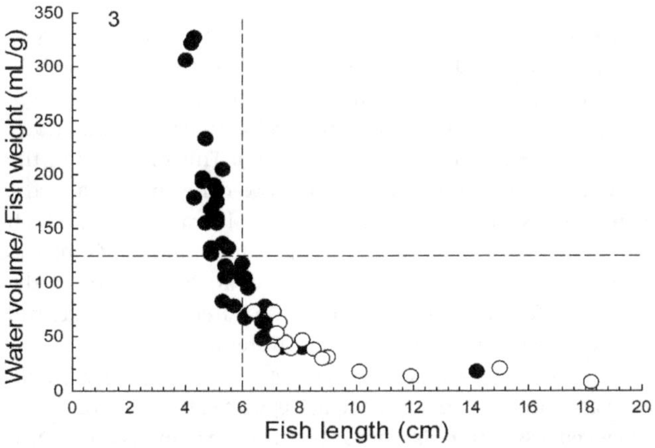

Figure 6. Transport water volume (ml) per gram of fish weight versus total body length (cm). Black circles, fish that were alive at the end of the experiment; white circles, fish that died during the experiment. Vertical and horizontal dashed lines respectively indicate maximum suggested length for the transport of *A. saxatilis* (6 cm) to ensure survival, and maximum water volume/fish weight relationship (125 ml/g fish) to minimize transport costs.

other for dead fish) are fitted to the weight/length versus length data. This is a common pattern for several species, with implications for production/management/fisheries (e.g., Safran 1992).

Mortality data indicate that the maximum length of fish to be transported should range 6–8 cm for better survival. It should be noted, however, that some measures to decrease stress during transport (e.g. addition of buffers or cooling of the bags) may contribute to a higher survival rate of larger fish. Interestingly, 7.5–11.25 cm *A. saxatilis* are commonly advertised (e.g., http://www.freshmarine.com and http://www.bluezooaquatics.com, both accessed on July 1, 2016 and May 16, 2018). The suggestion of a size limit for the trade of a particular species is important to improve sustainable management of aquarium fisheries (Wabnitz et al. 2003).

The amount of water used for the transport of ornamental fish is important. Even though there is a tendency to use as little water as possible (to minimize transport costs), the survival and well-being of fish need to be ensured. Since the small fish in our experiment ranged between 4–6 cm, 70–80 ml water/g fish (< 6 cm) would ensure their survival according to our analysis. In order to guarantee a safety margin, we suggest a 125 mL water/g fish (or 8 g/l) to transport the small sergeant major fish (up to 6 cm) for up to 72 h with no mortality (drawn lines in Fig. 6).

Studies that address the effects of fish density during transport are more common for food fish. Normally, fish are transported in great numbers as fingerlings, packed inside large plastic bags. According to some studies, appropriate densities (species-specific values) for fish transport inside the bags are as follows: 300 g/l for *Brycon cephalus* (Günther, 1869) (Carneiro and Urbinati 2002), 206 g/l for *Brycon amazonicus* (Spix and Agassiz, 1829) (Abreu et al. 2008), 68 g/l for *Pimelodus maculatus* Lacepède, 1803 (Braun and Nuñer 2014), 78 g/l for *Colossoma macropomum* (Cuvier, 1816) (Gomes et al. 2003). High densities are preferred, in order to reduce costs, but as long as they do not increase mortality or morbidity rates. Interestingly, for *B. cephalus*, high densities can actually reduce stress due to reduced aggression and establishment of a structure of social dominance (Urbinati et al. 2004). However, as marine ornamentals are transported individually, studies that indicate best transport densities – in weight of fish/liter of water, or volume of water for a given length of fish – are relevant. Much lower densities in g/l of fish are needed for ornamentals, such as suggested here for *A. saxatilis*: 8g/l.

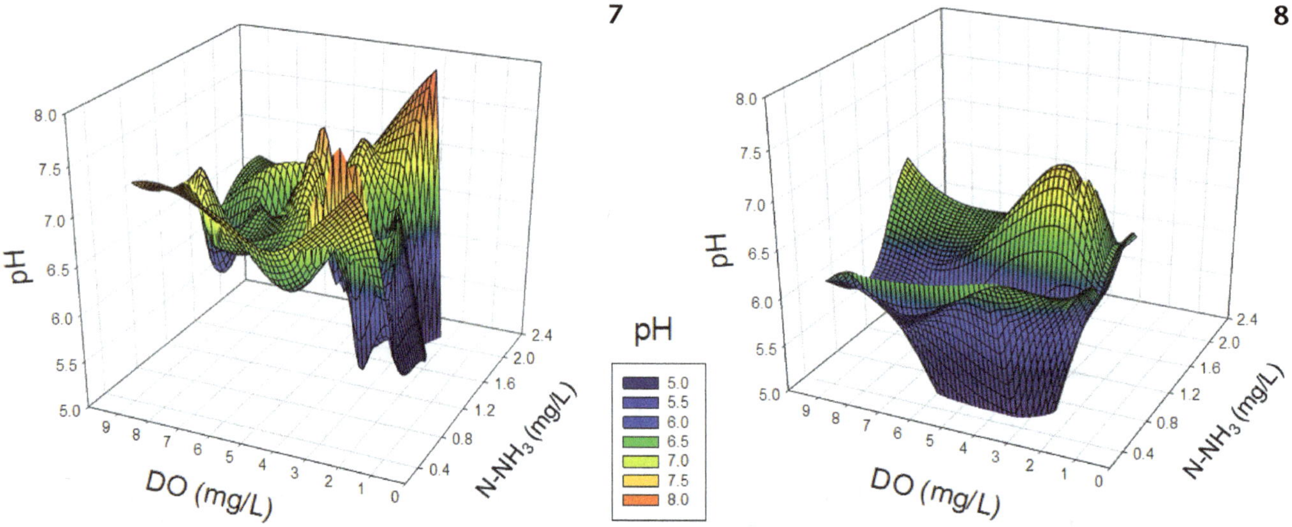

Figures 7–8. 3-D mesh plots showing the relationship among the 3 water parameters, whole set of data. NH_3-N (mg/L) versus DO(mg/L) versus pH, for live (7) and dead (8) fish. Legend illustrates color codes for pH interpolation.

Mortality data are actually the primary essential information in the development of protocols for the management of marine ornamental fish (Bruckner 2005). The causes of the high late mortality rates of marine ornamental fish are controversial (Rubec et al. 2001). The improvement in collection procedures, handling and transport tends to decrease mortality throughout the production chain (Bruckner 2005). The proposal of specific transport protocols for commercially relevant species can thus profit from the analysis of the causes of death of fish in transport simulation experiments.

Water parameters changing upon transport: what explains mortality?

The determination of a size limit or an adequate proportion between water volume and fish weight is relevant to establish a transport protocol for any given species. Nonetheless, it is also important to understand the reasons for fish mortality or morbidity during transport. Given that physiological parameters will obviously be altered in dead fish, it is interesting to ascertain if water parameters can give a clue on the reason for the deaths. If the proportion of water volume to fish mass is lower for larger fish – and it was, <75 mL/g – then the analysis of the water may give a hint.

Fish that died first (24 h) experienced acidosis (and acid water), and fish that died later (48 h) had water hypoxia. However, in either case, they produced less ammonia, and were much larger in size than the fish that survived. The transport times tested here were all "long", i.e., longer than eight hours (Davis 2006, Sampaio and Freire 2016). During these "long transport" times tested, both a reduction in pH and increase in ammonia were observed in live fish, consistent with the literature (Sampaio and Freire 2016). Given that mortality was clearly related to size, it was interesting to examine the relationship between each parameter and fish length, separately for live and dead fish. The only significant correlations to body length were those of water ammonia for the dead fish (-0.546, $p = 0.0433$), and water pH for the live fish (-0.512, $p = 0.0005$). This means that, the larger the size of the dead fish, the lower the ammonia in the water. This was not expected from the literature data, which often relates mortality to increased levels ammonia (Gomes et al. 2009, Tondolo et al. 2013, Parodi et al. 2014). Further, the longer the fish, the lower the pH of the water; that is, larger fish (in less water) acidify the transport water more.

Given that single parameters, when related to body length, were also not particularly conclusive, since larger fish that died produced less ammonia, these same three parameters (DO, NH_3 and pH) were again analyzed, but in a pairwise manner. In this analysis, Pearson correlations were calculated, also separately for live and dead fish. Significant correlations occurred only for live fish. There was a positive correlation between pH and DO (0.377, $p = 0.0139$), and a negative correlation between pH and ammonia (-0.473, $p = 0.0016$). DO drop follows pH decreases, and increased ammonia occurs concomitantly with pH drop for live fish. Over the hours of transport of *A. saxatilis*, the metabolism of fish that survived consumed oxygen (yields CO_2, which acidifies the water), NH_3 was putatively released (from protein metabolism, which alkalinizes the water); pH was observed to drop if there was higher input of CO_2 than NH_3 into the water (Sampaio and Freire 2016). Compatibly, as seen in Figs 1–3, ammonia levels increased in the water of dead fish, but less so than in live fish.

The situation of *A. saxatilis* seems to be of internal acidosis from CO_2/lactate production, resulting in water acidification (Sampaio and Freire 2016). Water acidification can result from metabolic CO_2 production and release to the water, whereas lower NH_3 excretion may have resulted from metabolic depression from acid accumulation; intracellular low pH is suggested to trigger metabolic depression (Guppy et al. 1994). Metabolic depression is also consistent with dead fish showing lower ammonia levels, and variable DO in the water after 48 hours. Although DO levels were ≤ 2.1 mg/L in the water of four dead fish, for the other 10 dead fish DO levels remained between 2.5 and 8.5 mg/L. Acidosis may have led to later (48 hours) increased oxygen uptake, similar to the "excess post-exercise oxygen consumption"(Wood 1991), apparent in the water of the fish that died later, between 24 and 48 hours. Indeed, exhaustive exercise and severe stress are metabolically similar situations (Wood 1991, Milligan 1996, Kieffer 2000). This happens in all fish, but small fish have much more water/g of fish, so the effects are present in lower magnitude. In addition, larger fish need more time to recover from exhaustive exercise, and thus, potentially, from the initial stages of the stress of confinement inside the bag (Gingerich and Suski 2012).

The analysis in the three-dimensional graph ("mesh-plot") integrated the three water parameters, and more clearly showed the importance of the drop in pH to explain mortality of the largest fish. When the three water parameters were analyzed together, for live fish, it is apparent that the pH of the water decreased in association with a reduction in DO (Fig. 7). However, for the water of the dead fish, a "pH-trap" or "pH-valley" seems apparent (Fig. 8). Given the low volume of water available, most (large) fish died within 24 hours.

Fish and other aquatic animals that breath water do not accumulate CO_2 in the extracellular fluid and therefore do not have the bicarbonate buffer system of animals that breath air (Schmidt-Nielsen 1996, Hill et al. 2012). This is a result of the high solubility of CO_2 in water. Thus, it can be expected that the generation of lactate by anaerobic metabolism and/or CO_2 accumulation in the water released by fish entails metabolic acidosis, which ends up acidifying the water and causing fish death. More importantly, fish, especially those from coastal or intertidal habitats, display Hb with significant Bohr and Root effects. Through these effects, O_2 is efficiently delivered to tissues, as low pH reduces both affinity for Hb and the carrying capacity of blood oxygen (Rummer and Brauner 2015). Given their small size, it was not possible to measure lactate or blood pH, which would have been extremely relevant to confirm these hypotheses. For example, in cod fingerlings, *Gadus morhua* Linnaeus, 1758, water pH decreased sharply in a closed system (Treasurer 2012). In the case of the clownfish, *Amphiprion ocellaris* Cuvier, 1830, a relationship between a decrease in pH and mortality was noted at pH <6.3 (Chow et al. 1994). A transport simulation experiment with the giant miss, *Lates calcarifer* (Bloch, 1790), associated higher mortality to increased CO_2 in the water (Paterson et al. 2003). As a practical recommendation, we suggest in such cases the use of substances that have a buffering effect (Paterson et al. 2003, Treasurer 2012). However, this should be tested, as the buffer may worsen the condition of the transport water by facilitating the production and release of ammonia (Paterson et al. 2003, Treasurer 2012). A further test should involve the use of buffers along with ammonia quenchers (Chow et al. 1994).

Abudefduf saxatilis inhabits tide pools (Barreiros et al. 2004, Cunha et al. 2008, Freitas et al. 2009). Fish from such environmentally unstable habitats are more tolerant to variations in NH_3, DO, pH, salinity, and temperature (Richards 2011, Freire et al. 2011). Considering the three parameters measured in this study (DO, ammonia and pH), we conclude that 1) the release of ammonia in the package was not enough to cause mortality; 2) pure oxygen added to the shipping container, a procedure commonly used by traders, can prevent a hypoxia; 3) low pH was correlated with mortality (especially within the first 24 hours of transport, of larger fish). Further research, however, is required to elucidate if this is due to acidosis of the fish (lactate build up) and/or increased CO_2 in the water.

Physiological parameters

The analysis aimed to examine the relationship between mortality rate and water deterioration. However, it is also important to verify the concomitant physiological changes occurring in live *A. saxatilis* fish during long transport. In general, the physiological parameters measured in this study (osmolality, muscle water content, and glucose) remained stable in fish that survived the long transport simulations.

The non-significant variability in the values of plasma osmolality and muscle water content reflects the ability of fish to regulate extracellular osmolality and tissue water, typical of fish in general, especially in situations of unchanged water salinity, as they were transported in closed vials containing full strength seawater. This stability is even more likely in a marine group of teleosts related to unstable environments such as tide pools, as is the case of the Pomacentridae (Marshall and Grosell 2006, Evans and Claiborne 2009).

The high variability in blood glucose levels is probably a result of individual variations, potentialized by handling and confinement stress. Stress effects through cortisol could potentially affect glycogen and glucose metabolism in these fish (Mommsen et al. 1999). In addition, these fish were collected from the environment and were directly packaged, so it was not possible to know when they fed for the last time prior to being captured. Still, fasting for 24 h should bring glucose levels back to baseline. For better understanding of glycemia as a marker of transport stress, more research is needed to evaluate fed/fasted fish, followed by transport simulations (Mommsen et al. 1999, Sampaio and Freire 2016).

Plasma osmolality or glucose, and the activities of metabolic enzymes, are part of a secondary response to stress (see review in Sampaio and Freire 2016). The activity of the branchial carbonic anhydrase (CA) was not altered by the transport simulation conducted here, despite the evidence of acid release by fish into the water, resulting in a reduction of water pH, even in live fish. The branchial CA is very relevant for osmoregulation, respiration, and acid base regulation (Marshall and Grosell 2006). The CA is a useful marker of exposure to pollutants, is sensitive to a wide variety of molecules (see Lionetto et al. 2012), but has not been used to assess metabolic disorders during fish transport.

Likewise, the enzymes of oxidative stress in fish are commonly used to measure the stress caused by drugs, heavy metals and other pollutants (Roche and Bogé 1996, Farombi et al. 2007). Studies to evaluate the stress of transport are recent, and use these markers also to verify the effects of using anesthetics during transport (e.g., Parodi et al. 2014, Salbego Joseânia 2014, Zeppenfeld et al. 2014). The specific activities of glutathione-S-transferase (GST) and catalase (CAT) were not different among the fish of the reference group and the three experimental groups (24, 48, and 72 h). However, the activity of the hepatic superoxide dismutase (SOD) decreased during the transport time. The decrease in SOD activity in transported fish can be related to a state of reduced metabolism in these fish, which is compatible with the stabilization in ammonia production rate and pH along time, apparent in Fig. 2, for live fish. Further research should be conducted within 24–72 hours of transport for a better understanding of the activation/suppression of the antioxidant response. For example, *Rhamdia quelen* (Quoy & Gaimard, 1824) transported for six hours in the presence of the essential oil of *Lippia alba* showed no change in liver SOD, but showed significant decrease in the activities of CAT and GST, compared to the control group (Salbego et al. 2014). With or without the use of anesthetics during transport, activation or inhibition of the antioxidant response should be more intensely studied, especially when fish reach their final destination. Special care should be taken when fish are transferred from their restricted transport vials, which may be hypoxic, to a stable aquarium, saturated with oxygen. Probably for that reason, marketers adopt protocols of acclimation of newly arrived fish: they drip water from the new system to the packaging of fish along periods of up to 4 hours. This procedure prevents sudden increases in dissolved oxygen which could foster the generation of oxygen free radicals.

Finally, marine ornamentals are sold according to size, which determines their market value. Conservatively, it is important to note that the maximum size of *A. saxatilis* for trade should be about 6 cm. For other species of MOF, the recommended maximum size for transport and appropriate relationships between water volume and fish weight should be established, in order to minimize costs, reduce stress and increase survival rates. Physiological and biochemical parameters evaluated in this study were not effective to characterize the stress of transport on individuals of this species. The assessment of changes in water quality during transport enabled a better understanding of the parameters that can affect fish homeostasis. With this approach, we propose that more studies be conducted on the physiology and water quality of the culture and transport of marine ornamental fish, aiming at their conservation. Ideally, more species should be cultivated, and successfully transported, to reduce their capture rates from nature.

ACKNOWLEDGEMENTS

The authors are thankful to Tony V.M. Sampaio for help in collecting fish, and Conselho Nacional de Desenvolvimento Científico e Tecnológico for a research grant awarded to CAF (CNPq 306630/2011–7). The authors also thank the two anonymous reviewers, for all the suggestions that resulted in correction and improvement of this article.

LITERATURE CITED

Abreu JS de, Sanabria-Ochoa AI, Gonçalves FD, Urbinati EC (2008) Stress responses of juvenile matrinxã (*Brycon amazonicus*) after transport in a closed system under different loading densities. Ciência Rural 38: 1413–1417. https://doi.org/10.1590/S0103-84782008000500034

Aebi H (1984) [13] Catalase in vitro. Methods in Enzymology 105: 121–126. https://doi.org/10.1016/S0076-6879(84)05016-3

Alencastro LA, Degner RL, Larkin SL (2005) Hobbyists preferences for marine ornamental fish: a discrete choice analysis of ecolabeling and selected product attributes. SPC Live Reef Fish Information Bulletin 15: 19–22.

Barreiros JP, Bertoncini Á, Machado L, et al. (2004) Diversity and seasonal changes in the ichthyofauna of rocky tidal pools from Praia Vermelha and Sao Roque, Santa Catarina. Brazilian Archives of Biology and Technology 47: 291–299.

Bradford MM (1976) A rapid and sensitive method for the quantitation of microgram quantities of protein utilizing the principle of protein-dye binding. Analytical Biochemistry 72: 248–254. https://doi.org/10.1016/0003-2697(76)90527-3

Braun N, Nuñer AP de O (2014) Stress in *Pimelodus maculatus* (Siluriformes: Pimelodidae) at different densities and times in a simulated transport. Zoologia 31: 101–104. https://doi.org/10.1590/S1984-46702014000100012

Brinn RP, Marcon JL, McComb DM, et al. (2012) Stress responses of the endemic freshwater cururu stingray (*Potamotrygon* cf. *histrix*) during transportation in the Amazon region of the Rio Negro. Comparative Biochemistry and Physiology 162: 139–145. https://doi.org/10.1016/j.cbpa.2011.07.004

Bruckner A (2005) The importance of the marine ornamental reef fish trade in the wider Caribbean. Revista de Biologia Tropical 53: 127–137.

Carneiro PCF, Urbinati EC (2002) Transport stress in matrinx, *Brycon cephalus* (Teleostei: Characidae), at different densities. Aquaculture International 10: 221–229. https://doi.org/10.1023/A:1022166411089

Chow PS, Chen TW, Teo LH (1994) Physiological responses of the common clownfish, *Amphiprion ocellaris* (Cuvier), to factors related to packaging and long-distance transport by air. Aquaculture 127: 347–361. https://doi.org/10.1016/0044-8486(94)90237-2

Cunha EA, Carvalho RAA, Monteiro-Neto C, et al. (2008) Comparative analysis of tidepool fish species composition on tropical coastal rocky reefs at State of Ceará, Brazil. Iheringia, Série Zoologia 98: 379–390. https://doi.org/10.1590/S0073-47212008000300013

Cunha MA da, Silva BF da, Delunardo FAC, et al. (2011) Anesthetic induction and recovery of *Hippocampus reidi* exposed to the essential oil of *Lippia alba*. Neotropical Ichthyology 9: 683–688. https://doi.org/10.1590/S1679-62252011000300022

Davis KB (2006) Management of physiological stress in finfish aquaculture. North American Journal of Aquaculture 68: 116–121. https://doi.org/10.1577/A05-007.1

Evans D, Claiborne J (2009) Osmotic and ionic regulation in fishes. In: Evans D (Ed.) Osmotic and ionic regulation: cells and animals. Boca Raton, CRC Press, 295–366.

Farombi EO, Adelowo OA, Ajimoko YR (2007) Biomarkers of oxidative stress and heavy metal levels as indicators of environmental pollution in African cat fish (*Clarias gariepinus*) from Nigeria Ogun River. International Journal of Environmental Research and Public Health 4(2): 158–165.

Freire CA, Welker AF, Storey JM, et al. (2011) Oxidative Stress in Aquatic Ecosystems. In: Abele D, Vázquez-Medina JP, Zenteno-Savín T (Eds) Oxidative Stress in Aquatic Ecosystems. Chichester, John Wiley & Sons, 524 pp.

Freitas MC de, Vieira RHS dos F, Araújo ME de (2009) Impact of the construction of the harbor at pecém (Ceará, Brazil) upon reef fish communities in tide pools. Brazilian Archives of Biology and Technology 52: 187–195. https://doi.org/10.1590/S1516-89132009000100024

Gao R, Yuan Z, Zhao Z, Gao X (1998) Mechanism of pyrogallol autoxidation and determination of superoxide dismutase enzyme activity. Bioelectrochemistry and Bioenergetics 45: 41–45. https://doi.org/10.1016/S0302-4598(98)00072-5

Gasparini JL, Floeter SR, Ferreira CEL, Sazima I (2005) Marine ornamental trade in Brazil. Biodiversity and Conservation 14:2883–2899. https://doi.org/10.1007/s10531-004-0222-1

Gingerich AJ, Suski CD (2012) The effect of body size on post-exercise physiology in largemouth bass. Fish Physiology and Biochemistry 38:329–340. https://doi.org/10.1007/s10695-011-9510-3

Gomes LC, Brinn RP, Marcon JL, et al. (2009) Benefits of using the probiotic Efinol((R))L during transportation of cardinal tetra, *Paracheirodon axelrodi* (Schultz), in the Amazon. Aquaculture Research 40: 157–165.

Gomes LC, Roubach R, Araujo-Lima CARM, Chippari-Gomes AR, Lopes NP, Urbinati EC (2003) Effectof fish density during transportation on stress and mortality of juvenile tambaqui *Colossoma macropomum*. Journal of the World Aquaculture Society 34: 76–84.

Guppy M, Fuery CJ, Flanigan JE (1994) Biochemical principles of metabolic depression. Comparative Biochemistry and Physiology, Part B 109: 175–189.

Hill RW, Wyse GA, Anderson M (2012) Endocrine and neuroendocrine physiology. In: Animal Physiology. Sunderland, Sinauer Associates, vol. 2, 419–454.

Hur JW, Park I-S, Chang YJ (2007) Physiological responses of the olive flounder, *Paralichthys olivaceus*, to a series stress during the transportation process. Ichthyology Research 54: 32–37.

Job S (2005) Integrating marine conservation and sustainable development : Community-based aquaculture of marine aquarium fish. Diversity 13: 1999–2004.

Keen JH, Habig WH, Jakoby WB (1976) Mechanism for the several activities of the glutathione S transferases. Journal of Biological Chemistry 251: 6183–6188.

Kieffer JD (2000) Limits to exhaustive exercise in fish. Comparative Biochemistry and Physiology, Part A 126: 161–179. https://doi.org/10.1016/S1095-6433(00)00202-6

Lionetto MG, Caricato R, Giordano ME, et al. (2012) Carbonic anhydrase as pollution biomarker: an ancient enzyme with a new use. International Journal of Environmental Research and Public Health 9: 3965–3977.

Marshall WS, Grosell M (2006) Ion transport, osmoregulation, and acid-base balance. In: Evans DH, Claiborne JB (Eds) The Physiology of Fishes. Boca Raton, CRC Press, 3rd ed., 624 pp.

Milligan CL (1996) Metabolic recovery from exhaustive exercise in rainbow trout. Comparative Biochemistry and Physiology, Part A 113(1): 51–60.

Mommsen TP, Vijayan MM, Moon TW (1999) Cortisol in teleosts:dynamics, mechanisms of action, and metabolic regulation. Reviews in Fish Biology and Fisheries 9: 211–268. https://doi.org/10.1023/A:1008924418720

Padilla DK, Williams SL (2004) Beyond ballast water: Aquarium and ornamental trades as sources of invasive species in aquatic ecosystems. Frontiers in Ecology and the Environment 2: 131–138.

Parodi TV, Cunha MA, Becker AG, et al. (2014) Anesthetic activity of the essential oil of *Aloysia triphylla* and effectiveness in reducing stress during transport of albino and gray strains of silver catfish, *Rhamdia quelen*. Fish Physiology and Biochemistry 40: 323–334. https://doi.org/10.1007/s10695-013-9845-z

Paterson BD, Rimmer MA, Meikle GM, Semmens GL (2003) Physiological responses of the Asian sea bass, *Lates calcarifer* to water quality deterioration during simulated live transport: Acidosis, red-cell swelling, and levels of ions and ammonia in the plasma. Aquaculture 218: 717–728. https://doi.org/10.1016/S0044-8486(02)00564-1

Richards JG (2011) Physiological, behavioral and biochemical adaptations of intertidal fishes to hypoxia. The Journal of Experimental Biology 214: 191–199. https://doi.org/10.1242/jeb.047951

Roche H, Bogé G (1996) Fish blood parameters as a potential tool for identification of stress caused by environmental factors and chemical intoxication. Marine Environmental Research 41: 27–43. https://doi.org/10.1016/0141-1136(95)00015-1

Rubec PJ, Cruz F, Pratt V, et al. (2001) Cyanide-free net-caught fish for the marine aquarium trade. Aquarium Sciences and Conservation 3: 37–51. https://doi.org/10.1023/A:1011370106291

Rummer JL, Brauner CJ (2015) Root effect haemoglobins in fish may greatly enhance general oxygen delivery relative to other vertebrates. Plos One 10(10): e0139477. https://doi.org/10.1371/journal.pone.0139477

Salbego J, Becker AG, Gonçalves JF, et al. (2014) The essential oil from *Lippia alba* induces oxidative stress in the silver catfish (*Rhamdia quelen*) after transportation. Neotropical Ichthyology 12(4): 811–818

Safran P (1992) Theoretical analysis of the weight-length relationship in fish juveniles. Marine Biology 112: 545–551.

Sampaio FDF, Freire CA (2016) An overview of stress physiology of fish transport: changes in water quality as a function of transport duration. Fish and Fisheries 17(4): 1055–1072. https://doi.org/10.1111/faf.12158

Sampaio FDF, Freire Carolina A, Sampaio TVM, et al. (2015) The precautionary principle and its approach to risk analysis and quarantine related to the trade of marine ornamental fishes in Brazil. Marine Policy 51: 163–168. https://doi.org/10.1016/j.marpol.2014.08.003

Sampaio FDF, Ostrensky A (2013) Brazilian environmental legislation as tool to conserve marine ornamental fish. Marine Policy 42: 280–285. https://doi.org/10.1016/j.marpol.2013.03.009

Schmidt-Nielsen K (1996) Animal Physiology – adaptation and environment. Cambridge, Cambridge University Press, 617 pp.

Schreck CB, Jonsson L, Feist G, Reno P (1995) Conditioning improves performance of juvenile chinook salmon, *Oncorhynchus tshawytscha*, to transportation stress. Aquaculture 135: 99–110. https://doi.org/10.1016/0044-8486(95)01018-1

Semmens BX, Buhle ER, Salomon AK, Pattengill-Semmens CV (2004) A hotspot of non-native marine fishes: evidence for the aquarium trade as an invasion pathway. Marine Ecology Progress Series 266: 239–244. https://doi.org/10.3354/meps266239

Shuman CS, Hodgson G, Ambrose RF (2004) Managing the marine aquarium trade: is eco-certification the answer? Environmental Conservation 31: 339–348. https://doi.org/10.1017/S0376892904001663

Tenningen M, Vold A, Olsen RE (2012) The response of herring to high crowding densities in purse-seines: Survival and stress reaction. ICES Journal of Marine Science 69: 1523–1531. https://doi.org/10.1093/icesjms/fss114

Tondolo JSM, de Amaral LP, Simões LN, et al. (2013) Anesthesia and transport of fat snook centropomus parallelus with the essential oil of *Nectandra megapotamica* (Spreng.) Mez. Neotropical Ichthyology 11(3): 667–674. https://doi.org/10.1590/S1679-62252013000300020

Treasurer JW (2012) Changes in pH during transport of juvenile cod *Gadus morhua* L. and stabilisation using buffering agents. Aquaculture 330–333: 92–99. https://doi.org/10.1016/j.aquaculture.2011.12.021

Urbinati EC, de Abreu JS, da Silva Camargo AC, Landinez Parra MA (2004) Loading and transport stress of juvenile matrinxã (*Brycon cephalus*, Characidae) at various densities. Aquaculture 229: 389–400. https://doi.org/10.1016/S0044-8486(03)00350-8

Wabnitz C, Taylor M, Green E, Razak T (2003) From ocean to aquarium: the global trade in marine ornamental species. Cambridge, UNEP/Earthprint, Biodiversity Series 17, ISBN: 92-807-2363-4, 64 pp. Available online at: https://archive.org/details/fromoceantoaquar03wabn

Weigle SM, Smith LD, Carlton JT, Pederson J (2005) Assessing the risk of introducing exotic species via the live marine species trade. Conservation Biology 19: 213–223. https://doi.org/10.1111/j.1523-1739.2005.00412.x

Wood BYCM (1991) Acid-Base and ion balance, metabolism, and their interactions, after exhaustive exercise in fish. Journal of Experimental Biology 160 (1): 285–308.

Wood E (2001a) Global advances in conservation and management of marine ornamental resources. Aquarium Sciences and Conservation 3: 65–77. https://doi.org/10.1023/A:1011391700880

Wood E (2001b) Collection of coral reef fish for aquaria: global trade, conservation issues and management strategies. United Kingdon, Marine Conservation Society, 80 pp.

Wright KA, Woods CMC, Gray BE, Lokman PM (2007) Recovery from acute, chronic and transport stress in the pot-bellied seahorse *Hippocampus abdominalis*. Journal of Fish Biology 70: 1447–1457. https://doi.org/10.1111/j.1095-8649.2007.01422.x

Zeppenfeld CC, Toni C, Becker AG, et al. (2014) Physiological and biochemical responses of silver catfish, *Rhamdia quelen*, after transport in water with essential oil of *Aloysia triphylla* (L'Herit) Britton. Aquaculture 418–419: 101–107. https://doi.org/10.1016/j.aquaculture.2013.10.013

Author Contributions: FDFS did field and lab work, generating the data of this manuscript, analyzed the data and wrote the first draft of the manuscript; HCSA helped with generating data on antioxidant enzymes and contributed to the final version of the manuscript; FLB helped in generating data on antioxidant enzymes and in discussing data; LFF helped in the idea of the project and text edition of the manuscript, CAF supervised the whole study, helped in all data analysis and text edition.
Competing Interests: The authors have declared that no competing interests exist.

15

Actual status of fishing reserves of the Yesil River

Zhumagazy Kurzhykayev[1], Kuanysh Syzdykov[2], Ainur Assylbekova[2], Dinara Sabdinova[2], Viktor Fefelov[1]

[1] *Fisheries Research and Production Center, North Kazakhstan Branch. Armandostar Street 2B, Nur-Sultan, Kazakhstan.*
[2] *S. Seifullin Kazakh Agro Technical University. Zhenis Avenue 62, Nur-Sultan, Kazakhstan.*
Corresponding author: Ainur Assylbekova (gamily-05@mail.ru)

ABSTRACT. In this article, the study and prediction of the state of fish resources in the Yesil River was carried out to determine the boundary reference points of the reserve and management decisions in case they are exceeded, to ensure sustainable fisheries and preserve biological diversity. Ichthyological studies, retrospective analyses and rankings of the effects of the water supply of the Yesil River on fish stocks have been carried out, critical values of the biomass of fishing reserves have been determined, as criteria for biologically safe fishing values and targets for management of fish stocks at reaching the boundary values of 24.03 tonnes have been established. According to the results of studies of basic biological indicators, it was determined that roach, bream, pike and perch populations each had an LC50 above LM50, which indicates their low utilization in the fishery industry. A strategy for management of fish stocks, recommendations for the conservation of fishery resources and their rational use has been developed as a result of these findings. The list and status of the monitored indicators for the careful management of fish stocks, as well as a list and indicators of the boundary indicators for the state of fish stocks, was established. These parameters included the fish size of 50% of the sampled fish and 50% of maturity achieved, where the LC50 was observed in the fish in question from 19 to 45.4 cm and LM50 from 12.3 to 41 cm, respectively. The minimum and maximum river levels that ranged from 111 up to 159 cm, as well as the critical value of the commercial reserve, corresponded to the boundary value of 24.03 tonnes. The results of this work will serve as the basis for making future managerial decisions, necessary for sustainable fisheries and conservation of the biological diversity of the Yesil River.

KEY WORDS. Biomass of fishing reserve, catch, fish productivity, ichthyological studies.

INTRODUCTION

The Yesil River is an important fishery reservoir in Northern Kazakhstan, not only socially, but also for preserving the biological diversity of the region. It is necessary to evaluate fish productivity and possible catches of fish in the different scenarios of water availability of the main fishing reservoirs of the country. For each of these scenarios, the boundary reference points of the fishery and the list of managerial decisions for its optimization should be defined.

For the Yesil River, boundary reserve guidelines have been defined to develop a strategy for cautious inventory management and sustainable fisheries, in accordance with the principles of responsible fishery management as outlined by the FAO. This will result in compliance with international norms and the conservation of fish stocks as the critical limits are approached (Koutsikos et al. 2012).

The stated research relates to ecological and fishery spheres and fully complies with the priority tasks of the "Concept on the Transition of the Republic of Kazakhstan to the Green Economy" (approved by the Decree of the President of the Republic of Kazakhstan of 30 May 2013 #577).

For the Yesil River, strategies (known as management decision schemes) have been developed for each of the possible scenarios of the state of the reserves. It should be remembered that, while recommending a particular catch factor for each fish species and setting limits on its catch, we do not take into account non-commercial catch. As a result, even if the recommended withdrawal rates are met, valuable fish species may be overfished. Therefore, a precautionary approach is necessary.

Fishes are the important indicator of an aquatic ecosystem and occupy a remarkable position from a socioeconomic point of view (Bera 2014). Due to their content of proteins, fats, vitamins, phosphorous and other compounds, the elements present in ichthyofauna have been used as a diet for humans since time immemorial. They also contribute to the economies in society (Ashok 2014).

In accordance with the chosen research direction, the following tasks were set: 1) Ichthyological studies, retrospective analyses and ranking of the impact of the water supply of the Yesil River on fish stocks; 2) Determination of critical values of the biomass of the commercial reserve, as criteria for biologically safe fishing values and targets for management of fish stocks when boundary values have been reached; 3) Development of strategies for management of fish stocks, recommendations for the conservation of fishery resources and their rational use; 4) Setting up lists and status of monitored indicators for careful management of fish stocks and 5) Setting lists and indicators of boundary guidelines for the state of fish stocks.

MATERIAL AND METHODS

The material was collected as a result of field trips in the period from 2016 to 2017. The volume of the collected material is given in Table 1.

Table 1. Volume of the collected material.

Hydrological regime (samples)	Biological analysis of fish (pic.)	Mass measures (pic.)	Network (pic.)	Fishing reserve (tons)
40	1076	2380	38	60

The observation stations were located along the entire length of the Yesil River. Sampling sites were chosen in order to conduct ichthyological research and determine commercial fish stock. The sites were found using echo-sounders and recorded using navigators. According to the Fig. 1, sampling was carried out in summer and autumn periods. Sampling in the summer period was carried out mainly from the riverbed of Esil River in the Akmola region. The major criterion for the determination of sample sites is, first of all, the fact that the water in these areas warms up evenly and creates favorable conditions for the rapid growth and development of zooplankton and benthic groups of organisms (the main food of ichthyofauna) and also creates favorable conditions for the growth and development of juvenile fish.

In autumn, determination of sampling sites was based on the fact that, in the North Region, rivers have wider channels, deeper sections and greater concentration of wintering pits and the bulk of the fish in the pre-winter period is concentrated in these pits in the riverbed. Hydrochemical properties are most favorable in these areas.

For the sampling, echo-sounders were used to determine the "dumps" – transitions from shallow to deep waters. Fish are mainly concentrated in these areas, which are favorable for feeding. These places were recorded by navigators. Subsequently, sampling was conducted at these sites. For the research, five areas were selected on the summer expedition and five areas on the autumn expedition (Fig. 1).

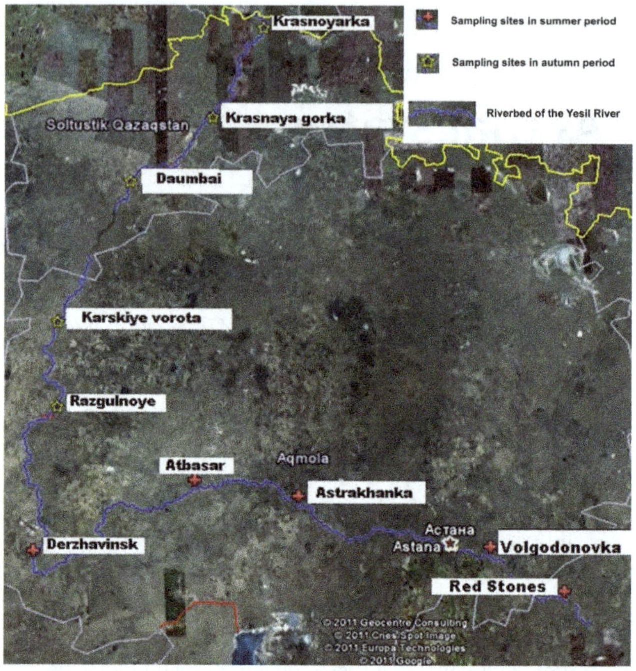

Figure 1. Yesil River with indicated sampling sites on the territory of the Republic of Kazakhstan.

Ichthyological analysis included the determination of linear dimensions, weight, sex and maturity stages of fish, according to standard methods (Pravdin 1966). Fish age was determined by the generally accepted method which is as follows: when scales were examined under magnification, concentric formations located around the centre in circles were observed. The number of such circles formed within one year is an annual ring. The number of annual rings corresponds to the age of the fish. Age determination could also be carried out by looking at the gill cover, ear bones (otoliths) and vertebrae. On the gill, the annual rings of the otoliths, after appropriate treatment (drying, degreasing, lightening and even boiling), are clearly visible.

The calculation of the number and industrial volume of fish catches was carried out according to the standard method of networking (Kushnarenko and Lugarev 1983, Sechin 1986), since the use of alloy networks, due to high weed density and a lack of necessary depths, was not possible. The basis was the formula: $N = (Q{\times}S)/k$, where: Q is the number of fish in catches in pcs., S is the registration area of the reservoir in hectares, k is the correction factor obtained by multiplying the three main coefficients for each size group (network): $K_i = P{\times}K{\times}C$, where: P is the coefficient of probability of meeting fish with fishing gear, K is the network catchability factor and C is the catch area. These coefficients are calculated on the basis of experimental data (Kushnarenko and Lugarev 1983, Sechin 1986).

The coefficient P is calculated on the basis of the formula, proposed by Sechin (1986). The coefficient of network efficiency is assumed equal to 0.5.

The catch area was calculated using the formula: $C = V \times t \times g \times (2 \times b + 3.14 \times Vt)$, where: V is the radial yaw rate (m/min.), t is the network setup time in minutes, g is the number of fixed networks and, b is the network length in meters.

In this case, the key figure is V. It was borrowed from the directory (Radakov and Protasov 1964). Yaw rate for perch and pike is 0.04; for bream and roach it is 0.05.

The industrial reserve was calculated according to the following formula: $M = N \times m$, where M is the ichthyomass of age/population, N is the age/population number, defined by the formula (1) and, m is the average weight of individuals at the age/population.

The LC50 and LM50 indices were determined in accordance with the recommendations of the FAO (Mees 2006). In connection with the lack of fishing on the Yesil River, a simulation of the fishery was carried out, i.e. for analysis, a separate order of networks with characteristics as close as possible to commercial networks was used. Of course, it is impossible to predict what networks the future nature user will use. In this regard, the order of 40 mm was used, which provides some underestimation of the LC50. This can also be attributed to our precautionary approaches.

All calculations were carried out on a PC using the MS Excel software (Zhivotovsky 1991, Korosov and Gorbach 2007, Buyul and Ceofel 2005).

RESULTS

The estimation of boundary reference points by biological indicators was done for the main commercial species. Pike-perch, tench and burbot were not evaluated due to the low populations of these species.

Roach is an aboriginal species in the Ob-Yertish basin, to which the Yesil River belongs. In the riverbed, the roach is one of the most numerous species distributed throughout the water area and occupies various biotopes. As such, it is one of the main fish species targeted in amateur fishing (Averintsev 1929, Alyoshina 2009). The biological parameters of the roach are not high, which is explained by the low fodder of the Yesil River (Table 2)

The catch is dominated by individuals aged 2+ to 4+ years. Beginning in 2010, seven-year-olds appeared in the catches and, since 2011, eight-year-old individuals of roach have been be found. This allows researchers to draw conclusions about the positive dynamics of roach age composition. The age range is limited to the eighth year. The maximum age of males is 7+ years. The well-being of the species in the Yesil River further confirms this thesis.

The average size of a roach relates to individuals which have already spawned more than once. The LC50 is significantly higher than the LM50. Moreover, the difference in these indicators clearly indicates an extremely low use of roaches in the fishery.

Bream is an acclimatizer and was not observed in the Yesil River until 1970 (Berg 1948, 1949a). The appearance of this species in the ichthyofauna of the river is associated with acclimatization work at the Sergeevsky and Vyacheslavsky reservoirs. At present, this species has spread widely throughout the entire river and has become a common species in the Yesil River system. Bream, possessing unpretentiousness in breeding conditions and eating a wide range of food, is common in almost all biotopes of the river (Berg 1949b, Borisov and Ovsyannikov 1954). The biological characteristics of the bream are not high, which is explained by the low fodder of the Yesil River (Table 2).

Table 2. Basic biological parameters of fish. $L_{average}$ – average length of fish (cm); $M_{average}$ – average mass of fish (g); L_{C50} – the length at which 50 percent of the fish is selected in the fishery (cm); L_{M50} – the length at which 50 percent of the fish reach puberty (g).

	Year	$L_{average}$	$M_{average}$	L_{C50}	L_{M50}
Roach	2016	14.6 ± 0.03	74.7 ± 0.03	20 ± 0.01	12 ± 0.05
	2017	14.5 ± 0.04	73.4 ± 0.01	20.1 ± 0.03	12.3 ± 0.02
Bream	2016	18.3 ± 0.02	164.6 ± 0.03	21 ± 0.01	13 ± 0.04
	2017	18.9 ± 0.03	152.3 ± 0.04	20.4 ± 0.03	12.8 ± 0.03
Pike	2016	40.3 ± 0.04	702.7 ± 0.04	47 ± 0.06	41 ± 0.05
	2017	38.7 ± 0.06	563.4 ± 0.03	45.4 ± 0.04	41.2 ± 0.05
Perch	2016	16.0 ± 0.02	88.5 ± 0.02	19 ± 0.03	13 ± 0.04
	2017	15.9 ± 0.03	88.4 ± 0.02	19.2 ± 0.01	12.4 ± 0.02

The catch is dominated by individuals aged 4+ and 5+ years. In the catches from 2006 to 2008, there were no older age groups; starting in 2010, eight-year-old bream species appeared and beginning in 2014, nine and ten-year-olds were caught in the catches, facts which allow one to conclude that the age composition is positive. The age range is limited to the tenth year. The maximum age of males is 7+ years (Veselov 1977). This also confirms the thesis about the well-being of the species in the Yesil River.

The average size of the bream of individuals that have already spawned more than once rises slightly. The LC50 is significantly higher than the LM50. Moreover, the difference in these indicators clearly indicates an extremely low use of bream in the fishery.

In the Yesil River, the pike is represented by unproductive populations. In the commercial shoals, younger age groups dominate. In experimental catches, it is represented by single specimens. Despite this, it is one of the most common species in the river system of Yesil, being found throughout the river (Vyshegorodtsev 2002, Zinoviev and Mandritsa 2003). Biological indicators of pike are not high (Table 2).

For the entire period of work (from 2005 to 2008 and from 2010 to 2016), the age limit of the pike caught in the river was 6+ years. In the population of the pike from the Yesil River in 2016, the age group dominates: 1+. Since 2010, 6+ years of age have been recorded in scientific catches, but the absence of older age groups prevents any discussion on the positive dynamics of the age composition. In addition, the dominance of individuals of younger age groups indicates a high pressure on the pike population in the Yesil River. The maximum age of males is 5+ years. This also confirms the thesis about the high pressure on the pike population in the Yesil River. The average size of the pike relates to individuals who have already spawned more than once. LC50 is higher than LM50.

Perch is an aboriginal species in the Yesil basin (Il'mast 2005, Kafanova 1984). Along with the roach, this species is the most abundant in the ichthyofauna of the river system of Yesil. This species is distributed throughout the water area and occupies various biotopes; it is one of the main objects of amateur fishing. The biological parameters of the perch are not high (Table 2).

The catch is dominated by individuals aged 3+ years. In the catches from 2006, 2008 and 2010, there were no older age groups. Starting in 2011, seven-year-olds appeared and since 2012, eight and nine-year-old perch species appeared as well, suggesting a positive dynamics in age composition. The age range is limited to the ninth year. The maximum age of males is 7+ years. This further confirms the thesis about the well-being of the species in the Yesil River.

The average size of perch falls on individuals who have already spawned more than once. The LC50 is significantly higher than the LM50. Moreover, the difference in these indicators clearly indicates the extremely low use of perch in the fishery.

In the Yesil River, a special ichthyocenosis was formed that is sufficiently well adapted to existing habitat conditions. There are a number of species that form the backbone of the community and fishery: roach, perch, bream and pike (Lebedev 1967, Lebedev et al. 1969, Lyagina 1984). In some areas, they are joined by tench and pike perch, which are not resistant and often exhibit unprecedented fluctuations in numbers. These species are of secondary importance, both in the cenotic and in the utilization plan.

In the Yesil River, the maturation of 50% of the roach population occurs when the length reaches 12–12.3 cm. Therefore, the indicator of the stability of the state of the population is the seizure of individuals exceeding the length of 12.3 cm. While the basis of catches in the simulation of commercial fishing was the specimen of 20–20.1 cm, this means a boundary reference point will not be reached soon. LC50 > LM50. The results of our studies indicate that a higher seizure factor of the commercial population than 0.311 (Malkin and Borisov 2000) may be applied.

Maturation of 50% of the bream population occurs when the length reaches 12.8–13 cm (Nikolsky 1971, 1974). Proceeding from this, the indicator of the stability of the state of the population is the seizure of individuals exceeding the length of 13 cm. While the basis of catches in the simulation of commercial fishing was individuals of 20.4–21 cm, this means the boundary landmark will not be soon reached. LC50> LM50. A higher capture factor of the commercial population than 0.311 (Malkin and Borisov 2000) may be applied.

For the pike population, maturation of 50% of individuals is typical when the length reaches 41–41.2 cm (Nikolsky 1980). Proceeding from this, the indicator of the stability of the state of the population is the seizures of individuals exceeding the length of 41.2 cm. While the baseline of catches in the simulation of commercial fishing was 45.4–47 cm, this means a boundary reference point will soon be reached. LC50 = LM50. When predicting the volume of seizures, the coefficient of seizure of the fishing population should be 0.266 (Malkin and Borisov 2000).

Maturation of 50% of the perch population occurs when the length reaches 12.4–13 cm (Silivrov 2008). Proceeding from this, the indicator of the stability of the state of the population is the seizure of individuals exceeding the length of 13 cm. While the basis of catches in the simulation of commercial fishing was made up of individuals 19–19.2 cm, this means the boundary landmark will not soon be reached. LC50> LM50. Thus, the results of the studies indicate that a higher seizure factor of the commercial population than 0.311 (Malkin and Borisov 2000) may be applied.

The water regime of the river is characterized by the high spring water (85–96% of the annual runoff) and a long low-water period. The spring rise of the level usually begins in mid-April, reaching a maximum in late April, early May. The duration of high water in the upper part of the river is 1.0–1.5 months and increases downstream to 2–3 months. The annual volume of runoff in the high water period may exceed the runoff of the shallow years by a hundredfold (Chugunova 1959).

Figure 2 shows the dynamics of the average annual water level in the river (Turgenevka and Volgodonovka) and the commercial fish stock for 2008–2017.

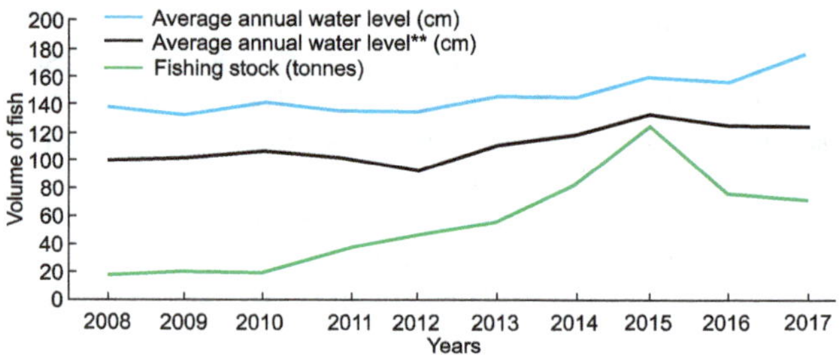

Figure 2. Dynamics of hydrological indices and industrial volume of fish in the Yesil River for the years 2008–2017.

Analysis of the available data shows the dependence of the commercial fish stock in the Yesil River on the average annual water level. Data on the average annual water level in the river for 2017 covers only an eight-month period.

Based on the available data from 2008 to 2017 (Table 3), a high correlation existed between the average annual water level in the river and the commercial fish stock. Thus, the relationship between the average annual water level in the river and the indicators for the commercial productivity of populations is present and its reliability is high.

Table 3. Correlation (R) of average annual water level and biomass of commercial fish stocks in the Yesil River.

Index	Water level										
	Law water					Medium water		Law water			
	2008	2009	2010	2011	2012	2013	2014	2015	2016	2017	R
Annual average level* (cm)	137	133	140	135	135	144	144	159	157	174	0.694
Annual average level** (cm)	99	101	106	100	93	111	118	131	125	125	0.851
Industrial stock, t.	19.3	19.4	19.3	35.1	45.5	56.6	81.1	123.5	75.0	71.3	–

*Turgenevka vil., **Volgodonovka vil., 2017: data on the water level for eight months.

In the period from 2008 to 2010, the biomass of commercial fish stocks was in a stable state and was at a low level during this period. From 2011 to 2015, there was an increase in the commercial fish stock in the river, which lasted until 2015. In 2016 and 2017, there was a decrease in total biomass, probably in connection with the spring overspills of the Yesil River, as a result of which a significant part of the commercial shoals of fish remained in floodplain reservoirs.

To determine the effect of water availability on the number of commercial stocks and determine the critical values of the water regime for the state of fish stocks, the average annual water level in the Yesil River was chosen.

In the range of investigated years, the water content of the Yesil River was generally low (low water period) and only the last five years can it be characterized as medium and high water. The critical value of the average annual level is 144 cm in the area of Turgenevka village and 111 cm in the area of Volgodonovka village. In addition, it should be noted that the critical water content for the Yesil River is also an increase in the water level of more than 159 cm in the area of the village of Turgenevka and 131 cm in the area of the village of Volgodonovka.

However, in view of the absence of fishing gear and the prospects for its development on the river, it is hardly possible to propose any specific management measures. The only thing that can be offered is the optimization of water consumption, including the use of alternative sources of water supply and the strengthening of work to save the young from reservoirs separated from the river. To implement this, local executive bodies should plan financing for said management.

In general, the following indicators of sustainable development were adopted for the Yesil River: the equality of replenishment and seizures for pike; and for bream, roach and perch, an excess of seizure over replenishment (Table 4).

In the Yesil River, in the age structure of the roach population, a gradual expansion of the age range has been observed beginning in 2010 and, in the last year, a noticeable increase in the number of juveniles occurred. The catch is dominated by individuals aged 2+ to 4+ years, accounting for a total of 65%. The most productive part of the shoal has been in a stable state for several years. In such circumstances, a target in evaluating the potential for roach extraction should be a small excess of the fishing over the replenishment process.

In the Yesil River, in the age structure of the bream population, from the year 2010, eight-year-old individuals appeared

Table 4. Status of indicators of sustainable development. TAC: total allowable catch, MAC: maximum allowable catch.

Types of fish	Target TAC (MAC)
Roach	Replenishment < fishing
Bream	Replenishment < fishing
Pike	Replenishment = fishing
Perch	Replenishment < fishing

and, beginning in 2014, nine and ten-year-old individuals were noted in the catches, which led to positive dynamics within the age composition. The catch is dominated by individuals aged 3+ to 5+ years, accounting for a total of more than 59%. The most productive part of the shoal has been in a stable state for several years. In such circumstances, a target in evaluating the potential for bream production should be a small excess of the fishing over the replenishment process.

The state of commercial pike stocks in the Yesil River can be estimated as satisfactory. In the age structure of the population, individuals today are from 1+ to 3+ years old, accounting for more than 65%. The most productive part of the shoal has been in a stable state for several years. In addition, the dominance of individuals of younger age groups indicates a high pressure on the pike population in the Yesil River. In this situation, the target for the assessment of the possibilities of extraction should be an equality between fishing and replenishment processes.

Seven-year-old individuals appeared in the Yesil River in the age structure of the perch population, starting in 2011 and, starting in 2012, eight and nine-year-old individuals were caught in the catches, which led to positive dynamics within the age composition of the species. The catch is dominated by individuals aged 3+ to 4+ years, totalling more than 52%. The most productive part of the shoal has been in a stable state for several years. In such circumstances, a target in evaluating the possibilities for perch production should be a small excess of the fishing over the replenishment process.

Other secondary fishery objects do not play a significant role here and the pressure on them is quite sparing, bordering on insignificant.

The critical values of the biomass of the commercial stock of the population (or the entire commercial shoal) of fish in the Yesil River were determined on the basis of analysis of the long-term dynamics of the fish population in the reservoir and are equal to the biomass of the year preceding the achievement of the minimum values of fish biomass or the next lowest biomass value (Table 5).

Table 5. Total biomass of commercial fish stocks by years in tonnes.

Years	2008	2009	2010	2011	2012	2013	2014	2015	2016	2017
Roach	5.58	5.55	5.60	6.33	10.46	8.56	12.95	18.49	11.96	13.39
Bream	4.06	4.03	4.20	7.30	10.88	13.83	15.71	21.04	17.61	13.56
Tench	–	–	–	–	0.88	1.70	3.36	2.31	1.95	2.52
Perch	4.32	4.30	4.40	6.71	11.96	14.60	21.60	29.86	25.93	24.27
Zander	–	–	–	–	–	1.99	3.37	3.03	1.57	3.33
Pike	5.35	5.47	5.10	14.71	11.33	15.87	24.13	48.74	14.66	14.22
Burbot	–	–	–	–	–	–	–	–	1.28	–
Total	19.31	19.35	19.30	35.05	45.51	56.55	81.12	123.47	74.96	71.29

In the period from 2008 to 2010, the biomass of commercial fish stocks was in a stable state and was at a low level during this period. Since 2011, there has been an increase in the commercial fish stock in the river, which lasted until 2015. In 2016 and 2017, there was a decrease in total biomass, probably in connection with the spring overspills of the Yesil River, as a result of which a significant part of the commercial shoal of fish remained in floodplain reservoirs.

Figure 3 shows the fluctuations in the biomass of commercial fish during the period under study and, in particular, a significant increase in 2015 and then a sharp decline in the biomass of the commercial pike shoal. The remaining species of studied fish did not undergo a significant change in the studied parameters.

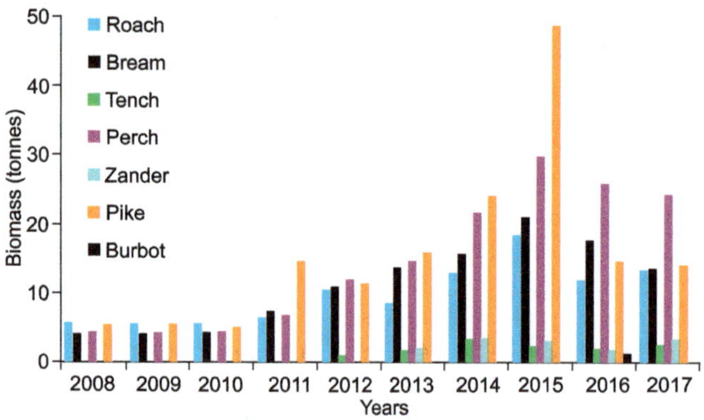

Figure 3. Total biomass of commercial fish stocks in tonnes.

It should be noted that populations of commercial fish stocks increase their biomass quite rapidly (Fediy 1964, Fulton 1906). In some cases, the biomass of the commercial part of the population may increase more than two times in just one year (Schultz 2003).

Table 6 shows the fishing stock for the last two years and the critical values of the fishing stock.

Table 6. The commercial reserve of the Yesil River over the last two years and the critical biomass of the commercial reserve.

Types of fish	Maximum fishing stock (ton)	Minimum fishing stock (ton)	Fishing stock in 2016 (ton)	Fishing reserve in 2017 (ton)	Critical biomass values of the commercial reserve (ton)
Roach	18.49	5.55	11.96	13.39	5.58
Bream	21.04	4.03	17.61	13.56	4.06
Tench	3.36	0.88	1.95	2.52	1.70
Perch	29.86	4.30	25.93	24.27	4.32
Zander	3.37	1.57	1.57	3.33	1.99
Pike	48.74	5.10	14.66	14.22	5.10
Burbot	1.28	1.28	1.28	–	1.28
Total	126.14	22.71	74.96	71.29	24.03

According to population-fishing indicators, the ichthyocenosis of the Yesil River is stable in its main fish species (roach, bream, perch and pike). The indicator of Lin's fishing stock in 2016 approached a critical value, even higher than the last year. The commercial pike perch in the current year was above the critical value; however, this species is recommended to be used only for reproductive purposes.

In accordance with the conducted studies and also taking into account the absence of commercial fishing on the Yesil River, we consider establishing the following list of boundary landmarks: a) $L_{C50} = L_{M50}$; b) critical water content; c) critical biomass values of the fishing stock.

Table 7 shows the list and status of the monitored indicators of the boundary indicators of the state of the fish stocks of the Yesil River.

Table 7. List and status of monitored indicators of the boundary reference points of the state of fish stocks in the reservoir of the Yesil River.

Indicators	Indicator value		
	Type of fish	L_{C50}	L_{M50}
Indicators L_{C50} and L_{M50}	Roach	20.00	12.3
	Bream	20.40	12.8
	Pike	45.40	41.0
	Perch	19.00	12.4
Critical water content	Minimum level (v.Turgenevka)	144.00	
	Maximum level (v.Turgenevka)	159.00	
	Minimum level (v.Volgodonovka)	111.00	
	Maximum level (v.Volgodonovka)	131.00	
Critical biomass values of the fishing stock	Roach	5.58	
	Bream	4.06	
	Tench	1.70	
	Perch	4.32	
	Zander	1.99	
	Pike	5.10	
	Burbot	1.28	
	Total	24.03	

DISCUSSION

For the purpose of assessing the state of fish stocks, developing their approaches to rational management and sustainability of use, research was carried out for the first time in this direction on the Yesil River, which attained the status of a water basin of international significance for the first time in 2016 and 2017.

Studies were carried out at eight stations, the ranking of water content was undertaken, the effect of the hydrological regime on fishery indicators was assessed, boundary reference points were drawn and critical biomass values of the commercial reserve for the population of the main fish species were determined.

Research conducted by scientists of the Kazakh Scientific Research Institute of Fisheries LLP has revealed that 13 species of fish from four families – carp, pike, perch and burbot, live in the Yesil River. Ichthyophagous are represented by the following species: xhtiophags – common pike and common pike-perch; ichthyobenthophagous: common perch, common ruff and burbot; euryphagous – Siberian roach, ide and tench; benthophagous – common minnow and bream. Ordinary stakes, Siberian and silver dace are, respectively, planktophag, planktobentophagous and phyto-benthophagous.

The research conducted in the Yesil River indicates that, throughout the area of the study, the reservoir is a typical perch and roach watercourse, since these species predominate in the ichthyofauna.

The data of our studies indicates that in the riverbed, the roach is one of the most numerous species distributed throughout the water area and occupies various biotopes, serving as one of the main objects of amateur fishing. Bream is an introduced species. The biological characteristics of the bream are not high, which is explained by the low fodder of the Yesil River. Pike is represented by low-productive populations. In the commercial shoals, younger age groups dominate. Perch represents aboriginal populations. Along with the roach, it is the most abundant in the ichthyofauna of the river system of Yesil. Distributed throughout the water area, perch occupy various biotopes and are also one of the main objects of sports and amateur fishing.

The results of our studies indicate that, after comparing pike, zander and bream populations to those of roach and perch, a higher seizure factor of the commercial population may now be applied.

The critical values of the biomass of the commercial stock of the population (or the entire commercial shoal) of fish in the Yesil River vary considerably. Especially large fluctuations in the fishing stock are observed in the pike population. The data indicates a significant decrease in the productive shoal of fish (pike) with significant mass and size classes. Overfishing can undermine the qualitative-quantitative state of the natural, healthy population of these species of fish.

The research data shows dominance amongst peaceful fish, like the roach and amongst predators, such as the perch.

We established, in the course of research, that the water level of the Yesil River was mostly low; in other words, low-water periods were observed. However, over the past four years, the water level of the Yesil River has increased significantly and is now considered as medium- and high-water, which significantly improves the habitat of the ichthyofauna. The reason for the improvement in the state of the environment is the fact that, in Akmola and Derzhavinka Regions in recent years, annual precipitation has increased from 230 to 320 mm, respectively and in winter, the snow cover increases, which favorably affects the flow of the riverbed in the spring. At the same time, we noted that, in 2017, biomass of ichthyofauna decreased due to increasing water level. The bulk of small fish were distributed over estuaries. This is also an indirect cause of the redistribution of ichthyofauna. In the summer, in the coastal areas, there is a large concentration of the food supply for many fish species, in particular roach, perch and, to some extent, bream. Predatory fish approach the coastal zone in the same way, following the fry and other aquatic organisms.

FINAL REMARKS

The analysis of the obtained data showed the dependence of the commercial fish stock in the Yesil River on the average annual water level. For the period from 2008 to 2017, there is a high degree of correlation here and a high level of reliability of the data.

In the range of years investigated, the water content of the Yesil River was generally low (with low-water periods) and only the last four years can it be characterized as medium and high-waters. We consider the level of 144 cm in the area of Turgenevka village and 111 cm in the area of Volgodonovka village to be critical values for the average annual levels. In addition, it should be noted that the critical water content for the Yesil River is also an increase in the water level of more than 159 cm in the vicinity of Turgenevka village and 131 cm in the vicinity of Volgodonovka village.

Critical values of the biomass of commercial fishing stock for the species studied were calculated. Several approaches were used to determine this indicator. Their choice was determined by methodological logic. As a result of this analysis, it was revealed that the situation with commercial reserves in the Yesil River is rare with few exceptions.

It should be noted that the definition of critical biomass of commercial fishing stock was carried out on the basis of available data for 2008–2017. This time interval is clearly insufficient to produce rigid conclusions on quantitative indicators. It is necessary to return annually to their revision for at least the next three years.

The materials obtained can form the basis for the concept of sustainable management of the river's fishery resources.

However, in view of the absence of fishing gear and the prospects for its development on the river, it is hardly possible to propose any specific management measures. The only thing that can be offered is the optimization of water consumption, including the use of alternative sources of water supply and the strengthening of work to save the young from reservoirs separated from the river. To implement the latter, local executive bodies should plan financing.

Recommendations on the strategy of fisheries management include: a) optimization of water consumption (regulation of the water level in the river Yesil); b) strengthening the work to save young fish from reservoirs separated from the river (especially in the years of high water); c) strengthening of control over sport (amateur) fishing.

ACKNOWLEDGEMENTS

For successful carrying out of research, special gratitude is due to employees of the Committee of Forestry and Fauna, and also to corresponding territorial divisions of Akmola and North Kazakhstan areas of the Ministry of Agriculture of the Republic of Kazakhstan. On hydro-chemistry and hydrology of the Yesil River, special thanks to Kazhydromet staff for providing relevant information on the years of research and on the months of each of the specific years of research. Also, the work would be impossible without the participation of the scientific, administrative and management and technical staff of the Northern branch of the Kazakh Research Institute of Fisheries and the Department of Hunting and Fisheries of the Kazakh Agrotechnical University named after S. Seifullin.

LITERATURE CITED

Alyoshina OA (2009) Plankton communities of the Ishim River and its tributaries as an indicator of their ecological state (within the Tyumen Region). Bulletin of the Tyumen State University 3: 223–232.

Ashok K (2014) Studies on ichthyofaunal diversity with special reference to monthly and seasonal variation of fish landing in glacial fed mountainous Goriganga River of Kumaun Himalaya, Uttarakhand, India. Research Journal of Animal, Veterinary and Fishery Sciences 2(4): 1–2. Available online at: http://www.isca.in/AVFS/Archive/v2/i4/1.ISCA-RJAVFS-2014-009.pdf

Averintsev SV (1929) On the issue of studying fish races. Bulletin of Fisheries 6: 24–25.

Bera A (2014) Ichthyofaunal diversity and water quality in the Kangsabati Reservoir, West Bengal, India. Advances in Zoology 2014: e674313. https://doi.org/10.1155/2014/674313

Berg LS (1948) Fishes of fresh waters of the USSR and neighboring countries. Tom I. Moscow, Publishing House of the Academy of Sciences of the USSR, 467 pp.

Berg LS (1949a) Fishes of fresh waters of the USSR and neighboring countries. Tom II. Moscow, Publishing House of the Academy of Sciences of the USSR, 459 pp.

Berg LS (1949b) Fishes of Fresh Waters of the USSR and Neighboring Countries. Tom III. Moscow, Publishing House of the Academy of Sciences of the USSR, 456 pp.

Borisov PG, Ovsyannikov NS (1954) Determinant of commercial fishes of the USSR. Moscow, 179 pp.

Buyul A, Ceofel P (2005) The art of information processing. St. Petersburg, DiSoft, 608pp.

Chugunova NI (1959) Guide to the study of the age and growth of fish. Moscow, Chugunova, 164 pp.

Fediy SP (1964) The faunistic complex of fish as an indicator of the sanitary state of fishing reservoirs In: "Radioactive Isotopes in Biology and Methods of Sanitary Hydrobiology". Leningrad, Science, 184–190.

Fulton TW (1906) On the rate of growth of Fishes. The Twenty fourth Annual Report of the Fishery Board for Scotland 3: 179–274.

Il'mast NV (2005) Introduction to Ichthyology. Petrozavodsk, Karelian Research Center of the Russian Academy of Sciences, 148pp.

Kafanova VV (1984) Methods of determining the age and growth of fish. Tomsk, Tomsk State University, 54 pp.

Korosov AV, Gorbach VV (2007) Computer processing of biological data. Petrozavodsk, Petrozavodsk State University, 76 pp.

Koutsikos N, Zogaris S, Vardakas L, Tachos V, Kalogianni E, Sanda R, Chatzinikolaou Y, Giakoumi S, Economidis P, Economou A (2012) Recent contributions to the distribution of the freshwater ichthyofauna in Greece. Mediterranean Marine Science, 13(2): 268–277. https://doi.org/10.12681/mms.308

Kushnarenko AI, Lugarev ES (1983) Estimating fish abundance by catch with passive fishing gear. Ichthyological Questions T23.6: 921–926.

Lebedev NV (1967) Elementary fish populations. Moscow, 212 pp.

Lebedev VD, Lebedev VD, Spanovskaya KA, Savvaitova LI, Sokolov EA, Tsepkin M (1969) Fish. Moscow, 346 pp.

Lyagina TN (1984) About intraspecies variability of roach *Rutilus rutilus* (L.) (Cyprinidae). Ichthyology Issues T24.5: 718–732.

Malkin EM, Borisov VM (2000) Methodical recommendations for monitoring the status of fish stocks and abundance estimates of fishes on the basis of biostatistical data. Moscow, Russian Federal Research Institute of Fisheries and Oceanography, 32 pp.

Mees CC (2006) Managing fishing effort in multispecies fisheries. Stock assessment for fishery management. A framework guide to the stock assessment tools of the Fisheries Management Science Programme. FAO Fishering Technical Paper 487: 189–194.

Nikolsky GV (1971) Private Ichthyology. Moscow, Enlightenment, 441 pp.

Nikolsky GV (1974) Theory of the dynamics of a herd of fish. Moscow, Food Industry, 447 pp.

Nikolsky GV (1980) Structure of the species and patterns of fish variability. Moscow, Food Industry, 182 pp.
Pravdin IF (1966) Guide to the study of fish. Moscow, Food Industry, 376 pp.
Radakov DV, Protasov VR (1964) Speed of movement and some features of the view of fish. Moscow, Science, 48 pp.
Schultz SK (2003) Field Guide to Freshwater Fish. Wiley, Amozon, 253 pp.
Sechin UT (1986) Methodological guidelines for assessing the abundance of fish in freshwater reservoirs. Moscow, All-Russian Research Institute of Freshwater Fisheries, 50 pp.
Silivrov SP (2008) Interpopulation variability of the pike *Esox lucius* (L.) in the reservoirs of the Middle and Southern Urals. Siberian Ecological Journal 1: 77–87.
Veselov EA (1977) The determinant of freshwater fish of the USSR. Moscow, Enlightenment, 238 pp.
Vyshegorodtsev AA (2002) Brief dictionary of the ichthyologist. Krasnoyarsk, KrasNU, 230 pp.
Zhivotovsky LA (1991) Population biometry. Moscow, Science, 271 pp.
Zinoviev EA, Mandritsa SA (2003) Methods of research of freshwater fish. Perm, The Perm University, 113 pp.

Author Contributions: ZK, KS, AA, DS and VF contributed equally to this article.
Competing Interests: The authors have declared that no competing interests exist.

16

The advertisement call of the phytotelm-breeding *Melanophryniscus xanthostomus* (Anura: Bufonidae)

Sarah Mângia[1], **Diego José Santana**[1], **Lucas Ribeiro Mariotto**[2]

[1] *Universidade Federal de Mato Grosso do Sul, Instituto de Biociências, Laboratório de Zoologia, Av. Costa e Silva, s/n, Cidade Universitária, 79070-900, Campo Grande, MS, Brazil*
[2] *Universidade de São Paulo, Faculdade de Filosofia Ciências e Letras de Ribeirão Preto, Departamento de Biologia. Avenida Bandeirantes, 3900, Monte Alegre, 14040-901, Ribeirão Preto, SP, Brazil*
Corresponding author: Lucas Ribeiro Mariotto (lucasrmariotto@gmail.com)

ABSTRACT. Vocalizations are often useful for understanding taxonomic relationships among anuran species. Despite this usefulness, vocalizations are described in only nine of 29 in *Melanophryniscus* Gallardo, 1961. Here we describe the advertisement call of *Melanophryniscus xanthostomus* Baldo, Bornschein, Pie, Ribeiro, Firkowski & Morato, 2015 of a population from Serra Dona Francisca, municipality of Campo Alegre, state of Santa Catarina, Brazil. The advertisement call (of three males, total of 17 calls) comprises two segments (the first with short and single notes, followed by a multi-pulsed note), with a duration of 12.194–20.986 s, and dominant frequency of 3101–3618 Hz (first and second segments combined). The advertisement call of *M. xanthostomus* differs from its congeners mostly by the higher number of pulses in the second segment (294–1033; from 16 to 321 in the other *Melanophryniscus* species), except *Melanophryniscus krauczuki* Baldo & Basso, 2004 (1018–1502 pulses in the second segment). This is the first described call of a phytotelm breeding *Melanophryniscus*, but it presents the same prototype (a compound call formed by two segments, the first composed of short notes followed by a long trill) of its congeners not-phytotelm breedings. This might indicate the advertisement call of *Melanophryniscus* as a conserved trait and thus can be considered diagnostic for the genus.

KEY WORDS. Atlantic Forest, bioacoustic, call description, natural history, vocalization.

INTRODUCTION

Vocalizations are imperative to solve major taxonomic problems in many distinct anuran groups (Haddad and Pombal-Jr 1998, Carvalho and Giaretta 2013, Pansonato et al. 2014). The most common vocalization among anuran repertoire is the advertisement call (Wells 2007), which is species-specific and therefore useful for species identification (Gerhardt and Davis 1988). The advertisement calls of many species-groups have a prototype pattern, well diffused among species – e.g., Microhylidae, Bufonidae (Heyer 1971, Martin 1972) –, and can be the synapomorphic characteristic for lineages.

Melanophryniscus Gallardo, 1961 toads (29 valid species) are distributed in South America (Argentina, Bolivia, Brazil, Paraguay, and Uruguay) (Frost 2018). In phylogenies, the genus has been recovered as the sister taxon of all other bufonids (e.g., Frost et al. 2006, Van Bocxlaer et al. 2010, Peloso et al. 2012), and its monophyly is supported by morphological, biochemical, behavioral, and molecular evidence for adults (McDiarmid 1971, Daly et al. 2007, Peloso et al. 2012) and tadpoles, based in a comparative description of the larvae of 23 *Melanophryniscus* species (Larson et al. 2003, Baldo et al. 2014). The genus is traditionally grouped into three species groups (*M. tumifrons*, *M. stelzneri*, and *M. moreirae* species groups) based mainly on morphology and coloration patterns (Caramaschi and Cruz 2002). Some species of *Melanophryniscus* described after 2003 have not been assigned to any group.

The toad *Melanophryniscus xanthostomus* Baldo, Bornschein, Pie, Ribeiro, Firkowski & Morato, 2015 was recently described based on individuals from Serra do Quiriri, Campo Alegre, Condomínio Vale dos Lagos and Reserva Particular do Patrimônio Natural Caetezal, Joinville and Morro do Boi, Corupá, municipalities from the state of Santa Catarina, Brazil, and was not assigned to any species groups. Some information about the natural history of *M. xanthostomus* is provided in the original description of the species (Bornschein et al. 2015), including data of vocalization sites, and period that males were found calling (from September to February). However, the advertisement call was not described.

Given the rapid increase of the species description in the genus, the search for useful characteristics, as vocalizations, is crucial important to help resolve their taxonomy and to test evolutionary hypothesis. However, from the 29 species of *Melanophryniscus*, only nine have the advertisement call described (see Caldart et al. 2013, Duré et al. 2015). Because anuran advertisement calls are species-specific (Gerhardt and Davis 1988) and therefore, are useful in the distinction of the species and an important character for taxonomy (Duellman and Trueb 1986), in the present study, we described the advertisement call of *M. xanthostomus* for a population from Serra Dona Francisca, Campo Alegre, Santa Catarina, Brazil. We also compared the calls of *M. xanthostomus* with the nine other species of *Melanophryniscus*.

MATERIAL AND METHODS

We recorded advertisement calls of three males (FURB 22851 - SVL 18.6 mm, FURB 22822 - SVL 19.7 mm, FURB 22713 - SVL 16.9 mm, total of 17 calls, Figs 1–3) at Serra Dona Francisca (-26°12'52.54"S, 49°13'04.92"W), Campo Alegre, Santa Catarina, on November 3rd, 2013. Calls were recorded with a Marantz PMD661 digital recorder coupled with a YOGA HT-81 directional microphone. Recordings were made around midnight (air temperature 18.8 °C, humidity 65%). We digitalized the recordings at 44.1 kHz, resolution of 16 bits. The three specimens were collected and deposited on the Zoology Collection of Universidade Regional de Blumenau (FURB), Santa Catarina, Brazil, under the numbers above.

We analyzed calls in RAVEN PRO 1.5 for Mac (Bioacoustics Research Program 2012) and constructed audio spectrograms in R using the package seewave (Sueur et al. 2008) with the following parameters: FFT window width = 256, Frame = 100, Overlap = 75, and flat top filter. We analyzed acoustic parameters normally used for species of *Melanophryniscus*: dominant frequency (Hz), call duration (sec), call interval (sec), first segment duration (sec), second segment duration (sec), interval between first and second segment (sec), number of short notes, duration of short notes (sec), interval between short notes (sec), note rate of the first segment (the ratio of the absolute number of notes and the absolute duration of the segment), pulse number of the second segment, and pulse rate of the second segment (the ratio of the absolute number of pulses and the absolute duration of the segment). Terminology of call descriptions follows Köhler et al. (2017).

Figures 1–3. Specimens of *Melanophryniscus xanthostomus*: dorsal view (1) and ventral view (2) from FURB 22851 specimen and FURB 22713 specimen in life (3).

We compared the advertisement call of *M. xanthostomus* with nine congeners, although those descriptions did not include all the call parameters that we analyzed (see references in Table 1).

RESULTS

The advertisement call of *M. xanthostomus* is a compound call formed by two segments (Table 1, Fig. 4–6). The first segment is composed by short and single notes (13 ± 4; range 7–20 notes), with a duration of 0.011–0.057 s (0.027 ± 0.010 s), and interval duration of 0.242–0.774 s (0.325 ± 0.109 s) between the short notes, with a pulse rate of 0.002–0.004 s (0.003 ± 0.001 s). The duration of the first segment is 1.774–6.794 s (4.115 ± 1248.98 s). The second segment consists of a multi-pulsed note (a long trill) of 6.278–18.318 s of duration (14.189 ± 2979.77 s), with 294–1033 pulses (685 ± 179.59 pulses), and pulse rate of 0.04–0.06 pulses per second (0.05 ± 0.01 pulses/s). The advertisement call (first and second segments combined) presents a duration of 12.194–20.986 s (18.577 ± 2443.70 s), and dominant frequency of 3101–3618 Hz (3395 ± 180.32 Hz). The interval duration between the first and the second segments is 9.258–221.673 s (82.949 ± 78878.03 s).

The advertisement call of *M. xanthostomus* is emitted with both segments in sequence, not only the first or the second

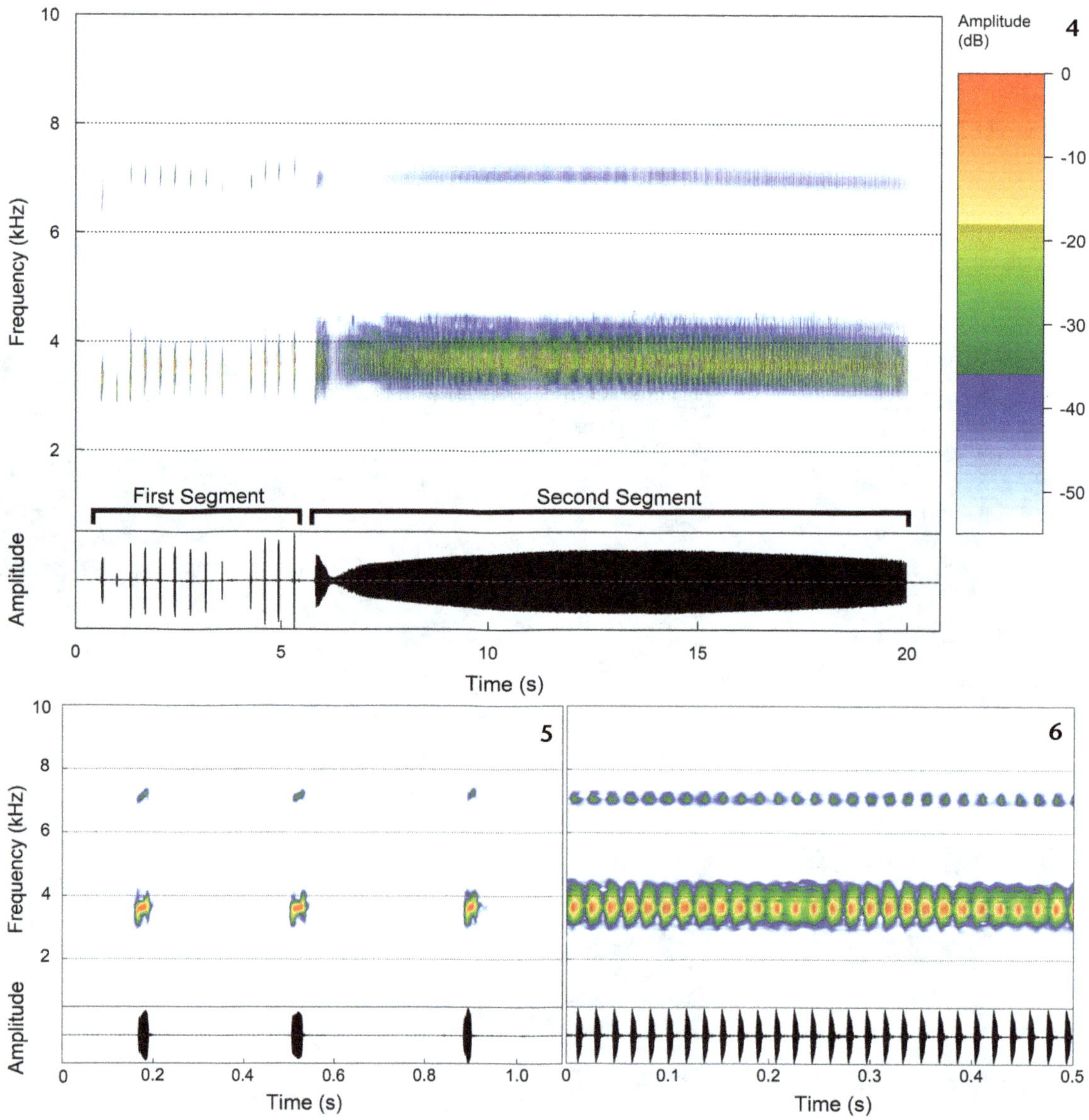

Figures 4–6. Advertisement call of *Melanophryniscus xanthostomus* (LM380, SLV 18.6 mm): (4) oscillogram and spectogram of one call; (5) oscillogram and spectogram of three notes from the first segment; (6) oscillogram and spectogram of 28 pulses from the second segment.

Table 1. Comparison of temporal and spectral parameters of the advertisement call of *Melanophryniscus xanthostomus* and nine other congeneric species. Values are means with the range in parentheses. (a) Air temperature; (w) water temperature.

Call parameters	*M. xanthostomus*[1]	*M. atroluteus*[2]	*M. atroluteus*[3]	*M. cupreuscapularis*[4]	*M. dorsalis*[5]	*M. dorsalis*[6]	*M. klappenbachi*[7]
Calls (n)	17 calls, 3 males	12 calls, 1 male	14 calls, 2 males	20 calls, 4 males	3 calls, 1 males	7 calls, 2 males	11 calls
Temperature (°C)	18.8 (a)	17.0 (w)	21.5 (a)	22.0 (a)	21.0 (a)	21.0 (a)	22.0–24.0 (a)
Dominant frequency (Hz)	3395 (3101–3618)	3000	(1900–3100)	2270 (2176–2357)	(2600–3200)	(2300–3200)	1900
Call duration (s)	18.577 (12.194–20.986)	7.523 (5.090–10.350)	3.700 (3.400–4.200)	3.023 (1.120–6.660)	–	–	0.507 (0.292–0.652)
Inter-call interval (s)	82.949 (9.258–221.673)	8.489 (2.922–7.954)	–	2.100 (1.350–2.780)	–	–	–
1st segment duration (s)	4.115 (1.774–6.794)	4.143 (2.575–5.360)	2.500 (1.800–3.200)	2.688 (0.780–6.390)	3.700 (3.500–4.100)	1.890 (1.000–2.300)	–
2nd segment duration (s)	14.189 (6.278–18.318)	3.012 (1.832–4.303)	1.100 (0.500–1.400)	0.23 (0.19–0.27)	2.37 (2.0–2.6)	1.50 (0.6–2.2)	–
Interval between segments (s)	0.265 (0.054–0.609)	0.026 (0.003–0.079)	–	0.100 (0.040–0.160)	–	–	–
Number of notes of 1st segment	13 (7 – 20)	20.6 (15–25)	17.6 (11–26)	10 (6–17)	19 (18–20)	13 (6–18)	3–4
1st segment notes duration (s)	0.027 (0.011–0.057)	0.102 (0.006–0.174)	–	0.020 (0.010–0.040)	0.054 (0.030–0.065)	0.042 (0.020–0.050)	–
Inter-notes interval of 1st segment (s)	0.325 (0.242–0.774)	0.091 (0.006–0.229)	(0.085–0.100)	0.270 (0.090–3.150)	0.153 (0.130–0.190)	0.140 (0.080–0.170)	–
Notes per s of 1st segment	0.003 (0.002–0.004)	–	(8–9)	–	–	(4–7)	–
Number of pulses 2nd segment	685.000 (294.000–1033.000)	222.380 (139.000–321.000)	87.600 (45.000–116.000)	20.400 (16.000–24.000)	161.000 (152.000–173.000)	122.000 (54.000–162.000)	43.000 (26.000–53.000)
Pulses per s of 2nd segment	47.960 (40.550–56.700)	75.440 (74.310–76.800)	(85.000–95.000)	88.950 (96.770–84.030)	(62.000–74.000)	(74.000–78.000)	86.000 (80.340–88.910)

Continues

Table 1. Continued.

Call parameters	*M. krauczucki*[8]	*M. montevidensis*[9]	*M. pachyrhynus*[10]	*M. rubriventris*[11]	*M. rubriventris*[12]	*M. rubriventris*[13]	*M. stelzneri*[14]	*M. stelzneri*[15]
Calls (n)	5 calls, 1 male	14 calls, 2 males	12 calls, 6 males	12 calls, 4 males	12 calls, 4 males	12 calls, 4 males		
Temperature (°C)	17(w)	24(a)	16–17(a) 16–18(w)	19(a)	19(a)	19(a)		20.5(a)
Dominant frequency (Hz)	3300	(2100–2800)	2668 (2261–2932)	1st seg. 1788 (1704–1846) 2nd seg. 1734 (1653–1789)	1st seg. 1721 (1515–1845) 2nd seg. 1687 (1535–1783)	1st seg. 1841 (1754–1944) 2nd seg. 1782 (1705–1869)	(1600–2600)	FH (2200–2600) 2H (4600–5200) 3H (7000–7600)
Call duration (s)	32.699 (25.013–36.646)	–	37.070 (6.640–75.200)	2.710 (1.400–3.250)	3.150 (2.030–4.170)	2.130 (1.320–3.790)	7.300 (4.500–9.300)	
Inter-call interval (s)	21.988 (18.196–25.737)	–	–	–	–	–	–	
1st segment duration (s)	2.031 (1.128–3.160)	1.980 (1.000–4.500)	4.440 (1.850–7.670)	0.912 (0.296–1.489)	1.438 (0.606–2.318)	0.617 (0.231–1.449)	4.200 (2.500–6.300)	
2nd segment duration (s)	30.455 (23.784–33.408)	1.580 (1.200–2.000)	25.270 (4.790–45.750)	1.365 (0.691–1.648)	1.413 (0.558–2.251)	1.323 (0.772–2.125)	2.100 (1.800–2.400)	0.500
Interval between segments (s)	0.258 (0.085–0.827)	–	–	0.435 (0.212–0.929)	0.299 (0.111–0.404)	0.187 (0.210–0.524)	–	
Number of notes of 1st segment	8.6 (6–12)	17 (7–28)	17 (8–28)	3.8 (2–7)	4.7 (2–8)	2.9 (1–6)	26 (12–36)	
1st segment notes duration (s)	0.009 (0.005–0.023)	0.0313 (0.021–0.039)	0.017 (0.010–0.026)	0.039 (0.042–0.048)	0.044 (0.033–0.043)	0.044 (0.035–0.050)		0.110
Inter-note interval of 1st segment (s)	0.217 (0.147–0.837)	0.103 (0.078–0.130)	0.234 (0.082–0.322)	0.285 (0.209–0.496)	0.316 (0.171–0.987)	0.268 (0.157–0.254)	(0.110–0.130)	0.090
Notes per s of 1st segment	–	–	3.900 (2.900–4.300)	–	–	–		(5.000–6.000)
Number of pulses 2nd segment	1298.500 (1018–1502)	147 (100–192)	818 (164–1382)	–	–	–		
Pulses per s of 2nd segment	43.670 (42.350–44.950)	(85.000–95.000)	32.800 (30.200–34.300)	54.000 (51.000–58.000)	48.000 (45.000–61.000)	57.000 (54.000–64.000)	(65.000–85.000)	70.000

Speciemens from: [1]Campo Alegre, Santa Catarina, Brazil (present study); [2]Misiones, Argentina (Baldo and Basso 2004); [3]São José dos Ausentes, Rio Grande do Sul, Brazil (Kwet and Miranda 2001); [4]Corrientes, Argentina (Duré et al. 2015); [5]Laguna, Santa Catarina, Brazil (Kwet et al. 2005); [6]Torres, Rio Grande do Sul, Brazil (Kwet et al. 2005); [7]Paraguay (Kurth et al. 2013); [8]Misiones, Argentina (Baldo and Basso 2004); [9]La Paloma, Uruguay (Kwet et al. 2005); [10]São Sepé, Rio Grande do Sul, Brazil (Caldart et al. 2013); [11]Tiraxi, Jujuy, Argentina (Ferrari and Vaira 2008); [12]Abra Colorada, Jujuy, Argentina (Ferrari and Vaira 2008); [13]Canto Del Monte, Argentina (Ferrari and Vaira 2008); [14]Calamuchita, Córdoba, Argentina (Kwet and Miranda 2001); [15]Santa Rosa de Calamuchita, Córdoba, Argentina (Barrio 1964).

segment separately. In November 2013, we observed six males calling on bromeliads at ground level, in two forest fragments at 1027 m altitude. Most toads can be found in bromeliads closer to the ground, which were usually under 1 m high, as well as inside those that fell in the leaf litter. Also we found one female near an egg clutch at the fence of pitfall trap. In January 2014, we registered only one male calling during the fieldworks. No individuals were capture in pitfall traps in the forest fragment. Other species calling along with *M. xanthostomus* in the forest were *Fritizana* sp., *Cycloramphus bolitoglossus* (Werner, 1897) and *Adenomera araucaria* Kwet & Angulo, 2002.

DISCUSSION

The natural history of the phytotelma-using *Melanophryniscus* species is very different from their congeners, because of the reproductive mode and tadpole morphology (Langone et al. 2008, Baldo et al. 2014), and the type of habitats they inhabit (other species occurs and lay the eggs in freestanding water) (Kwet et al. 2010, Maneyro et al. 2017). Concerning its call parameters, this species showed some temporal and spectral characteristics that suggest some acoustic adaptations for forest environments, such as described in the Acoustic Adaptation Hypothesis (AAH) (Morton 1975, Erdtmann and Lima 2013). According this hypothesis, calls in forest environments will: (1) be longer in length, (2) have a lower repetition rate, (3) have lower minimal, maximal and dominant frequencies and (4) have a smaller frequency bandwidth (Morton 1975).

The advertisement call of *M. xanthostomus* differs from its congeners by the higher number of pulses in the second segment (294–1033; from 16 to 321 in the other *Melanophryniscus* species), except from *Melanophryniscus krauczuki* Baldo & Basso, 2004 (1018–1502 pulses in the second segment) and *M. pachyrhynus* (Miranda-Ribeiro, 1920) (164–1382 pulses in the second segment). The lowest pulse rate of the second segment (40.55–56.70 pulses per second) differs *M. xanthostomus* from most of its congeners (62 to 95 pulses per second in *M. atroluteus* (Miranda-Ribeiro, 1920), *M. cupreuscapularis* Céspedez & Alvarez, 2000, *M. dorsalis* (Mertens, 1933), *M. klappenbachi* Prigioni & Langone, 2000, *M. montevidensis* (Philippi, 1902), *M. stelzneri* (Weyenbergh, 1875), except from *M. krauczuki, M. pachyrhynus*, and *M. rubriventris* (Vellard, 1947) (42.35–44.95, 3020–34.30, 45.00–64.00 pulses per second, respectively). The total call and the second segment durations of *M. xanthostomus* (12.19–20.99 s, 6.28–18.32 s, respectively) are only shorter than *M. krauczuki* (25.01–36.65 s, 23.78–33.41 s). *Melanophryniscus pachyrhynus* reaches higher values of call and second segment durations (6.64–75.20 s, 4.79–45.75 s) than *M. xanthostomus*, however, the values overlap. The higher number of notes in the first segment (7–20) of the advertisement call of *M. xanthostomus* differs from *M. klappenbachi* (3–4) and *M. rubriventris* (2–7). The dominant frequency is higher (3395 Hz) and the amplitude range is lower (± 500 Hz) than other species of *Melanophryniscus*, except from *M. krauczuki* (dominant frequency 3300 Hz), *M. cupreuscapularis* (amplitude ± 200 Hz) and *M. rubriventris* (amplitude between 100-300 Hz). The interval between calls is longer in *M. xanthostomus* (82.949 s) than other species. The other acoustic parameters, such as first segment duration and interval between segments, varies among species of the genus, but all the values overlap. Even with some acoustic traits that corroborate with the AAH, we observed that the advertisement calls are similar among *Melanophryniscus* species.

Species of *Melanophryniscus* have an advertisement call composed of short notes followed by a long trill (e.g., Kwet et al. 2005, Caldart et al. 2013, Duré et al. 2015), as we observed to *M. xanthostomus*. Most species, including *M. xanthostomus*, have the first segment composed by one group of short notes, except *M. cupreuscapularis* and *M. dorsalis* (from Laguna, Santa Catarina, Brazil), which have two or more groups of short notes, separated by a higher distance than that among notes (Kwet et al. 2005, Duré et al. 2015). *Melanophryniscus xanthostomus* emit the advertisement call with both segments in sequence as described for other species of the genus (see Duré et al. 2015), except *M. pachyrhynus*, that also emit calls with the second segment (the trill) alone (Caldart et al. 2013). The functional differences of the two segments of the advertisement call of *Melanophryniscus* species might be investigate in future bioacoustic and behavior studies of these species.

Thus, it is remarkable that the advertisement call of *M. xanthostomus* presents the same prototype (a compound call formed by two segments, the first composed of short notes followed by a long trill) of its congeners not-phytotelm breedings. Biological concerns, such as species recognition, sexual selection, physiological traits or body size (Gerhardt 1991, Bevier et al. 2008), may explain the similarities of advertisement calls among *Melanophryniscus* species, while some evidences showed contrary pattern expected for AAH. Some amphibians species showed larger body mass and lower call frequencies in open-land environments (Bevier et al. 2008), such as we detected for *M. xanthostomus* (higher dominant frequency associated to forest environment). But the open land and stream inhabitant *M. krauczuki* (Baldo and Basso 2004) also showed higher dominant frequency, suggesting other environment variables may be associated, such as stream noise. However, the lack of phylogenetic hypothesis for the genus and other described calls from phytotelm breeding *Melanophryniscus*, makes it hard to test this hypothesis at this moment. Thus, future phylogenetic comparative studies, testing the AAH, could bring new explanations for the evolutinary acoustic patterns of *Melanophryniscus*. Once this is the first described call of a phytotelm breeding *Melanophryniscus*, we highlight the need to know the call of the other species with the same reproductive mode in order to confirm this proposal.

ACKNOWLEDGMENTS

We are thankful to all friends involved directly and indirectly to this work, specially to Carlos E. Conte, Renato C.

Nali and Daniel Son for suggestions and the help during the fieldworks. We are thankful to H.L. Doerrier and C.T. Ferriolli for the English editing. To Conselho Nacional de Desenvolvimento Científico e Tecnológico (CNPq processes 132559/2012-9, 167888/2014-5 and 311492/2017-7), Coordenação de Aperfeiçoamento de Pessoal de Nível Superior (CAPES), and Fundação Boticário for the grants. To the SISBIO (#35005) and FUNDEMA (#013/12-GEMAP) for the research licenses.

LITERATURE CITED

Baldo D, Basso NG (2004) A new species of *Melanophryniscus* Gallardo, 1961 (Anura: Bufonidae), with comments on the species of the genus reported for Misiones, Northeastern Argentina. Journal of Herpetology 38: 393–403.

Baldo D, Candioti FV, Haad B, Kolenc F, Borteiro C, Pereyra MO, Zank C, Colombo P, Bornschein MR, Sisa FN, Brusquetti F, Conte CE, Nogueira-Costa P, Almeida-Santos P, Pie MR (2014) Comparative morphology of pond, stream and phytotelm-dwelling tadpoles of the South American Redbelly Toads (Anura: Bufonidae: *Melanophryniscus*). Biological Journal of the Linnean Society 112: 417–441. https://doi.org/10.1111/bij.12296

Barrio A (1964) Peculiaridades del canto nupcial de *Melanophryniscus stelzneri* (Weyenbergh) (Anura: Brachicephalidae). Physis 24: 435– 437.

Bevier CR, Gomes FR, Navas CA (2008). Variation in call structure and calling behavior in treefrogs of the genus *Scinax*. South American Journal of Herpetology 3(3): 196-206. https://doi.org /10.1670/144-03A

Bornschein MR, Firkowski CR, Baldo D, Ribeiro LF, Belmonte-Lopes R, Corrêa L, Morato SAA, Pie MR (2015) Three New Species of Phytotelm-Breeding *Melanophryniscus* from the Atlantic Rainforest of Southern Brazil (Anura: Bufonidae). PlosOne 10(12): e0142791. https://doi.org/10.1371/journal.pone.0142791

Caldart VM, Santos TG, Maneyro R (2013) The advertisement and release calls of *Melanophryniscus pachyrhynus* (Miranda-Ribeiro, 1920) from the central region of Rio Grande do Sul, southern Brazil. Acta Herpetologica 8(2): 115–122.

Caramaschi U, Cruz CAG (2002) Taxonomic status of Atelopus pachyrhynus Miranda-Ribeiro, 1920, redescription of *Melanophryniscus tumifrons* (Boulenger, 1905), and descriptions of two new species of *Melanophryniscus* from the state of Santa Catarina, Brazil (Amphibia, Anura, Bufonidae). Arquivos do Museu Nacional 60(4): 303-314.

Carvalho TR, Giaretta AA (2013) Bioacoustics reveals two new syntopic species of Adenomera Steindachner (Anura: Leptodactylidae: Leptodactylinae) in the Cerrado of central Brazil. Zootaxa 3731(3): 533–551.

Daly JW, Wilham JM, Spande TF, Garraffo HM, Gil RR, Silva GL, Vaira M (2007) Alkaloids in bufonid toads (*Melanophryniscus*): temporal and geographic determinants for two Argentinian species. Journal of Chemical Ecology 33(4): 871–887. https://doi.org/10.1007/s10886-007-9261-x

Duellman WE, Trueb L (1986) Biology of Amphibians. Baltimore, Johns Hopkins University Press, 670 pp.

Duré MI, Schaefer EF, Kehr AI (2015) Acoustic Repertoire of *Melanophryniscus cupreuscapularis* (Céspedez and Álvarez 2000) (Anura: Bufonidae): Advertisement, Encounter, and Release Calls. Journal of Herpetology 49(1): 53–59.

Erdtmann LK, Lima AP (2013) Environmental effects on anuran call design: what we know and what we need to know. Ethology Ecology & Evolution 25(1): 1-11. https://doi.org/10.1080/03949370.2012.744356

Ferrari L, Vaira M (2008) Análisis comparativo de la estructura del canto de anuncio de tres poblaciones de *Melanophryniscus rubriventris* (Vellard, 1947) (Anura: Bufonidae). Cuadernos de Herpetologia 22: 25–34.

Frost DR (2018) Amphibian Species of the World: an Online Reference. New York, American Museum of Natural History, v. 6.0. http://research.amnh.org/herpetology/amphibia/index.html

Frost DR, Grant T, Faivovich J, Bain RH, Haas A, Haddad CFB, de Sá RO, Channing A, Wilkinson M, Donnellan SC, Raxworthy CJ, Campbell JA, Blotto BL, Moler P, Drewes RC, Nussbaum RA, Lynch JD, Green DM, Wheeler WC (2006) The Amphibian Tree of Life. Bulletin of the American Museum of Natural History 297: 1-291.

Gerhardt HC (1991). Female mate choice in treefrogs: static and dynamic acoustic criteria. Animal Behaviour 42(4): 615-635. https://doi.org/10.1016/S0003-3472(05)80245-3

Gerhardt HC, Davis MS (1988) Variation in the coding of species identify in the advertisement calls of *Litoria verreauxi* (Anura: Hylidae). Evolution 42: 556–563.

Haddad CF, Pombal-Jr JP (1998) Redescription of *Physalaemus spiniger* (Anura: Leptodactylidae) and description of two new reproductive modes. Journal of Herpetology 557–565.

Heyer WR (1971) Mating calls of some frogs from Thailand. Fieldiana Zoology 58: 61–82.

Köhler J, Jansen M, Rodríguez A, Kok PJR, Toledo LF, Emmrich M, Glaw F, Haddad CFB, Rödel MO, Vences M (2017) The use of bioacoustics in anuran taxonomy: Theory, terminology, methods and recommendations for best practice. Zootaxa 4251: 1–124. https://doi.org/10.11646/zootaxa.4251.1.1

Kurth M, Hörnes D, Esser S, Rödder D (2013) Notes on the acoustic repertoire of *Melanophryniscus klappenbachi* Prigioni & Langone, 2000. Zootaxa 3626: 597–600.

Kwet A, Miranda T (2001) Zur Biologie und taxonomie der Schwarz-kro te *Melanophryniscus atroluteus* (Miranda-Ribeiro, 1920). Herpetofauna 23: 19–27.

Kwet A, Maneyro R, Zillkens A, Mebs D (2005) Advertisement call of *Melanophryniscus dorsalis* (Mertens, 1933) and *M. montevidensis* (Phillipi, 1902), two parapatric species from southern Brazil and Uruguay, with comments on morphological variation in the *Melanophryniscus stelzneri* group (Anura: Bufonidae). Salamandra 41: 3–20.

Kwet A, Lingnau R, Di-Bernardo M (2010) Pró-Mata: Anfíbios da Serra Gaúcha, sul do Brasil. Tübingen, Brasilien-Zentrum, University of Tübingen, 148 pp.

Langone JA, Segalla MV, Bornschein M, de Sá RO (2008) A new reproductive mode in the genus *Melanophryniscus* Gallardo, 1961 (Anura: Bufonidae) with description of a new species from the state of Paraná, Brazil. South American Journal of Herpetololy 3(1): 1–9.

Larson PM, de Sá RO, Arrieta D (2003) Chondrocranial, hyobranchial and internal oral morphology in larvae of the basal bufonid genus *Melanophryniscus* (Amphibia: Anura). Acta Zoologica 84: 145-154.

Maneyro R, Loebmann D, Tozetti A, Fonte LFM (2017) Anfíbios das planícies costeiras do extreme sul do Brasil e Uruguai. São Paulo, Anolis Books, 176 pp.

Martin WF (1972) Evolution of vocalization in the genus Bufo. In: Blair WF (Ed.) Evolution in the genus Bufo. Austin, University of Texas Press, 279–309.

McDiarmid RW (1971) Comparative morphology and evolution of frogs of the neotropical genera *Atelopus*, *Dendrophryniscus*, *Melanophryniscus* and *Oreophrynella*. Bulletin of the Los Angeles County Museum of Natural History Science 12: 1-66.

Morton E S (1975) Ecological sources of selection on avian sounds. The American Naturalist 109(965): 17-34. https://doi.org/10.1086/282971

Pansonato A, Mudrek JR, Simioni F, Alves Martins I, Strüssmann C (2014) Geographical variation in morphological and bioacoustic traits of *Pseudopaludicola mystacalis* (Cope, 1887) and a reassessment of the taxonomic status of *Pseudopaludicola serrana* Toledo, 2010 (Anura: Leptodactylidae: Leiuperinae). Advances in Zoology 2014: 1–13. https://doi.org/10.1155/2014/563165

Peloso PLV, Faivovich J, Grant T, Gasparini JL, Haddad CFB (2012) An extraordinary new species of *Melanophryniscus* (Anura, Bufonidae) from Southeastern Brazil. American Museum Novitates 3762: 1-32.

Sueur J, Aubin T, Simonis C (2008) Equipment review: seewave, a free modular tool for sound analysis and synthesis. Bioacoustics 18: 213–226.

Van Bocxlaer I, Loader SP, Roelants K, Biju SD, Menegon M, Bossuyt F (2010) Gradual adaptation toward a range-expansion phenotype initiated the global radiation of toads. Science 327: 679-682.

Wells KD (2007) The ecology and behavior of amphibians. University of Chicago Press, 1400 pp.

Author Contributions: LRM conducted the field works, recording the vocalizations and collecting the specimens, SM, DJS and LRM analyzed the data and wrote the paper.

Competing Interests: The authors have declared that no competing interests exist.

17

Morphological study of the gastrointestinal tract of the snow trout, *Schizothorax esocinus* (Actinopterygii: Cypriniformes)

Mohd Y. Bhat[1], **Ashok Channa**[1], **Bilal A. Paray**[2], **Mohammed K. Al-Sadoon**[2], **Irfan A. Rather**[3,4]

[1]*Department of Zoology, University of Kashmir. Srinagar 190006, India.*
[2]*Department of Zoology, College of Science, King Saud University. P.O. Box 2455, Riyadh, 11451, Saudi Arabia.*
[3]*Department of Biological Sciences, Faculty of Science, King Abdulaziz University. Jeddah, 21589, Saudi Arabia.*
[4]*Centre of Excellence in Bionanoscience Research, King Abdulaziz University. Jeddah, 21589, Saudi Arabia*
Corresponding authors: Mohd Yousuf Bhat (myousufpp@Yahoo.co.in), *Irfan A. Rather* (erfaan21@gmail.com)

ABSTRACT. The present study aimed to investigate the macroscopic structure of the gastrointestinal tract (GIT) of *Schizothorax esocinus* Heckel, 1838. The surface architecture of the buccopharynx, oesophagus and the entire intestinal tract of *S. esocinus* has been examined under scanning electron microscope (SEM) after fixing in 2.5% glutaraldehyde buffered with 0.1 M sodium cacodylate at pH 7.3 for 18–48 hours and post-fixation for two hours at room temperature in 1% osmium tetra oxide buffered at pH 7.3 with 0.1 M cacodylate. The mucosal surface of buccopharynx, esophagus, intestinal bulb, and intestine reveal prominent longitudinal major or primary mucosal folds which are further subdivided into the series of irregular and well-circumscribed folds called minor or secondary folds. However, in the intestinal bulb and intestine, the longitudinal major or primary folds themselves form wavy or zigzagging patterns along the mucosal surface. The fine structure of the surface epithelium further shows that the apical surfaces of the epithelial cells are ped with finger-print like microridges, arranged in various patterns and regularly spaced. The rectal mucosa, on the other hand, displays a highly irregular type of major mucosal folds. The separation can't be seen between major mucosal folds. A thin film of mucous spread over the mucosal folds and the numerous pores through which mucous cells release their content has also been noted along the rectal mucosa. This investigation suggests the possible role of different digestive organs in relation to feeding, digestion, storage, absorption, and various other physiological processes, thereby providing a knowledge necessary to the understanding of pathological or physiological alterations in both aquaculture and natural environment.

KEY WORDS. Buccopharynx, esophagus, intestinal tract.

INTRODUCTION

Ultrastructural studies are important to understand the relationship between physiological and biochemical functions and molecular mechanisms of an animal (Kalhoro et al. 2018). The morphological studies of the gastrointestinal tract (GIT) of teleost fish received attention and have shown a significant diversity of morphological and functional characters (Kapoor et al. 1976, Fange and Grove 1979). Moreover, due to several morphological variations among the GIT of many fish species, nutritional requirements have been the focus of research in the field of fish ecology. Scanning electron microscope (SEM) has been used to describe the alimentary canal of teleosts morphologically to provide a detailed information on the structure and function of various anatomical structures in relation to their different feeding adaptations (Grau et al. 1992, Yashpal et al. 2006). There are several studies on the morpho-physiological characteristics of the GIT of fish (Murray et al. 1994, Park and Kim 2001). Carrassón and Matallanas (1994) used SEM to study the digestive tract in *Dentex dentex* (Linnaeus, 1758); while the same technique was used by Mir and Channa (2010) to examine the intestinal tract of *Schizothorax curvifrons* Heckel, 1838.

The snow trout, *Schizothorax esocinus* Heckel, 1838 (Cypriniformes: Cyprinidae), is a native freshwater teleost fish that inhabit cold streams, rivers, and lakes (Menon 1999). In the Jammu and Kashmir state of India, it is mainly found in the inland waters viz. Mansbal lake, Dal lake, Lidder stream, river Jhelum (Kullander et al. 1999), The population of *Schizothorax* Heckel, 1838 have declined drastically in their natural habitats due to many environmental factors like habitat destruction, eutrophication, climate change and chemical pollution (Mir and Channa 2009). Snow trout are highly preferred food fishes among local people because of its nutritional value and taste that fetches high market price (Singh and Paul 2010). Details of the digestive tract of this benthopelagic species is of vital physiological significance to better understanding of the features of this tract in relation to its function as it is the site of temporary storage of food besides digestive and absorptive process. *Schizothorax esocinus* lacks true stomach, as observed in other cyprinids, the proximal portion of the intestine is of greater diameter than the rest of the intestine and acts as a temporary storage organ for the ingested food (Mir and Channa 2010). There is a dearth of information regarding the scanning electron microscopy of the digestive tract of fishes of Kashmir in general and *Schizothorax* in particular. The purpose of the present study is to investigate the detailed ultrastructure of mucosal modifications of the entire digestive tract of naturally feeding fish, *S. esocinus* by using SEM. The study will provide an insight in understanding the possible role of different digestive organs in relation to feeding, digestion, storage, absorption, defecation as well as in the physiological processes, acting as a control for investigations in relation to pathology, pollution and stress conditions in both natural and polluted water bodies besides its application in fish culture.

MATERIAL AND METHODS

All procedures were carried out according to the ethical principles of animal experimentation adopted by the Ethics Committee on Animal Use of the University of Kashmir, India. Live specimens of snow trout, *S. esocinus*, weighing 300 to 350 g (total length ranging from 25 to 32 cm) were collected from different freshwater sources of Kashmir valley. After being anesthetized with tricaine methanesulfonate, MS 222, solution by using a dosage of 100 mg L^{-1}, the fishes were euthanized via a longitudinal incision through their ventral side. After dissection, various segmental tissue fragments of the alimentary tract viz. buccopharynx, esophagus, intestinal bulb, intestine and rectum of 2 cm long were removed and immediately processed for scanning electron microscopic studies. The esophagus, intestinal bulb, intestine, and rectum were incised and split longitudinally to expose the mucosal surface and rinsed in normal saline to remove excess mucus and other adhered particles. The tissues were then fixed in 2.5% glutaraldehyde buffered with 0.1 M sodium cacodylate at pH 7.3 for 18–48 hours. Following this primary fixation, the samples were then thoroughly rinsed three times in 1 M cacodylate buffer and post-fixed for two hours at room temperature in 1% osmium tetraoxide buffered at pH 7.3 with 0.1 M cacodylate (secondary fixation). The post-fixed tissues were washed three times in buffer, dehydrated in ascending grades of ethanol and transferred to a 100% amyl acetate bath. The tissues were then critical-point dried through liquid CO_2 in Hitachi HCP-2 dryer and mounted on aluminum stubs with silver paint. These were sputter-coated with gold in a Hitachi HUS-5GB high vacuum evaporator. The coated specimens were examined in a Hitachi S-3000H scanning electron microscope.

RESULTS

Scanning electron microscope observations of the mucosal surface of the buccopharynx, esophagus, intestinal bulb and intestine of snow trout, *S. esocinus* reveal prominent longitudinal mucosal folds or major mucosal folds which left deep furrows in between (Fig. 1). The apical surfaces of the stratified epithelial cells of buccopharynx were sculped with a network of well-defined and prominent linear elevations called microridges as revealed by the fine structure of the surface epithelium (Figs 2, 3).

The primary longitudinal and secondary mucosal folds in the epithelium of the esophagus are visible along with goblet cells (Figs 4, 5). Though extensively branched and connected, the microridges were observed to be regularly spaced bearing shallow groove between them which help in holding the mucous. These irregular elevations were most prominent in the anterior portion of the esophagus (Fig. 5). The mucous secreting oval shaped goblet cells, extensively scattered mucin droplets and numerous pores, representing openings for mucous release, were also noted on the surface of the epithelial cells of mucosal folds (Fig. 5).

The SEM observation on the anterior intestine shows the organization of longitudinal mucosal folds (Fig. 6). A zigzagging or chevron-like pattern of arrangement of primary mucosal folds (major mucosal folds) which course along the long axis of the intestine bulb and intestine were also observed (Fig. 6). Moreover, the angles of the zigzag patterns were noted to be quite distant. The round-topped major mucosal folds of the intestinal bulb showing microridges were observed to be larger in size than the intestine (Figs 7, 8). The space between the primary longitudinal folds was large in the intestine bulb. These folds were lined with columnar cells bearing microvilli on their apical surfaces and appeared as a honeycomb polyhedral epithelial cells in the form of pentagonal or hexagonal elevations with the presence of gastric pits in between. When compared with the intestine bulb, the intestine has less distinct separation and the continuous cavity runs parallel throughout the entire length. Smaller infoldings of the mucosa (secondary folds) were noted to be present on the sides of the main primary folds (Fig. 9). The rectum, on the other hand, exhibits a highly irregular pattern of major mucosal folds. The secondary mucosal folds (minor mucosal folds) and well-defined cavities were not observed in the present study, however, irregular mucosal folds were seen (Fig. 10). The rectal mucosal folds were covered by a thin film of mucus and were beset with distinct pores through which the mucous cells discharge their contents.

DISCUSSION

The morpho-anatomical features of the digestive tract of Indian freshwater teleosts, in relation to their feeding habits, are well documented (Sinha and Moitra 1975, Sinha and Chakrabarti

1984, 1985). According to these authors, densely coiled intestine, modification of mucosal folds and complexity of the intestinal villi are the main features associated with the herbivorous mode of feeding. It has been observed in the present study that the mucosal folds in the various regions of the alimentary tract of the fish studied get modified in different ways. The buccopharynx and esophagus display a well-elaborated pattern of primary or major folds, which are further subdivided into many secondary

Figures 1–3. Scanning electron micrograph of buccopharynx showing: (1) primary longitudinal (MF_1) and secondary folds (MF_2); (2) goblet cell (GC) and stratified epithelial cells (SEC); (3) microridges (MR). Scale bars: 1 = 200 µm, 2 = 20 µm, 3 = 10 µm.

Figures 4–5. Scanning electron micrograph of esophagus showing: (4) primary longitudinal (MF_1) and secondary mucosal folds (MF_2); (5) stratified epithelial cells (SEC), goblet cells (GC), Pores of goblet cells (PG) and microridges (MR). Scale bar: 4 = 500 µm, 5 = 10 µm.

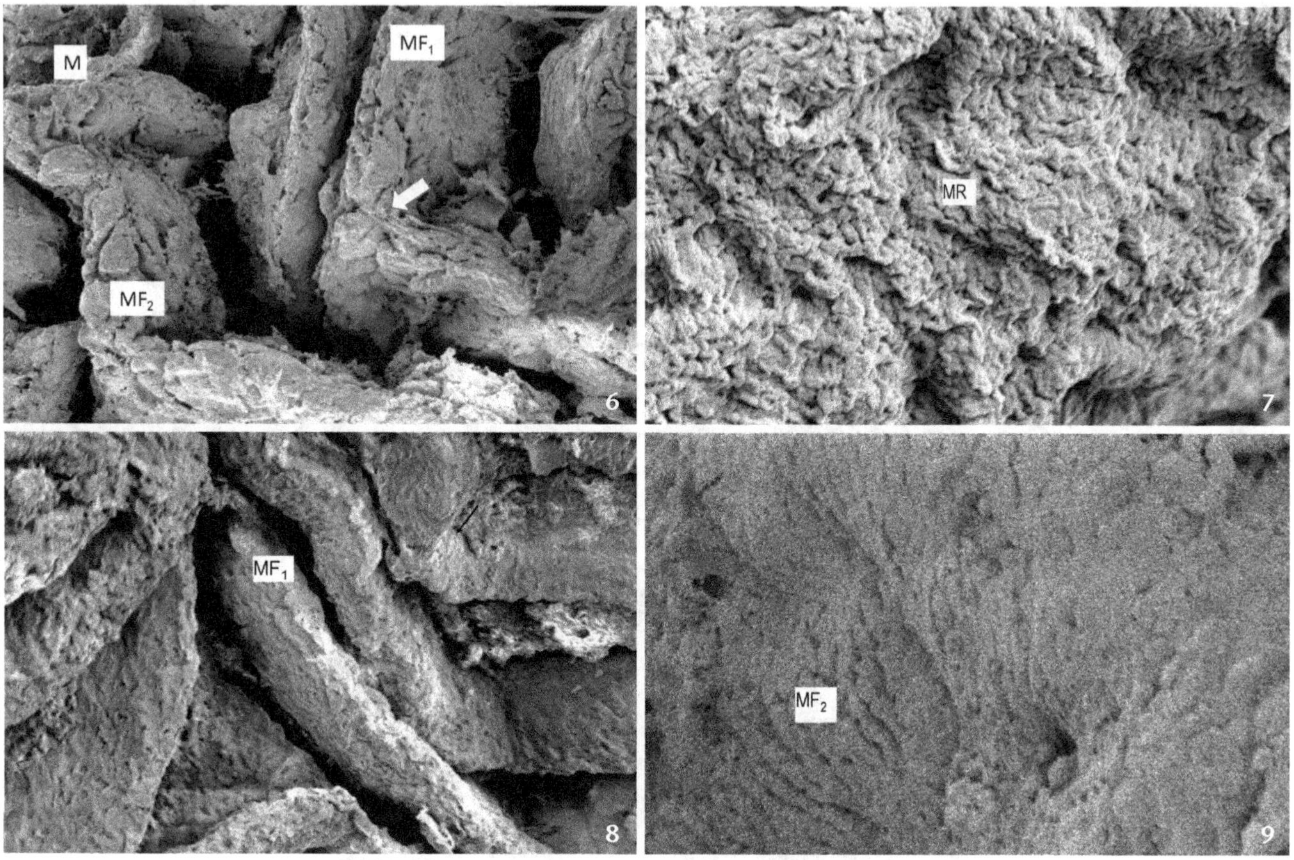

Figures 6–9. Scanning electron micrograph of intestinal bulb showing: (6) primary or major (MF_1) and secondary mucosal folds (MF_2), covered with thin film of Mucin (M). The zig-zag arrangement of primary folds is clearly visible in this region (←); (7) microridges (MR); (8) primary or major mucosal folds (MF_1) and the zig-zag invagination (←); (9) secondary folds (MF_2). Scale bar: 6 = 500 μm, 7 = 10 μm, 8 = 100 μm, 9 = 20 μm.

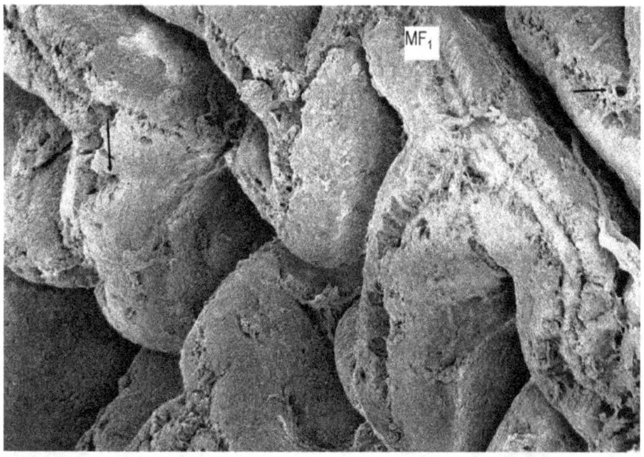

Figure 10. Scanning electron micrograph of rectum showing irregular mucosal folds (MF_1), mucous (↓) and pores (→). Scale bar: 200 μm.

folds. These primary and secondary folds aid in swallowing. Similar mucosal foldings have been observed in carp, *Catla catla* (Hamilton, 1822) by Sinha and Chakrabarti (1985) and in *Serola dumerili* (Risso, 1810) by Grau et al. (1992). The apical surfaces of the stratified epithelial cells of buccopharynx and the luminal surface of the stratified epithelial cells of the esophagus are beset with a complex pattern of microridges, which increase the surface area and provide the channel for mucus transport. According to Harding (1973), the microbridge on the epithelial surfaces are subjected to mechanical insult. The presence of microridges may, therefore, represent a mechanical adaptation which in the buccopharynx and esophagus would withstand the trauma resulting from ingested materials. The microridges of the buccopharynx and esophagus by virtue of their nature seem to spread and hold mucous film secreted by the adjacent mucous cells over the soft mucous membrane. These observations are compatible with the studies made earlier (Ezeasor and Stokoe 1980, Sinha and Chakrabarti 1985, Murray et al. 1994).

The secretion of mucous serves as a lubricant in transmitting food from buccopharynx to esophagus thereby enabling the esophagus to act as a transit tube for ingested food from the oropharyngeal cavity to the intestinal bulb. The secretion of mucous by the mucous cells to the exterior has been reported by several authors in various teleosts (Khanna 1964, Martin and Blaber 1984, Humbert et al. 1984, Sinha and Chakrabarti 1985, 1986, Chakrabarti and Sinha 1987, Tibbets 1997, Park and Kim 2001, Podkowa and Goniakowska-Witalinska 2003). Some authors have reported taste buds in the oesophageal wall of many teleosts (Ezeasor and Stokoe 1980, Fishelson et al. 2004). Sis et al. (1979) demonstrated taste buds only in the cranial portion of the esophagus of the channel catfish, *Ictalurus punctatus* (Rafinesque, 1818). However, during the present study, no taste buds were observed in the esophagus. This is in correlation with the findings of Elbal and Angulleiro (1986) and Grau et al. (1992) in different teleosts.

Fine anatomical features of the mucosal surface of the intestinal tract in *S. esocinus* as revealed by present SEM study correlated with the other teleosts (Sinha and Chakrabarti 1985, 1986, Chakrabarti and Sinha 1987, Kapoor and Khanna 1994, El-Shammaa et al. 1995), it has been observed that the mucosal folds in various regions of the intestine get modified in different ways. The major mucosal folds noted to be very high, consist of zigzagging pattern that course along the long axis of the entire intestine and form deep cavities between them. The major mucosal folds along with the minor ones around the cavities not only serve to increase the total surface area but also retain ingested food for longer periods in the stomach less fish studied. The deep cavities in the intestinal region are an adaptational feature meant mainly for storage. It is well known that the primary function of this region is absorption and secretion. Nevertheless, partial retention of the semi-digested food also occurs while the same passes through the zigzag course of this region. The zigzagging pattern of the major mucosal folds has been reported by McVay and Kann (1940) and Caceci (1984). According to Curry (1939), the tops of the mucosal folds in the intestinal bulb are flat, round or pointed. However, during the present study, the tops of the mucosal folds in the entire intestine are flat and rounded. Caceci (1984) also observed flat and round folds in the intestinal mucosa of goldfish, *Carassius auratus* (Linnaeus, 1758). The luminal surface of the entire intestine in *S. esocinus* is lined with well-developed columnar epithelial cells. The luminal plasma membrane of the columnar epithelial cells of the anterior portion of the intestine sculped into microridges. Sinha and Chakrabarti (1985) report that the microridges represent a mechanical adaptation in the intestine. According to various authors, the microridges are not only restricted to the intestinal surface but have been found to occur also on the epidermal cells of the body (Hawkes 1974, Lanzing and Higginbotham 1974, Harris and Hunt 1975, Hunter and Nayudu 1978, Bereiter-Hahn et al. 1979, Whitear 1990), epithelial cells of the gills (Mattey et al. 1980, Eiras-Stofella and Charvet-Almeida 1998) and the stratified epithelial cells of buccopharynx and oesophagus (Mallatt 1979, Sis et al. 1979, Clarke and Witcomb 1980, Ezeasor and Stokoe 1980, Humbert et al. 1984, Sinha and Chakrabarti 1985, MacDonald 1987, Grau et al. 1992, Murray et al. 1994).

The rectum reveals irregular mucosal foldings that increase the surface area of the rectal mucosa. The rectum plays a very little role in food storage, digestion and absorption as evidenced by the low mucosal folds as well as the absence of well-defined cavities and the minor mucosal folds, thereby bearing a close relationship with its functional aspects. A thin film of mucous secretion has been observed in the rectal mucosa, which aids in easy defecation. The occurrence of few mucous cells in between the epithelial cells of the various regions of the digestive tract of various teleosts has been reported by different authors (Khanna 1964, 1968, Sinha 1975, Park and Kim 2001). According to the authors mentioned above, the presence of copious mucin discharged by the mucous cells in the different regions of the alimentary canal of various teleosts is an adaptation for adequate lubrication of food for facilitating conduction posteriorly along the alimentary canal.

On the basis of our results, the surface architecture of the mucosa of the digestive tract of *S. esocinus* throws light on the modifications of the mucosal epithelium and morphological characteristics of the various cells lining the alimentary canal which are useful in correlating the morphological features with function. These findings provide insights into the structure and functioning of the digestive system in this fish. Further detailed studies should be conducted on the ultrastructure level of the gastrointestinal tract of *S. esocinus*.

ACKNOWLEDGMENT

The authors would like to express their sincere appreciation to the Deanship of Scientific Research at the King Saud University, Riyadh, Saudi Arabia for funding this Research Group project no RG-1437-005.

LITERATURE CITED

Bereiter-Hahn J, Osborn M, Weber K, Vöth M (1979) Filament organization and formation of microridges at the surface of fish epidermis. Journal of Ultrastructure Research 69(3): 316–330. https://doi.org/10.1016/S0022-5320(79)80050-7

Caceci T (1984) Scanning electron microscopy of goldfish, *Carassius auratus*, intestinal mucosa. Journal of Fish Biology 25(1): 1–12. https://doi.org/10.1111/j.1095-8649.1984.tb04845.x

Carrassón M, Matallanas J (1994) Morphometric charateristics of the alimentary tract of deep-sea Mediterranean teleosts in relation to feeding habits. Marine Biology 118(2): 319–322. https://doi.org/10.1007/BF00349799

Chakrabarti P, Sinha GM (1987) Mucosal surface of the alimentary canal in *Mystus vittatus* (Bloch). A scanning electron microscopic study. Proceedings of the National Academy of Sciences, India Section B 53(4): 317–322.

Clarke AJ, Witcomb DH (1980) A study of the histology and morphology of the digestive tract of the common eel, *Anguilla anguilla*. Journal of Fish Biology 16(2): 159–170. https://doi.org/10.1111/j.1095-8649.1980.tb03695.x

Curry E (1939) The histology of the digestive tube of the carp (*Cyprinus carpio communis*). Journal of Morphology 65(1): 53–78. https://doi.org/10.1002/jmor.1050650104

Eiras-Stofella DR, Charvet-Almeida P (1998) Ultrastructure (SEM) of the gills of *Prochilodus scrofa* Steindachner (Pisces, Teleostei). Revista Brasileira de Zoologia 15(2): 279–287. https://doi.org/10.1590/S0101-81751998000200001

Elbal MT, Agulleiro B (1986) A histochemical and ultrastructural study of the gut of *Sparus auratus* (Teleostei). Journal of submicroscopic Cytology 18(2): 335–347.

El-Shammaa MA, El-bargeesy GA, Wali YR, El-Gharbawy SM (1995) Anatomical histological studies on the alimentary canal of *Clarias lazera* fish (Karmout). Egyptian Journal of Anatomy 18: 101–122.

Ezeasor DN, Stokoe WM (1980) Scanning electron microscopic study of gut mucosa of the rainbow trout, *Salmo gairdneri* Rich. Journal of Fish Biology 17(5): 529–539. https://doi.org/10.1111/j.1095-8649.1980.tb02784.x

Fange R, Grove D (1979) Digestion, in: Hoar WS, Randall DJ, Brett JR (Eds) Fish Physiology. London, Academic Press, vol. 7, 162–241.

Fishelson L, Delarea Y, Zverdling A (2004) Taste bud form and distribution on lips and in the oropharyngeal cavity of cardinal fish species (Apogonidae, Teleostei), with remarks on their dentition. Journal of Morphology 259(3): 316–327. https://doi.org/10.1002/jmor.10186

Grau A, Crespo S, Sarasquete MC, Gonzalez de Canales ML (1992) The digestive tract of the amberjack *Seriola dumerili*, Risso: A light and scanning electron microscope study. Journal of Fish Biology 41(2): 287–303. https://doi.org/10.1111/j.1095-8649.1992.tb02658.x

Harding CV (1973) Corneal epithelial surfaces in elasmobranchs and teleosts as seen with the scanning electron microscope. Biological Bulletin 5: 345–351.

Harris JE, Hunt S (1975) The Fine Structure of the Epidermis of Two Species of Salmonid Fish, the Atlantic Salmon (*Salmo salar* L.) and the Brown Trout (*Salmo trutta* L.) Cell and Tissue Research 163(4): 535–543. https://doi.org/10.1007/BF00218498

Hawkes JW (1974) The structure of fish skin. I. General organization. Cell and Tissue Research 149(2): 147–158. https://doi.org/10.1007/BF00222270

Humbert W, Kirsch R, Meister MF (1984) Scanning electron microscopic study of the oesophageal mucous layer in the eel, *Anguilla anguilla* Linn. Journal of Fish Biology 25(1): 117–122. https://doi.org/10.1111/j.1095-8649.1984.tb04856.x

Hunter CR, Nayudu PL (1978) Surface folds in superficial epidermal cells of three species of teleost fish. Journal of Fish Biology 12(2): 163–166. https://doi.org/10.1111/j.1095-8649.1978.tb04160.x

Kalhoro H, Tong S, Wang L, Hua Y, Volatiana JA, Shao QJ (2018) Morphological study of the gastrointestinal tract of *Larimichthys crocea* (Acanthopterygii: Perciformes). Zoologia 35: e25171 https://doi.org/10.3897/zoologia.35.e25171

Kapoor BG, Khanna B (1994) The alimentary canal of teleosts: A brief survey of structure and function. In: Singh HR (Eds) Advances in Fish Biology. Delhi, Hindustan Publishing, vol. 1, 12–24.

Kapoor BG, Smith H, Verighina IA (1976) The alimentary canal and digestion in teleosts. Advances in Marine Biology 13: 109–239. https://doi.org/10.1016/S0065-2881(08)60281-3

Khanna SS (1964) Histology of the digestive tract of a teleost, *Clarias batrachus* (Linn.). Journal of the Zoological Society of India 16: 53–58.

Khanna SS (1968) The structure and distribution of the taste buds and the mucous secretion cells in the buccopharynx in some Indian teleosts (Pisces). Studies, DSB Government collage, Nainital (India). 5: 143–149.

Kullander SO, Fang F, Delling B, Ahlander E (1999) The fishes of Kashmir Valley. In: Nyman L (Ed.) Impacts on the aquatic environment. Jhelum, Kashmir Valley, Lenart Nyman Ed., 99–163.

Lanzing WJR, Higginbotham DR (1974) Scanning microscopy of surface structures of *Tilapia mossambica* (Peters) scales. Journal of Fish Biology 6(3): 307–310. https://doi.org/10.1111/j.1095-8649.1974.tb04547.x

Mac Donald NL (1987) An electron microscopic examination of the gastrointestinal epithelium in Dover sole, *Solea solea* (Linn.). Journal of Fish Biology 31(1): 27–36. https://doi.org/10.1111/j.1095-8649.1987.tb05211.x

Mallatt J (1979) Surface morphology and functions of pharyngeal structures in larval lamprey, *Petromyzon marinus*. Journal of Morphology 162(2): 249–274. https://doi.org/10.1002/jmor.1051620206

Martin TJ, Blaber SJM (1984) Morphology and histology of the alimentary tracts of Ambassidae (Cuv.) (Teleostei) in relation to feeding. Journal of Morphology 182(3): 295–305. https://doi.org/10.1002/jmor.1051820305

Mattey DL, Morgan M, Wright DE (1980) A scanning electron microscope study of the pseudobranchs of two marine teleosts. Journal of Fish Biology 16(3): 331–343. https://doi.org/10.1111/j.1095-8649.1980.tb03709.x

McVay JA, Kann HW (1940) The digestive tract of *Carassius auratus*. Biological Bulletin 78: 53–67.

Menon AGK (1999) Checklist freshwater fishes of India. Records of the Zoological Survey of India Occasional Paper 175: 366

Mir IH, Channa A (2009) Gills of the snow trout, *Schizothorax curvifrons* Heckel: A SEM study. Pakistan Journal of Biological Sciences 12(23): 1511–1515. https://doi.org/10.3923/pjbs.2009.1511.1515

Mir IH, Channa A (2010) A Scanning Electron Microscopic Examination of the Intestinal Tract of the Snow Trout, *Schizothorax curvifrons* Heckel. Journal of Fisheries and Aquatic Science 5(5): 386–393. https://doi.org/10.3923/jfas.2010.386.393

Murray HM, Wright GM, Goff GP (1994) A study of the posterior oesophagus in the winter flounder *Pleuronectes americans* and the yellowtail flounder, *Pleuronectes ferruginea*. Morphological evidence for pregastric digestion? Canadian Journal of Zoology 72: 1191–1198. https://doi.org/10.1139/z94-160

Park JY, Kim IS (2001) Histology and mucin histochemistry of the gastrointestinal tract of the mud loach, in relation to respiration. Journal of Fish Biology 58(3): 861–872. https://doi.org/10.1111/j.1095-8649.2001.tb00536.x

Podkowa D, Goniakowska-Witalinska L (2003) Morphology of the air-breathing stomach of the catfish *Hypostomus plecostomus*. Journal of Morphology 257(2): 147–163. https://doi.org/10.1002/jmor.10102

Singh NO, Paul AK (2010) Fitting of allometric model with expected-value parameters for different species of snow trout from Jhelum River, Kashmir. The Indian Journal of Animal Scince 80(1): 85–88

Sinha GM (1975) A histochemical study of the mucous cells in the buccopharyngeal region of four Indian freshwater fishes in relation to their origin, development, occurrence and probable functions. Acta Histochemica 53(2): 217–223.

Sinha GM, Chakrabarti P (1984) Mechanisms of mucus release in the alimentary canal of a freshwater major carp *Labeo rohita* (Ham.): A light and scanning electron microscopic study. Proceedings of the National Academy of Sciences, India Section B 50(1): 27–31.

Sinha GM, Chakrabarti P (1985) On topological characteristics of the mucosal surface in buccopharynx and intestine of an Indian freshwater major carp, *Catla catla* (Ham.): A light and scanning electron microscopic study. Zoologische Jahrbücher. Abteilung für Anatomie und Ontogenie der Tiere 113: 375–389.

Sinha GM, Chakrabarti P (1986) Scanning electron microscopic studies on the mucosa of the digestive tract in *Mystus aor* (Ham.). Proceedings of the National Academy of Sciences 52: 267–273.

Sinha GM, Moitra SK (1975) Functional morpho-histology of the alimentary canal of an Indian freshwater major carp, *Labeo rohita* (Ham.) during its different life history stages. Anatomischer Anzeiger 138(3): 222–239.

Sis RF, Ives PJ, Jonas DM, Lewis DH, Haensly WE (1979) The microscopic anatomy of the oesophagus, stomach and intestine of the channel catfish, *Ictalurus punctatus*. Journal of Fish Biology 14(2): 179–186. https://doi.org/10.1111/j.1095-8649.1979.tb03508.x

Tibbetts IR (1997) The distribution and function of mucous cells and their secretions in the alimentary tract of *Arrhamphus sclerolepis krefftii*. Journal of Fish Biology 50(4): 809–820. https://doi.org/10.1111/j.1095-8649.1997.tb01974.x

Whitear M (1990) Causative aspects of microridges on the surface of fish epithelia. Journal of Submicroscopic Cytology and Pathology 22: 211–220.

Yashpal M, Kumari U, Mittal S, Mittal AK (2006) Surface architecture of the mouth cavity of a carnivorous fish *Rita rita* (Ham., 1822) (Siluriformes, Bagridae). Belgian Journal of Zoology 136(2): 155–162.

Author Contributions: MYB and AC conducted the experiments; BAP and MKAS designed the experiments; MYB, BAP and IAR analyzed the data; MYB, BAP, IAR wrote and revised the paper.
Competing Interests: The authors have declared that no competing interests exist.

18

A new cecidogenous species of many-plumed moth (Alucitidae) associated with *Cordiera* A. Rich. ex DC. (Rubiaceae) in the Brazilian Cerrado

Gilson R.P. Moreira[1], Cristiano M. Pereira[2], Vitor O. Becker[3], Alexandre Specht[4], Gislene L. Gonçalves[5]

[1]*Departamento de Zoologia, Instituto de Biociências, Universidade Federal do Rio Grande do Sul. Avenida Bento Gonçalves 9500, 91501-970 Porto Alegre, RS, Brazil.*
[2]*Programa de Pós-graduação em Biologia Animal, Departamento de Zoologia, Instituto de Biociências, Universidade Federal do Rio Grande do Sul. Avenida Bento Gonçalves 9500, 91501-970 Porto Alegre, RS, Brazil.*
[3]*Reserva Serra Bonita, Instituto Uiraçú. Caixa Postal 001, 45880-970 Camacan, BA, Brazil.*
[4]*Embrapa Cerrados. Rodovia BR 020, km 18, Caixa Postal 08223, 73310-970 Planaltina, DF, Brazil.*
[5]*Departamento de Recursos Ambientales, Facultad de Ciencias Agronómicas, Universidad de Tarapacá. Casilla 6-D, Arica, Chile.*
*Corresponding author: Gilson R.P. Moreira (*gilson.moreira@ufrgs.br*)*

ABSTRACT. Larvae of many-plumed moths (Alucitidae), especially in the world-wide genus *Alucita* Linnaeus, 1758 are known as borers or gall-inducers on flowers, fruits and shoots of a few dicotyledonous families, including Bignoniaceae, Caprifoliaceae and Rubiaceae. However, there is no study available on the biology of the monotypic, Neotropical genus *Prymnotomis* Meyrick, 1931 except for its original description that was based on a single male, the holotype of *Prymnotomis crypsicroca* Meyrick, 1931 from Espirito Santo, Brazil. We describe here a second species for this genus, *Prymnotomis cecidicola* **sp. nov.** whose larvae induce galls on *Cordiera elliptica* (Cham.) Kuntze (Rubiaceae), a dioecious plant with dimorphic inflorescences found in the Brazilian Cerrado, Planaltina City, Federal District. Adults, larvae, pupae and galls are illustrated under light and scanning electron microscopy. Galls are green, spherical, unilocular and develop individually on *C. elliptica* flower buds. During development they look like fruits in shape and colour but are larger, do not have style scars when on female plants, and are induced also in male inflorescences. Pupation occurs outside the gall within a silk cocoon, presumably in the litter. A preliminary analysis of DNA barcode sequences including putative members of other alucitid lineages and Neotropical BINs (Barcode Index Number) supports *Prymnotomis cecidicola* **sp. nov.** as an independent phylogenetic unit, with 12 to 18% divergence. Its nearest-neighbour was the BIN cluster 5 (BOLD:AAA0842) that includes specimens from Costa Rica.

KEY WORDS. Alucitid moths, Brazilian Savanna, insect galls, *Prymnotomis*, taxonomy.

INTRODUCTION

Alucitidae is a small, worldwide family of apoditrysian moths with nine valid genera and ca 216 species (Gielis 2003, Nieukerken et al. 2011). Several additional alucitid species have been described for the family recently, particularly in the tropics (e.g., Vargas 2011, Heppner 2015, Ustjuzhanin and Kovtunovich 2016, Ustjuzhanin et al. 2018), and many are supposed yet to be discovered in these regions. A total of 27 species of alucitids are found in the Neotropics, of which only six have been recorded in Brazil (Gielis 2003, Vargas 2011, Heppner 2015). These micromoths are well known by their specialized fore- and hindwings, which are multiply divided into lobes that look like bird feathers. Larvae are usually borers or gall-inducers, associated with flowers, fruits and shoots of a few plant families including the Bignoniaceae, Caprifoliaceae and Rubiaceae (Dugdale et al. 1998). In contrast to the adults, their immature stages are poorly known, especially in the Neotropical region where the host plants for a few species have been documented, particularly within *Alucita* Linnaeus, 1758 the most speciose, worldwide genus (e.g. Vargas 2011). Four other endemic genera have been described for the region: *Hexeretmis* Meyrick, 1929, *Prymnotomis* Meyrick, 1931, *Paelia* Walker, 1866, and *Alinguata* Fleming, 1948; to the best of our knowledge, none of the host plants or the immature stages of these genera have been identified so far (see also Lima 1945 and Pastrana 2004). Taxonomy of *Hexeretmis* and *Alinguata* was revised by Heppner (2010a,b), based on the adults.

Prymnotomis was proposed by Meyrick (1931) to include *P. crypsicroca* Meyrick, that was described briefly based on the general morphology of only one adult male from Espírito Santo, Brazil. It has remained monotypic genus since its original description, and there is no additional information about its biology. This study concerns a second, new species of *Prymnotomis* that induces conspicuous, spherical galls on inflorescences of *Cordiera elliptica* (Cham.) Kuntze (Rubiaceae), a native shrub of the Brazilian Savanna (Cerrado Biome). A preliminary comparison of genitalia structures suggested that it is congeneric with but is not *P. crypsicroca*. In this study we describe and illustrate the adult, larva, pupa and the gall under light and scanning electron microscopy, and provide information on the natural history of this new species. An analysis of DNA barcode sequences including putative members of other alucitid lineages, as well as Neotropical BINs (Barcode Index Number; Ratnasingham and Hebert 2013) is also provided to estimate the phylogenetic position and genetic distances of the new species.

MATERIAL AND METHODS

Adult specimens were reared by V.O. Becker from galls collected during October of 1981, 1982, and 1983 at the Centro Nacional de Pesquisa Agropecuária dos Cerrados (Embrapa Cerrados), Planaltina City, Federal District, Brazil (15°36′26.4″S; 47°42′52.4″W), and maintained in small plastic pots under room temperature in the Laboratório de Entomologia of the same Institution. They were checked daily for the emergence of adults, which were pin-mounted and dried. Immatures used for descriptions were dissected from additional galls that were collected by C.M. Pereira and A. Specht at same locality during September 2018, and brought to the Laboratório de Morfologia e Comportamento de Insetos (LMCI), Departamento de Zoologia, Universidade Federal do Rio Grande do Sul (UFRGS), Porto Alegre, Rio Grande do Sul, Brazil, where they were reared in a similar manner. They were fixed in Dietrich's fluid and preserved in 75% ethanol. Additional larvae used for DNA extraction were preserved in100% ethanol at -20 °C.

For descriptions of adults, genitalia were dissected and cleared in a 10% potassium hydroxide (KOH) solution, stained with either eosin or Chlorazol black E and slide-mounted in Canada balsam, following Robinson (1976). Last instar larvae were prepared similarly for the study of chaetotaxy. Observations were performed with the aid of a Leica® M125 stereomicroscope. Structures selected to be drawn were previously photographed with a Sony® Cyber-shot DSC-H10 digital camera attached to the stereomicroscope. Vectorized line drawings were then made with the software Corel Photo-Paint® X7, using the corresponding digitalized images as a guide.

Additional specimens were used for scanning electron microscope analyses. They were dehydrated in a Bal-tec® CPD030 critical-point dryer, mounted with double-sided tape on metal stubs, coated with gold in a Bal-tec® SCD050 sputter coater and examined and photographed in a JEOL® JSM6060 scanning electron microscope at the Centro de Microscopia e Microanálise (CMM) of UFRGS.

Terminology used in descriptions followed Heppner (1987), Patočka and Turčani (2005) and Landry and Landry (2004) for the larva, pupa and adults.

DNA was extracted from larvae of four specimens (CMP 008-01A, D) of *Prymnotomis* sp. nov. using the PureLink Genomic DNA extraction kit (Thermo Fisher Scientific, Carlsbad, California). Extracted DNA was resuspended in 80 mL of Tris: EDTA (10 mm Tris-HCl, 1 mm EDTA, pH 5 8.0). DNA barcoding PCR was conducted using primers LCO1490 and HCO2198 (Folmer et al. 1994), which amplify ca. 500 bp of the mitochondrial gene cytochrome oxidase (COI). PCR reactions were conducted using 2 µL of the extracted DNA. The thermal cycler profile consisted of 35 cycles of 94 °C for 45 s, 48 °C for 45 s and 72 °C for 45 s. Excess dNTP and primers were removed and the amplified DNA concentrated using exonuclease I and FastAP thermosensitive alkaline phosphatase (Thermo Fisher Scientific). Samples were sequenced in both directions using BigDye Terminator v. 3.1 Cycle Sequencing kit (Thermo Fisher Scientific) following standard procedure according to manufacturer's instructions, and analysed in an ABI3730XL (Thermo Fisher Scientific, Waltham, USA) automatic sequencer. The new data were deposited in BOLD Systems (http://www.boldsystems.org/) under the project MISA. The barcode sequences were aligned using CodonCode Aligner (CodonCode Corp, Massachusetts, USA).

To explore the phylogenetic position of the new species within the family we used the COI data generated for *Prymnotomis cecidicola* sp. nov. with a published dataset of 40 Alucitidae (Table 1). Specifically, the genera *Alucita* and *Pterotopteryx* Hannemann, 1959 were included in the analysis, as well as five BIN clusters (BOLD: AAH5751, AAG9907, AAJ6491, AAU0280 and AAA0842) that were identified in the Neotropics, publicly available in BOLD (Table 1). The tree was rooted with *Isonomeutis amauropa* Meyrick, 1888 (Copromorphidae) according to the phylogeny proposed by Mutanen et al. (2010). We used the maximum likelihood algorithm, which was performed in PHYML v. 3.0 (Guindon et al. 2010) using 1000 replicates of heuristic search, with random addition of sequences and TBR branch swapping. The Tamura-Nei substitution model was selected based on the Akaike information criterion run in MEGA v. 6 (Tamura et al. 2013). Monophyly confidence limits were assessed with the bootstrap method at a 50% cut-off after 1000 iterations. Pairwise genetic distances between the new species and other Alucitidae were quantified using the Kimura 2-parameter model in MEGA v. 6.

Museum collections. AMNH – American Museum of Natural History, New York, NY, USA; DZUP – Coll. Padre Jesus S. Moure, Departamento de Zoologia, Universidade Federal do

Table 1. Sample information for specimens used in this study.

Species	Acession number			
	Sample ID	Genbank	BOLD COI-5P	BIN clusters
Ingroup			–	–
Alucita adriendenisi	CGWC-3887	–	LOWCE127-06	–
Alucita cancellata	TLMF Lep 03910	JN307745	PHLSA260-11	–
Alucita debilella	TLMF Lep 03912	JN307747	PHLSA262-11	–
Alucita desmodactyla	TLMF Lep 09154	KP253214	PHLAI592-13	–
Alucita grammodactyla	JBA-05-0004	–	LTOL071-06	–
Alucita hexadactyla	TLMF Lep 03909	JN307744	PHLSA259-11	–
Alucita lalannei	jflandry2557	–	MECC537-06	–
Alucita montana	BIOUG21945-E01	–	SMTPL7338-15	–
Prymnotomis cecidicola sp. nov.	CMP 008-04A	MK919497	MISA041-19	–
Prymnotomis cecidicola sp. nov.	CMP 008-04D	MK919498	MISA042-19	–
Pterotopteryx dodecadactyla	TLMF Lep 08735	KM573355	PHLAH931-12	–
alucitBioLep01	BioLep698	HQ936333	BLPDT350-10	BOLD:AAA0842
alucitBioLep01	BioLep698	HM411224	BLPDM2386-10	BOLD:AAA0842
alucitBioLep01	BioLep698	HQ555946	BLPDQ816-10	BOLD:AAA0842
alucitBioLep01	BioLep698	HM402693	BLPDN1595-10	BOLD:AAA0842
alucitBioLep01	BioLep698	HM402110	BLPDN1038-10	BOLD:AAA0842
alucitBioLep01	BioLep698	HQ555644	BLPDQ346-10	BOLD:AAA0842
Alucita	BioLep698	–	LTOL788-07|	BOLD:AAA0842
Lepidoptera	BioLep698	–	BLPDW575-11	BOLD:AAA0842
alucitBioLep01	BioLep698	HM403333	BLPDN2246-10	BOLD:AAA0842
alucitBioLep01	BioLep696	JN296976	BLPEA475-11	BOLD:AAU0280
alucitBioLep01	BioLep696	JN295546	BLPDX574-11	BOLD:AAU0280
Lepidoptera	–	–	NOUD2153-12	BOLD:AAG9907
Lepidoptera	–	–	BLPEF6714-14	BOLD:AAG9907
Lepidoptera	–	–	BLPEF6357-14	BOLD:AAG9907
alucitBioLep01	BioLep693	HQ934518	BLPDR306-10	BOLD:AAG9907
Lepidoptera	–	–	MHMYS2880-13	BOLD:AAG9907
alucitBioLep01	BioLep693	HM375200	BLPDF869-09	BOLD:AAG9907
Alucitidae	–	–	LMEMB642-09	BOLD:AAG9907
Lepidoptera	–	–	BLPEE4426-14	BOLD:AAH5751
Lepidoptera	–	–	BLPEE3748-14	BOLD:AAH5751
Lepidoptera	–	–	BLPEE3457-14	BOLD:AAH5751
Lepidoptera	–	–	BLPEE4540-14	BOLD:AAH5751
alucitBioLep01	BioLep694	–	BLPDY391-11	BOLD:AAH5751
alucitBioLep01	BioLep694	–	BLPDL1914-10	BOLD:AAH5751
alucitBioLep01	BioLep694	–	BLPDY535-11	BOLD:AAH5751
Lepidoptera	–	–	BLPEE3749-14	BOLD:AAH5751
Lepidoptera	–	–	BLPEE4316-14	BOLD:AAH5751
alucitBioLep01	BioLep694	HQ934494	BLPDR282-10	BOLD:AAH5751
Lepidoptera	–	–	BLPEE3667-14	BOLD:AAH5751
Lepidoptera	–	–	BLPEE3751-14	BOLD:AAH5751
Lepidoptera	–	–	–	–
	–	–	BLPEE3747-14	BOLD:AAH5751
Outgroup	–	–	–	–
Isonomeutis amauropa	MM11203	GU828850	LEFIA1190-10	–

Paraná, Curitiba, PR, Brazil; LMCI – Coll. Laboratório de Morfologia e Comportamento de Insetos, Universidade Federal do Rio Grande do Sul, Porto Alegre, RS, Brazil; NM – Natural History Museum, Vienna, Austria; NHMUK – Natural History Museum, London, United Kingdom; VOB – Coll. Vitor O. Becker, Reserva Serra Bonita, Camacan, BA, Brazil.

RESULTS

Molecular data

Sequencing of COI resulted in an average amplicon size of 500 bp. The aligned data matrix had 683 characters, of which 227 (33%) were phylogenetically informative. Maximum likelihood analysis recovered an optimal ML ln likelihood tree = 5937 with nucleotide frequencies of A = 31.9%, C = 15.2%, G = 14.6%, and T = 38.3%. In the preliminary barcode tree all Neotropical specimens clustered, including *Prymnotomis cecidicola* sp. nov. and the five BIN taxa. Specimens of *Alucita* and *Pterotopteryx* grouped in a second clade, the Nearctic + Palearctic (Fig. 1). The nearest neighbour of *P. cecidicola* was the BIN cluster 5 BOLD:AAA0842 (ca 12% genetic distance), which included specimens from Costa Rica (data from BOLD). Pairwise genetic distance of the new taxon to BIN clusters and lineages within Alucitidae ranged from 12 to 18%, with the highest divergence to *Pterotopteryx dodecadactyla* (Hübner, [1813]) (Table 2).

Table 2. Genetic distance between *Prymnotomis cecidicola* sp. nov. and members of Alucitidae based on 683 base pairs of the DNA barcode sequences using the Kimura 2-parameter model. BIN clusters are identified in Fig. 1.

	1	2	3	4	5	6	7	8
1 *Prymnotomis cecidicola* sp. nov.	–							
2 *Alucita* spp.	0.16	–						
3 BIN cluster1 (BOLD:AAH5751)	0.16	0.17	–					
4 BIN cluster2 (BOLD:AAG9907)	0.14	0.15	0.08	–				
5 BIN cluster3 (BOLD:AAJ6491)	0.14	0.14	0.09	0.03	–			
6 BIN cluster4 (BOLD:AAU0280)	0.17	0.16	0.15	0.12	0.12	–		
7 BIN cluster5 (BOLD:AAA0842)	0.12	0.15	0.17	0.13	0.14	0.14	–	
8 *Pterotopteryx dodecadactyla*	0.18	0.14	0.17	0.15	0.15	0.17	0.13	–

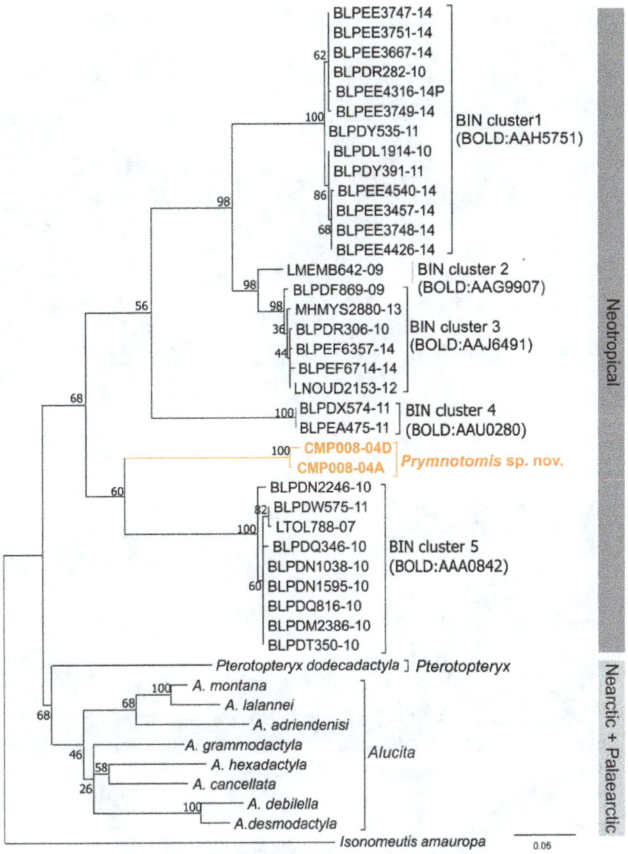

Figure 1. Maximum likelihood tree based on COI sequences for 41 Alucitidae species and lineages – ln likelihood = 5937. The phylogenetic position of *Prymnotomis cecidicola* sp. nov. (CMP 008) is indicated in orange. Bootstrap values are indicated for nodes with more than 50% support (1000 replications).

Taxonomy

Prymnotomis cecidicola Moreira & Becker, sp. nov.

Figs 2–48, 51, 54

Diagnosis. *Prymnotomis cecidicola* shares with the closely related *Hexeretmis* Meyrick the porrect maxillary palpi, forewing cleft only to 1/5 from termen, and pattern of wing venation. However, it differs from this genus in general appearance in coloration pattern and by presenting deeper hindwing clefts (ca 1/3 of wing length), as pointed out by Meyrick (1931). *P. cecidicola* can be separated from the congeneric *P. crypsicroca* Meyrick by the smaller size of the latter, whose forewing length of the type specimen measures ca 6 mm, and by the less contrasting coloration of *P. crypsicroca*, especially in relation to the hindwing. The basal process of the valva of the male genitalia is upturned, finger-like, covered with sparse filiform setae in *P. crypsicroca*, while in *P. cecidicola* it is turned down, looking like a claw, bearing short, stout spines on the base.

Description of adults (Figs 2–19). **Male.** Forewing length (mean ± standard error) = 9.34±0.45 mm; n = 6. Body with most scales greyish brown, interspersed by either isolated or small patches of either entirely light grey or bicoulored scales, mostly with whitish beige at the base and pale greyish brown at the apex (Fig. 2). Palpi and thoracic legs a little lighter at distal portion of each segment (Figs 7, 8). Abdominal segments also lighter ventrally. Head bearing paired tufts of scales on posterior vertex that project mesally, forming a collar (Fig. 6). Similar to description of *P. crypsicroca* provided by Meyrick (1931): maxillary palpus small; labial palpus well developed, porrected, with middle segment more than double the other two in length (Fig. 7); antennae filiform, reaching ca 2/3 forewings in length, with flagellomeres ventrally ciliated; forewings mostly dark grey; a wavy whitish line on costal edge marking alternate dark fuscous spaces; a short and fine whitish, transversal bar on end of cell; three wavy whitish parallel lines crossing the terminal lobes (Fig. 2), proximal one weakly defined, restricted to basal

Figures 2–8. Pinned-dried adult of *Prymnotomis cecidicola* sp. nov. (2) dorsal view, and corresponding morphology (3–7) in detail: (3) right forewing apical angle (indicated by rectangle in Fig. 2); (4) left hind wing discal cell area (marked by asterisk in Fig. 2); (5) right hind wing outer margin (delimited by brackets in Fig. 2); (6,7) head in dorsal and lateral views, respectively; (8) left hind leg, anterior view. Scale bars: 2.5, 1, 1, 1, 0.5, 0.5, 2 mm, respectively.

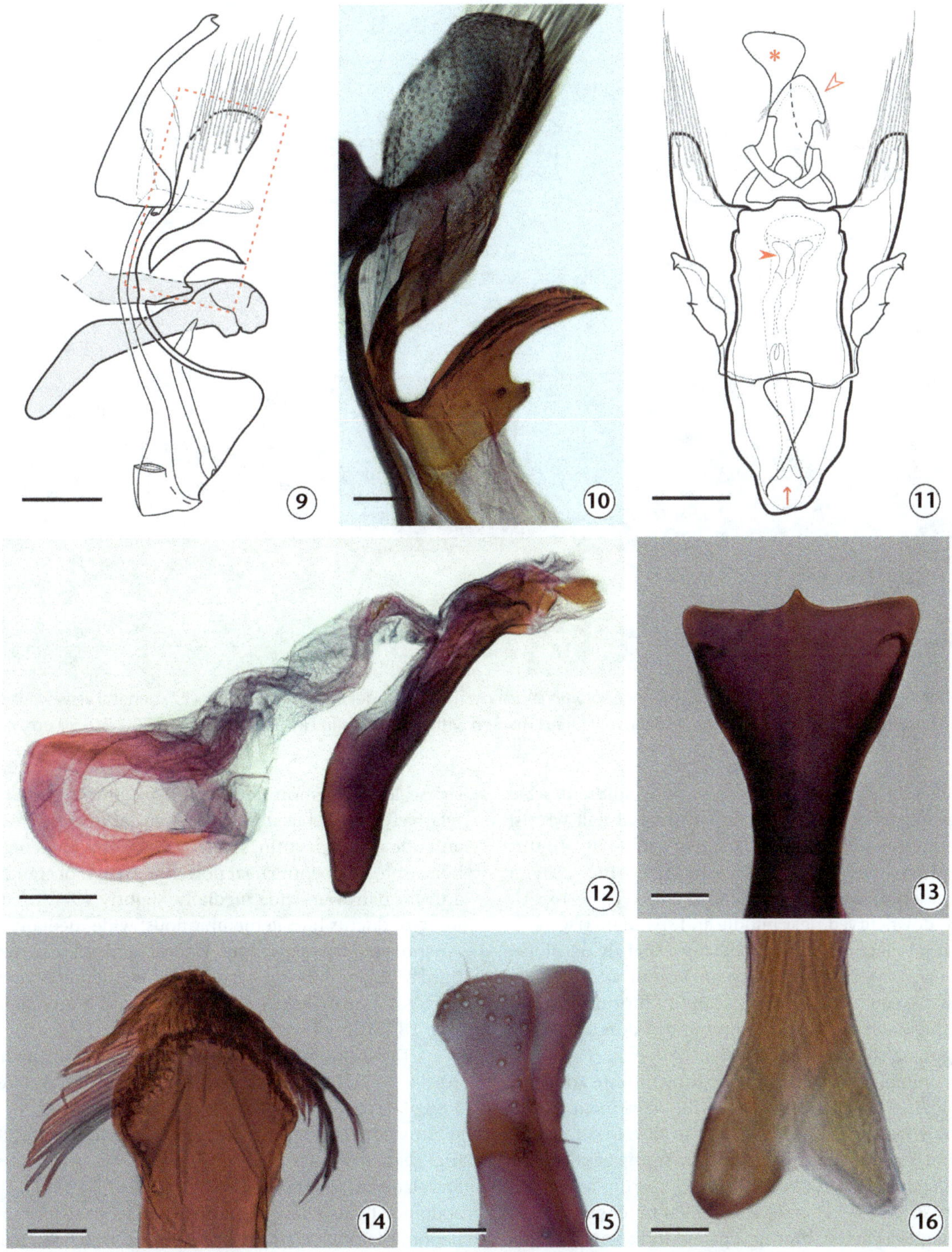

Figures 9–16. Male genitalia morphology of *Prymnotomis cecidicola* sp. nov. under light microscopy: (9, 11) general, lateral and ventral views, respectively (aedeagus omitted in Fig. 11); (10) valva, lateral (area marked with rectangle in Fig. 9); (12) aedeagus, lateral; (13) apex of uncus, ventral (indicated by asterisk in Fig. 11); (14) distal portion of gnathos (pointed by open arrow in Fig. 11); (15) arms of juxta, lateral (indicated by closed arrow in Fig. 11); (16) base of juxta (pointed by seta in Fig. 11). Scale bars; 200, 100, 200, 200, 100, 100, 70, 100 µm, respectively.

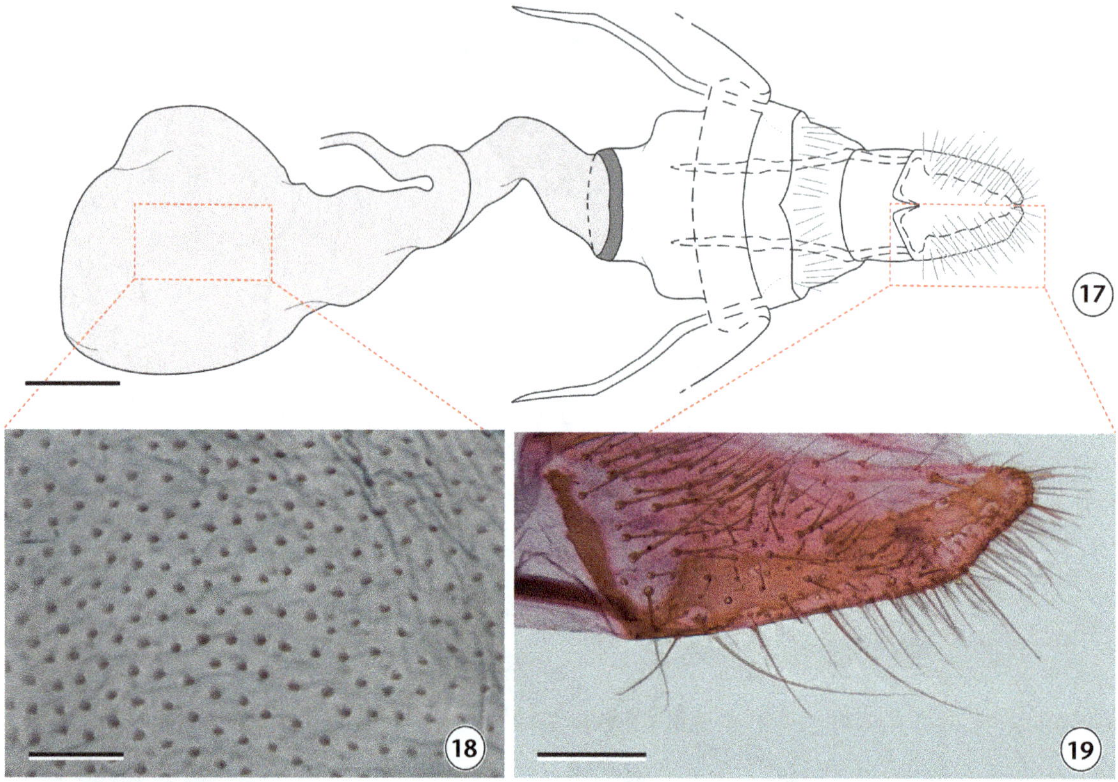

Figures 17–19. Female genitalia morphology of *Prymnotomis cecidicola* sp. nov. under light microscopy: (17) general view, ventral; (18, 19) detail of corpus bursae and papillae anales, respectively (areas marked with rectangles in Fig. 17). Scale bars: 200, 30, 50 μm, respectively.

intersections (Fig. 3); fringe concolours with adjacent scales; hind wings mostly dark grey distally, with basal half whitish, densely permeated on base with dark grey scales (Fig. 4); three fine white wavy lines on lobes and associated fringes (Fig. 5), similar to description of forewings. Genitalia (Figs 9–16): Uncus narrow, downcurved near middle and spatulate (Figs 9, 11, 13), with distal margin showing medially a slightly developed, pointed process. Median arm of gnathos strait and flattened, with a bow-shaped, distal margin bearing filiform setae (Figs 9, 11, 14). Tegumen short, compact, rounded dorsally (Fig. 9). Juxta straight, narrow and flat, forked on both ends (Figs 11, 16) with a pair of long, cylindrical arms bearing minute and sparse setae on widened distal ends (Figs 11, 15). Valva membranous, with cucullus well developed, widened in distal half, bearing long filiform setae on distal, rounded margin; basal process well developed and sclerotized, claw-like, bearing a few stout spines on base (Figs 9–11). Aedeagus similar in length to valva, tubiform, slightly curved, bearing a pair of indistinctly shaped, sclerotized plates apically. Vesica with elongated and narrow plates of indistinct shape. Coecum penis ca two-thirds of total length of aedeagus (Figs 9, 12). **Female.** Virtually no differences from male regarding size and coloration. Genitalia (Figs 17–19): Papillae anales elongated, narrowly rounded apically, bearing sparse, long and short setae (Fig. 19). Eighth tergum broad with relatively long setae at distal on ventral margin. Posterior and anterior apophyses thin, similar to each other in length. Ostium bursae broad, opening near posterior margin of eighth sternum; antrum narrower, wide medially, slightly sclerotized on distal margin; ductus bursae membranous, wide medially, similar to anterior apophyses in length. Ductus seminalis inserted on distal third of ductus bursae. Corpus bursae membranous, ovoid, ca 2/3 ductus bursae in length, wall covered with minute microtrichea (Fig. 18), without signum.

Description of immature stages. **Last instar larva** (Figs 20–34, 51): Head capsule width (mean ± standard error) = 1.88±0.01 mm; body length = 15.24±1.31 mm, n = 5. Body slightly cuneiform, proportionally wider in the middle abdominal segments (Fig. 20). Head tan-brown, with frontoclypeal area, labrum and mandibles darker (Figs 21–23, 51). Thorax and abdomen light yellow (Fig. 51), turning into reddish prior to pupation. Prothoracic shield slightly melanized, with faint patches of light brown spots located laterally. Anal plate and thoracic legs not melanized. **Head:** subrectangular, with lateral margins convex, semiprognathus; a sculptured area latero-dorsally on posterior margin (Figs 24, 27). Frontoclypeus subtriangular, with adfrontal sutures extending to apex of epicranial notch (Fig. 21).

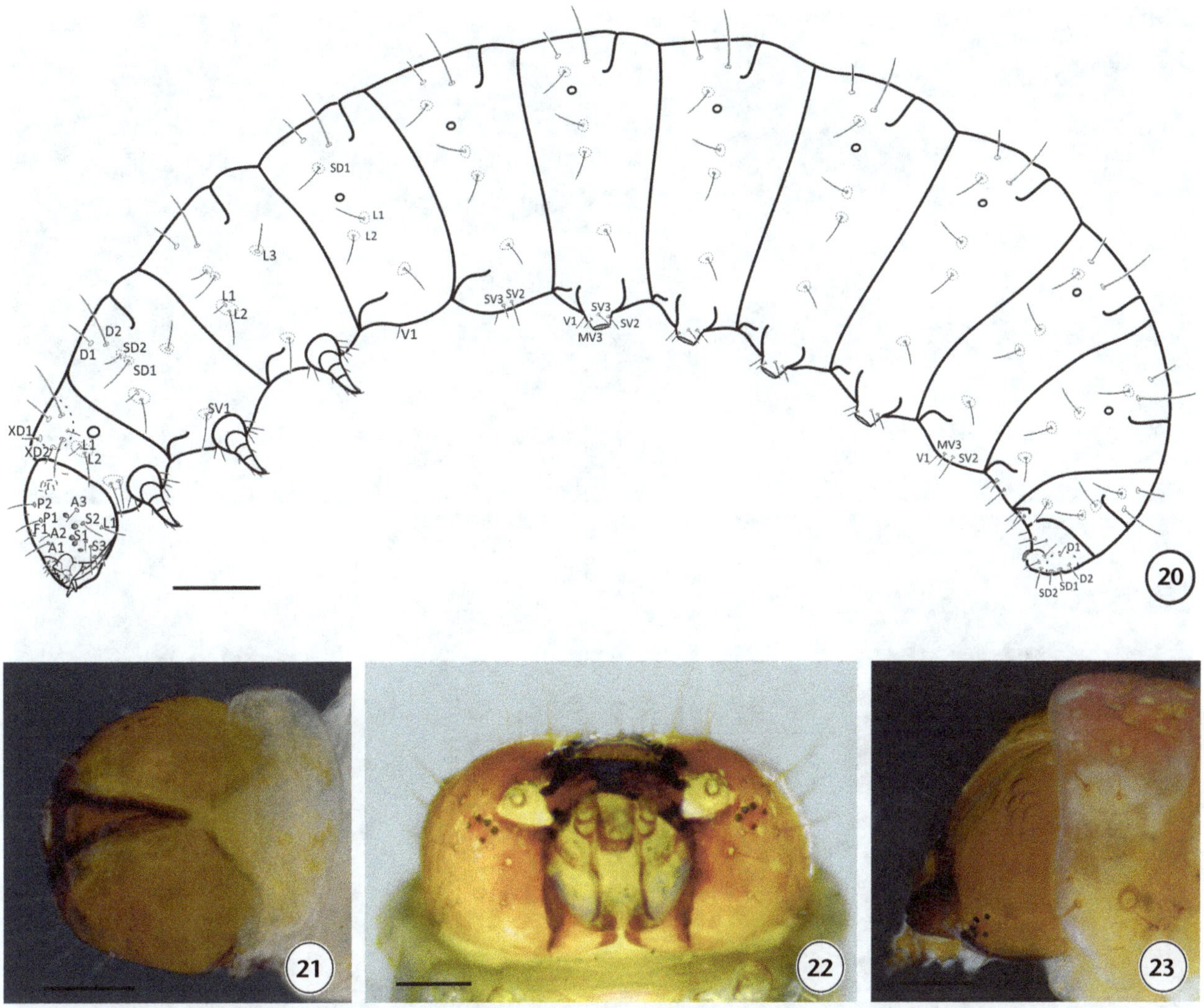

Figures 20–23. Last larval instar of *Prymnotomis cecidicola* sp. nov. under light microscopy: (20) chaetotaxy, lateral view; (21–23) head under dorsal, ventral and lateral views, respectively. Scale bars: 1 mm; 100, 100, 100 μm, respectively.

Five poorly developed, laterally located stemmata (Figs 22, 24). Labrum (Fig. 25) slightly bilobed, with three pairs of setae on distal margin, two pairs laterally on proximal margin and another pair mesally. Antenna (Fig. 26) 2-segmented; basal segment with five sensilla on distal margin, two short and stout, one minute, and two long with more than 3x the length of the others; distal segment much thinner and shorter, bearing one short sensilla on distal margin. Mandible well developed with four cusps along distal margin and two setae basally on external surface. Maxilla (Fig. 28) with palpus and galea well developed, stipes bearing well-developed flap-like, distally forked protrusions that project mesally (Fig. 29). Spinneret short, conical (Figs 28, 29). Labial palpus (Fig. 29) bi-segmented; distal segment thinner and much shorter, both with well-developed apical seta. Chaetotaxy (Fig. 20): MD group trisetose; F unisetose; C group bisetose; A group trisetose, with A1 longer than A2 and A3; AF group bisetose; S group trisetose, with S2 and S3 longer than S1; L unisetose, SS group trisetose. **Thorax (T) and abdomen (A):** Integument covered with microtrichia (Figs 30, 33, 34). Thoracic legs well developed (Fig. 30), with stout tarsal claw bearing a tooth on ventral basis (Fig. 31). Circular spiracles (Fig. 32) with slightly elevated peritreme, laterally on T1, and A1–8. Abdominal pseudopodia short (Figs 33, 34), on A3-6 and A10, with crochets arranged on uniserial and uniordinal, as a penellipse. Chaetotaxy (Fig. 20): T1 with D, XD and SD group bisetose, all on the dorsal shield; D2 longer than D1; XD similar to each other in length; SD2 shorter than SD1; L group bisetose, both on the same pinacula, with L1 shorter and latero-dorsal to L2; SV bisetose, with SV1 shorter than SV2. T2-3 with D group bisetose, similar to prothorax; SD bisetose; L1 trisetose, with L1-2 anterior to L3, and posterior

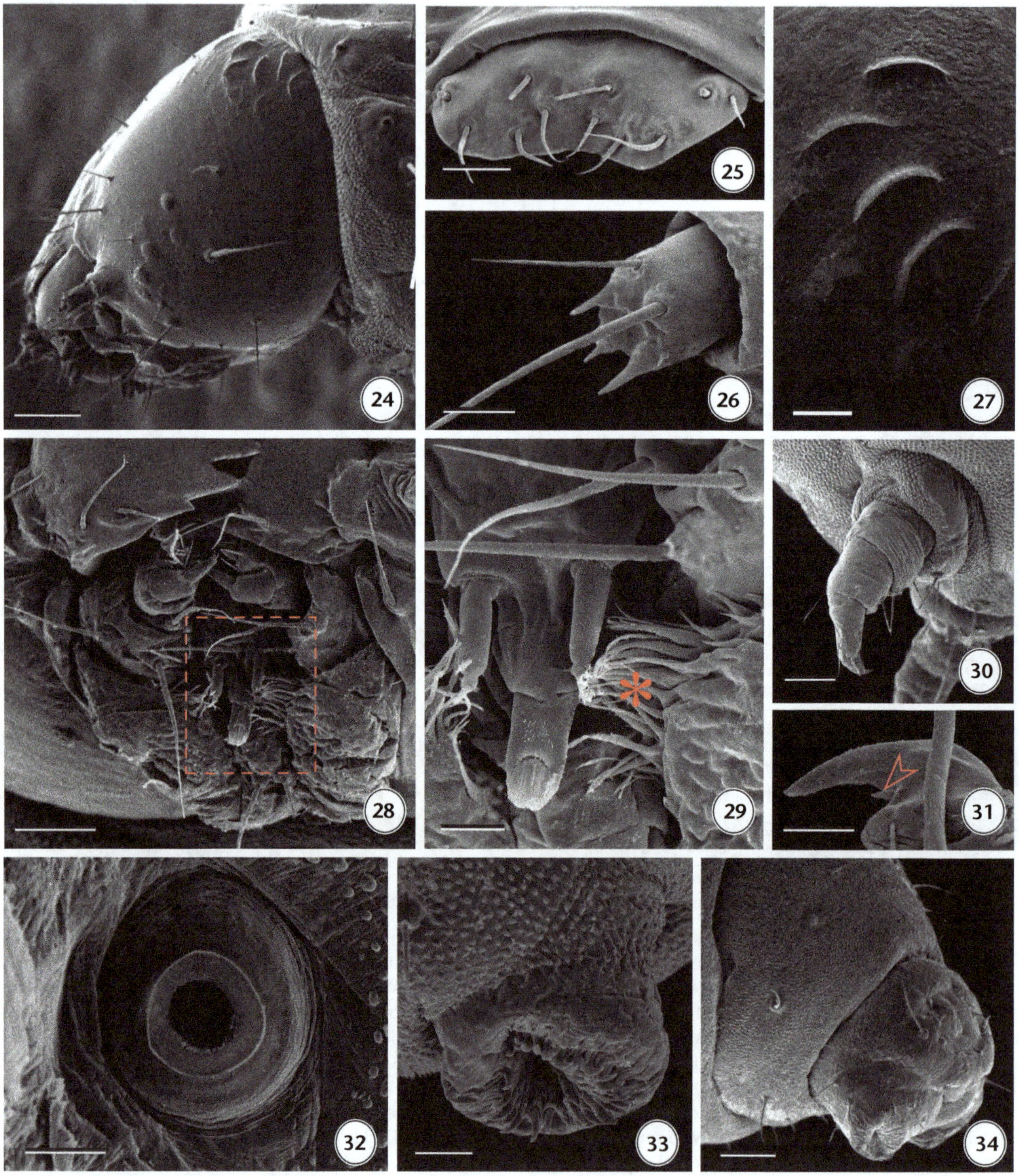

Figure 24–34. Morphology of *Prymnotomis cecidicola* sp. nov. last larval instar under scanning electron microscopy: (24) head, lateral view; (25) labrum, dorsal; (26) antenna, lateral; (27) latero-dorsal area of head in detail; (28) maxillae and labium, antero-ventral; (29) labium in detail (area marked by rectangle in Fig. 28; asterisk indicates associated flap-like protrusions of stipes); (30) mesothoracic leg, postero-lateral; (31) tarsal claw in detail, posterior (open arrow indicates basal spine); (32) prothoracic spiracle, lateral; (33) proleg of fifth abdominal segment, lateral; (34) last two abdominal segments, lateral. Scale bars: 200, 50, 50, 50, 100, 25, 100, 25, 50, 50, 200 μm; 0.5, 0.5, 0.5 mm, respectively.

to SD; SV and V unisetose. Abdominal segments (A) with only short setae that are more or less aligned on the middle region of each segment, which are herein tentatively named. A1 with D and L groups bisetose; SD and SV unisetose; V present. A2 with D and L groups bisetose; SV trisetose; SD unisetose; V present. A3-6 with D and L groups bisetose; SD unisetose; SV trisetose, with SV2 and SV3 on proleg; MV3 and V present. A7-9 with D, SV and L groups bisetose; SD unisetose; MV3 and V present. A10 with D and SD groups bisetose, located within the anal plate; L bisetose; SV trisetose; V present.

Pupa (Figs 35–48). Body cylindrical, yellowish brown, mean length (± standard error) = 9.6 ± 1.74 mm; maximum width = 4.08 ± 0.07 mm; n = 5. Head with vertex deprived of setae and without a differentiated gall-cutter (Figs 38–41). Frons wide, posteriorly expanded mesally to the eyes. Clypeus subtrapezoidal, also without setae. Antennae filiform, reaching the distal end of the middle legs. Mandibles small, rounded, latero-posterior to the clypeus. Maxillary palpi small, rounded, latero-posterior to the eyes. Proboscis well developed, reaching A2. Prothorax fairly developed, deprived of setae, bearing postero-laterally a wide open spiracle (Fig. 43). Hindwings concealed by forewings; the latter with six well marked longitudinal lobes on external surface (Fig. 44), extending to sixth abdominal segment. Protho-, meso- and methatoracic legs reaching the third, fifth, and seventh abdominal segments, respectively. Thoracic and abdominal setae extremely reduced in size (Fig. 45); one pair dorsally on mesothorax and A1; two pairs of such setae on metathorax and A2-7, one dorsal and the other lateral, dorsally to spiracles. Abdominal spiracles rounded, with slightly elevated peritreme (Fig. 46), laterally on A2–7; spiracle on A8 partially closed (Fig. 47). Distal margin of the last abdominal segment with six pairs of stout, distally hooked setae (Fig. 48).

Material examined. All specimens examined came from galls associated with *Cordiera elliptica* (Cham.) Kuntze (Rubiaceae) at Embrapa Cerrados, as already described. Adults were reared by V.O. Becker, from galls collected during October 1982-83

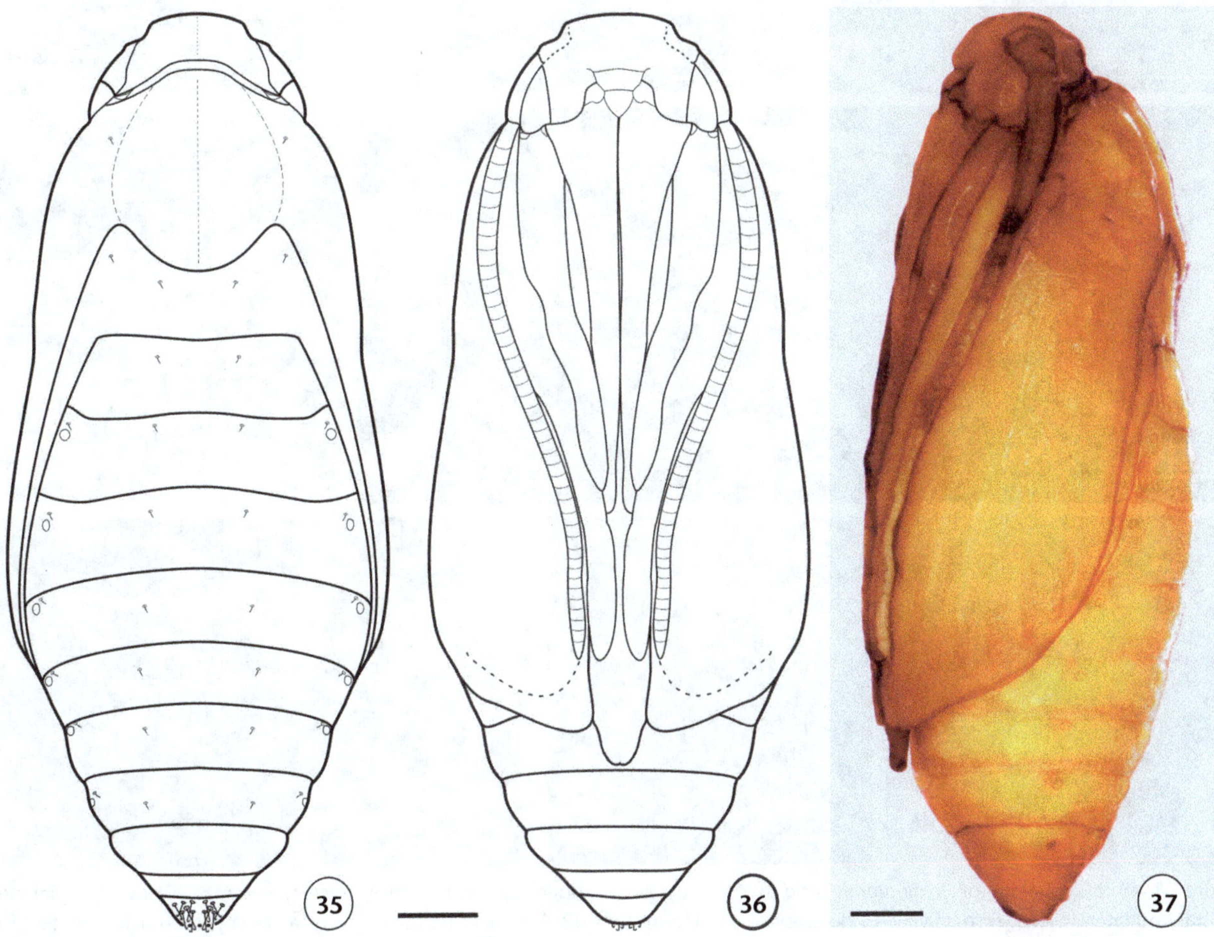

Figures 35–37. *Prymnotomis cecidicola* sp. nov. pupa under light microscopy, in dorsal (35), ventral (36) and lateral (37) views. Scale bar: 0.5 mm.

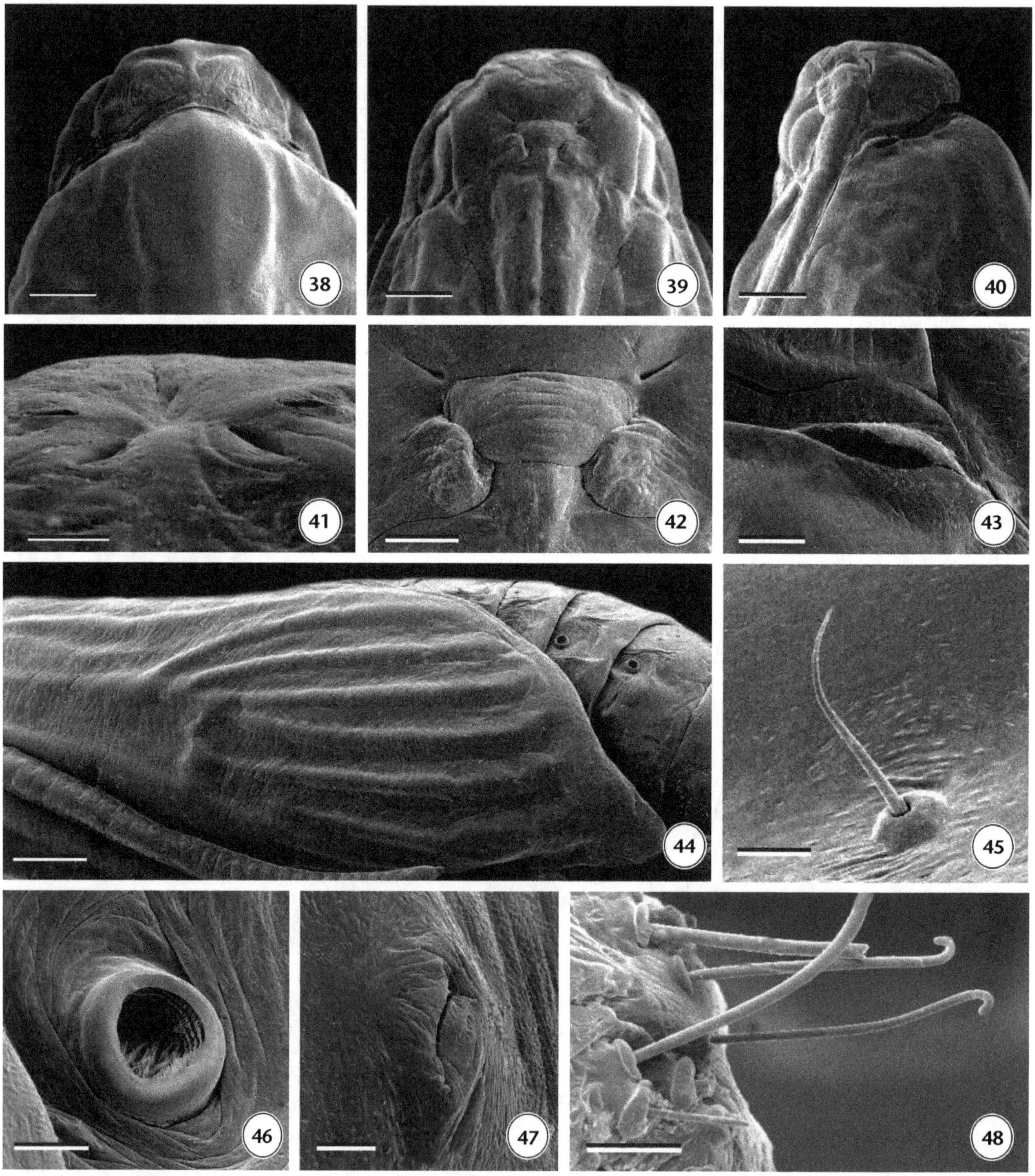

Figure 38–48. Morphology of *Prymnotomis cecidicola* sp. nov. pupa under scanning electron microscopy: (38–40) head, under dorsal, ventral and lateral views, respectively; (41) vertex of head, anterior; (42) buccal appendages, ventral; (43) prothoracic spiracle, dorsal; (44) left forewing, lateral; (45) mesothoracic seta, dorsal; (46, 47) fourth and eight abdominal spiracles, respectively, lateral; (48) dorsal hooks of last abdominal segment, lateral. Scale bars: 500, 100, 200, 100, 500, 25, 50, 50, 50 μm, respectively.

(LMCI 313 series). Immatures were either dissected or reared by C.M. Pereira and A. Specht, from galls collected by C.M. Pereira (LMCI 349 series) on 19.x.2018. Additional galls were collected by G.R.P. Moreira & J. Fochezato (LMCI 346 series) on 1-4.xii.2018.

Type material. Holotype male, BRAZIL: DF, Planaltina, 1100 m, 14.i.1985, ex *Cordiera elliptica*(Cham.) Kuntze (Rubiaceae) (Becker, 57100) (VOB). Paratypes: 11 males, 4 females, 3.xii.1984-8.i.1985, same data as holotype; 1 female, same data as holotye, but 20.ii.1976, at light (Becker, 19564); 2 male, 1 female, same data as holotype, but 15-30.xii.1982 (Becker, 40740); 1 male, same data as holotype, but 15.x.1982, at light (Becker, 40613); 11 males, 1 female, same data as holotype, but 14.xi-23.xii.1983, at light (Becker, 41731); 1 male, same data as holotype, but 5.i.1984, at light (Becker, 56053). Additional paratypes, same data as holotype: 3 male (VOB 57100, 3.xii.1984, deposited LMCI 313-869; VOB 41731, 26.xii.1983, deposited at LMCI 313-874; LMCI 313-879/VOB 40740, donated to DZUP/DZ 33.402), and 3 ♀♀ (VOB 57100, 19.xii.1984, deposited LMCI 313-870; VOB 41731, 5.xii.1983, deposited at LMCI 313-876; VOB 41731, 14.xii.1983, donated to DZUP/DZ 33.412).

Additional material examined, not type specimens: 1 female, GO, Ipameri, 10.X.1988 (Becker, 59682); MG, Nova Lima, 850m: 2 males, 25-27.xii.1982; 2 males, 1-10.i.1985; 1 female, 30.xii.1988; all at light (Becker, 50290, 55772, 60523) (VOB, USNM, NHMUK). Pinned-dried adults with genitalia preparations mounted in Canada balsam on slides – 2 males (VOB 40740, 17.i.1983, deposited at LMCI 313-867; VOB 40740, 15.xii.1982, deposited at LMCI 313-880); 2 females (VOB 57100, 6.xii.1984, deposited at LMCI 313-872; VOB 40740, 30.xii.1982, deposited at LMCI 313-878). Immatures fixed in Dietrich's fluid and preserved in 70% ethanol – three last instar larvae (VOB 1519, 2.x.1984, deposited at LMCI 313-865A); five pupae (VOB 1519, 2.x.1984, deposited at LMCI 313-865B); twelve mature galls (CMP 008-01, donated LMCI 349-1); five empty, senescent galls (LMCI 346-01). Also, two last instar larvae, preserved in 100% ethanol at -20 °C, used for DNA extraction (CMP 008-04, donated to LMCI 349-2). Two last instar larvae preparations, also mounted in Canada balsam on a slide (VOB 1519, donated to LMCI 313-865C).

Other species examined (pinned-mounted adults): *Prymnotomis crypsicroca* Holotype male, BRAZIL: ES, [Baixo] Guandú, ES, 1920 (Hoffmann) (NM, Vienna) (g.s. NM 13322). *Hexeretmis pontopora* Meyrick, 1??? Holotype males, BRAZIL: PA, Taperinha, 11-20.vi.1927 (Zerny) (NM, Vienna) (g.s. NM 13321. *Alinguata neblina* Fleming, Alotype male, VENEZUELA: [Aragua], Rancho Grande, 3.vii.1946 (*Fleming*) (AMNH) (g.s. VOB).

Distribution. *Prymnotomis cecidicola* is known from the Brazilian Savanna, within the biome called "Cerradão" (= Cerrado *stricto sensu*; for a description, see Parron et al. 1998) where their host-plant and associated galls were found. The adults of this species do not come readily to light, as shown by the few specimens studied. The third author collected all over the Cerrado region of Brazil for over 30 years. At the type locality (Planaltina) he collected this species regularly in places very close to the host plant where the galls were common.

Host plant. Galls of *P. cecidicola* have been found only in association with *Cordiera elliptica* (Cham.) Kuntze (Rubiaceae) (Fig. 49), a plant native to the Brazilian Savanna (Cerrado Biome), with distribution ranging from northeast Bahia to southeast São Paulo state. It is a dioecious shrub (1.5 to 3.0 m high), with thin, cylindrical, glabrous branches, having dimorphic inflorescences; these are fasciculate on male plants, but only solitary flowers appear on female ones (Matsuoka LG, UFG, unpublished data). Also according to this author, flowers appear during the dry season (from July to September), fruits maturing later on, during the peak of the rainy season (from November to December), which was confirmed by our field observations. At the type locality, *C. elliptica* is commonly known as "marmelinho" and "marmela-da-de-pinto", being found scattered on vegetation, particularly along trails. The fruits are consumed by native people either fresh or as a home-made jelly.

Etymology. The species name is derived from the Greek *kekis – idos* = gall + the Latin *co – col* = with; to be treated as feminine.

Natural history. Mature galls of *P. cecidicola* (Figs 50–53) measure on average (± standard error) 1.80 ± 0.34 cm (n = 5) in diameter. They are green when active, spherical, unilocular and develop individually on *C. elliptica* inflorescences of both male and female plants. Thus we infer they are induced early on flower buds, since male flowers are dehiscent. *Cordiera elliptica* fruits are also green and spherical during development, but a little smaller (maximum diameter = 1.49±0.14 cm; n = 5) and turn brown when mature (Fig. 55). Contrary to *P. cecidicola* galls that have a smooth surface, *C. elliptica* fruits show conspicuous style scars distally (Fig. 50). Empty galls of *P. cecidicola* dry up, turning black, remaining attached to the plant (Fig. 56).

Larvae of *P. cecidicola* assume an arched position within their galls by placing the body around the fecal pellets left inside the gall. These are packaged and positioned centrally, as a sphere, firmly attached to the gall wall (Fig. 51), except when leaving the gall, when fecal pellets are left aside (Fig. 53). This sphere also contains the larval exuviae packaged within the feces. It is apparently increased in size and modelled periodically, being covered each time by a fine, blackish, silk net.

The body of full-grown *P. cecidicola* larvae progressively changes to red before pupation, when they leave the gall through a circular orifice made laterally on the gall wall (Fig. 52). Invariably when offered sandy soil and dried-broken leaves at the bottom of the rearing plastic pots in the laboratory, they built a flat, semi-rectangular, tied, silk-woven cocoon with debris attached (Fig. 54). Exit orifices were present on the surface of empty galls found in the field. Since we did not encounter cocoons attached to the host plant bearing empty galls under these conditions, we presume pupation occurs in the litter, which should be explored further.

Figure 49–56. Natural history of *Prymnotomis cecidicola* sp. nov. on *C. elliptica*: (49) host plant at the type locality; (50) young fruit and gall on female plant (gall is marked by asterisk; closed arrow indicates style scar of the fruit); (51) dissected gall showing last instar larva (seta indicates sphere made of faeces and exuviae attached to the gall wall); (52) external aspect of empty gall, showing larval exit orifice (pointed by open arrow); (53) dissected empty gall, showing sphere of packaged faeces and exuviae with fecal pellets left aside (pointed by closed arrow) by the larva before leaving for pupation; (54) fresh cocoon made by last larval instar in association with sand grains and dead-broken leaves under laboratory conditions; (55) mature fruit; (56) senescent empty gall. Scale bars: 3, 3.5, 4, 1, 3.5, 3.5 mm, respectively.

Cordiera elliptica plants bearing galls of *P. cecidicola* were found scattered in the field, density varying from one to four galls per plant. Field collections suggested that at the type locality *P. cecidicola* is a univoltine species. Galls and associated larvae were noticeable during the end of the dry season (August) up to beginning of the rainy season (October), thus coinciding in phenology with the reproductive phase of the host plant, described above. Full-grown larvae and pupae were obtained in late October. Emergence of adults under laboratory conditions was recorded from November to January.

DISCUSSION

This study sheds light on the biology of *Prymnotomis*, a poorly known genus of Neotropical many-plumed moth. Lack of morphological and molecular data on the other three alucitid genera endemic to the region makes a broad discussion about the descriptions presented here difficult. However, field collections in Costa Rica, Puerto Rico and French Guiana that resulted in several barcoded specimens (clustered in five BINs) allowed us to compare *Prymnotomis* with other Neotropical material from a genetic perspective. As expected, our sequence clustered with these BINs and was most closely related to cluster #5, presenting ca 12% genetic distance. Thus a revision is needed in Alucitidae, also to reduce the gap of lineage coverage in the analysis, which likely influences the large genetic divergence between taxa. The most distant affinity was found between *Prymnotomis* and *Pterotopteryx*, a Paleartic genus, not represented in the Neotropics, and that contrary to *Prymnotomis* shows deeply divided wings. *Prymnotomis* is expected to be more closely related morphologically to *Hexeretmis* and *Paelia*, according to Meyrick (1931). This should be further explored, also taking into account morphological characteristics of immature stages, their host plants and larval feeding habitats not only for these genera but also including *Alinguata*, also endemic to the Neotropics. Their original descriptions were based mostly on coloration and relative depth of wing lobes.

The unusual flap-like protrusion on stipes described here for the larva of *P. cecidicola* has been found in other alucitds, and also in the closely related copromorphid and carposinid lineages (Heppner 1987). We confirmed the suggestion of this author that these structures are functionally associated with the spinneret. As described by him, the tip of the spinneret is nested within these flaps in *P. cecidicola*. Additional observation demonstrated that they are used to retain partially a brownish dark, liquid substance over the spinneret's tip, which is used to soak the silk strands continuously when they emerge during the weaving process. This substance then solidifies, sealing the silk threads of the net that is used, for example, to cover the faecal pellet described here, and that thus remains isolated from the larva within the gall. We also noted under laboratory conditions that an orifice artificially made on the gall wall is immediately covered by the larva in this manner. Whether this substance is regurgitated and/or produced by an exocrine gland associated with its buccal apparatus should be further explored.

The absence of a differentiated cocoon-cutter and abdominal spines on the pupa of *P. cecidicola* and the presence of curved-pointed hooks on the terminal portion of the abdomen suggest that adult emergence in this species occurs inside the cocoon, which should be further examined. The emergence of the adult on the pupation site apparently appeared earlier in Lepidoptera evolution, within the Gellechioidea (e.g., Powell 1973, Becker 1982, Luz et al. 2014). Also interesting are the six raised lobes that appear externally on the *P. cecidicola* pupa forewing, particularly under scanning electron microscopy. Further studies should explore whether this characteristic is unique to and how variable it is within the Alucitidae, in order to rank its value as a diagnostic character for the family in this stage. From an ontogenetic perspective, such lobes supposedly correspond to early divisions in the wing that are present on the adults (for a description of the corresponding position related to wing veins in adult alucitids, see Dugdale et al. 1998).

ACKNOWLEDGEMENTS

We thank Centro de Pesquisas Agropecuárias dos Cerrados – Embrapa Cerrados for allowing us to collect the specimens in areas under their care and for use of laboratory facilities. Special thanks are due to Julia Fochezato (UFRGS), Amabílio José Aires de Camargo and Fernando Souza Rocha (Embrapa Cerrados) for providing assistance with field work. We also acknowledge the staff members of CMM/UFRGS and Thales O. Freitas (UFRGS) for the use of facilities and assistance with scanning electron microscopy and molecular analyses, respectively. Thanks are also due to Lafayette Eaton for editing the text. We are especially grateful to Héctor Vargas (UTA, Chile) and an anonymous reviewer for important suggestions the improved substantially the second version of the manuscript, and also to John Heppner (UFL, USA) for providing literature. G.L. Gonçalves was supported by project 16/2551-0000485-4 PRONEX, granted by FAPERGS/CNPq. G.R.P. Moreira and A. Specht were supported by fellowships from CNPq (process 311991/2018-1 and 306601/2016-8, respectively). C.M. Pereira was supported by Ph.D. scholarship from CNPq.

LITERATURE CITED

Becker VO (1982) Stenomine moths of the Neotropical genus *Timocratica* (Oecophoridae). Bulletin of the British Museum of Natural History 45: 211–306.

Dugdale JS, Kristensen NP, Robinson GS, Scoble MJ (1998) The smaller Microlepidoptera-grade superfamilies. In: Kristensen NP (Ed.) Lepidoptera: Moths and butterflies. Berlin, Walter de Gruyter, vol. 1, 217–232.

Folmer O, Black M, Hoeh W, Lutz R, Vrijenhoek R (1994) DNA primers for amplification of mitochondrial cytochrome c oxidase subunit I from diverse metazoan invertebrates. Molecular Marine Biology and Biotechnology 3: 294–299. https://doi.org/10.1371/journal.pone.0013102

Gielis C (2003) Pterophoroidea & Alucitoidea (Lepidoptera). In: World catalogue of insects. Stenstrup, Apollo Books, 198 pp.

Guindon S, Dufayard JF, Lefort V, Anisimova M, HordijkW, Gascuel O (2010) New algorithms and methods to estimate Maximum-Likelihood phylogenies: assessing the performance of PhyML 3.0. Systematic Biology 59: 307–21. https://doi.org/10.1093/sysbio/syq010

Heppner JB (1987) Copromorphidae, Alucitidae, Carposinidae (Copromorphoidea). In: Stehr FW (Ed.) Immature Insects. Dubuque, Kendall/Hunt Publishing Company, vol. 1, 399–402.

Heppner JB (2010a) Review of the genus *Alinguata* from Venezuela (Lepidoptera: Alucitidae). Lepidoptera Novae 3: 57–59.

Heppner JB (2010b) Review of the Neotropical genus *Hexeretmis* (Lepidoptera: Alucitidae). Lepidoptera Novae 3: 263–268.

Heppner JB (2015). A new *Alinguata* species from Ecuador (Lepidoptera: Alucitidae). Lepidoptera Novae 8: 41–42.

Landry B, Landry J-F (2004) The genus *Alucita* in North America, with description of two new species (Lepidoptera: Alucitidae). The Canadian Entomologist 136: 553–579. https://doi.org/10.4039/n03-095

Lima AC (1945) Insetos do Brasil. 5° Tomo. Lepidópteros, 1ª Parte. Escola Rio de Janeiro, Nacional de Agronomia, 379 pp.

Luz FA, Gonçalves GL, Moreira GRP, Becker VO (2014) Three new cecidogenous species of *Palaeomystella* Fletcher (Lepidoptera, Momphidae) from the Brazilian Atlantic Rain Forest. ZooKeys 433: 97–127. https://doi.org/10.3897/zookeys.433.7379

Meyrick E (1931) Orneonidae. Exotic Microlepidoptera 4: 86–87.

Mutanen M, Wahlberg N, Kaila L (2010) Comprehensive gene and taxon coverage elucidates radiation patterns in moths and butterflies. Proceedings of the Royal Society B 277: 2839–2848. https://doi.org/10.1098%2Frspb.2010.0392

Nieukerken EJ van, Kaila L, Kitching IJ, Kristensen NP, Lees DC, Minet J, Mitter C, Mutanen M, Regier JC, Simonsen TJ, Wahlberg N, Yen S-H, Zahiri R, Adamski D, Baixeras J, Bartsch D, Bengtsson BÅ, Brown JW, Bucheli SR, Davis DR, De Prins J, De Prins W, Epstein ME, Gentili-Poole P, Gielis C, Hättenschwiler P, Hausmann A, Holloway JD, Kallies A, Karshol O, Kawahara A, Koster JC, Kozlov MV, Lafontaine JD, Lamas G, Landry J-F, Lee S, Nuss M, Park K-T, Penz C, Rota J, Schmidt BC, Schintlmeister A, Sohn JC, Solis MA, Tarmann GM, Warren AD, Weller S, Yakovlev VR, Zolotuhin VV, Zwick A (2011) Order Lepidoptera. In: Zhang Z-Q (Ed), Animal biodiversity: An outline of higher-level classification and survey of taxonomic richness. Zootaxa 3148: 212–221. http://www.mapress.com/zootaxa/2011/f/zt03148p221.pdf

Patočka J, Turčani M (2005) Lepidoptera pupae: Central European species. Stenstrupp, Apollo Books, 542 pp.

Parron LM, Sousa-Silva JC, Camargo AJA (1998) Reservas ecológicas da Embrapa Cerrados: caracterização e zoneamento. Planaltina, Centro de Pesquisa Agropecuária dos Cerrados/EMBRAPA, 76 pp.

Pastrana JA (2004) Los Lepidópteros Argentinos: sus plantas hospedadoras y otros substratos alimenticios. Buenos Aires, Sociedad Entomológica Argentina, 334 pp.

Powell JA (1973) A systematic monograph of New World ethmiid moths (Lepidoptera: Gelechioidea). Smithsonian Contributions to Zoology 120: 1–302. https://doi.org/10.5479/si.00810282.120

Ratnasingham S, Hebert PDN (2013) A DNA-based registry for all animal species: The Barcode Index Number (BIN) system. PloS ONE 8(7): e66213. https://doi.org/10.1371/journal.pone.0066213

Robinson GS (1976) The preparation of slides of Lepidoptera genitalia with special reference to the Microlepidoptera. Entomologist's Gazette 27: 127–132.

Tamura K, Stecher G, Peterson D, Filipski A, Kumar S (2013) MEGA6: molecular Evolutionary Genetics Analysis version 6.0. Molecular Biology and Evolution 30: 2725–2729. https://doi.org/10.1093/molbev/mst197

Ustjuzhanin P, Kovtunovich V (2016) The Alucitidae (Lepidoptera) of Malawi with descriptions of five new species. Zootaxa 4126: 533–547. https://doi.org/10.3897/nl.39.8559

Ustjuzhanin P, Kovtunovich V, Sáfián S, Maicher V, Tropek R (2018) A newly discovered biodiversity hotspot of many-plumed moths in the Mount Cameroon area: first report on species diversity, with description of nine new species (Lepidoptera, Alucitidae). ZooKeys 777: 119–139. https://doi.org/10.3897/zookeys.777.24729

Vargas HA (2011) A new species of *Alucita* L. (Lepidoptera: Alucitidae) from Northern Chile. Neotropical Entomology 40: 85–88. https://doi.org/10.1590/S1519-566X2011000100012

Author Contributions: GRPM: designed the study, analyzed the data, and wrote the manuscript excepting the molecular analyses; CMP: collected and reared the immature stages, took the pictures, drew the structures, and organized the plates; VOB: designed the study, collected and reared the adults, and analyzed the genitalia; AS: collected and reared the immature stages, and analyzed the corresponding data; GLG: conducted the molecular analyses, interpreted and wrote the corresponding part.
Competing Interests: The authors have declared that no competing interests exist.

19

Hybridism between *Cyanocorax chrysops* and *Cyanocorax cyanopogon* (Aves: Corvidae) in Brazil

Cristiane Apolinario[1], **Luís Fábio Silveira**[1]

[1] *Museu de Zoologia da Universidade de São Paulo. Avenida Nazaré 481, Ipiranga, 04263-000 São Paulo, SP, Brazil.*
Corresponding author: Cristiane Apolinario (cris.snt4@gmail.com)

ABSTRACT. *Cyanocorax chrysops* (Vieillot, 1818) and *Cyanocorax cyanopogon* (Wied, 1821) are widespread jays. They are considered sister species according to recent molecular phylogenies. When analyzing museum specimens of both species, we found two individuals with intermediate plumage characteristics. They were collected near the range limits of both species, and we classified them as hybrids. Based on the sites where these specimens were found, we discuss the possible factors leading the interbreeding and compare our findings with information available in the literature. This is the first documented case of natural hybridization in jays in Brazil and the second case involving sister species within the New World jays.

KEY WORDS. Contact zones, hybrids, interbreeding, jays.

INTRODUCTION

Hybridization is a phenomenon of basic relevance in biology (McCarthy 2006), and zoologists have often considered it a sporadic event (Mayr 1963). Hybridization, the mating between two different species that results in offspring, might be considered a rare phenomenon because it is geographically restricted and because hybrids represent between 0.1 and 5% of populations (Mallet 2005, Arnold 1997, 2006). Nevertheless, it has been suggested that about 10% of all animal species hybridize naturally (Mallet 2005). With the advance of molecular techniques, the number of reports on natural hybridization and hybrid speciation has increased in the past decades (Brelsford 2011).

Hybridization can occur in a contact zone, there is, in areas where the ranges of two populations overlap; and those can become a hybrid zone, where hybridization occurs regularly (McCarthy 2006). A hybrid can be defined as an individual mixing the physical traits of individuals of two different but not necessarily related taxa. Hybrids can express the traits of their parents in three ways: combination, when a trait resembles only one of the parents; intermediacy, when the traits seem to express a mix between the parents; and heterotic, when the hybrid presents a trait well outside of the range of parental variation (McCarthy 2006).

McCarthy (2006) listed all reported cases of hybridization in the family Corvidae. In this family, natural hybridization is rare (Pitelka et al. 1956). Even so, one of the most complex case of hybridization involves the interbreeding between *Corvus corone corone* Linnaeus, 1758 and *Corvus corone cornix* Linnaeus, 1758 with an extensive hybrid zone (Picozzi 1976, Rolando 1993, Haas et al. 2009, Wolf et al. 2010, Brodin et al. 2013).

New World jays do not hybridize as often or as extensively as Old-World ones. Some records mention occasional hybridization involving congeneric species, for instance *Cyanocorax morio* (Wagler, 1829) and *Cyanocorax formosus* (Swainson, 1827) (Pitelka et al. 1956), two morphologically distinct species, whereas others involve hybridization between sister species, for example *C. formosus* and *Cyanocorax colliei* (Vigors, 1829) (McCarthy 2006, Anjos and de Juana 2018) and *Cyanocitta cristata* (Linnaeus, 1758) and *Cyanocitta stelleri* (Gmelin, 1788) (Williams and Wheat 1971). Hybridization in captivity has also been observed, comprising almost half of the known cases, more often when there is no conspecific individual to pair up with (Hardy and Wheat 1982), for instance between *Cyanocorax yncas* (Boddaert, 1783) and *C. cristata* (Pulich and Dellinger, 1981).

The Neotropical jays belonging to the genus *Cyanocorax* Boie, 1826 comprise a very distinct group of New World corvids. *Cyanocorax chrysops* (Vieillot, 1818), the Plush-crested Jay, is a polytypic species widely distributed in South America, ranging from northern Brazil to northern Argentina, with four subspecies currently accepted: *C. c. chrysops* (Vieillot, 1818), *C. c. diesingii* Pelzeln, 1856, *C. c. insperatus* Pinto and Camargo, 1961 and *C. c. tucumanus* Cabanis, 1883. This species is found in various types of lowland forests and temperate rainforest, also inhabiting patches of forest in open areas, riparian forests and even disturbed areas. *Cyanocorax cyanopogon* (Wied, 1821), the White-naped Jay, in turn, inhabits mostly areas of 'caatinga' and 'cerrado', but it is also seen in secondary woodland, riparian forests, and in the borders of tropical deciduous forests. It is Brazilian endemic, distributed from the state of Maranhão to the state of Minas Gerais, also occurring in southeastern Pará, eastern Mato Grosso and Goiás (Madge and Burn 1994, Anjos et al. 2009, 2019a, b). Regarding the preferences of habitat occupancy, *C. chrysops* prefers to occupy the understory and the middle level, exhibiting some variation in habitat use (Uejima et al. 2012). On the contrary, the unique information available about the preferences of *C. cyanopogon* is the use of the ground when foraging (Barros et al. 2014). The information about the reproductive behavior of both species is very limited (Uejima et al. 2012).

Cyanocorax chrysops and *C. cyanopogon* are considered sister species (Bonaccorso et al. 2010) and due their morphological similarity they have been previously treated as conspecific (Pinto 1954, Pinto and Camargo 1961, Blake and Vaurie 1962). In addition, Pinto (1954) described a taxon based on a single specimen from Alagoas, *Cyanocorax chrysops interpositus*, considering it as an intermediate form between the two species. Later, Hardy (1969) questioned the validity of *C. c. interpositus*, arguing that the specimens assigned to this subspecies were actually molting, worn, and subadults of *C. cyanopogon*, discarding its intermediate position both in geographic and genetic sense. There are no reliable records or specimens confirming the existence of intermediate individuals between the two so far, but we found two museum specimens showing intermediate plumage patterns between *C. chrysops* and *C. cyanopogon* from two different sites in Brazil, which we identified as hybrids. Thus, our objectives were to describe these hybrid specimens and discuss the possible causes behind the interbreeding and the existence of contact zones.

MATERIAL AND METHODS

We analyzed 80 specimens of *C. chrysops* and 137 of *C. cyanopogon* housed at Museu de Zoologia da Universidade de São Paulo (MZUSP) and Museu Paraense Emílio Goeldi (MPEG) (see Supplementary Material S1) in order to describe the plumage pattern of each hybrid and then compare it with the plumage of their parents. We analyzed plumage coloration following Smithe (1975) and Munsell (1994) color catalogues (hereafter S and M, respectively). In addition, we also analyzed records of both species on the Brazilian online database WikiAves (http://wikiaves.com.br) aiming to delimit a more precise range limit of both taxa, trying to identify putative contact zone areas.

RESULTS

We found two specimens, one from Três Lagoas, state of Mato Grosso do Sul (MZUSP 64191), and a second from Comendador Gomes, state of Minas Gerais (MZUSP 103009), showing intermediate plumage characters between *C. cyanopogon* and the nominate form *C. c. chrysops*. The most remarkable plumage characters exhibited by these particular specimens were: (1) back and wings dark grayish brown, (2) tail indigo blue or dark grayish brown + indigo blue and (3) nape light bluish gray/dull violaceous blue or white/lavender blue (Table 1). The main difference between the two is that the specimen from Três Lagoas has yellowish underparts whereas the specimen from Comendador Gomes shows white underparts, with each one being more similar to one of its parental species, resembling *C. c. chrysops* and *C. cyanopogon*, respectively. Following the classification given by McCarthy (2006), the hybrids showed both, intermediate traits (characters 1 and 3 in both hybrids and character 2 in the hybrid 2) and a combined trait (character 2 in the hybrid 1).

There are specimens of both species from Três Lagoas, *C. c. chrysops* (MZUSP 64192) and *C. cyanopogon* (MZUSP 73779) (Figs 1–3) in addition to records from WikiAves database; WA12379 (*C. c. chrysops*) and WA 3230616 (*C. cyanopogon*). The existence of these specimens and records demonstrates that parental forms come into contact in this region, one of the prerogatives to designate a true hybrid according to McCarthy (2006). Otherwise, in Comendador Gomes, a place near the São Paulo border, there are no specimens or records of both species rather than the hybrid.

DISCUSSION

Identifying the causes of hybridization requires understanding how ecology, demography and phenotype influence mate choice in a particular species (Willis 2013). Many factors could be involved in hybridization, but environment and behavior are known to be the main drivers causing this phenomenon (McCarthy 2006, Randler 2006); however, a close genetic relationship is considered an important factor leading to it (Randler 2006). In fact, it is been argued that species with small genetic distances are more prone to hybridize (Gholamhosseini et al. 2013), and sister taxa tend to interbreed more than non-sister taxa due their similar morphology (Randler 2002, 2004, 2006). There is no doubt about the close relationship between *C. chrysops* and *C. cyanopogon* based on morphology and molecular studies (Bonaccorso et al. 2010). Not only morphological similarity, but similarity in mating signals might be also a significant factor inducing interspecific pairing and hybridization (Gholamhosseini et al. 2013) and even the voice repertoire of the two seems to be very similar (Anjos 2019a), which would facilitate recognition of mating signals. The case of *C. chrysops* and *C. cyanopogon* match all these statements, revealing that multiple and related factors are involved and could be used to explain why they hybridize.

McCarthy (2006) pointed to breeding range overlap between *C. chrysops* and *C. cyanopogon* in southern Pará, but no specimen was mentioned or was found by us to substantiate his statement. Willis (1992), in turn, believed that the two taxa meet in western São Paulo, while Madge and Burn (1994) asserted that

Table 1. Comparison of plumage characters of the hybrids between *Cyanocorax chrysops chrysops* and *Cyanocorax cyanopogon.*

Character	*C. c. chrysops*	Hybrid 1 (MZUSP 64191)	Hybrid 2 (MZUSP 103009)	*C. cyanopogon*
Nape	Light sky blue (S 168D)/Campanula (S 71)	Light bluish Gray (M 10B 8/1)/Dull violaceous blue (S 170B)	White (M 2.5Y 8/1)/Lavender blue (S 170D)	White (M 2.5Y 8/1)
Superciliar spot	Smalt blue (S 170)/Light Sky blue (S 168D)	Smalt blue (S 170)/Light Sky blue (S 168D)	Light Sky blue (S 168D)	Light Sky blue (S 168D)
Malar spot	Campanula (S 71)	Campanula (S 71)	Smalt blue (S 70)	Cyanine blue (S 74)
Infraocular spot	Smalt blue (S 170)	Smalt blue (S 170)	Smalt blue (S 170)	Smalt blue (S 170)
Abdomen and underparts	Pale Yellow (M 2.5Y 8/3)	Pale Yellow (M 2.5Y 8/3)	White (M 2.5Y 8/1)	White (M 2.5Y 8/1)
Central rectrices and outer vanes of lateral rectrices	Indigo Blue (S 73)	Indigo Blue (S 73)	Dark grayish brown (S 20) + Indigo blue (S 173)	Sepia (S 119)
Rectrices tips	Pale Yellow (M 2.5Y 8/3)	Pale Yellow (M 2.5Y 8/3)	White (M 2.5Y 8/1)	White (M 2.5Y 8/1)
Back, rump and wing coverts	Indigo blue (S 73)	Dark grayish Brown (S 20)	Dark grayish Brown (S 20)	Sepia (S 119)

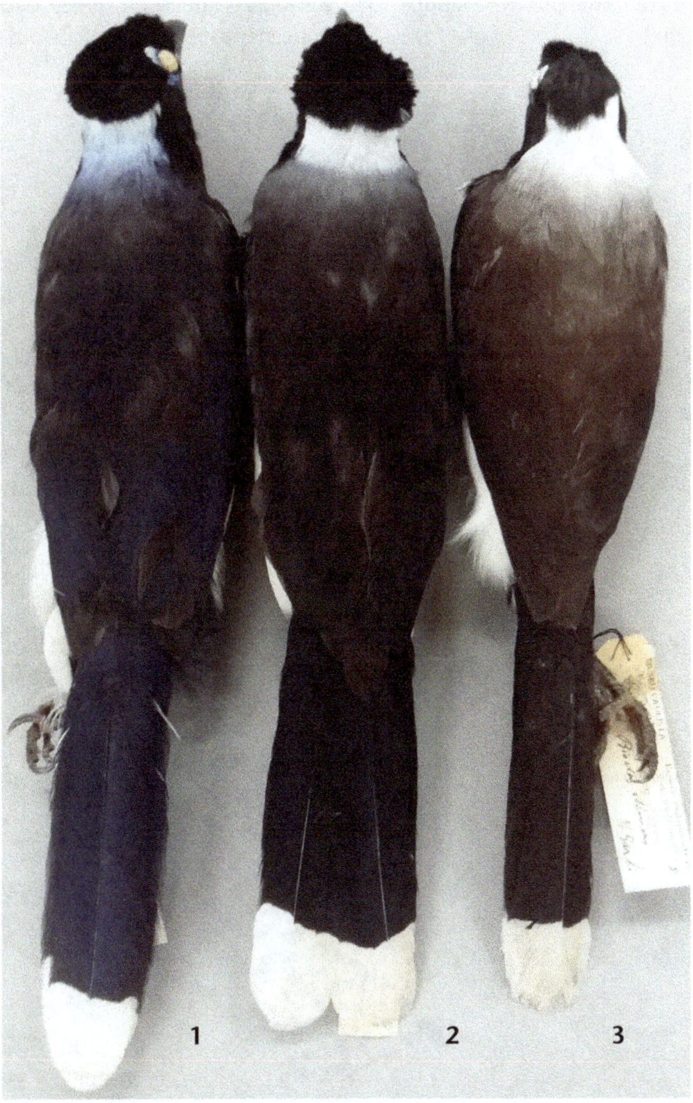

Figures 1–3. Hybrid MZUSP 64191 (2) between *C. c. chrysops* MZUSP 26041 (1) and *C. cyanopogon* MZUSP 27910 (3). Back and wings are brownish, resembling *C. cyanopogon*. The crest feathers have the typical *C. c. chrysops* shape, but nape coloration is lighter, almost completely white like *C. cyanopogon*.

the two species meet in eastern Mato Grosso. Our findings agree with Willis (1992) since one of the specimens was collected at the border between the states of Mato Grosso do Sul and São Paulo (Fig. 4). Moreover, it is possible to identify two conspicuous areas where the ranges of both species overlap: southern Minas Gerais and eastern Mato Grosso do Sul, the very same areas where we found the hybrid specimens. However, eastern Mato Grosso and Pará could be potential contact areas due the proximity of the range limits of both taxa, especially for *C. c. diesingii*, which shows an easternmost record very close to the westernmost records of *C. cyanopogon*, and because *C. c. diesingii* inhabits patches of open drier vegetation (Pacheco and Olmos 2005, Lees et al. 2008, Anjos et al. 2009, Whittaker 2009, Olmos et al. 2011, Santos et al. 2011), environments very similar to those inhabited by *C. cyanopogon*.

We are not able to ascertain if there is a stable hybrid zone, and there is no information about an area where individuals of *C. chrysops* and *C. cyanopogon* interbreed regularly. Possible reasons for the paucity of records of intermediate individuals could come from the great similarity of the parental species, which are often erroneously identified: individuals of *C. chrysops* are misidentified as *C. cyanopogon* and vice-versa. When the parental taxa are similar, the phenotype of the offspring produced by hybridization might be also more similar to one of the parents, making it difficult to identify visually as intermediate so they could just have remained unnoticed. On the other hand, the lack of intermediate individuals could just reflect the rarity of the interbreeding, as hybridization between other species of *Cyanocorax* is rare. For instance, *C. chrysops* is sympatric with other congeneric species, *C. cyanomelas* (Vieillot, 1818) and *C. caeruleus*

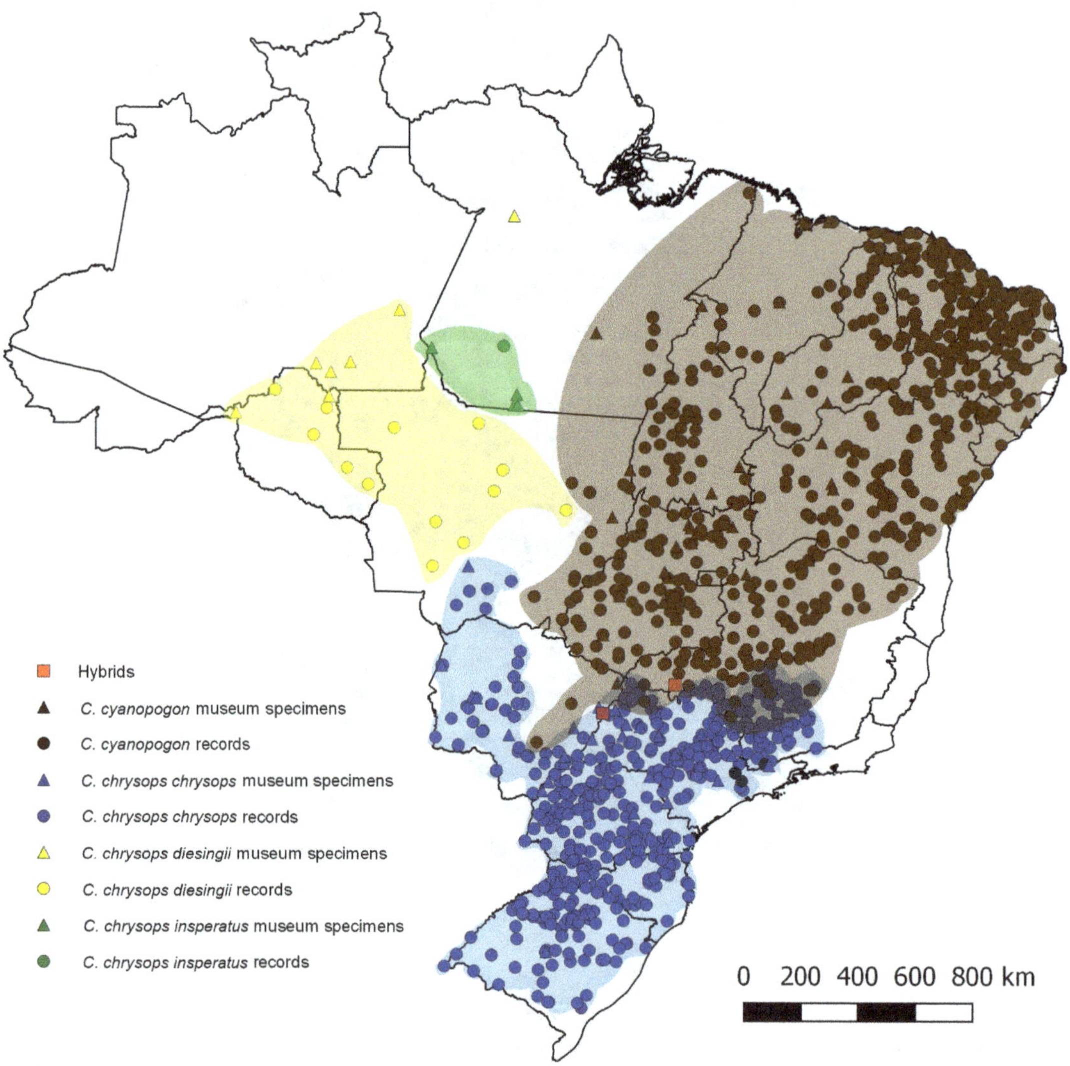

Figure 4. Distribution of *C. chrysops* and *C. cyanopogon* based on specimens analyzed and WikiAves' photographic records. Triangles represent the specimens and circles the photos analyzed.

(Vieillot, 1818) in part of its range (Goodwin 1976, Madge and Burn 1994) and there is no evidence of hybridism between these species, which have a dissimilar morphology when compared to *C. chrysops*, also demonstrating that sympatry by itself may not be a factor leading to hybridization. Thus, without any other clear indication and based on data we have, we suppose that hybridization between the two taxa occurs only occasionally in contact zones.

Hybridization is more common in areas where two related species meet and one of them is rarer. Thus, restricted mate choice drives the interbreeding (Hubbs 1955, Mayr 1967, Short 1969, Gillespie 1985, Grant and Grant 1997, Wirtz 1999, 2000). In the absence of conspecifics, individuals have to "choose" between mating heterospecifically or not mating at all (Randler 2006), a hypothesis known as "making the best of a bad job" (Baker 1996). The occurrence of *C. cyanopogon* in Mato Grosso do Sul is rare (Godoi et al. 2013), and there are only three municipalities where this species was recorded in WikiAves database. Otherwise, there are several records of *C. chrysops* and museum specimens from the same state, proving that it is the commonest species there. On the other hand, in southwestern Minas Gerais both species seem to be uncommon, as *C. cyanopogon* is considered rare in northern São Paulo (Anjos 2019a), and *C. chrysops* is known by a handful of records in the same region.

Deforestation and habitat modification affect the range of the species, and those inhabiting non-forest environments are expanding their distribution throughout Brazil, for instance *Patagioenas picazuro* (Temminck, 1813) (Willis and Oniki 1987), *Ramphastos toco* Statius Müller, 1776 (Sick 1997), *Rhynchotus rufescens* (Temminck, 1815) (Willis and Oniki 2002), and *Cyanocorax cristatellus* (Temminck, 1823) (Lopes 2008). One of the effects of range expansions is that it increases the possibility of taxa coming into secondary contact due the lack of a natural barrier. For instance, an invading species, initially rare locally, might mate heterospecifically because conspecific partners are at low frequency (Rheindt and Edwards 2011, Duckworth and Semenov 2017). The same occurs in transitional zones, which often show disparities in the relative abundances of species (Jansson et al. 2007, Lepais et al. 2009, Larson et al. 2013). *Cyanocorax cyanopogon* is expanding its range southward due the habitat modification, reaching the state of Espírito Santo (Anjos 2019a). The range of *C. chrysops*, in turn, does not seem to be expanding. This species is not very vulnerable to habitat fragmentation, being present in urban areas in southern and southeastern Brazil (Anjos 2019b). *Cyanocorax cyanopogon* and *C. chrysops* have a very diverse diet (Madge and Burn 1994, Anjos et al. 2009, Barros et al. 2014) and exhibit plasticity and differences in habitat use (Uejima et al. 2012, Barros et al. 2014). Therefore, they probably do not compete for resources when co-occurring, so that these features also contribute to the ability of individuals to persist in disturbed or transitional areas.

The two sites where the hybrids were collected are transitional areas from drier vegetation formations (Cerrado, southwestern Minas Gerais and eastern Mato Grosso do Sul) to areas of humid forests (western São Paulo), and this may explain the differences in abundance of both species, easing the possibility of hybridization. Moreover, with the advance of deforestation, we detected putative new contact areas. Both species were recorded on some urban surroundings in the cities of Araxá (WA 1333137 and WA 745949), Belo Horizonte (WA 2006717 and WA 47372) in Minas Gerais, and Meridiano (WA 1702233 and WA 775584) in São Paulo state, showing that *C. chrysops* and *C. cyanopogon* can also meet in very modified environments and new events of hybridization are likely to occur.

In summary, the best predictors to explain the occurrence of hybrids between *C. chrysops* and *C. cyanopogon* are those from morphological and behavior similarity coupled with a putative rarity of conspecific pairs. Our findings highlight the need to conduct further studies in areas of range overlap to investigate the rate and frequency of hybridization and its consequences for the genetic integrity of the species involved.

ACKNOWLEDGEMENTS

We thank Fundação de Amparo à Pesquisa do Estado de São Paulo (FAPESP) (grant 2013/22026-1). LFS is supported by Conselho Nacional de Desenvolvimento Cientifico e Tecnológico research productivity fellowship (process #302291/2015-6). We are thankful to Alexandre Aleixo (MPEG) for the specimens' loan and Gustavo Bravo for assisting with the procedures associated with the loan. Thiago Costa and two anonymous reviewers helped us to improve an earlier version of this manuscript.

LITERATURE CITED

Anjos L, Debus S, Madge S, Marzluff JM (2009) Family Corvidae. Species accounts. In: del Hoyo J, Elliott A, Christie DA (Eds) Handbook of the birds of the world. Barcelona, Lynx Edicions, vol. 14, 566–640.

Anjos L, de Juana E (2018) Black-throated Magpie-jay (*Cyanocorax colliei*). In: del Hoyo J, Elliott A, Sargatal J, Christie DA, de Juana E (Eds) Handbook of the birds of the world alive. Barcelona. Lynx Edicions, https://www.hbw.com/node/60724 [Accessed: 20/11/2018]

Anjos L (2019a) White-naped Jay (*Cyanocorax cyanopogon*). In: del Hoyo J, Elliott A, Sargatal J, Christie DA, de Juana E (Eds) Handbook of the birds of the world alive. Barcelona, Lynx Edicions, https://www.hbw.com/node/60722 [Accessed: 10/04/2019]

Anjos L (2019b) Plush-crested Jay (*Cyanocorax chrysops*). In: del Hoyo J, Elliott A, Sargatal J, Christie DA, de Juana E (Eds) Handbook of the birds of the world alive. Barcelona, Lynx Edicions, https://www.hbw.com/node/60721 [Accessed: 10/04/2019]

Arnold ML (1997) Natural hybridization and evolution. New York, Oxford University Press, 232 pp.

Arnold ML (2006) Evolution through genetic exchange. New York, Oxford University Press, 272 pp.

Baker MC (1996) Female buntings from hybridizing populations prefer conspecific males. The Wilson Bulletin 108: 771–775. https://www.jstor.org/stable/4163756

Barros RAM, Costa CA, Pascotto MC (2014) Diet and feeding behavior of the White-naped Jay, *Cyanocorax cyanopogon* (Wied, 1821) (Aves, Passeriformes, Corvidae) in a disturbed environment in central Brazil. Brazilian Journal of Biology 74(4): 899–905. https://doi.org/10.1590/1519-6984.09313

Blake ER, Vaurie C (1962) Corvidae. In: Mayr E, Greenway JC (Eds) Peters' check-list of birds of the world. Cambridge, Museum of Comparative Zoology, vol. 15, 204–282.

Bonaccorso E, Peterson AT, Navarro-Sigüenza AG, Fleischer RC (2010) Molecular systematics and evolution of the *Cyanocorax* jays. Molecular Phylogenetics and Evolution 54(3): 897–909. https://doi.org/10.1016/j.ympev.2009.11.014

Brelsford A (2011) Hybrid speciation in birds: allopatry more important than ecology? Molecular Ecology 20: 3705–3707. https://doi.org/10.1111/j.1365-294X.2011.05230.x

Brodin A, Haas F, Hansson B (2013) Gene flow across the European crow hybrid zone – a spatial simulation. Journal of avian Biology 44(3): 281–287. https://doi.org/10.1111/j.1600-048X.2012.00017.x

Duckworth RA, Semenov GA (2017) Hybridization associated with cycles of ecological succession in a passerine bird. The American Naturalist 190(4): E94–E105. https://doi.org/10.1086/693160

Gillespie GD (1985) Hybridization, introgression, and morphometric differentiation between mallard (*Anas platyrhynchos*) and grey duck (*Anas superciliosa*) in Otago, New Zealand. Auk 102: 459–469. https://www.jstor.org/stable/4086641

Gholamhosseini A, Vardakis M, Aliabadian M, Nijman V, Vonk R (2013) Hybridization between sister taxa versus non-sister taxa: a case study in birds. Bird Study 60(2): 195–201. https://doi.org/10.1080/00063657.2013.770815

Godoi MN, Morante-Filho JC, Módena ÉS, Faxina C, Tizianel FAT, Bocchese R, Pivatto MAC, Nunes AP, Posso SR (2013) Birds of upper Paraná River basin in the state of Mato Grosso do Sul, Brazil. Revista Brasileira de Ornitologia 21(3): 176–204.

Goodwin D (1976) Crows of the world. Ithaca, Cornell University Press, 354 pp.

Grant PR, Grant BR (1997) Hybridization, sexual imprinting, and mate choice. The American Naturalist 149: 1–28. https://doi.org/10.1086/285976

Haas F, Pointer MA, Saino N, Brodin A, Mundy NI, Hansson B (2009) An analysis of population genetic differentiation and genotype-phenotype association across the hybrid zone of carrion and hooded crows using microsatellites and MC1R. Molecular Ecology 18(2): 294–305. https://doi.org/10.1111/j.1365-294X.2008.04017.x

Hardy JW (1969) A taxonomic revision of the New World jays. The Condor 71(4): 360–375. https://www.jstor.org/stable/1365735

Hardy JW, Wheat P (1982) Vocalisations and other characteristics of Blue and Steller's jays and their hybrids. Bird Behavior 4(1): 21 29. https://doi.org/10.3727/015613882791560567

Hubbs CL (1955) Hybridization between fish species in nature. Systematic Zoology 4: 1–20. https://www.jstor.org/stable/2411933

Jansson G, Thulin C-G, Pehrson Å (2007) Factors related to the occurrence of hybrids between brown hares *Lepus europaeus* and mountain hares *L. timidus* in Sweden. Ecography 30: 709–715. https://doi.org/10.1111/j.2007.0906-7590.05162.x

Larson EL, Becker CG, Bondra ER, Harrison RG (2013). Structure of a mosaic hybrid zone between the field crickets *Gryllus firmus* and *G. pennsylvanicus*. Ecology and Evolution 3(4): 985–1002. https://doi.org/10.1002/ece3.514

Lees AC, Davis BJ, Oliveira AVG, Peres CA (2008) Avifauna of a structurally heterogeneous forest landscape in the Serra dos Caiabis, Mato Grosso, Brazil: a preliminary assessment. Cotinga 29: 149–159. http://e-space.mmu.ac.uk/id/eprint/617294

Lepais O, Petit RJ, Guichoux E, Lavabre JE, Alberto F, Kremer A, Gerber S (2009) Species relative abundance and direction of introgression in oaks. Molecular Ecology 18: 2228–2242. https://doi.org/10.1111/j.1365-294X.2009.04137.x

Lopes LE (2008) The range of the Curl crested Jay: lessons for evaluating bird endemism in the South American Cerrado. Diversity and Distributions 14(4): 561–568. https://doi.org/10.1111/j.1472-4642.2007.00441.x

Madge S, Burn H (1994) Crows and Jays: A guide to the crows, jays, and magpies of the world. Boston, Houghton Mifflin Co., 256 pp.

Mallet J (2005) Hybridization as an invasion of the genome. Trends in Ecology & Evolution 20: 229–237. https://doi.org/10.1016/j.tree.2005.02.010

Mayr E (1963) The breakdown of isolating mechanisms (hybridisation) In: Mayr E (Ed.) Animal species and evolution. Cambridge, Belknap Press, 110–135.

Mayr E (1967) Artbegriff und Evolution. Hamburg, Paul Parey, 617 pp.

McCarthy EM (2006) Handbook of avian hybrids of the world. New York, Oxford University Press, 602 pp.

Munsell (1994) Soil, color charts, revised edition. New Windsor, Macbeth Division of Kollmorgan Instruments Corporation.

Olmos F, Silveira LF, Benedicto GA (2011) A Contribution to the ornithology of Rondônia, southwest of the Brazilian Amazon. Revista Brasileira de Ornitologia 19(44): 200–229.

Pacheco JF, Olmos F (2005) Birds of a latitudinal transect in the Tapajós-Xingu interfluvium, eastern Brazilian Amazonia. Revista Brasileira de Ornitologia 13(22): 29–43.

Picozzi N (1976) Hybridization of carrion and hooded crows *Corvus c. corone* and *Corvus c. cornix*, in northeastern Scotland. Ibis 118(2): 254–257. https://doi.org/10.1111/j.1474-919X.1976.tb03072.x

Pinto O (1954) Resultados ornitológicos de duas viagens científicas ao estado de Alagoas. Papéis do Departamento de Zoologia 12(1): 1–98.

Pinto OM, Camargo EA (1961) Resultado ornitológico de quatro recentes expedições do Departamento de Zoologia no nordeste do Brasil, com a descrição de seis novas subespécies. Arquivos de Zoologia de São Paulo 11: 193–284.

Pitelka FA, Selander RK, del Toro MA (1956) A hybrid jay from Chiapas, Mexico. The Condor 58(2): 98–106. https://www.jstor.org/stable/1364976

Pulich WM, Dellinger RM (1981) An example of a hybrid Green jay × Blue jay. The Wilson Bulletin 93(4): 538–540. https://www.jstor.org/stable/4161546

Randler C (2002) Avian hybridization, mixed pairing and female choice. Animal Behaviour 63: 103–119. https://doi.org/10.1006/anbe.2001.1884

Randler C (2004) Frequency of bird hybrids: does detectability make all the difference? Journal of Ornithology 145: 123–128. https://doi.org/10.1007/s10336-004-0022-0

Randler C (2006) Behavioural and ecological correlates of natural hybridization in birds. Ibis 148(3): 459–467. https://doi.org/10.1111/j.1474-919X.2006.00548.x

Rheindt FE, Edwards SV (2011) Genetic introgression: an integral but neglected component of speciation in birds. Auk 128: 620–632. https://doi.org/10.1525/auk.2011.128.4.620

Rolando A (1993) A study on the hybridization between Carrion and Hooded Crow in Northwestern Italy. Ornis Scandinavica 24(1): 80–83. https://www.jstor.org/stable/3676414

Santos MPD, Silveira LF, Silva JMC (2011) Birds of Serra do Cachimbo, Para state, Brazil. Revista Brasileira de Ornitologia 19(2): 244–259.

Short LL (1969) Taxonomic aspects of avian hybridization. Auk 86(1): 84–105. https://www.jstor.org/stable/4083543

Sick H (1997) Ornitologia brasileira. Rio de Janiero, Editora Nova Fronteira, 862 pp.

Smithe FB (1975) Naturalist's color guide. New York, American Museum of Natural History, 229 pp.

Uejima AMK, Boesing AL, Anjos LD (2012) Breeding and foraging variation of the Plush-crested Jay (*Cyanocorax chrysops*) in the brazilian Atlantic Forest. The Wilson Journal of Ornithology 124(1): 87–95. https://doi.org/10.2307/41932804

Whittaker A (2009) Pousada Rio Roosevelt: a provisional avifaunal inventory in south-western Amazonian Brazil, with information on life history, new distributional data and comments on taxonomy. Cotinga 31: 20–43.

Williams O, Wheat P (1971) Hybrid jays in Colorado. Wilson Bulletin 83: 343–346.

Willis EO, Oniki Y (1987) Invasions of deforested regions of São Paulo state by the Picazuro Pigeon, *Columba picazuro* Temminck, 1813. Ciência e Cultura 39: 1064–1065.

Willis EO (1992) Zoogeographical origins of eastern Brazilian birds. Ornitología Neotropical 3: 1–15.

Willis EO, Oniki Y (2002) Birds of Santa Teresa, Espírito Santo, Brazil: do humans add or subtract species? Papéis Avulsos de Zoologia 42: 193–264. https://doi.org/10.1590/S0031-10492002000900001

Willis PM (2013) Why do animals hybridize? Acta Ethologica 16(3): 127–134. https://doi.org/10.1007/s10211-013-0144-6

Wirtz P (1999) Mother species-father species: unidirectional hybridization in animals with female choice. Animal Behaviour 58: 1–12. https://doi.org/10.1006/anbe.1999.1144

Wirtz P (2000) Einseitige hybridisierung im tierreich. Naturwissenschaftliche Rundschau 53: 172–175.

Wolf JB, Bayer T, Haubold B, Schilhabel M, Rosenstiel P, Tautz D (2010) Nucleotide divergence vs. gene expression differentiation: comparative transcriptome sequencing in natural isolates from the carrion crow and its hybrid zone with the hooded crow. Molecular Ecology 19: 162–175. https://doi.org/10.1111/j.1365-294X.2009.04471.x

Author Contributions: CA and LFS contributed equally to this article.

Competing Interests: The authors have declared that no competing interests exist.

20

Molecular sexing and intersexual differences in the morphometry of the Hangnest Tody-Tyrant *Hemitriccus nidipendulus* (Passeriformes: Rhynchocyclidae)

Reinaldo T. Medeiros[1], Flávia G. Chaves[2,3], Maurício B. Vecchi[4], Denise M. Nogueira[5], Maria Alice S. Alves[4]

[1]*Graduação em Ciências Biológicas, Universidade Veiga de Almeida. Rua Ibituruna 108, 20271-020 Rio de Janeiro, RJ, Brazil.*
[2]*Programa de Pós-graduação em Ecologia e Evolução, Instituto de Biologia Roberto Alcantara Gomes, Universidade do Estado do Rio de Janeiro. Rua São Francisco Xavier 524, 20550-011 Rio de Janeiro, RJ, Brazil.*
[3]*Instituto Nacional da Mata Atlântica. Avenida José Ruschi 4, 29650-000 Santa Teresa, ES, Brazil.*
[4]*Departamento de Ecologia, Instituto de Biologia Roberto Alcantara Gomes, Universidade do Estado do Rio de Janeiro. Rua São Francisco Xavier 524, 20550-011 Rio de Janeiro, RJ, Brazil.*
[5]*Departamento de Genética, Instituto de Ciências Biológicas e da Saúde, Universidade Federal Rural do Rio de Janeiro. Rodovia BR 465, km 7, 23897-000 Seropédica, RJ, Brazil.*
Corresponding Author. Maria Alice S. Alves (masaalves19@gmail.com)

ABSTRACT. Variation in the morphometry of individuals in a population may result from natural or sexual selection. In the present study we investigated morphometric differences between males and females of a bird species endemic to the Brazilian Atlantic Forest, the Hangnest Tody-Tyrant, *Hemitriccus nidipendulus* (Wied, 1831), with no apparent sexual dimorphism. All individuals (n = 56) were sexed by PCR amplification of the intronic fragments of the *CHDZ/CHDW* alleles obtained from blood samples, and 12 morphometric measurements were recorded. The molecular configuration was similar to that described for other species of Passeriformes, with males presenting a single band of approximately 360 base pairs (bp), and females with two bands of 360 and 400 bp. Males had significantly larger tarsi and wings than the females, while the females had two larger beak measurements. This indicates that differential selection pressures may be modeling the morphometry of the Hangnest Tody-Tyrant. In the males, larger tarsi and wings may be beneficial for the exploration of the habitat or the acquisition of potential mates, while females with larger beaks may be able to exploit larger food items and forage more efficiently. However, these hypotheses need to be tested empirically in future studies.

KEY WORDS. Atlantic Forest, gender differences, molecular sexing, restinga, sexual monomorphism.

INTRODUCTION

Birds may exhibit sexual dimorphism in their plumage, morphometry or behavior (Fuchs and Montalti 2016), reflecting the influence of natural selection (Selander 1966) and/or sexual selection (Andersson 1994, Székely et al. 2004). In the presence of intersexual differences in size and shape, the occupancy of different niches within a given habitat is possible. This may lead each sex to exploit a different part of the habitat, resulting in spatial segregation. In this context, morphometric traits may also vary systematically between juveniles and adults, and between males and females (Senay et al. 2017).

Although about half of all bird species are dimorphic, their sex differences become apparent only in their adult stage (Griffiths et al. 1998). Considering that, a fundamental problem for the evaluation of the role of sexual dimorphism in the ecology of these organisms are the difficulties in assigning immature individuals to the correct sex. One of the most reliable and widely-used sexing methods is the amplification by Polymerase Chain Reaction (PCR) of an intronic segment of the chromo-helicase DNA binding gene (*CHD*), which is linked to the bird sex chromosomes ZZ/ZW (Griffiths et al. 1998). The differences in the length of the two alleles can be used to identify the individual's gender. Females have two distinct fragments (*CHDZ/CHDW*), whereas the males have only one (*CHDZ*). In most bird species, the *CHDW* fragment, found only in the female, has the longest amplicon, although in some species, such as the Rock Pigeon, *Columba livia* Gmelin, 1789 (Columbidae), *CHDZ* is the longest amplicon (Griffiths et al. 1998). The size and sequence of the amplified fragments vary among different species, but are normally homogeneous in individuals of the same species, although some intraspecific polymorphisms have been described (Dawson et al. 2001).

The sequences of these intronic regions have been described in only a few Brazilian bird species up to now. For example, the *CHDZ/CHDW* intronic fragments of only one of the 67 species of Rhynchocyclidae Berlepsch, 1907, the Ochre-bellied Flycatcher *Mionectes oleagineus* (Lichtenstein, 1823), have been described (Caio S. Ferreira unpublished data). In this species, the *CHDZ* fragment has approximately 360 base pairs (bp) and the *CHDW* fragment, 390 bp. The rhynchocyclid genus *Hemitriccus* Cabanis & Heine, 1859 has 22 species (Piacentini et al. 2015) and, given its phylogenetic proximity to *Mionectes*, a similar configuration of the *CHDZ/CHDW* intronic fragments would be expected in *Hemitriccus*, including the Hangnest Tody-Tyrant, *Hemitriccus nidipendulus* (Wied, 1831).

Rhynchocyclidae comprises exclusively Neotropical species, which are often classified as a subfamily of the Tyrannidae Vigors, 1825. Based on an extensive analysis of the DNA sequence data of the tyrannid clade, however, Tello et al. (2009) proposed that a group of rhynchocycline flycatchers (including *Hemitriccus*) should be considered a distinct family, an arrangement that has been upheld since 2010 by the Brazilian Ornithological Records Committee (see Piacentini et al. 2015 for the current checklist of the birds of Brazil). With no apparent sexual dimorphism, the Hangnest Tody-Tyrant is an insectivorous species endemic to the Brazilian Atlantic Forest (Bencke et al. 2006, Vale et al. 2018), where it occurs in coastal "restingas" and humid evergreen forests, particularly in dense and secondary habitats (Sick 1997, Clock 2018). The analysis of morphometric differences between the sexes in phylogenetically close species has identified variation in body dimensions, such as the wing and tail, with males showing larger measurements than females in both the Rhynchocyclidae (Botero-Delgadillo 2010) and Tyrannidae (Cueto et al. 2015). Differences in beak dimensions have also been found in some rhynchocyclids (Botero-Delgadillo 2010).

In the present study, the *CHDZ/CHDW* gene fragments of the Hangnest Tody-Tyrant were investigated to confirm the sex of individuals and to evaluate potential morphometric measurements that could be used to reliably identify the sex of individuals without the need for molecular analyses, and in particular, parameters that can be applied to other species of the same family. As males are typically larger than females, and have larger beaks, as observed in other rhynchocyclid species (Botero-Delgadillo 2010) and most other birds (Székely et al. 2007), we hypothesized that a similar pattern may be found in the Hangnest Tody-Tyrant.

MATERIAL AND METHODS

Individuals were sampled in the restinga habitat (sandy coastal plain associated with the Atlantic Forest) of the Massambaba region in the state of Rio de Janeiro, Brazil, which is considered an Important Bird Area (IBA RJ08: "Restinga de Massambaba and Ilha de Cabo Frio") of the Brazilian Atlantic Forest domain, due to the presence of the local endemic, Restinga Antwren, *Formicivora littoralis* Gonzaga & Pacheco, 1990 (Bencke et al. 2006). In the present study, we sampled three locations in the Costa do Sol State Park: Ponta das Coroinhas (22°55'3"S; 42°16'34"W) and Brejo do Espinho (22°55'40"S; 42°16'12"W), in the municipality of Arraial do Cabo, and Praia do Vargas (22°56'14"S; 42°17'30"W), in the municipality of Araruama. The distance between sample sites ranged from 2 to 6 km, and all the sites are covered with shrubby and dense restinga vegetation.

The Hangnest Tody-Tyrant individuals were captured regularly using mist-nets (12 x 2.5 m, 36 mm mesh) between October 2011 and September 2014. The nets were set randomly along trails and in natural gaps in the vegetation in the morning (6:00–10:00 am) and afternoon (3:00–6:00 pm). We banded each captured individual with a metal ring (CEMAVE license number: 14210), measured it, and collected 10–50 µl of blood from the tarsal vein with a disposable needle (13 x 4.5 mm). Using a Pesola dynamometer (precision of 0.5 g), a ruler (precision 1 mm), and a caliper (precision 0.1 mm), one of the authors (RTM) took the following morphometric measurements: weight, total length, tail length, wing length, nostril-beak tip, exposed culmen, beak depth at the nostril and the base, beak width at the nostril and the base, tarsus length and length of the head to beak tip. Each individual was released near the site of its capture, usually less than one hour after being processed. The present study was conducted under the approval of Institutional Ethic Commitee at UERJ (CEUA – Comissão de *Ética* no Uso de Animais) #065/2018.

The DNA was extracted using the salting-out method (Fitz-Simmons et al. 1995). All individuals were sexed by amplifying the highly conserved ZW chromosome-linked *CHD* gene, using the primers P2 and P8 (Griffiths et al. 1998). The *CHDW* and *CHDZ* alleles were amplified by PCR in a 10 µl reaction volume containing 10 ng of DNA, 1X buffer (75 mM Tris-HCl pH 9.0, 20 mM(NH4)2SO_4, 0.01% Tween 20), 0.25 U Taq DNA polymerase (Thermoprime plus, Advanced Biotechnologies), 200 µM of dNTPs (Gibco), 2 mM of $MgCl_2$ and 1 µM of each primer, P2 (5'-TCT GCA TCG CTA AAT CCT TT- 3') and P8 (5'-CTC CCA AGG ATG AGR AAY TG-3'). The alleles were amplified in a Hybaid Touchdown thermal cycler (Thermo Hybaid, Ashford, Middlesex, UK). The PCR protocol used here was an initial denaturation at 94 °C for 2 min, followed by 40 cycles of 94 °C for 45 s, 50 °C for 45 s and 72 °C for 45 s, with a final extension of 5 min at 72 °C. PCR products were analyzed by electrophoresis in 2.5% agarose gel in 1X TBE buffer at 70V for three hours, stained with 5% ethidium bromide solution (10 mg/ml) and visualized in a transilluminator with UV light.

Student's t test for independent samples, run in Statistica 10, was used to compare the morphometric measurements between the two sexes, and a standard discriminant analysis was used to evaluate the proportion of individuals whose sex was identified correctly, and which variable contributed most to the identification of the sex of the individuals. Discriminant analysis is sensitive to sample size (Hair et al. 2005, Dechaume-Moncharmont et al. 2011), requiring a minimum of 20 individuals for each predictor variable (Hair et al. 2005), but smaller sample sizes has been used successfully in a number of studies (reviewed in Dechaume-Moncharmont et al. 2011). Therefore, we performed this test, even with the small sample size for females (n = 16). We used a jackknife validation, following Dechaume-Moncharmont et al. (2011), to estimate the proportion of individuals sexed correctly by the discriminant analysis. This procedure was run in the R program (R Core Team 2011).

RESULTS

Fifty-six adult Hangnest Tody-Tyrant were captured and banded. The molecular genetic analysis revealed that 40 of these individuals were males and 16 were females. The intronic *CHDZ* fragment had approximately 360 bp, while the *CHDW* fragment had 400 bp (Fig. 1). Based on this molecular analysis, we compared the morphometric traits of the two genders. Males had significantly longer wings and tarsi than females (Table 1), while larger values were recorded for the nostril-tip and beak depth of females (at the base). The discriminant analysis showed that males and females differed in their morphometry (Wilks' Lambda = 0.60, F = 2.54, p < 0.01) with tarsus length being the morphometric trait that most contributed to the differentiation of the sexes (Table 2). The jackknife cross-validation process revealed that 60% of the females (10 of 16 individuals) and 90% of the males (36 of 40 individuals) were sexed correctly based on their morphometric traits.

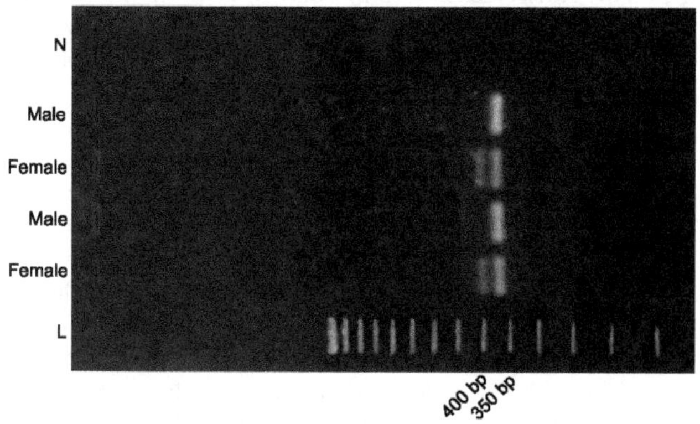

Figure 1. The DNA fragments of the intronic region of the *CHDZ/CHDW* alleles of the Hangnest Tody-Tyrant individuals captured in the Costa do Sol State Park, in southeastern Brazil. The fragments were amplified by Polymerase Chain Reaction (PCR) and electrophoresed in 2.5% agarose gel. The first column on the left (L) is a 50 bp DNA ladder, followed by the *CHDZ/CHDW* amplicons of the females, which present two bands (*CHDZ* = ~360 bp and *CHDW* = 400 bp), while males had only one band (*CHDZ*), with the negative control (N), consisting of only water.

DISCUSSION

The amplification of the intronic region of the *CHD* gene of the Hangnest Tody-Tyrant revealed that the *CHDW* allele is the largest amplicon, as observed in most bird species (Griffiths et al. 1998). A similar pattern has been recorded in a second rhynchocyclid, the Ochre-bellied Flycatcher (Caio S. Ferreira unpublished data), supporting that phylogenetically close species have a similar configuration of the *CHDZ/CHDW* alleles, when amplified with the P2/P8 primers (Griffiths et al. 1998). The male and female Hangnest Tody-Tyrant were morphometrically different, with males having larger body size parameters (tarsus and wing length), while the females had larger beaks (nostril-tip and beak height at the base). Theoretically this difference may reflect different selection pressures influencing the morphometry of each sex. In males, larger tarsi and wings can be beneficial for longer and faster flight, allowing a more efficient exploration of habitats, even though smaller wings may enhance maneuverability (Anderson and Norberg 1981), which may favor territorial defense, for example, and might be advantageous if females prefer more acrobatic males (Hakkarainen et al. 1996). It is important to note that differences in tarsal morphometry, as observed in the present study, may be associated with gender-based niche partitioning, given the importance of the tarsus in tyrannid foraging behavior (Fitzpatrick 1985). An enhanced tarsus may be beneficial for ambushing arthropods by pouncing, which would be important for an insectivorous bird such as the Hangnest Tody-Tyrant.

In the females, morphometric traits may have evolved in response to the need to avoid intraspecific competition with males for food resources. In this specific case, larger beaks may facilitate the capture of a more diverse array of arthropods by the females, in particular, larger prey. Unfortunately, this hypothesis cannot be tested empirically due to the lack of data on the diet of the Hangnest Tody-Tyrant. Selander (1966) argued that gender-based differences in size should be reflected in differences in the diets of the two sexes. However, variation in beak morphometry may also be related to mate choice and sexual selection, as observed in four subspecies of the Swamp Sparrow *Melospiza georgiana* (Latham, 1790), in which females mated preferentially with males that had larger beaks (Olsen et al. 2013).

Morphometric differences between males and females have been recorded in a number of other Neotropical birds, such as the Creamy-bellied Thrush, *Turdus amaurochalinus* Cabanis, 1850, a migrant that overwinters in the same habitat as the Hangnest Tody-Tyrant. The males of this thrush have larger wings than the females, which could allow them to fly faster and arrive first at their breeding grounds (Silva et al. 2011). In resident species, larger wing size has been recorded in male Chestnut-capped foliage-gleaner *Clibanornis rectirostris* (Wied, 1831) in the Serra do Cipó National Park, in the Brazilian Cerrado savanna (Faria et al. 2007), and in the male Restinga Antwren in the coastal restinga (Chaves and Alves 2013). The males of both

Table 1. Mean, standard deviation (sd), t-test result, and corresponding probabilities of morphometric measurements (mm) and weight (g) of 56 individuals of the Hangnest Tody-Tyrant (40 males and 16 females) in Costa do Sol State Park, southeastern Brazil.

Measurements	Males ± sd (range)	Females ± sd (range)	Student's t-test	Probability (p)
Weight	7.90 ± 0.68 (6.5–9.7)	7.58 ± 0.55 (6.5–8.5)	0.71	0.47
Total length	90.4 ± 2.97 (85–97)	89.3 ± 3.01 (83–94)	0.88	0.38
Wing length*	45.2 ± 1.27 (43–48)	44.4 ± 1.54 (41–47)	2.00	0.04
Length of tail	39.4 ± 1.31 (35–39)	38.9 ± 0.86 (33–36)	0.10	0.91
Tarsus length*	19.0 ± 0.55 (18–19.8)	18.5 ± 0.43 (17.8–19.4)	3.16	0.002
Exposed culmen	11.2 ± 0.51 (9.9–12.3)	11.4 ± 0.52 (10.2–11.9)	-1.06	0.29
Nostril-tip*	8.1 ± 0.33 (7.5–8.9)	8.3 ± 0.44 (7.9–8.9)	-2.10	0.04
Beak depth in nostril	2.8 ± 0.14 (2.5–3.2)	2.9 ± 0.15 (2.8–3.4)	-1.72	0.08
Beak width in nostril	3.7 ± 0.19 (3.4–4.1)	3.8 ± 0.19 (3.5–4.2)	-0.79	0.43
Beak height in base*	3.1 ± 0.13 (2.8–3.4)	3.2 ± 0.21 (3.0–3.9)	-2.38	0.02
Beak width in base	8.0 ± 0.72 (7.0–9.0)	8.0 ± 0.65 (7.1–9.7)	0.02	0.97
Head to beak tip	29.3 ± 0.44 (28.4–30)	29.3 ± 0.41 (28.2–29.9)	-0.72	0.47

*Significant results.

Table 2. Discriminant analysis for all morphometric measurements of the Hangnest Tody-Tyrant (Males = 40 and females = 16) captured in Costa do Sol State Park, southeastern Brazil.

Measurements	Wilk's lambda	p-level
Weight	0.60	0.96
Total length	0.62	0.18
Wing length	0.63	0.11
Length of tail	0.61	0.53
Tarsus length*	0.67	0.02
Exposed culmen	0.60	0.67
Nostril-tip	0.65	0.06
Beak depth in nostril	0.62	0.20
Beak width in nostril	0.61	0.50
Beak height in base	0.61	0.36
Beak width in base	0.60	0.92
Head to beak tip	0.60	0.69

*Significant results.

species defend their territories to guarantee access to potential mates. We were unable to confirm the existence of territorial behavior in the Hangnest Tody-Tyrant, but when playback was used to detect the species, these birds tended to answer promptly, possibly reflecting potential territoriality (unpublished data).

Gender differences in beak morphometry have also been found for the Restinga Antwren (Chaves and Alves 2013), reflecting habitat partitioning, with males foraging higher up in the vegetation than females (Chaves et al. 2017), even though the two sexes had a similar diet. In addition to body size, male Restinga Antwren had larger beaks than females in two parameters – nostril-tip and beak width at the nostril (Chaves and Alves 2013). The Restinga Antwren occurs in the same habitat (sandy coastal plain) in which the Hangnest Tody-Tyrant was sampled in the present study. Here, the Hangnest Tody-Tyrant females had longer and deeper beaks than the males, the opposite pattern to that observed in the previous study, and in all other studies of rhynchocyclid morphometry. In the Ochre-bellied Flycatcher, for example, males were heavier and had longer bodies, wings, and tails than the females (Caio S. Ferreira unpublished data), but no significant differences in beak parameters, while males Olive-striped Flycatchers, *Mionectes olivaceus* Lawrence, 1868, studied in Colombia by Botero-Delgadillo (2010), had deeper beaks than the females, as well as larger wings and tails.

Although clear gender differences were observed in the morphometry of the Hangnest Tody-Tyrant, the larger beak of the females did contradict our hypothesis. Further research should focus on the diet of the species, and in particular, the possible partitioning of feeding niches between the sexes. It will also be important to investigate the morphometry of other rhynchocyclids to determine in which species the females have larger beaks than the males, and the possible ecological implications of this morphometric difference. Despite the relatively small sample size of our study, our results indicate that body size measurements can be useful for sexing the Hangnest Tody-Tyrant, given the correct classification of 90% of males. Higher confidence level can be achieved, however, by applying the P2 and P8 primers, which can provide for this species a 100% certainty of individual sexing.

ACKNOWLEDGMENTS

We thank Luzia Bogel and Reinaldo Fortes for logistical support, and to members of the Laboratório de Ecologia de Aves – UERJ for help during fieldwork. To Stephen Ferrari for his valuable help with revising the language. We are also grateful to CEMAVE/ICMBio and INEA for research licenses (numbers 1237 and 013/2009, respectively), Conselho Nacional de Desenvolvimento Científico e Tecnológico for the fellowships granted to M.A.S.A. (CNPq processes 308792/2009-2, 305798/2014-6 and 306579/2018-9), Fundação de Amparo à Pesquisa do Estado do Rio de Janeiro (M.A.S.A. CNE processes E-26/102.837/2012,

E-26/203191/2015 and E-26/202835/2018, F.G.C. process E-26/201.724/2015, M.B.V. process E-26/201.778/2017), and Fundação de Amparo à Pesquisa e Inovação do Espírito Santo for the current fellowship to F.G.C. (BPIG-I, INMA).

LITERATURE CITED

Andersson M (1994) Sexual selection. Princeton University Press, Princeton, 624 pp.

Anderson M, Norberg RA (1981) Evolution of reversed sexual size dimorphism and role partitioning among predatory birds, with a size scaling of flight performance. Biological Journal of Linnean Society 15: 105–130. https://doi.org/10.1111/j.1095-8312.1981.tb00752.x

Bencke GA, Maurício GN, Develey PF, Goerck JM (2006) Áreas Importantes para a Conservação das Aves no Brasil. Parte I – Estados do Domínio da Mata Atlântica. SAVE Brasil, São Paulo, 494 pp.

Botero-Delgadillo E (2010) Criterios morfométricos y cualitativos para la determinación de la edad y el sexo en *Mionectes olivaceus* (Tyrannidae). Hornero. 25: 9–16.

Chaves FG, Alves MAS (2013) Gender-related diet composition and morphometry of the Restinga Antwren, *Formicivora littoralis* (Aves: Thamnophilidae). Zoologia. 30(6): 601–606. https://doi.org/10.1590/S1984-46702013005000006

Chaves FG, Vecchi MB, Alves MAS (2017) Intersexual differences in the foraging behavior of *Formicivora littoralis* (Thamnophilidae), an endangered Neotropical bird. Studies on Neotropical Fauna and Environment. 52(3): 179–186. https://doi.org/10.1080/01650521.2017.1335275

Clock B (2018) Hangnest Tody-tyrant (*Hemitriccus nidipendulus*). In: Del Hoyo J, Elliott A, Sargatal J, Christie DA, de Juana E (Eds) Handbook of the Birds of the World Alive. Lynx Edicions, Barcelona. https://www.hbw.com/node/57276

Cueto VR, Bravo SP, Trujillo-Arias N, Cabanne GS (2015) Sex determination by morphometry of adult White-crested *Elaenia* (*Elaenia albiceps chilensis*). Revista Brasileira de Ornitologia 23(1): 18–24.

Dawson DA, Darby S, Hunter FM, Krupa AP (2001) A critique of avian CHD-based molecular sexing protocols illustrated by a Z-chromosome polymorphism detected in Auklets. Molecular Ecology Notes 1(3): 201–204. https://doi.org/10.1046/j.1471-8278.2001.00060.x

Dechaume-Moncharmont FX, Monceau K, Cezilly F (2011) Sexing birds using discriminant function analysis: a critical appraisal. The Auk 138(1): 78–86. https://doi.org/10.1525/auk.2011.10129

Faria LP, Carrara L, Rodrigues M (2007) Dimorfismo sexual de tamanho no fura-barreira *Hylocryptus rectirostris* (Wied) (Aves, Furnariidae). Revista Brasileira de Zoologia 24(1): 207–212. https://doi.org/10.1590/S0101-81752007000100027

Fitzpatrick JW (1985) Form, Foraging behavior, and adaptive radiation in the Tyrannidae. Ornithological Monographs 36: 447–470. https://doi.org/10.2307/40168298

FitzSimmons NN, Moritz C, Moore SS (1995) Conservation and dynamics of microsatellite loci over 300 million years. Molecular Biology Evolution 12(3): 432–440. https://doi.org/10.1093/oxfordjournals.molbev.a040218

Fuchs DV, Montalti D (2016) Do morphometric measurements allow sex discrimination in Mockingbirds (*Mimus* sp.)? All Results Journal, Biology 7(3): 34–40.

Griffiths R, Double MC, Orr K, Dawson RJG (1998) A DNA test to sex most birds. Molecular Ecology 7(8): 1071–1075. https://doi.org/10.1046/j.1365-294x.1998.00389.x

Hair JF, Anderson RE, Tatham RL, Black WC (2005) Análise multivariada de dados. Bookman, Porto Alegre, 688 pp.

Hakkarainen H, Huhta E, Lahti K, Lundvall P, Mappes T, Tolonen P, Wiehn J (1996) A test of male mating and hunting success in the krestel: the advantages of smallness? Behavioral Ecology and Sociobiology 39(6): 375–380. https://doi.org/10.1007/s002650050303

Olsen BJ, Greenberg R, Walters JR, Fleischer RC (2013) Sexual dimorphism in a feeding apparatus is driven by mate choice and not niche partitioning. Behavioral Ecology 24(6): 1327–1338. https://doi.org/10.1093/beheco/art071

Piacentini VDQ, Aleixo A, Agne CE, Maurício GN, Pacheco JF, Bravo GA, Brito GRR, Naka LN, Olmos F, Posso S, Silveira LF, Betini GS, Carrano E, Franz I, Lees AC, Lima LM, Pioli D, Schunck F, Raposo FA, Bencke GA, Cohn-Haft M, Figueiredo LFA, Straube FC, Cesari E (2015) Annotated checklist of the birds of Brazil by the Brazilian Ornithological Records Committee. Revista Brasileira de Ornitologia 23(2): 91–298.

R Core Team (2011) R: a language and environment for statistical computing. Available online at: http://www.R-project.org, [Accessed on 18/12/2018]

Selander RK (1966) Sexual dimorphism and differential niche utilization in birds. The Condor 68(2): 113–151. https://doi.org/10.2307/1365712

Senay C, Harvey-Lavoie S, Macnaughton CJ, Bourque G, Boisclair D (2017) Morphological differentiation in northern pike (*Esox lucius*): the influence of environmental conditions and sex on body shape. Canadian Journal of Zoology 95(6): 383–391. https://doi.org/10.1139/cjz-2016-0159

Sick H (1997) Ornitologia brasileira. Nova Fronteira, Rio de Janeiro, 862 pp.

Silva KVA, Hajdu GL, Alves MAS (2011) Sex determination in *Turdus amaurochalinus* (Passeriformes: Muscicapidae): morphometrical analysis supported by CHD gene. Revista de Biología Tropical 59(2): 789–794.

Székely T, Freckleton RP, Reynolds JD (2004) Sexual selection explains Rensch's rule of size dimorphism in shorebirds. Proceedings of the National Academy of Sciences 101(33): 12224–12227. https://doi.org/10.1073/pnas.0404503101

Székely T, Lislevand T, Figuerola J (2007) Sexual size dimorphism in birds. In: Blanckenhorn W, Fairbairn D, Székely T (Eds). Sex, size and gender roles. Oxford University Press, Oxford, 27–37.

Tello JG, Moylea RG, Marchesea DJ, Cracraft J (2009) Phylogeny and phylogenetic classification of the tyrant flycatchers, cotingas, manakins, and their allies (Aves: Tyrannide). Cladistics 25(5): 429–467. https://doi.org/10.1111/j.1096-0031.2009.00254.x

Vale MM, Tourinho L, Lorini ML, Reis HR, Figueiredo MSL (2018) Endemic birds of the Atlantic Forest: traits, conservation status, and patterns of biodiversity. Journal of Field Ornithology 89(3): 193–206. https://doi.org/10.1111/jofo.12256

Author Contributions: MASA, MBV and RTM designed the experiments; RTM, MBV, FGC, and DMN conducted the experiments; RTM, FGC, MBV, MASA and DMN analyzed the data and wrote the paper.

Competing Interests: The authors have declared that no competing interests exist.

21

Mortality of primates due to roads and power lines in two forest patches in Bangladesh

Hassan Al-Razi[1,2], Marjan Maria[1,2], Sabir Bin Muzaffar[3]

[1]*Faculty of Life and Earth Science, Department of Zoology, Jagannath University. Dhaka, Bangladesh.*
[2]*Bangladesh Slow Loris Research and Conservation Project. 531/2, Shahidbagh, Mosque Road, Dhaka, Bangladesh.*
[3]*Department of Biology, United Arab Emirates University. Al Ain, United Arab Emirates.*
Corresponding author: Hassan Al-Razi (chayan1999@yahoo.com)

ABSTRACT. Primates are in imminent risk of extinction due to different types of anthropogenic activities. Mortality due to road accidents and electrocution from power lines are among the major direct anthropogenic threats to the survival of primates. We collected primate mortality data from 2015 to 2017 at Lawachara National Park and Satchari National Park in northeastern Bangladesh. We recorded 27 fatalities in five species of primates caused by road accidents (n = 15) and electrocution (n = 12). Most mortality records were for *Trachypithecus phayrei* (Blyth, 1847) (n = 8) while the lowest recorded mortality was for *Macaca mulatta* (Zimmermann, 1780) (n = 3). Ninety percent of primates in Bangladesh are threatened and populations are gradually declining. Our results suggest that roads and power supply lines are major sources of primate mortality that should be managed in these two forests. We strongly suggest avoiding construction of roads and power supply lines inside forests. Furthermore, control of the speed limit of vehicles inside the forests, use of insulated power lines, maintenance of natural canopy bridges and preparation of artificial canopy bridges are strongly recommended.

KEY WORDS. Bengal slow loris, conservation, electrocution, fragmentation, road accident, Phayre's langur.

Primates are one of the largest groups of seed dispersers in tropical forests (Stevenson 2011). They are vital to the reproductive cycles of many plant species (Beaune et al. 2013, Levi and Peres 2013, Arroyo-Rodríguez et al. 2015) and are also central to the livelihood and culture of many societies (Estrada et al. 2017). Although primates are recognized as essential components of ecosystems, about 60% of the primates are globally threatened and an astounding 73% of Asian species are threatened (Estrada et al. 2017). Habitat loss caused by agriculture, logging, wood harvesting, livestock farming and infrastructure development are some of the primary threats to primates (IUCN 2017).

Forest fragmentation associated with forest loss can create small and isolated subpopulations, raising extinction probabilities arising from demographic, environmental, and genetic factors (Frankham et al. 2002, Goosem 2007, Muzaffar et al. 2007). Infrastructure, such as pipelines, roads, railways, and transmission lines have the potential to impact wildlife in many ways, most notably by reducing access to resources and increasing mortality rates (Laurence et al. 2009, Benítez-López et al. 2010, Jenkins et al. 2010, Dean et al. 2019). With rapid development in many developing countries, mortality due to road accidents and electrocution from power lines is among the major direct human causes of mortality of terrestrial animals worldwide (Drews 1995, Fedigan and Zohar 1997, Forman and Alexander 1998, Printes 1999, Chhangani 2004, Lokschin et al. 2007, Parker et al. 2008, Cáceres et al. 2010, Pragatheesh 2011, Umapathy et al. 2011).

Asia as a continent has undergone development at unprecedented rates during the last 50 years (Muzaffar et al. 2011, Islam and Sato 2012, FAO 2015). Within Asia, Bangladesh is a small, densely populated country with a total population of over 160 million. Forested lands are heavily fragmented and few forest patches remain scattered in the northeast, southeast and southwest of the country, representing about 9.82% of the total land area (IUCN 2015). These patches are surrounded by human habitation, roads and railway lines that often pass through these forests. Thus, primates here are at high risk of mortality from road accidents on highways and from electrocution from electric power supply lines that pass through the remaining forest habitats. In this study the threats of primates arising from road accidents and electrocution were characterized.

We selected two forest patches of northeastern Bangladesh, namely Lawachara National Park (LNP) and Satchari National Park (SNP) for this study. Both national parks consist of mixed evergreen forest types planted in the 1930s and 1940s (Kabir and Muzaffar 2002). Lawachara National Park is situated in Kamalganj Upazila, Maulvi Bazar District and covers a 1250 ha area (Mollah and Kunda 2004). The canopy height of the forested areas in LNP varies from 10 to 30 m (Muzaffar et al. 2007). This forest is fragmented by Sreemangal-Bhanugach highway and Dhaka-Sylhet railway line (Fig. 1). In addition, an electric power supply line also passes through the forest approaching the highway. The second site, Satchari National Park, is within Chunarughat Upazilla, Habiganj District and has an area of 243 ha. The current extent of mixed evergreen forest is about 150 ha (Choudhury et al. 2004). The canopy height of this forest varies from 15 to 35 m. This forest patch is dissected by the old Dhaka–Sylhet highway and an electric power supply line that passes through the forest approaching the highway (Fig. 1).

We collected primate mortality data from both primary (direct observation) and secondary sources (published articles, interviews, social media uploads and newspaper reports) from 2015 to 2017 at these two national parks. Firstly, we collected primary data by direct observation during field work from July, 2015 to December 2017 during which we spent a total of 37 days in LNP and 68 days in SNP. Secondly, we collected secondary data from published articles. From March 2018 to April 2018 we searched the internet using Google and Google Scholar to find all related articles regarding primate mortality in SNP and LNP. Thirdly, we searched news published by national and local newspapers, online news portals, wildlife conservation groups and social media outlets. Finally, we collected data from interviews with forest staff, local people, as well as data from the record books of the range offices of SNP and LNP (Table 1). We conducted a questionnaire survey over a seven-day period in February 2018 in SNP. In LNP we conducted the questionnaire survey between May 2018 to June 2018. The questionnaire was developed in advance and approved by the Forest Department prior to the study. After data collection we cross-checked all data to avoid duplication of reports. In order to test possible differences between the types of accidents both within and between

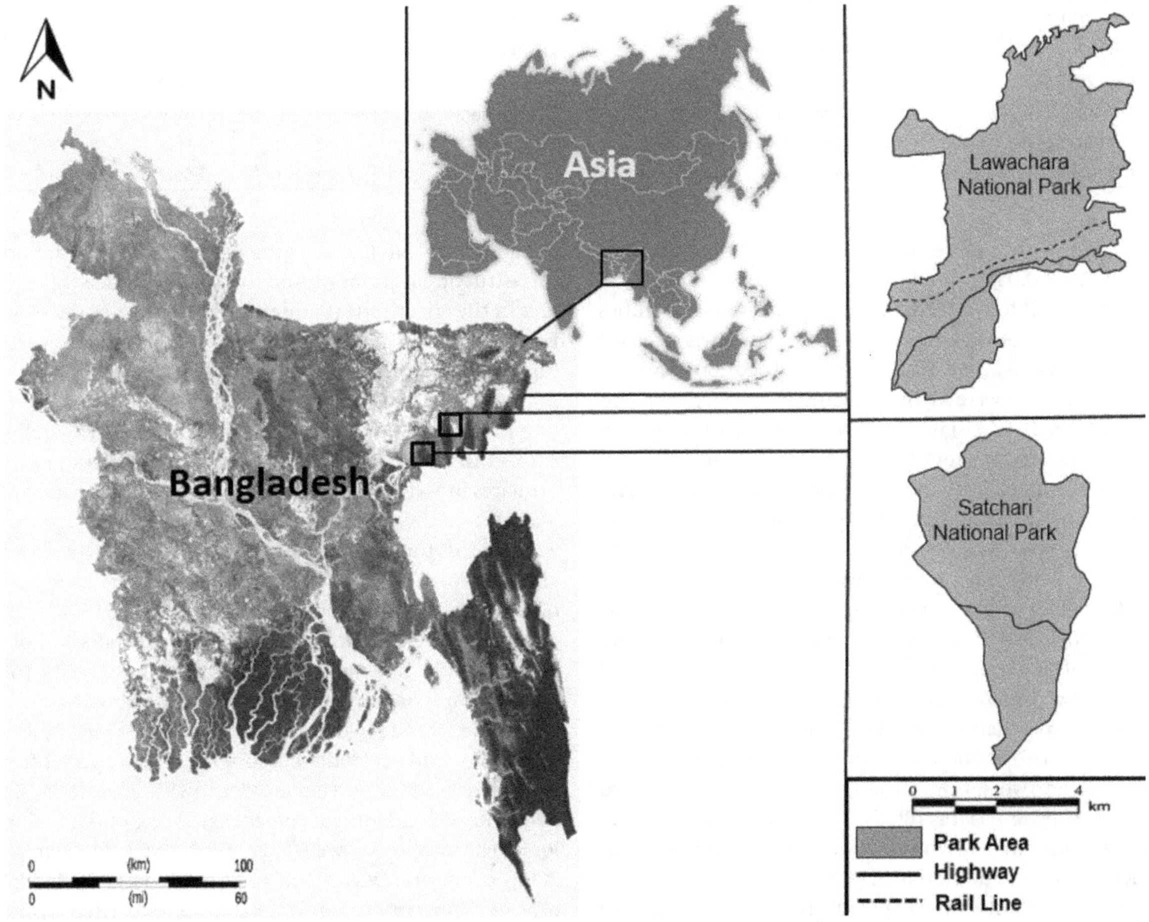

Figure 1. Map of two forest patches in Bangladesh, showing the locations of linear infrastructures inside the forest.

Table 1. Mortality of primates between 2015 and 2017 in Lawachara National Park (LNP) and Satchari National Park (SNP). (Source of data: I, Interview; SM, Social Media; N, Newspaper; PA, Published Article; DO, Direct Observation).

Incidence month	Affected species	Type of incidence	# mortality	Place	Source of data
January, 2015	*Nycticebus bengalensis*	Roadkill	1	SNP	I, SM
March, 2015	*Macaca leonina*	Roadkill	1	SNP	I
August, 2015	*Macaca leonina*	Roadkill	1	LNP	I
October, 2015	*Trachypithecus phayrei*	Electrocution	1	LNP	I
December, 2015	*Nycticebus bengalensis*	Roadkill	1	SNP	N, I
March, 2016	*Trachypithecus phayrei*	Roadkill	1	LNP	N
March, 2016	*Trachypithecus phayrei*	Roadkill	2	SNP	PA, N
June, 2016	*Trachypithecus pileatus*	Electrocution	2	SNP	DO
July, 2016	*Trachypithecus pileatus*	Electrocution	1	SNP	I, DO
August, 2016	*Trachypithecus pileatus*	Electrocution	1	SNP	SM
July, 2016	*Macaca mulatta*	Roadkill	1	SNP	DO
August, 2016	*Trachypithecus phayrei*	Roadkill	1	LNP	N, I
October, 2016	*Trachypithecus phayrei*	Roadkill	1	LNP	N, DO
October, 2016	*Nycticebus bengalensis*	Roadkill	1	LNP	N, I
November, 2016	*Macaca leonina*	Roadkill	1	LNP	I
December, 2016	*Trachypithecus phayrei*	Electrocution	1	LNP	N, I
February, 2017	*Nycticebus bengalensis*	Electrocution	2	LNP	I
May, 2017	*Macaca mulatta*	Roadkill	1	LNP	N, I
May, 2017	*Macaca leonina*	Roadkill	1	LNP	N
June, 2017	*Macaca mulatta*	Roadkill	1	SNP	N
July, 2017	*Trachypithecus pileatus*	Electrocution	1	SNP	I
October, 2017	*Trachypithecus pileatus*	Electrocution	1	SNP	DO
October, 2017	*Trachypithecus phayrei*	Electrocution	1	LNP	SM
December, 2017	*Trachypithecus pileatus*	Electrocution	1	SNP	I

parks we use the chi-square test (with values of significance determined at $\alpha < 0.05$).

A total of 27 (Mean = 9 ± SD 4 per year) individual fatalities of five primate species were recorded, of which 15 were caused by road accidents and 12 were caused by electrocution (Table 1). Most mortality records were for the Phayre's langur, *Trachypithecus phayrei* (Blyth, 1847) (n = 8) while the lowest recorded mortality was for the Rhesus Monkey, *Macaca mulatta* (Zimmermann, 1780) (n = 3). Fourteen individuals of primates died in SNP and 13 individuals in LNP (Table 1). The rate of mortality from road accidents (57.14%) was higher than the rate of mortality from electrocution (42.86%) in LNP, whereas mortality from electrocution (53.85%) was higher than mortality from road accidents (46.15%) in SNP ($\chi^2 = 0.83$, $p < 0.05$). In the case of electrocution, mortality rate of langurs (66.66%) was higher than mortality rates of other primates (16.66%). In contrast, mortality rate of langurs (33.33%) was lower than the mortality rates of other primates (83.33%) due to road accident in both study sites ($\chi^2 = 0.99$, $p < 0.0001$). Between two species of langur, mortality of Capped langur, *Trachypithecus pileatus* (Blyth, 1843), was higher in SNP (n = 6) than in LNP (n = 1) whereas mortality of Phayre's langur was higher in LNP (n = 6) than in SNP (n = 2).

Primate mortality rate appeared to be higher in SNP (13 individuals/2.5 km of road = 5.2 deaths per km of road) compared to LNP (2.15 deaths per km of road). The impact of infrastructure is proportional to its length since this affects the area of the effect zone (Forman 2000, Benítez-López et al. 2010). Furthermore, primate mortality rate in SNP was particularly high in 2016. In 2016, 12 of 18 natural canopy bridges over the highway were cut down to introduce an electric power supply line in the middle of the forest approaching the highway. We suspect that this was the likely cause of the higher mortality of primates in SNP in 2016. Removal of the natural canopy bridges likely forced the primates to use the road or electric power supply wire to cross the road making them vulnerable to road accidents and electrocution. It may be added that there were no records of electrocution of primates in SNP before 2016. Electrocution of primates were recorded only after the installation of electric power supply lines inside the forest, highlighting that such installations constitute a major threat to primates.

Rhesus and pig-tailed macaques were only recorded to have died from road accidents because these two species frequently use the ground for movement (Southwick et al. 1976, Bernstein 1967). Road accidents are considered as one of the major threats for Bengal slow loris as well (Choudhury 1992, Radhakrishna et al. 2006, Kumar and Devi 2010, Das et al. 2015). Only Radhakrishna et al. (2010) reported an incident of mortality from electrocution in this species near Siju Wildlife Sanctuary in Garo Hills, northeastern India. In this study, we recorded two incidents of mortality due to electrocution of Bengal slow loris from LNP. Among the two species of langur, mortality of capped langur occurred only due to electrocution. On the other hand, among eight cases of mortality of Phayre's langur, five occurred due to electrocution and three were due to road accidents (Fig. 2). From direct observation and interview of forest staff we found that both Capped langurs and Phayre's langurs frequently use the power supply lines to cross roads. Mortality in langurs occurred primarily due to short-circuiting of two electric parallel power lines connected by overhanging tails. In comparison, the other species have relative short tails (macaques) or almost no tail (e.g. slow loris). This is possibly why the macaques would be less vulnerable to electrocution even if they did use electric lines for movement. However, movement of the slow loris is slow and individuals may move from one wire to the next without releasing their grip on the first wire, causing fatal short-circuiting.

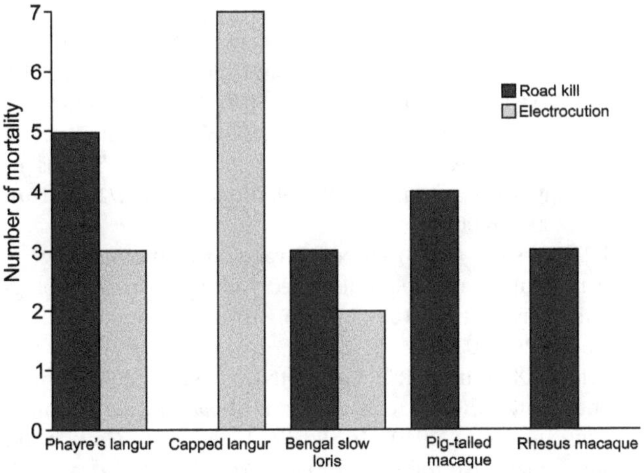

Figure 2. Type of mortality of different primate species in both study sites.

Ten primate species have been recorded from Bangladesh (IUCN 2015). Two species are globally Endangered, three are Vulnerable, and one is Near Threatened (Roos et al. 2014, IUCN 2015). In this study we recorded mortality of one Critically Endangered: *Trachypithecus phayrei*, and three Vulnerable species: *Trachypithecus pileatus*, *Nycticebus bengalensis* (Lacépède, 1800) and *Macaca leonina* (Blyth, 1863). Ninety percent of primates are threatened and the population of most of the primates are gradually declining in Bangladesh (IUCN 2015).

The Phayre's langur is surviving in few semi-evergreen forests of eastern parts of the country in small fragmented populations (IUCN 2015). The total population has declined by more than 80% in the last 20 years making it vulnerable to imminent extinction (Molur et al. 2003, IUCN 2015). Beside habitat destruction, logging, hunting, road accidents and electrocution are major threats of primates in Bangladesh (IUCN 2015). To mitigate the effects of road accidents and electrocution of primates and other wildlife in forest patches in Bangladesh, we strongly suggest avoiding construction of roads and power supply lines inside the last remaining forest patches. However, there is no direct law on the construction of roads and power supply lines inside the forest. As development progresses, roads and electric lines are constructed according to project needs. Stricter control of the speed limit of vehicles inside the forest should be imposed by creating speed breakers. To reduce fatalities from electrocution, we recommend the use of insulated power lines at least in the forested areas. Furthermore, maintenance of natural canopy bridges, and the preparation of artificial canopy bridges over the roads and electric power supply lines could further minimize mortality of primates and other arboreal mammals in forest patches of Bangladesh (Goosem 2004, Thurber and Ayarza 2005, Dean et al. 2019).

ACKNOWLEDGMENTS

First of all we are thankful the Forest Department of Bangladesh for permission (FD-22.01.0000.004.04.021.15.) to carry out work in forest and to provide information on road kills. The authors are also grateful to Professor Abdul Alim, Chairman, Department of Zoology, Jagannath University, for his support. We are thankful to Md. Anisur Rahman beat officer and Mahbub Hussain range officer of Satchari National Park for their support during the study. We are thankful to Sabit Hasan and Tanvir Ahmad for their support in field work. The field work of 2017 was funded by The Rufford Foundation. We thank David Thomson, Department of Biology, UAE University for reviewing the language of the paper.

LITERATURE CITED

Arroyo-Rodríguez V, Andresen E, Bravo SP, Stevenson PR (2015) Seed dispersal by howler monkeys: Current knowledge, conservation implications, and future directions. In: Youlatos D, Kowalewski MM, Garber PA, Cortes-Ortiz L, Urbani B, Youlatos D (Eds) Howler monkeys: Behavior, ecology and conservation. Developments in primatology: Progress and prospects. New York, Springer, 111–139.

Beaune D, Bretagnolle F, Bollache L, Bourson C, Hohmann G, Fruth B (2013) Ecological services performed by the bonobo (*Pan paniscus*): seed dispersal effectiveness in tropical forest. Journal of Tropical Ecology 29: 367–380. https://doi.org/10.1017/S0266467413000515

Benítez-López A, Alkemade R, Verweij PA (2010) The impacts of roads and other infrastructure on mammal and bird populations: a meta-analysis. Biological Conservation 143: 1307–1316. https://doi.org/10.1016/j.biocon.2010.02.009

Bernstein IS (1967). A field study of the pigtail monkey. Primates 8: 217-228.

Cáceres NC, Hannibal W, Freitas DR, Silva EL, Roman C, Casella J (2010) Mammal occurrence and road kill in two adjacent ecoregions (Atlantic Forest and Cerrado) in south-western Brazil. Zoologica 27: 709–717. https://doi.org/10.1590/S1984-46702010000500007

Chhangani AK (2004) Killing of Hanuman Langur in road accidents in Knumbhalgarh Wildlife Sanctuary, Rajasthan, India. Primate Report 69: 49-57

Choudhury AU (1992) The slow loris (*Nycticebus coucang*) in northeast India. Primate Report 34: 77–83.

Choudhury JK, Biswas SR, Islam M, Rahman SO, Uddin SN (2004) Biodiversity of Shatchari Reserved Forest, Habiganj. Dhaka, Bangladesh, IUCN Bangladesh Country Office, 34 pp.

Cuarón AD (1995) Pole bridges to avoid primate kills. Neotropical Primates 3: 74–75.

Das N, Nekaris KAI, Biswas J, Das J, Bhattacharjee PC (2015) Persistence and protection of the Vulnerable Bengal slow loris *Nycticebus bengalensis* in Assam and Arunachal Pradesh, north-east India. Oryx 49: 127–132. https://doi.org/10.1017/S0030605312001287

Dean WRJ, Seymour CL, Joseph GS, Foord SHA (2019) Review of the Impacts of Roads on Wildlife in Semi-Arid Regions. Diversity 11(81): 1–19. https://doi.org/10.3390/d11050081

Drews C (1995) Road kill of animals by public traffic in Mikumi National Park, Tanzania with notes on baboon mortality. African Journal of Ecology 33: 89–100. https://doi.org/10.1111/j.1365-2028.1995.tb00785.x

Estrada A, Garber PA, Rylands AB, Roos C, Fernandez-Duque E, et al. (2017) Impending extinction crisis of the world's primates: Why primates matter. Science Advances 3: e1600946. https://10.1126/sciadv.1600946

FAO (2015) Global Forest Resources Assessment 2015. Rome, FAO, Forestry Paper No. 1.

Fedigan LM, Zohar S (1997) Sex Differences in Mortality of Japanese Macaques: Twenty-One Years of Data From the Arashiyama West Population. American Journal of Physical Anthropology 102: 161–175. https://doi.org/10.1002/(SICI)10968644(199702)102:2<161::AID-AJPA2>3.0.CO;2-1

Forman RTT (2000). Estimate of the area affected ecologically by the road system in the United States. Conservation Biology 14: 31–35. https://doi.org/10.1046/j.1523-1739.2000.99299.x

Forman RTT, Alexander LE (1998) Roads and their major ecological effects. Annual Review of Ecology and Systematics 29: 207–232. https://doi.org/10.1146/annurev.ecolsys.29.1.207

Frankham R, Briscoe DA, Ballou JD (2002) Introduction to Conservation Genetics. Cambridge, Cambridge University Press, 642 pp.

Goosem M (2007) Fragmentation impacts caused by roads through rainforests. Current Science 93: 1587–1595.

Goosem M (2004) Linear infrastructure in tropical rainforests: mitigating impacts on fauna of roads and powerline clearings. In: Lunney D (Eds) Conservation of Australia's Forest Fauna. Sydney, Royal Zoological Society of New South Wales, 418–434 pp.

Islam K, Sato N (2012) Deforestation, Land Conversion and Illegal Logging in Bangladesh: The Case of the Sal (*Shorea robusta*) Forests. iForest-Biogeosciences and Forestry 5: 171–178. https://doi.org/10.3832/ifor0578-005

IUCN (2015) Red List of Bangladesh. Dhaka, Bangladesh, International Union for Conservation of Nature, vol. 2, 232 pp.

IUCN (2017) IUCN red list of threatened species. International Union for Conservation of Nature, v. 2017-3-4. Available online at: http:// www.iucnredlist.org. [Accessed: 26/12/2018]

Jenkins AR, Smallie JJ, Diamond M (2010) Avian collisions with power lines: a global review of causes and mitigation with a South African perspective. Bird Conservation International 20: 263–278. https://doi.org/10.1017/S0959270910000122

Kabir DS, Muzaffar SB (2002) The review of the present state of protected areas of Bangladesh. In: Ahmed MF, Tanveer SA, Badruzzaman ABM (Eds) Bangladesh environment 2002. Dhaka, Bangladesh, Bangladesh Poribesh Rokha Andolon, 389–403.

Kumar A, Devi A (2010) Status and conservation of slow loris (*Nycticebus bengalensis*) in northeast India. Ecotone 2: 18–20. http://www.scribd.com/doc/32168876/Ecotone2-1-2010

Laurence WF, Goosem M, Laurance SGW (2009) Impacts of roads and linear clearings on tropical forests. Trends in Ecology & Evolution 24: 659-669. https://doi.org/10.1016/j.tree.2009.06.009

Levi T, Peres CA (2013) Dispersal vacuum in the seedling recruitment of a primate-dispersed Amazonian tree. Biological Conservation 163: 99–106. https://doi.org/10.1016/j.biocon.2013.03.016

Lokschin LX, Printes RC, Cabral JNH, Buss G (2007) Power lines and howler's conservation (*Alouatta guariba clamitans*; Cabrera, 1940) in Porto Alegre, Rio Grande do Sul, Brazil. Netropical Primates 14: 76–80. https://doi.org/10.1896/044.014.0206

Mollah AR, Kunda DK (2004) Site Level Appraisal for Protected Area Co-Management: Lawachara National Park. Bangladesh, International Resources Group, Nature Conservation Management, 120 pp.

Molur S, Brandon-Jones D, Dittus W, Eudey A, Kumar A, Singh M, Feeroz MM, Chalise M, Priya P, Walker S (2003) Status of South Asian Primates: Conservation Assessment and Managment Plan Report. Coimbatore, India, Zoo Outreach Organization/CBSG-South Asia.

Muzaffar SB, Islam MA, Feeroz MM, Kabir M, Begum S, Mahmud S, Chakma S (2007) Habitat characteristics of the endangered Hoolock Gibbons (*Hoolock hoolock*) of Bangladesh: the role tree species richness. Biotropica 39: 539–545. https://doi.org/10.1111/j.1744-7429.2007.00298.x

Muzaffar SB, Islam MA, Kabir DS, Khan MH, Ahmed FU, Chowdhury GW, Aziz MA, Chakma S, Jahan I (2011) The endangered forests of Bangladesh: why the process of implementation of the Convention on Biological Diversity is not working. Biodiversity and Conservation 20: 1587-1601. https://doi.org/10.1007/s10531-011-0048-6

Parker L, Nijman V, Nekaris KAI (2008) When there is no forest left: fragmentation, local extinction, and small population sizes in the Sri Lankan western purple-faced langur. Endangered Species Research 5: 29–36. https://doi.org/10.3354/esr00107

Pragatheesh A (2011) Effect of human feeding on the road mortality of Rhesus Macaques on National Highway-7 routed along Pench Tiger Reserve, Madhya Pradesh, India. Journal of Threatened Taxa 3: 1656–1662. https://doi.org/10.11609/JoTT.o2669.1656-62

Printes RC (1999) The Lami biological reserve, Rio Grande do-Sul, Brazil and the danger of power lines to howlers in urban reserves. Neotropical Primates 7: 135-136.

Radhakrishna S, Datta-Roy A, Swapna N, Sinha A (2010). Population survey of the Bengal slow loris, *Nycticebus bengalensis*, in Meghalaya, northeast India. Primate Conservation 25: 105–110. https://doi.org/10.1896/052.025.0102

Radhakrishna S, Goswami AB, Sinha A (2006) Distribution and Conservation of *Nycticebus bengalensis* in Northeastern India. International Journal of Primatology 27: 971–982. https://doi.org/10.1007/s10764-006-9057-9

Roos C, Boonratana R, Supriatna J, Fellowes JR, Groves CP, Nash SD, Rylands AB, Mittermeier RA (2014) An updated taxonomy and conservation status review of Asian primates. Asian Journal of Primatology 4: 2–38.

Southwick CH, Sidiqi MF, Farooqui MY, Pal BC (1976) Effects of artificial feeding on aggressive behaviour of rhesus monkeys in India. Animal Behavior 24: 11–15.

Stevenson PR (2011) The abundance of large ateline monkeys is positively associated with the diversity of plants regenerating in Neotropical forests. Biotropica 43: 512–519. https://doi.org/10.1111/j.1744-7429.2010.00708.x

Thurber M, Ayarza P (2005) Canopy bridges along a rainforest pipeline in Ecuador. Kuala Lumpur, Malaysia, Society of Petroleum Engineers Asia Pacific Health, Safety and Environment Conference and Exhibition, 1–4.

Umapathy G, Hussain S, Shivaji S (2011) Impact of habitat fragmentation on the demography of lion-tailed macaque (*Macaca silenus*) populations in the rainforests of Anamalai Hills, Western Ghats, India. International Journal of Primatology 32: 889–900. https://doi.org/10.1007/s10764-011-9508-9

Author Contributions: HA and MM planned and designed the research project; HA and MM collected and managed the data; SBM and HA analyzed the data. All of the authors contributed to prepare the manuscript.

Competing Interests: The authors have declared that no competing interests exist.

22

Immature stages of the *Selaginella*-feeding *Euptychia mollina* (Nymphalidae: Satyrinae) from the Brazilian Amazon

André V. L. Freitas[1,2], Luísa L. Mota[1], Eduardo P. Barbosa[1], Junia Y. O. Carreira[1]

[1]*Departamento de Biologia Animal, Instituto de Biologia, Universidade Estadual de Campinas. Rua Monteiro Lobato 255, 13083-862 Campinas, SP, Brazil.*
[2]*Museu de Zoologia, Instituto de Biologia, Universidade Estadual de Campinas. Rua Charles Darwin, Bloco N, Barão Geraldo, 13083-863 Campinas, SP, Brazil.*
Corresponding author: André Victor Lucci Freitas (baku@unicamp.br)

ABSTRACT. The present paper describes the immature stages of the Neotropical satyrine butterfly *Euptychia mollina* (Hübner, [1813]) from the Brazilian Amazon Forest. Eggs were laid singly on the under surface of the fronds of its host plant, *Selaginella* sp. (Bryophyta: Selaginellaceae). Larvae are solitary in all instars, presenting a color pattern and shape that make them cryptic on its host plant. The pupa is short, smooth and varies from rusty brown to green. Despite the lack of a close phylogenetic relationship, larvae of *Euptychia* are very similar to those of the paleotropical satyrines *Ragadia* and *Acrophtalmia*, suggesting that camouflage is likely to be one of the factors explaining the similarities among them.

KEY WORDS. Camouflage, caterpillar, Euptychiina, Lepidoptera.

INTRODUCTION

In recent years, several aspects of Euptychiina have been extensively studied, including their taxonomy and phylogeny, natural history, biogeography and conservation (Peña et al. 2010, Marín et al. 2011, 2017, Freitas et al. 2012, Zacca et al. 2018 and references therein). As a result, the systematics of the subtribe has been revised from the species to the subtribal level, with significant reorganization in most of the large genera (see references above). Nevertheless, some important points remain under discussion, including the monophyly of the subtribe in relation to its nominal genus *Euptychia* Hübner, 1818 (Peña et al. 2010, Marín et al. 2017, Espeland et al. 2019).

As for other large genera of Euptychiina, *Euptychia* has also been the target of several studies in recent years, and since the checklist of Lamas (2014), the number of recognized species in the genus has increased from 13 to 34 (Freitas et al. 2012, Neild et al. 2014, 2015, Nakahara et al. 2014, 2015a,c,d, 2016, Fratello et al. 2015, Warren et al. 2018, see also Freitas et al. 2013 and Nakahara et al. 2015b), with at least 15 additional species awaiting to be described (S. Nakahara, unpublished data). Species of *Euptychia* are present in most Neotropical forests, although they are more common in lowland and submontane forests, where they can be locally common (AVLF personal observation). Despite their prevalence in some habitats, very little is known about their ecology, behavior and natural history. For example, *Euptychia* is well-known for being one of the few butterflies to use Lycopsida as larval host plants (Singer et al. 1971, Singer and Mallet 1986, DeVries 1987). However, although there are some images of immature stages available for *Euptychia westwoodi* A. Butler, 1867 (Hamm and Fordyce 2016, Janzen and Hallwachs 2018) and partial descriptions for four additional Costa Rican species (Singer et al. 1983, DeVries 1987), a detailed, complete description of the immature stages for any species of *Euptychia* is still lacking.

In the present paper, descriptions of the immature stages for *Euptychia mollina* (Hübner, [1813]) are presented. This is the first paper illustrating in detail all life stages for a species of *Euptychia* and also the first report on immature stages from a species outside of Costa Rica.

MATERIAL AND METHODS

A single egg was obtained through the observation of an oviposition event in the rural region of Belterra municipality, Pará State, Northern Brazil, on August 19, 2018 (2°35'47"S; 54°57'00"W). The oviposited plant was an unidentified *Selaginella* (Lycopsida: Selaginellaceae) that was brought to the laboratory and potted to be used to feed the larva. The egg was brought to laboratory and, after hatching, the larva was reared in a plastic container cleaned daily and provided with fresh plant material (see above) every two or three days (following Freitas 2007). Data were recorded on morphology and development time for all stages. Dry head capsules and the pupal case were retained in glass vials. Voucher specimens of the immature stages and adults (the ovipositing female and the emerged adult) were deposited in the Museu de Zoologia "Adão José Cardoso" (ZUEC-AVLF), Universidade Estadual de Campinas, Campinas, São Paulo, Brazil. Additional material was obtained from the Teles Pires River, Alta Floresta municipality, northern Mato Grosso State, Central Brazil, on June 20, 2000 (9°38'12"S; 55°56'24"W). Several eggs and larvae were also collected in the field on fronds of *Selaginella* sp. along a forest edge following a dirt road running parallel to the river.

Measurements were taken and general aspects of morphology were observed using a Leica® MZ7.5 stereomicroscope equipped with a micrometric scale. Scanning electron microscopy (SEM) was conducted using a JEOL® JSM-5800 microscope, and samples were prepared in accordance with the following protocol: Sample critical point dried using Bal-tec® – CPD030 equipment and attached with double stick tape to aluminum stubs; gold/palladium coated with a Bal-tec® – SCD050 sputter coater. Egg size is presented as height and diameter, and head capsule size is the distance between the most external stemmata (as in Freitas 2007). Terminology for early stages descriptions follows García-Barros and Martín (1995) for eggs and Stehr (1987) for larvae and pupae.

RESULTS

The morphological description and measurements of the immature stages below are based on material from Belterra, Pará. Additional data from Teles Pires river, Alta Floresta, Mato Grosso, are included at the end of this section.

Egg (Figs 1, 2). Green, round with poorly marked irregular cells. Height and diameter 0.60 mm. Duration 7 days (n = 1).

First instar (Figs 3, 4, 17–19). Head capsule width 0.40 mm; head scoli 0.08 mm. Head black, bearing a pair of short scoli on vertex (Figs 18, 19). Third stemma larger than other stemmata. Body light green with several small protuberances; caudal filaments short. Legs and prolegs light green. Setae light green; most dorsal and subdorsal setae clubbed at tip (see detail in Figs 17, 19). The chaetotaxy of the first instar is presented in Fig. 20. Maximum length 3 mm. Duration 8 days (n = 1).

Second instar (Figs 5, 6). Head capsule width 0.58 mm; head scoli 0.22 mm. Head black, with two short pointed scoli on vertex. Body green, with light green poorly marked longitudinal stripes and covered with small protuberances giving a rough appearance to the larva; caudal filaments short. Legs and prolegs light green. Maximum length 5 mm. Duration 5 days (n = 1).

Third instar (Figs 7, 8). Head capsule width 0.80 mm; head scoli 0.46 mm. Very similar to second instar in color and general shape. Maximum length 8 mm. Duration 7 days (n = 1).

Fourth instar (Figs 9, 10). Head capsule width 1.20 mm; head scoli 0.72 mm. Head black, with two diverging short pointed scoli on vertex. Body green, with several short dorsal and lateral white chalazae, with oblique dark/light markings converging dorsally; legs and prolegs green; caudal filaments short. Maximum length 13 mm. Duration 7 days (n = 1).

Fifth (last) instar (Figs 11, 12). Head capsule width 1.70 mm; head scoli 0.80 mm. Head green, bordered with light brown markings, with two diverging light brown short scoli on vertex and a frontal rounded bump just below the scoli. Body green, with several short dorsal and lateral white chalazae, with oblique dark/light markings converging dorsally; legs and prolegs green; caudal filaments short, green with a reddish tip. Maximum length 21 mm. Duration 10 days (n = 1).

Pupa (Figs 13–15). Short and smooth; mostly rusty brown, with short ocular caps; cremaster broad, dark in ventral portion; dorsal abdomen with a paired series of short subdorsal light brown protuberances and a conspicuous dorsal shelf on A4. Total length 7.5 mm. Duration 8 days (n = 1).

Additional rearing data. Measurements of the immature stages from Alta Floresta (Mato Grosso, Brazil): Egg: duration 6 days (n = 2), diameter 0.66–0.68 mm, height 0.66 mm (n = 2); first instar: Duration 3 days, maximum length 3.5 mm (n = 2), head width 0.40–0.42 mm (mean = 0.41 mm, SD = 0.009), scoli 0.08–0.10 mm (mean = 0.09 mm, SD = 0.010) (n = 8); second instar: Duration 5 days, maximum length 6 mm (n = 2), head width 0.52–0.58 mm (mean = 0.56 mm, SD = 0.023), scoli 0.22–0.30 mm (mean = 0.26 mm, SD = 0.025) (n = 9); third instar: Duration 9 days, maximum length 9 mm (n = 2), head width 0.78–0.86 mm (mean = 0.81 mm, SD = 0.026), scoli 0.44–0.54 mm (mean = 0.49 mm, SD = 0.033) (n = 8); fourth instar: Duration 14 days, maximum length 15 mm (n = 3), head width 1.10–1.20 mm (mean = 1.16 mm, SD = 0.040), scoli 0.56–0.72 mm (mean = 0.63 mm, SD = 0.062) (n = 7); fifth (last) instar: Duration 17 days, maximum length 19 mm (n = 3), head width 1.50–1.74 mm (mean = 1.63 mm, SD = 0.107), scoli 0.72–0.84 mm (mean = 0.78 mm, SD = 0.050) (n = 4); pupa: general color either mostly green or rusty brown; duration 55–56 days, length 10 mm (n = 2).

Behavior and natural history. *Euptychia mollina* (Fig. 16) is very common all year round in several habitats where its host plant *Selaginella* sp. occurs (Fig. 23), including forest gaps, sand forests with open canopy, secondary vegetation and forest edges (Figs 21–23). In some of these places, *E. mollina* can be locally abundant, such as in Alta Floresta, where dozens of individuals were observed flying in a 200 m transect. The eggs were placed singly on the under surface of the fronds of its host plant (*Selaginella* sp., see above). The larvae are solitary in all instars, and its color pattern and shape make it cryptic on fronds of *Selaginella* sp. Two Braconidae parasitoids (Hymenoptera) were obtained from two field-collected larvae from Mato Grosso in 2000. Adults have a low flight near the ground, rarely flying above 2 m. Males were not observed exhibiting territorial behavior.

Figure 1–16. Life stages of *Euptychia mollina*: (1–2) egg, lateral and dorsal; (3–4) first instar, lateral and dorsal; (5–6) second instar, lateral and dorsal; (7–8) third instar, lateral and dorsal; (9–10) fourth instar, lateral and dorsal; (11–12) fifth (last) instar, lateral and dorsal; (13–15) pupa, lateral, ventral and dorsal; (16) reared adult.

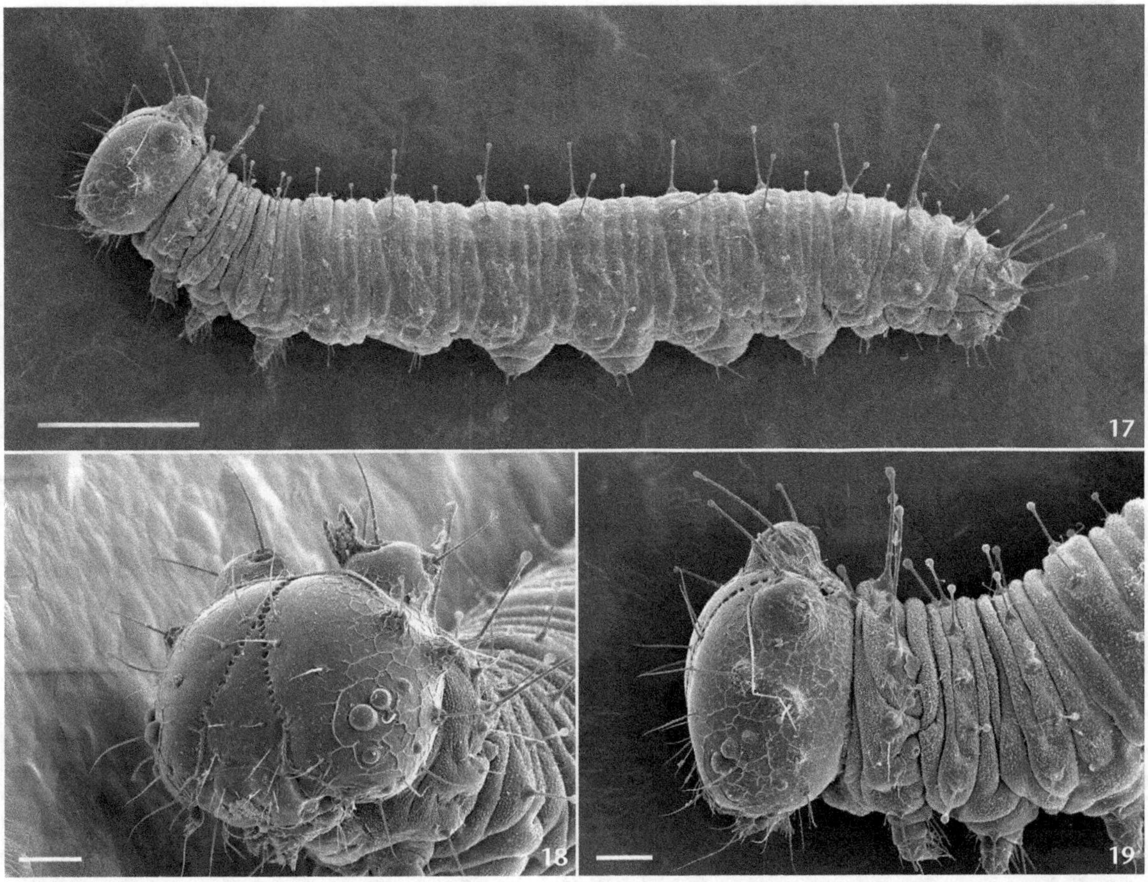

Figure 17–19. Scanning electron microscopy of first instar of *Euptychia mollina*. (17) general view; (18–19) frontal and lateral view of head, respectively. Scale bars: 17 = 0.5 mm, 18–19 = 0.1 mm.

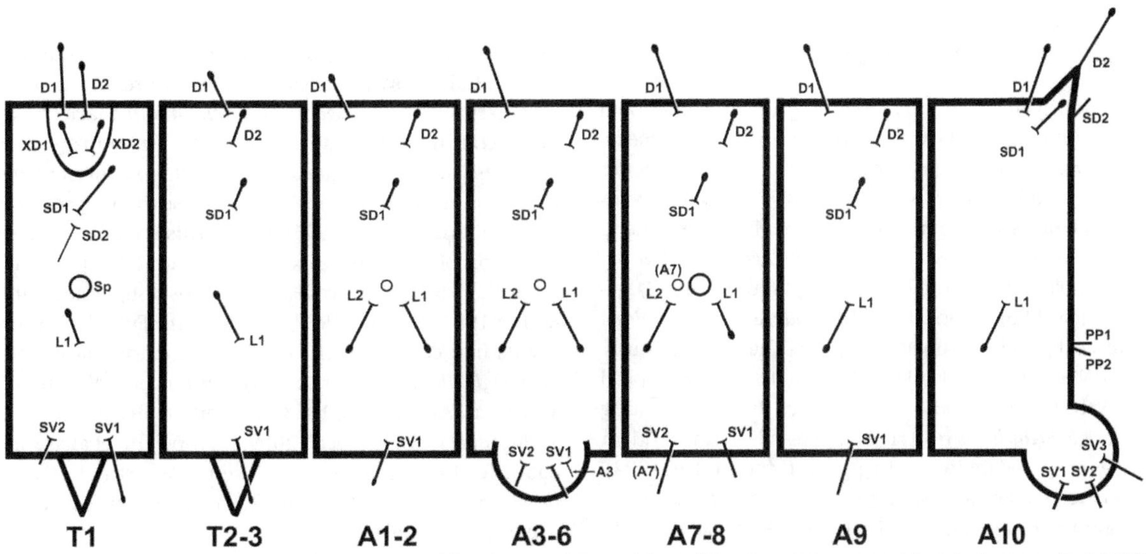

Figure 20. Body diagram showing larval chaetotaxy of first instar of *Euptychia mollina*. For chaetotaxy abbreviations see Stehr (1987) and Murray (2001).

Figure 21–23. Habitat of *Euptychia mollina* in Belterra, Pará, Brazil. (21) general view of the forest showing tall trees and an open canopy; (22) view of the trail where the ovipositing female was observed; (23) view of the larval host plant, *Selaginella* sp.

DISCUSSION

As previously mentioned, *Euptychia* species are well-known for using non-monocots as larval host plants, mainly *Selaginella*, but also *Neckeropsis* mosses (Bryophyta: Neckeraceae) (Singer et al. 1971, Singer and Mallet 1986, DeVries 1987). The use of Lycopsida as host plants is known in only two other butterfly genera, coincidently also in the tribe Satyrini, namely *Ragadia* Westwood, [1851] and *Acrophtalmia* C. and R. Felder, 1861 (Fukuda 1983, Igarashi and Fukuda 1996). Curiously, both *Ragadia* and *Acrophtalmia* are Paleotropical genera not related to each other (as previously suggested by Miller 1968 and Fukuda 1983) or to the Neotropical *Euptychia* (Wahlberg et al. 2009, Espeland et al. 2019), suggesting that the use of *Selaginella* as hostplants is a result of independent host shifts, as proposed for insects in general (Mound et al. 1994).

Nevertheless, despite the lack of a close phylogenetic relationship among *Euptychia*, *Ragadia* and *Acrophtalmia*, larvae in these three genera are surprisingly similar in shape and color pattern (Igarashi and Fukuda 1996 and this paper), although diverging in several other morphological characters. For example, while larvae of all known species of *Euptychia* present a pair of head scoli, larvae of both *Ragadia* and *Acrophtalmia* lack conspicuous head scoli (Igarashi and Fukuda 1996). The pupal shape is also quite distinct, with *Ragadia* and *Acrophtalmia* presenting pupae that are much more rounded and globose than those of *Euptychia* (Igarashi and Fukuda 1996). Given this scenario, independent evolution of camouflage explains the similar appearance of the caterpillars in all these cases, as previously suggested (Singer and Mallet 1986, DeVries 1987, Igarashi and Fukuda 1996), which results in a color pattern convergence among larvae that feed on *Selaginella*. Moreover, larvae of *Orsotriaena* Wallengren, 1858 and *Zipaetis* Hewitson, 1863, two genera phylogenetically close to *Ragadia* (Wahlberg et al. 2009, M. Espeland et al. unpublished data), feed on monocots and are quite distinct from *Ragadia* larvae (Sourakov and Emmel 2001, Kalesh and Valappil 2018, Saji and Kunte 2018).

As additional detailed descriptions of immature stages of Satyrini come to light, coupled with natural history observations, it becomes clear that camouflage is possibly one of the major strategies of larvae in this group to escape from predators (e.g. DeVries 1987, Freitas 2002, 2017, Freitas et al. 2016).

ACKNOWLEDGMENTS

Thanks to Keith Willmott for carefully revising and giving valuable suggestions on the last version of the manuscript; to Carla Penz and an anonymous reviewer for valuable comments in the submitted version of our manuscript. Keith S. Brown helped during field work in Alta Floresta in 2000. To Shinichi Nakahara for identifying adults from both rearing lots. André Rangel Nascimento identified the parasitoid. JYOC and LLM thanks Capes for a postgraduate fellowship. This study was financed in part by the Coordenação de Aperfeiçoamento de Pessoal de Nível Superior (CAPES – Finance Code 001). EPB thanks Fundação de Amparo à Pesquisa do Estado de São Paulo (FAPESP) for a post-doc fellowship (FAPESP 2016/15873-8). AVLF acknowledges support from FAPESP (Biota-Fapesp – grants 2011/50225-3), from the Conselho Nacional de Desenvolvimento Científico e Tecnológico (CNPq fellowship 302585/2011-7, 421248/2017-3), from the National Science Foundation (DEB-1256742) and from the United States Agency for International Development – USAID/the U.S. National Academy of Sciences (NAS), under the PEER program (Sponsor Grant Award Number: AID-OAA-A-11-00012) (Mapping and Conserving Butterfly Biodiversity in the Brazilian Amazon). This publication is part of the RedeLep 'Rede Nacional de Pesquisa e Conservação de Lepidópteros' SISBIOTA-Brasil/CNPq (563332/2010-7). Butterfly species are registered in the SISGEN (A4E702A).

LITERATURE CITED

DeVries PJ (1987) The butterflies of Costa Rica and their natural history. Papilionidae, Pieridae, Nymphalidae. Princeton, Princeton University Press, 327 pp.

Espeland M, Breinholt JW, Barbosa EP, Casagrande MM, Huertas B, Lamas G, Marín MA, Mielke OHH, Miller JY, Nakahara S, Tan D, Warren AD, Zacca T, Kawahara AY, Freitas AVL, Willmott KR (2019) Four hundred shades of brown: Higher level phylogeny of the problematic Euptychiina (Lepidoptera, Nymphalidae, Satyrinae) based on hybrid enrichment data. Molecular Phylogenetics and Evolution 131: 116–124. https://doi.org/10.1016/j.ympev.2018.10.039

Fratello SA, Nakahara S, Brévignon CR, Harvey DJ (2015) Two new species of *Euptychia* Hübner, 1818 (Lepidoptera: Nymphalidae: Satyrinae) from the Guiana Shield, with notes on *E. marceli* Brévignon, 2005 and *E. rufocincta* Weymer, 1911. Journal of the Lepidopterists' Society 69(4): 293–306. https://doi.org/10.18473/lepi.69i4.a5

Freitas AVL (2002) Immature stages of *Eteona tisiphone* (Nymphalidae: Satyrinae). Journal of the Lepidopterists' Society 56(4): 286–288.

Freitas AVL (2007) A new species of *Moneuptychia* Forster (Lepidoptera: Satyrinae: Euptychiina) from the highlands of Southeastern Brazil. Neotropical Entomology 36(6): 919–925. https://doi.org/10.1590/S1519-566X2007000600014

Freitas AVL (2017) Immature stages of the Neotropical satyrine butterfly *Taygetis acuta* (Nymphalidae: Euptychiina). Tropical Lepidoptera Research 27(1): 1–5.

Freitas AVL, Wahlberg N, Matos-Maravi PF, Marín MA, Mielke OHH (2012) *Euptychia boulleti* (Le Cerf) new combination (Nymphalidae: Satyrinae), a rare and endangered butterfly from southeastern Brazil. Neotropical Entomology 41: 461–467. https://doi.org/10.1007/s13744-012-0073-5.

Freitas AVL, Barbosa EP, Santos JP, Mielke OHH (2013) A new genus, *Atlanteuptychia* gen. nov., for *Euptychia ernestina* (Lepidoptera: Nymphalidae: Satyrinae). Zoologia (Curitiba) 30(6): 661–668. https://doi.org/10.1590/S1984-46702013005000005

Freitas AVL, Barbosa EP, Marín MA (2016) Immature Stages and Natural History of the Neotropical Satyrine *Pareuptychia ocirrhoe interjecta* (Nymphalidae: Euptychiina). Journal of the Lepidopterists' Society 70(4): 271–276. https://doi.org/10.18473/lepi.70i4.a4

Fukuda H (1983) Life history of two satyrid butterflies feeding on *Selaginellas*. Tyo to Ga 33: 132–144.

García-Barros E, Martín J (1995) The eggs of European satyrine butterflies (Nymphalidae): external morphology and its use in systematics. Zoological Journal of the Linnean Society 115: 73–115.

Hamm CA, Fordyce JA (2016) *Selaginella* and the satyr: *Euptychia westwoodi* (Lepidoptera: Nymphalidae) oviposition preference and larval performance. Journal of Insect Science 16(1): 39. https://doi.org/10.1093/jisesa/iew01.

Igarashi S, Fukuda H (1996) The life histories of Asian butterflies. Tokyo, Tokai University Press, 550 pp.

Janzen DH, Hallwachs W (2018) Dynamic Database for an Inventory of the Macrocaterpillar Fauna, and its Food Plants and Parasitoids, of the Area De Conservacion Guanacaste, Northwestern Costa Rica. Philadelphia, University of Pennsylvania. Available online at: http://janzen.sas.upenn.edu/caterpillars/database.lasso [Accessed: 24/10/2018]

Kalesh S, Valappil B (2018) *Zipaetis saitis* Hewitson, 1863 – Banded Catseye. In: Kunte K, Sondhi S, Roy P (Eds) Butterflies of India. Indian Foundation for Butterflies, v. 2.56. http://www.ifoundbutterflies.org/sp/951/Zipaetis-saitis [Accessed: 24/10/2018]

Lamas G (2004) Nymphalidae. Satyrinae. Subtribe Euptychiina. In: Heppner JB (Ed.) Atlas of Neotropical Lepidoptera. Gainesville, Scientific Publishers, vol. 5A, 217–223.

Marín MA, Peña C, Freitas AVL, Wahlberg N, Uribe SI (2011) From the Phylogeny of the Satyrinae Butterflies to the Systematics of Euptychiina (Lepidoptera: Nymphalidae): History, Progress and Prospects. Neotropical Entomology 40(1): 1–13. https://doi.org/10.1590/S1519-566X2011000100001

Marín MA, Peña C, Uribe SI, Freitas AVL (2017) Morphology agrees with molecular data: phylogenetic affinities of Euptychiina butterflies (Nymphalidae: Satyrinae). Systematic Entomology 42(4): 768–785. https://doi.org/10.1111/syen.12245

Miller LD (1968) The higher classification, phylogeny and zoogeography of the Satyridae (Lepidoptera). Memoirs of the American Entomological Society 24: [6] + iii + 174 pp.

Mound LA, Martin JH, Polaszek A (1994) The insect fauna of *Selaginella* (Pteridophyta: Lycopsida), with descriptions of three new species. Journal of Natural History 28: 1403–1415.

Murray DL (2001) Immature stages and biology of *Taygetis* Hübner (Lepidoptera: Nymphalidae). Proceedings of the Entomological Society of Washington 103: 932–945.

Nakahara S, Fratello SA, Harvey DJ (2014) A new species of *Euptychia* Hübner, 1818 (Lepidoptera: Nymphalidae: Satyrinae: Satyrini) from Mount Roraima, Guyana. Zootaxa 3881(1): 291–300. https://doi.org/10.11646/zootaxa.4371.1.1

Nakahara S, Hall JPW, Lamas G, Willmott KR (2015a) Seven new species and one new subspecies of *Euptychia* Hübner, 1818 (Lepidoptera: Nymphalidae: Satyrinae) from the tropical Andes. Tropical Lepidoptera Research 25: 63–79.

Nakahara S, Janzen DH, Hallwachs W, Espeland M (2015b) Description of a new genus for *Euptychia hilara* (C. Felder and R. Felder, 1867) (Lepidoptera: Nymphalidae: Satyrinae). Zootaxa 4012(2): 525–541. https://doi.org/10.11646/zootaxa.4184.2.7

Nakahara S, Le Crom J-F, Lamas G (2015c) A remarkable new *Euptychia* Hübner, 1818 (Lepidoptera: Nymphalidae: Satyrinae: Satyrini) from the Amazon basin of Peru and Colombia. Journal of Research on the Lepidoptera 48: 59–63.

Nakahara S, Llorente-Bousquets JE, Luis-Martínez A, Miller JY, Warren AD (2015d) Two new species of *Euptychia* Hübner, 1818 (Lepidoptera: Nymphalidae: Satyrinae) from Mexico and Guatemala. Journal of Research on the Lepidoptera 48: 51–57.

Nakahara S, Vega G, Willmott KR (2016) Description of a new species of *Euptychia* Hübner, 1818 (Lepidoptera: Nymphalidae: Satyrinae) from the western Andes. Zootaxa 4184: 358–366. https://doi.org/10.11646/zootaxa.4184.2.7.

Neild AFE, Nakahara S, Fratello SA, Harvey DJ (2014) A new species of *Euptychia* Hübner, 1818 (Nymphalidae: Satyrinae: Satyrini) from the Amazon basin and the Guianas. Tropical Lepidoptera Research 24: 4–9.

Neild AFE, Nakahara S, Zacca T, Fratello S, Lamas G, Le Crom J-F, Dolibaina DR, Dias FMS, Casagrande MM, Mielke OHH, Espeland M (2015) Two new species of *Euptychia* Hübner, 1818 from the upper Amazon basin (Lepidoptera, Nymphalidae, Satyrinae). ZooKeys 541: 87–108. https://doi.org/10.3897/zookeys.541.6297.

Peña C, Nylin S, Freitas AVL, Wahlberg N (2010) Biogeographic history of the butterfly subtribe Euptychiina (Lepidoptera, Nymphalidae, Satyrinae). Zoologica Scripta 39: 243–258. https://doi.org/10.1111/j.1463-6409.2010.00421.x

Saji K, Kunte K (2018) *Orsotriaena medus* Fabricius, 1775. – Medus Brown. In: Kunte K, Sondhi S, Roy P (Eds) Butterflies of India. Indian Foundation for Butterflies, v. 2.56. http://www.ifoundbutterflies.org/sp/564/Orsotriaena-medus [Accessed: 24/10/2018]

Singer MC, Mallet J (1986) Moss-feeding by a satyrine butterfly. Journal of Research on the Lepidoptera 24: 392.

Singer MC, Ehrlich PR, Gilbert LE (1971) Butterfly feeding on lycopsid. Science 172: 1341–1342.

Singer MC, DeVries PJ, Ehrlich PR (1983) The *Cissia confusa* species-group in Costa Rica and Trinidad (Lepidoptera: Satyrinae). Zoological Journal of the Linnean Society 79: 101–119.

Sourakov A, Emmel TC (2001) Notes on life histories of Satyrinae in the Solomon Islands (Lepidoptera: Nymphalidae). Tropical Lepidoptera 12 (Suppl. 1): 7–12.

Stehr FW (1987) Order Lepidoptera. In: Stehr FW (Ed.) Immature insects. Dubuque, Kendall/Hunt, 288–305.

Wahlberg N, Leneveu J, Kodandaramaiah U, Peña C, Nylin S, Freitas AVL, Brower AVZ (2009) Nymphalid butterflies diversify following near demise at the cretaceous/tertiary boundary. Proceedings of the Royal Society of London B 276: 4295–4302. https://doi.org/10.1098/rspb.2009.1303.

Warren AD, Davis KJ, Stangeland EM, Pelham JP, Grishin NV (2018) Illustrated Lists of American Butterflies. Available online at: http://www.butterfliesofamerica.com [Accessed: 24/10/2018]

Zacca T, Casagrande MM, Mielke OHH, Huertas B, Barbosa EP, Freitas AVL, Magaldi LM, Espeland M, Nakahara S, Willmott KR (2018) Systematics of the butterfly genus *Cissia* Doubleday, 1848 (Lepidoptera: Nymphalidae, Satyrinae) using an integrative approach. Arthropod Systematics and Phylogeny 76(2): 349–376.

Author Contributions: AVLF conceived the study and reared the immatures, EPB prepared the MEV material and chaetotaxy schemes; EPB and LLM prepared the pictures, AVLF and JYOC collected adults and immature stages in the field, AVLF, LLM, EPB and JYOC wrote the paper.
Competing Interests: The authors have declared that no competing interests exist.

23

Xenochlora meridionalis sp. nov. (Hymenoptera: Apidae), a new halictine bee from eastern Brazil as evidence of past connections between Amazonia and Atlantic Forest

Gabriel A.R. Melo[1], **Luiz R.R. Faria**[2], **Leandro M. Santos**[3]

[1]*Laboratório de Biologia Comparada de Hymenoptera, Departamento de Zoologia, Universidade Federal do Paraná. Caixa Postal 19020, 81531-980 Curitiba, PR, Brazil.*
[2]*Instituto Latino-Americano de Ciências da Vida e da Natureza, Universidade Federal da Integração Latino-Americana. Avenida Tarquínio Joslin dos Santos 1000, Jardim Universitário, 85870-901 Foz do Iguaçu, PR, Brazil.*
[3]*Instituto de Formación Docente Salomé Ureña (ISFOODSU). Santo Domingo, República Dominicana.*
Corresponding author. Gabriel A.R. Melo (garmelo@ufpr.br)

ABSTRACT. We describe a new species of the bee genus *Xenochlora* Engel, Brooks & Yanega, 1997, *X. meridionalis* sp. nov., based on a single female collected in the coastal forests of southeastern Brazil, in the state of Espírito Santo. The disjunct distribution exhibited by *Xenochlora*, with species in northern South America and in the Atlantic Forest of eastern Brazil, is discussed in light of current knowledge about other taxa with similar distribution pattern.

KEY WORDS. Augochlorini, Halictidae, Neotropical, taxonomy, systematics.

INTRODUCTION

Xenochlora Engel, Brooks & Yanega, 1997 is a rare genus of halictine bees until recently assumed to be restricted to the Amazon Basin and the Guiana shield (Engel et al. 1997, Santos and Melo 2013). The genus comprises four species: *X. ianthina* (Smith, 1861), *X. nigrofemorata* (Smith, 1879), *X. chalkeos* Engel, Brooks & Yanega, 1997 and *X. ochrosterna* Engel, Brooks & Yanega, 1997. Specimens of *Xenochlora* are rarely collected, and the entire genus is known only from a few females and a single male of *X. nigrofemorata* (Santos and Melo 2013). Nests of *Xenochlora* are built by females in dead wood, notably small dead branches hanging on the understory vegetation, and can be inhabited by single or multiple females (Smith 1861, Tierney et al. 2008). Multifemale nests are regarded to exhibit parasocial behavior, and a temporal switch between semisocial, eusocial and possibly communal phases was also proposed to occur within the genus (Tierney et al. 2008).

Within the Augochlorini, *Xenochlora* has been considered as closely related to *Megalopta* Smith, 1853 (Engel et al. 1997, Engel 2000, Tierney et al. 2012, Gonçalves 2016). In a recent molecular and morphological phylogenetic analysis of the extant Augochlorini, *Megalopta* was recovered as a paraphyletic group, as *Megalopta atra* Engel, 2006 appears in a clade with *Xenochlora*, forming the sister group of the other species of *Megalopta* (Gonçalves 2016). Reconstructing the phylogenetic history of *Xenochlora*, *Megalopta*, *Megaloptidia* Cockerell, 1900, *Megommation* Moure, 1943 and related genera is essential to understand the evolution of dim-light foraging in the tribe Augochlorini, which yet is possible to be interpreted in different ways (see Tierney et al. 2012, Gonçalves 2016).

In 2013, during fieldwork of a study on orchid bees at Reserva Natural Vale (Coswosk et al. 2018), one female of *Xenochlora* was unexpectedly caught in the scent traps. A preliminary comparison with available material suggested that it belonged to an undescribed species. More recently, Nemésio et al. (2016) recorded the genus in the same locality, Reserva Natural Vale, from the study of a female previously deposited in the Coleção Entomológica da Reserva Natural Vale (CERNV). They even considered the possibility of this individual to belong to a new undescribed species.

We herein describe a new species of *Xenochlora* from a large remnant of Atlantic Forest, collected in 2013. This is the first species occurring outside of Amazonian Formations at northern South America.

MATERIAL AND METHODS

The holotype was collected in the Reserva Natural Vale, a large remnant of Brazilian Atlantic forest, encompassing ca.

22,700 ha in northern Espírito Santo state, southeastern Brazil (Lopes and Mello-Silva 2014). This reserve and Reserva Biológica Sooretama (ca. 24,000 ha) are recognized as the last Tabuleiro Forest massif in northern Espírito Santo (Amorim 1984, Silva 2014). Tabuleiro forests (Floresta Ombrófila Densa de Terras Baixas *sensu* Veloso et al. 1991) have distinctive physiognomic and floristic characteristics (e.g., Peixoto et al. 2008) and share many plant taxa with the Amazon Forest, including several species with disjunct distribution between Amazon and Atlantic forests (reviewed in Rolim et al. 2016).

The terminology for the external morphology follows Eickwort (1969), Engel (2000) and Michener (2007), except for the scutum, referred to as mesoscutum, and the "basal area of propodeum", here the metapostnotum, following Brothers (1976). The following abbreviations are used: F1–F11 for the flagellomeres; T1–T6, metasomal terga; and S1–S8, metasomal sterna. The punctation density is described in relation to the interspaces between punctures, measured in terms of puncture diameter (pd). The information provided between quotation marks in the type material section is an exact transcription of the label associated with the specimen; the inverted bars (\) indicate different lines in the same label. The holotype is deposited in the Coleção Entomológica Padre Jesus Santiago Moure, Departamento de Zoologia, Universidade Federal do Paraná, Curitiba, Brazil (DZUP).

TAXONOMY

Xenochlora meridionalis sp. nov.

Figs 1–7

Diagnosis. The new species resembles *X. nigrofemorata* and *X. ianthina* but differs from them in having a greenish brassy upper frons and vertex; an upper gena without metallic reflexes; wings covered with uniformly brown and denser pilosity; plumose pilosity on parocular area and sides of mesoscutum less ramified; punctation on lower half of gena, adjacent to eye orbit, finer and sparser; shallow concavity of upper gena with striation relatively weak and restricted to portion adjacent to vertex and not as developed as striation on anterior corners of mesoscutum; central portion of mesoscutal disc, in the area devoid of metallic luster, with relatively finer and sparser punctation; and disc of T2 with denser punctation, punctures separated by 1–2 pd and about as coarse as those on marginal zone.

Description. Holotype female. Approximate body length: 9.0 mm; maximum head width: 3.15 mm; intertegular distance: 2.25 mm; forewing length, including tegula: 8.4 mm. Integumental Color. Mostly pale ferruginous, with greenish golden reflexes on clypeus, lower frons, gena, and mesosoma; upper frons, vertex, anterior one-fourth, lateral thirds and a narrow band along posterior margin of mesoscutum metallic greenish brassy to bronze. Antennae, tibiae and basitarsi of mid and hindlegs mostly ferruginous to light reddish-brown; apex of mandibles black. Wing membrane hyaline; veins and pterostigma pale reddish-brown to brown. Terga mostly ferruginous, with darker marginal zones, edge of T1–T4 reddish-brown. Pubescence. Mostly pale yellow; upper frons, vertex and most of mesoscutum with reddish-brown to brown pubescence; long erect setae on scutellum and metanotum mostly ferruginous to reddish-brown. Posterior surface and lower outer surface of midtibia, most of outer surface and basal third of inner surface of midbasitarsus, apex of femur, entire tibia except for reddish-brown plumose setae on basal half of its anterior margin, and most of basitarsus of hind leg with conspicuous dark brown to black setae; setae on inner surface of fore tibia, tarsi of fore and mid legs, outer apex and most of inner surface of basitarsus and entire distitarsus of hind leg mostly bright reddish-brown; mesotibial comb mostly reddish-brown with a few black simple setae on posterior basal edge. Wing microtrichia uniformly dark brown to black, relatively sparse on basal two-thirds of wings, distance between them about as long as microtrichium length. Short plumose pubescence on lower parocular area and lateral portions of disc of mesoscutum distinctly less plumose than that on anterior corner of mesoscutum. Tergal pilosity mostly dark brown to black, including that on lateroventral portions of sclerites; erect pilosity on anterior vertical portion of T1 and pilosity bordering medial cleft of T5 pale orange; lateroventral portions of terga with some orange erect setae; S1–S4 with pale yellow pubescence; S5 with mostly orange setae and some intermingled brown setae; tuft on apex of S6 orange. Integumental surface. Punctation on lower half of gena, adjacent to eye orbit, very fine and relatively sparse, punctures separated by 2–4 pd; shallow concavity of upper gena with coarser and denser punctation, its striation relatively weak and restricted to portion adjacent to vertex and not as developed as striation on anterior corners of mesoscutum; longitudinal striation radiating from posterior mandibular condyle well developed and distinctly stronger than that on concavity of upper gena; central portion of mesoscutal disc, in the area devoid of metallic luster, with relatively fine and sparse punctation, punctures separated by 3–5 pd; carinae on basal area of metapostnotum well-marked, medial carina strongest and lateral ones becoming progressively weaker; punctation on disc of T2 relatively dense, punctures separated by 1–2 pd and about as coarse as those on marginal zone.

Structure and measurements (in mm). Head slightly wider than long (3.17:2.91); labral elevation relatively broad, about 2.5x wider than long (0.35:0.14); clypeus about 3.7x wider than long (2.34:0.63); length of subantennal suture about 1.4x interalveolar distance (0.54:0.38) and slightly shorter than alveolo-orbital distance (0.54:0.60); length of scape, including radicle, shorter than distance between clypeus and mid ocellus (1.42:1.55); distance between posterior ocelli subequal to ocello-orbital distance (0.44:0.47); genal spine well developed and pointed (Fig. 5). Lateral corner of pronotal collar forming a slightly acute angle (Fig. 6); length of mesoscutum about 0.75x

Figures 1–4. *Xenochlora meridionalis* sp. nov., female holotype: (1) Habitus, lateral view; (2) Habitus, dorsal view; (3) Head, frontal view; (4) Head and mesosoma, posterodorsal view. Scale bars: 1–2 = 2 mm, 3–4 = 1 mm.

the intertegular distance (1.68:2.22); length of scutellum, along midline, about 1.7x the metanotum length (0.63:0.38); distal hamuli arranged as 3-1-1-1-1-3; inner spur of hind tibia with five teeth in addition to apex.

Type material. Holotype female, "BRA: *ES*, Linhares Reserva\Vale\23.24.iii.2013\2000m/replica 3/vanilina\L.R.R. Faria e eq. col." (DZUP).

Etymology. The new species is named for being the southernmost taxon in the genus *Xenochlora*, from the Latin *meridionalis*, meaning "southern".

DISCUSSION

The occurrence of *Xenochlora meridionalis* sp. nov. in the Atlantic forest supports the already well-established pattern of disjunct distributions between this forest formation of eastern Brazil and the Amazon basin for some plants and animals (e.g., Vanzolini and Williams 1970, Costa 2003, Martini et al. 2007, Fiaschi and Pirani 2009, Salles et al. 2010, Tonini et al. 2010, Batalha-Filho et al. 2013). Past connections between Amazon and Atlantic forests through central areas of South America are highly supported by independent evidence (e.g., Auler et al. 2004, Wang et al. 2004, Sobral-Souza et al. 2015).

In addition to the distributional pattern of *Xenochlora*, other bees also provide good examples of disjunct taxa where the Atlantic morphs closely resemble their Amazonian counterparts. Orchid bees (Apidae, Euglossini) deserve special attention in this scenario when taking into consideration the following reasons (i) they are closely related to forested areas (e.g., Dressler 1982), (ii) they are well sampled when compared to other groups of bees, mainly in the Atlantic forest (e.g., Sydney et al. 2010), and (iii) there have been recent revisions at species level for the fauna (e.g., Nemésio 2009, Faria and Melo 2012). Some examples where a species inhabits the Amazon Basin and its putative

Figures 5–7. *Xenochlora meridionalis* sp. nov., female holotype: (5) Head, lateral view; (6) Mesosoma, dorsal view; (7) Metasoma, dorsal view. Scale bars: 5 = 1 mm, 6–7 = 2 mm.

sibling is found in the Atlantic forest include (i) *Euglossa parvula* Dressler, 1982 and *Eg. pepei* Nemésio & Engel, 2012, (ii) *Eufriesea ornata* (Mocsáry, 1896) and *Ef. atlantica* Nemésio, 2008, and (iii) *Exaerete trochanterica* (Friese, 1900) and *Ex. salsai* Nemésio, 2011. It is important to highlight that phylogenetic analyses and divergence times of these pairs of species are not available, so their closer relationship is hypothesized based only in the taxonomic interpretations of the authors. But phylogenetically supported evidence also exists. When studying the phylogeography of some species of *Eulaema* Lepeletier, 1841, López-Uribe et al. (2014) found that the individuals from Atlantic forest are closely related to their counterparts in the Amazon Basin from both species complexes involving respectively *Eulaema meriana* (Olivier, 1789) and *Eulaema bombiformis* (Packard, 1869). They also found that the separation between the Amazonian and Atlantic forms are somewhat recent, ca. 1.0–2.0 Mya in both species complexes.

Nonetheless, unlike the examples presented above, the distribution patterns of some other bee taxa raise the following question: why some Amazonian groups have never crossed the forest bridges? It is remarkable in this context of faunal exchange that some particular genera of stingless bees as *Aparatrigona* Moure, 1951, *Duckeola* Moure, 1944, *Nogueirapis* Moure, 1953 and *Ptilotrigona* Moure, 1951, cleptoparasites as *Aglae* Lepeletier & Serville, 1825 and *Aglaomelissa* Snelling & Brooks, 1985 and other solitary bees as *Xylocopa* (*Xylocopina*) Hurd & Moure, 1963 have not been found in the Atlantic forest. Finding an answer to this question is certainly not a simple task, but the bionomics and phylogeny of taxa with disjunct or restricted distribution, both in Amazon basin or Atlantic forest, may shed light on the relation between past environmental filters and the functional traits of species with these distributional patterns. In this sense, the pattern exhibited by the stingless bees, whose newly founded nests depend of a nearby mother nest during their early beginning, suggests that the bridges between the Atlantic and Amazonian forests probably formed a fragmented landscape and not broadly continuous forested areas.

The southeastern-northwestern bridge (SE–NW; Bigarella et al. 1975) was proposed based on the distribution of taxa that inhabit western Amazonia and southeastern Atlantic Forest but not northeastern areas of this latter formation (e.g., Costa 2003) and have been regarded as the most important route between these forested biomes (Por 1992). This route seems to be the most ancient connection between southeastern Atlantic and Amazon forests and was established since the Andean uplift and the formation of the South American monsoon system (Ledo and Colli 2017). Batalha-Filho et al. (2013) presented data supporting two distinct pathways connecting both biomes, (i) more ancient connections (middle to late Miocene) through the current southern Cerrado and Mato Grosso and the transition towards the Chaco and palm savannas of Bolivia and Paraguay, and (ii) more recent connections (Pliocene and Pleistocene) through drier formations (Cerrado and Caatinga) in northeastern Brazil. Molecular dating of splits between sister lineages (one in Amazon and one in Atlantic forest) suggests the events related to the SE-NW bridge are more frequent and older than those related to the NE route (Ledo and Colli 2017).

Xenochlora meridionalis sp. nov. brings a new piece to the Amazon-Atlantic forest connection puzzle, and we hypothesize that *Xenochlora* reached the Atlantic forest through the SE–NW bridge, since so far it has been collected only in eastern Brazil. Considering the recent discovery of *Aglae* in forest fragments in Goiás, central Brazil (Silva et al. 2015), it is possible that more intensive sampling in this region may reveal the southernmost records for Amazonian species of *Xenochlora*. Furthermore, the general morphological similarity between *Xenochlora meridionalis* sp. nov. and their Amazon counterparts (mainly *X. nigrofemorata* and *X. ianthina*) suggests that the occurrence of this bee genus in the Atlantic Forest derives from a more recent faunal exchange, in the same way as with several pairs of species of orchid bees, dated from the late Pliocene-Pleistocene (see molecular dating of euglossine evolution in Ramírez et al. 2010 and López-Uribe et al. 2014).

ACKNOWLEDGMENTS

We acknowledge Maria Cecília Martins Kierulff for all the support during fieldwork, and Vale company/Reserva Natural Vale administration for allowing us to sample the bees of the preserve. J.A. Coswosk, E.D.G. Soares, R.A. Ferreira, A. Storari and F.F. Salles are acknowledged for all the help with fieldwork. We also thank Eduardo A.B. Almeida and Ângelo P. Pinto for reviewing the manuscript and for suggestions that helped improve it. SISBIO/ICMBio is acknowledged for the provided collecting permit (#21803). Partial support has been provided by the Conselho Nacional de Desenvolvimento Científico e Tecnológico (CNPq 309641/2016-0) to GARM. All authors revised the text and are solely responsible for accuracy in use of the English language.

LITERATURE CITED

Amorim HB (1984) Florestas nativas dos estados do Rio de Janeiro e Espírito Santo. Instituto Brasileiro do Desenvolvimento Florestal-IBDF, Brasilia, 204 pp.

Auler AS, Wang A, Edwards RL, Cheng H, Cristalli PS, Smart ML, Richards DA (2004) Quaternary ecological and geomorphic changes associated with rainfall events in presently semi-arid northeastern Brazil. Journal of Quaternary Science 19(7): 693–701. https://doi.org/10.1002/jqs.876

Batalha-Filho H, Fjeldså J, Fabre P-H, Miyaki CY (2013) Connections between the Atlantic and the Amazonian forest avifaunas represent distinct historical events. Journal of Ornithology 154(1): 41–50. https://doi.org/10.1007/s10336-012-0866-7

Bigarella JJ, Andrade-Lima D, Riehs PJ (1975) Considerações a respeito das mudanças paleoambientais na distribuição de algumas espécies vegetais e animais no Brasil. Anais da Academia Brasileira de Ciências 47(Supl.): 411–464.

Brothers DJ (1976) Modifications of the metapostnotum and origin of the "propodeal triangle" in Hymenoptera Aculeata. Systematic Entomology 1(3): 177–182. https://doi.org/10.1111/j.1365-3113.1976.tb00036.x

Costa LP (2003) The historical bridge between the Amazon and the Atlantic Forest of Brazil: a study of molecular phylogeography with small mammals. Journal of Biogeography 30(1): 71–86. https://doi.org/10.1046/j.1365-2699.2003.00792.x

Coswosk JA, Ferreira RA, Soares EDG, Faria LRR (2018) Responses of euglossine bees (Hymenoptera, Apidae, Euglossina) to an edge-forest gradient in a large Tabuleiro forest remnant in eastern Brazil. Neotropical Entomology 47(4): 447–456. https://doi.org/10.1007/s13744-017-0533-z

Dressler RD (1982) Biology of the orchid bees (Euglossini). Annual Reviews of Ecology and Systematics 13: 373–394. https://doi.org/10.1146/annurev.es.13.110182.002105

Eickwort GC (1969) A comparative morphological study and generic revision of the augochlorine bees (Hymenoptera: Halictidae). University of Kansas Science Bulletin 48: 325–524.

Engel MS (2000) Classification of the bee tribe Augochlorini (Hymenoptera: Halictidae). Bulletin of the American Museum of Natural History 250: 1–89. http://hdl.handle.net/2246/1598

Engel MS (2006) A new nocturnal bee of the genus *Megalopta*, with notes on other Central American species (Hymenoptera: Halictidae). Mitteilungen des Internationalen Entomologischen Vereins 31: 37–49.

Engel MS, Brooks RW, Yanega D (1997) New genera and subgenera of augochlorine bees (Hymenoptera: Halictidae). Scientific Papers of the Natural History Museum of the University of Kansas 5: 1–21. https://doi.org/10.5962/bhl.title.4042

Faria LRR, Melo GAR (2012) Species of *Euglossa* of the *analis* group in the Atlantic forest (Hymenoptera, Apidae). Zoologia 29(4): 349–374. https://doi.org/10.1590/S1984-46702012000400008

Fiaschi P, Pirani JR (2009) Review of plant biogeographic studies in Brazil. Journal of Systematics and Evolution 47(5): 477–496. https://doi.org/10.1111/j.1759-6831.2009.00046.x

Gonçalves RB (2016) A molecular and morphological phylogeny of the extant Augochlorini (Hymenoptera, Apoidea) with comments on implications for biogeography. Systematic Entomology 41(2): 430–440. https://doi.org/10.1111/syen.12166

Ledo RMD, Colli GR (2017) The historical connections between the Amazon and the Atlantic Forest revisited. Journal of Biogeography 44(11): 2551–2563. https://doi.org/10.1111/jbi.13049

Lopes JC, de Mello-Silva R (2014) Annonaceae da Reserva Natural Vale, Linhares, Espírito Santo. Rodriguésia 65(3): 599–635. https://doi.org/10.1590/2175-7860201465304

López-Uribe MM, Zamudio KR, Cardoso CF, Danforth BN (2014) Climate, physiological tolerance and sex-biased dispersal shape genetic structure of Neotropical orchid bees. Molecular Ecology 23(7): 1874–1890. https://doi.org/10.1111/mec.12689

Martini AMZ, Fiaschi P, Amorim AM, Paixão JM (2007) A hot-point within a hot-spot: a high diversity site in Brazil's Atlantic Forest. Biodiversity Conservation 16(11): 3111–3128. https://doi.org/10.1007/s10531-007-9166-6

Michener CD (2007) The Bees of the World. Johns Hopkins, Baltimore, 2nd ed., 953 pp.

Nemésio A (2009) Orchid bees (Hymenoptera: Apidae) of the Brazilian Atlantic Forest. Zootaxa 2041: 1–242.

Nemésio A, Santos Júnior JE, Buys SC (2016) A fauna de abelhas e vespas apoideas (Hexapoda: Hymenoptera: Apoidea) da Reserva Natural Vale, norte do Espírito Santo. In: Rolim SG, Menezes LFT, Srbek-Araujo AC (Eds) Floresta Atlântica de Tabuleiro: diversidade e endemismos na Reserva Natural Vale. Rupestre, Belo Horizonte, 329–340.

Peixoto AL, Silva IM, Pereira OJ, Simonelli M, Jesus RM, Rolim SG (2008) Tabuleiro forests north of the Rio Doce: their representation in the Vale do Rio Doce Natural Reserve, Espírito Santo, Brazil. In: Thomas WW (Ed) The Atlantic coastal forest of northeastern Brazil. The New York Botanical Garden Press, New York, 319–350.

Por FD (1992) Sooretama: the Atlantic rain forest of Brazil. SPB Academic, The Hague, 130 pp.

Ramírez SR, Roubik DW, Skov C, Pierce NE (2010) Phylogeny, diversification patterns and historical biogeography of euglossine orchid bees (Hymenoptera: Apidae). Biological Journal of the Linnean Society 100(3): 552–572. https://doi.org/10.1111/j.1095-8312.2010.01440.x

Rolim SG, Magnago LFS, Saiter FZ, Amorim AM, Abreu KMP (2016) São as florestas do norte do Espírito Santo e sul da Bahia as mais ricas em espécies arbóreas no domínio da Floresta Atlântica? In: Rolim SG, Menezes LFT, Srbek-Araujo AC (Eds) Floresta Atlântica de Tabuleiro: diversidade e endemismos na Reserva Natural Vale. Rupestre, Belo Horizonte, 91–100.

Salles FF, Nascimento J, Massariol F, Angeli K, Barcelos-Silva P, Rúdio J, Boldrini R (2010) First survey of mayflies (Ephemeroptera, Insecta) from Espírito Santo state, southeastern Brazil. Biota Neotropica 10(1): 293–307. https://doi.org/10.1590/S1676-06032010000100025

Santos LM, Melo GAR (2013) Taxonomic notes and description of the male of *Xenochlora nigrofemorata* (Smith, 1879) (Hymenoptera: Apidae: Halictinae). Zootaxa 3670: 371–377.

Silva AG (2014) A importância da Reserva Natural Vale para a conservação das florestas tropicais nativas do Norte do Espírito Santo, Brasil. Natureza Online 12(5): 206–211. http://www.naturezaonline.com.br/natureza/conteudo/pdf/02_SilvaAG_206-211.pdf

Silva DP, Varela S, Nemésio A, de Marco P Jr (2015) Adding biotic interactions into paleodistribution models: a host-cleptoparasite complex of Neotropical orchid bees. Plos One 10: e0129890. https://doi.org/10.1371/journal.pone.0129890

Smith F (1861) Descriptions of new genera and species of exotic Hymenoptera. Journal of Entomology 1: 146–155.

Sobral-Souza T, Lima-Ribeiro MS, Solferini VN (2015) Biogeography of Neotropical Rainforests: past connections between Amazon and Atlantic Forest detected by ecological niche modeling. Evolutionary Ecology 29(5): 643–655. https://doi.org/10.1007/s10682-015-9780-9

Sydney NV, Gonçalves RB, Faria LRR (2010) Padrões espaciais na distribuição de abelhas Euglossina (Hymenoptera, Apidae) da região Neotropical. Papéis Avulsos de Zoologia 50(43): 667–679. https://doi.org/10.1590/S0031-10492010004300001

Tierney SM, Gonzales-Ojeda T, Wcislo WT (2008) Nesting biology and social behavior of *Xenochlora* bees (Hymenoptera: Halictidae: Augochlorini) from Peru. Journal of the Kansas Entomological Society 81(1): 61–72. https://doi.org/10.2317/JKES-704.24.1

Tierney SM, Sanjur O, Grajales GG, Santos LM, Bermingham E, Wcislo WT (2012) Photic niche invasions: phylogenetic history of the dim-light foraging augochlorine bees (Halictidae). Proceedings of the Royal Society of London B 279(1729): 794–803. https://doi.org/10.1098/rspb.2011.1355

Tonini JFR, Carão LM, Pinto IS, Gasparini JL, Leite YLR, Costa LP (2010) Non-volant tetrapods from Reserva Biológica de Duas Bocas, state of Espírito Santo, southeastern Brazil. Biota Neotropica 10(3): 339–351. http://dx.doi.org/10.1590/S1676-06032010000300032

Vanzolini PE, Williams EE (1970) South American Anoles: The geographic differentiation and evolution of the *Anolis chrysolepis* species group (Sauria, Iguanidae). Arquivos de Zoologia 19(3-4): 1–124. https://doi.org/10.11606/issn.2176-7793.v19i3-4p125-298

Veloso HP, Rangel ALR, Lima JCA (1991) Classificação da vegetação brasileira, adaptada a um sistema universal. IBGE-DERMA, Rio de Janeiro, 124 pp.

Wang XF, Auler AS, Edwards RL, Cheng H, Cristalli PS, Smart PL, Richards DA, Shen CC (2004) Wet periods in northeastern Brazil over the past 210 kyr linked to distant climate anomalies. Nature 432: 740–743. https://doi.org/10.1038/nature03067

Author Contributions: GARM, LRRF and LMS planned the paper; LRRF and GARM wrote the manuscript, with contributions from LMS.

Competing Interests: The authors have declared that no competing interests exist.

24

Species richness and community composition of sphingid moths (Lepidoptera: Sphingidae) along an elevational gradient in southeast Peru

Clayton A. Sublett[1], Jerry L. Cook[1], John P. Janovec[1]

[1]*Sam Houston State University Natural History Museum, Sam Houston State University, Box 2116, Huntsville, TX, USA. Corresponding author: Jerry L. Cook (jcook@shsu.edu)*

ABSTRACT. A clear low-elevation skewed unimodal richness pattern is presented for hawkmoths in Southeast Peru. Several hypotheses offer plausible explanations for such a distribution. The effects of water-energy dynamics are partially supported by a strong correlation between temperature and species richness at higher elevations. Further, hypotheses of plant diversity influences on hawkmoth ranges are supported by species richness peaking in transitional habitats. Sphingid subfamilies do not appear to be influenced by habitat type or elevational factors, such as temperature. This may make subfamily analysis a poor means of characterizing sphingid community composition unless study sites vary in the level of disturbance. This study documents 134 species in 23 genera of Sphingidae from five Southeastern Peru sites from the 7,545 specimens collected for the study.

KEY WORDS. Biodiversity, community structure, elevation, Neotropical region.

INTRODUCTION

One major challenge to studying patterns and mechanisms of biodiversity is finding gradients sufficiently large enough to observe any patterns while limiting confounding factors, such as biogeographical differences, between sites (Beck and Chey 2008). It becomes harder to study local-scale influences on diversity without accounting for the role of historical differences as the size of the study area increases (Ricklefs 2004). Mountains offer one potential solution by providing environmental gradients, such as temperature, area, and habitat, within a limited area, allowing for studies of local biodiversity processes while minimizing confounding factors. Additionally, the abundance of elevational gradients allows for high reproducibility and comparison of patterns from a variety of habitats and latitudes (McCain 2007, Beck and Kitching 2009).

One common type of elevational diversity pattern is a mid-elevation diversity peak. Historically, such a pattern was attributed to sampling artifacts (McCoy 1990), but the description of a unimodal pattern in many recent studies (Brehm et al. 2007, McCain 2007, Beck and Chey 2008, Beck and Kitching 2009) has led to its acceptance as a legitimate altitudinal diversity pattern. Several hypotheses attempting to explain mid-elevational diversity peaks are based on water-energy (Hawkins et al. 2003, McCain 2007).

Beck and Chey (2008) suggested that diversity is limited by ambient energy (for which temperature can act as a proxy) at high elevations and water availability at lower elevations. This hypothesis is adapted from findings showing a shift from temperature to water as a limiting factor to diversity as latitude decreases, and is based on an indirect, trophic cascade effect (Hawkins et al. 2003). Beck and Chey (2008) found support for energy limitation at high elevations, but not for water limitation at lower elevations, for geometrid moths in Borneo. Beck and Kitching (2009) found support for temperature, and weak support for water, as predictive factors of peak species richness for sphingid moths. McCain (2007), for bats, and Hawkins et al. (2003), for plants, vertebrates, and invertebrates, however, found that both water and energy had strong predictive capabilities for diversity patterns.

The Sphingidae are probably one of the most well-known moth families, with the fauna of Borneo (Barlow and Woiwod 1989, Schulze et al. 2001, Schulze and Fiedler 2003, Beck et al. 2006a, Beck and Kitching 2009) and Central America (Haber and Frankie 1989, Powell and Brown 1990) being particularly well studied, which allows them to be confidently used in diversity studies (Schulze et al. 2001, Amorim et al. 2009). Compared to the amount of information on sphingids from Southeast Asia, relatively little work has been done on the communities of the South American Neotropics, and the species rich altitudinal gradient between the Andes mountains and Amazonian lowlands is an ideal location for diversity studies (Myers et al. 2000, Malhi et al. 2010). Studies in a variety of taxa show this to be a particularly diverse area (e.g., Myers et al. 2000 for plants and vertebrates, Spector 2002 for beetles, Orme et al. 2005 for birds, Jackson and Flowers 2006 for aquatic insects, Ignatov et al. 2011 for moths). Southeastern Peru is an example of an area in this Andes-Amazon transitional zone, and the site of the current study. Since little work has been done on hawkmoths in Peru, the present study will both add to our understanding of Peruvian sphingid diversity and the pool of knowledge available for studies of global diversity patterns.

This study aims to 1) describe the species richness and community composition of hawkmoths along an elevational gradient in Southeastern Peru, 2) describe whether species or genera are characteristic of a particular elevational range, and 3) investigate the predictive power of temperature in explaining any observed patterns. Although not specifically testing the "water-energy hypothesis" of McCain (2007) and Beck and Chey (2008), this study addresses the correlation of temperature and species richness.

MATERIAL AND METHODS

Detailed descriptions are given for the two sites not previously reported in the literature, Vitobabma and Culebrayoc. Short descriptions of the remaining three sites, Los Amigos, Atalaya, and Wayquecha, are included, but for more detailed information see Ignatov et al. (2011). Site descriptions are given in order of elevation (Fig. 1).

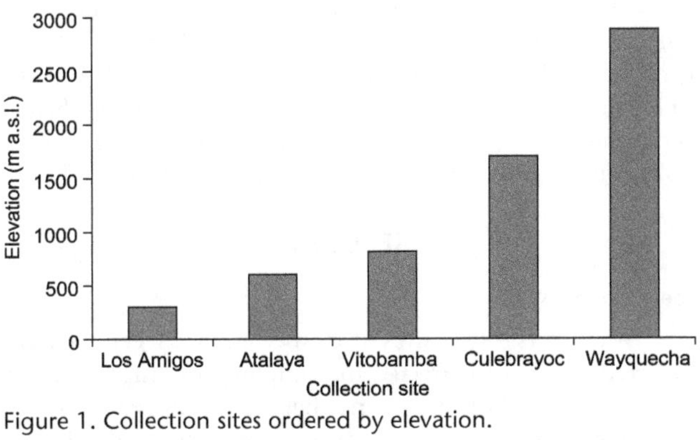

Figure 1. Collection sites ordered by elevation.

Los Amigos Biological Station is located in lowland rainforest at 300 m a.s.l. Atalaya is in tropical moist/pluvial forest at 600 m a.s.l. Wayquecha Cloud Forest Research Station is in a montane cloud forest at ~2,900 m a.s.l.

Vitobamba (13°18.596'S; 70°48.984'W; 818m a.s.l.) is a transitional forest between lowland tropical rainforest and montane rainforest. The collection site was in an area of low mountains on a former plantation that has transitioned secondary forest. The site was approximately 75 m from a paved road, which was recently under construction, and 400 m from a small river. Vegetation within the forest consisted of trees, shrubs, herbs, and lianas. The understory had patches of bamboo and semi-thick undergrowth; it was not thick enough to impede travel or obscure vision. About 50 m from the trap site, inside the forest, was a newly established vanilla plantation, about 50x30 m in size. Vegetation at the forest's edge immediately surrounding the light trap consisted of an invasive species of ginger. Collections at this site were made in 2010 from June 8th to June 15th, August 7th to August 15th, and October 3rd to October 11th. Average nightly temperatures ranged from 20.3 °C (June) to 19.0 °C (August). October is the start of the rainy season, so there was more precipitation, during the day and evening, than in June or August.

Culebrayoc (13°29.979'S; 70°53.977'W; 1,701 m a.s.l.) is a montane rainforest. The collection site was located within a low valley in the Andes Mountains. The site was approximately 50 m from a river and 100 m from the edge of primary forest. To one side of the trap was a flat, open, graveled area, which was under construction in September. Nearby was a patch of secondary forest consisting of bamboo, trees, and herbs; a small garden with coffee, corn, tomato, and a yuca relative was also near the light trap. Understory vegetation in both the secondary and primary forests tended to be thick, making it difficult to travel off trails. Collections at Culebrayoc were made in 2010 from July 7th to July 16th (except for July 13th), September 4th to September 14th (except for September 10th and 11th), and November 1st to November 9th. It was common for a strong breeze to pick up during the night, and occasionally a heavy fog was present for the first one to two hours of collecting. Average nightly temperatures ranged from 15.1 °C (July) to 16.3 °C (November), and there were no noticeable differences in precipitation among the three months.

Moths were collected nine nights each month, alternating between sites, from June to November of 2010, making three collections at each site. Collecting was conducted around the new moon, generally beginning four nights before the new moon and continuing four nights after the new moon. A 175-watt mercury-vapor bulb (Bioquip), powered by a Honda EU-1000 generator, was suspended in front of a 2 x 2 m white sheet to attract moths. Traps that emit light in the ultra violet spectra, such as those using mercury-vapor or black lights, have been shown to be more efficient at attracting sphingids than white lights (Nabli et al. 1999). Black lights are a weaker light source, and are preferable to MV bulbs when sampling needs to be restricted to a small spatial scale, such as a particular vegetation layer (Schulze et al. 2001, Axmacher and Fiedler 2004). A clear, plastic roof structure was installed at each site to protect the mercury-vapor bulb and allow collecting during precipitation. Schulze et al. (2001) found that at the Mount Kinabalu National Park (Sabah, East Malaysia) most sphingid diversity was concentrated in the canopy, with only one understory specialist. Since the goal of this study was to describe total sphingid diversity, and not just that from a single forest layer, collections were made 5m to 100m outside the tree line. In the understory, light is not able to penetrate the canopy. Collecting at the tree line allows the light to penetrate the understory and canopy, attracting moths from both vegetation layers.

Each night the light was turned on between 5:45 and 6:00 pm, and collections were made hourly from 7:00 pm to 3:00 am. Each hour all sphingids present on the sheet were collected and killed by injecting a 3:1 mixture of alcohol:ethyl acetate into the thorax. Specimens were stored in individual glassine envelopes labeled with the location, date, and time of capture. In addition, temperature was recorded every three hours from 6:00 pm till 3:00 am using a digital thermometer (RadioShak).

In the field, moths were stored in plastic containers containing silica gel and paradiclorobenzene (PDB). Upon returning to the field station, moths were dried, sorted into morphospecies, and stored in plastic bags with PDB. All species were identified to species, with the exception of a small number that consisted of distinct morphospecies in a given genus, but whose species identification could not be accurately assured. Identifications were aided by the reference collection at the San Marcos Museo de Historia Natural in Lima, Peru. Nomenclature follows Kitching and Cadiou (2000). Representatives of each morphospecies were deposited in the San Marco Museo de Historia Natural.

For all data analyses hawkmoth abundance data and temperature readings from previous collections made at the Los Amigos Biological Station, Atalaya, and Wayquecha Cloud Forest Research Station during 2004–2006 were used. Temperature data for Los Amigos was obtained from Atrium Biodiversity Information System (AABP Atrium, http://atrium.andesamazon.org), while data for Atalaya and Wayquecha was obtained from John P. Janovec (unpublished data). The procedure described in Ignatov et al. (2011) was used in order to facilitate the incorporation of their data into these analyses. Wei et al. (2010) and Walther and Morand (1998) determined that the performance of species richness estimators varies with sampling intensity, so standardized sampling intensity was used by only using June, August, and October data from Los Amigos and Atalaya and July, September, and November data from Wayquecha. This allowed comparisons to the three alternating months of sampling at each site during the same June-November sampling window.

It is well known that obtaining a complete inventory of any species rich taxonomic group is exceedingly difficult (Price et al. 1995), so observed species richness is a poor estimator of actual species richness (Wei et al. 2010). Non-parametric estimators and extrapolating species accumulation curves provide more reliable alternatives to estimating species richness than the observed number of species (Walther and Morand 1998, Wei et al. 2010). Three non-parametric, abundance-based estimators, Chao1, Jackknife1 (Jack1), and abundance-based coverage estimator (ACE), were used to estimate total species richness. Since each estimator gives a slightly different estimation of species richness, the average of the three estimates was used for all analyses. Chao1 and Jack1 use the number of singletons and doubletons (species with an abundance of one and two respectively) to estimate the total number of observed and unobserved species (Palmer 1991, Chao et al. 2005). ACE functions in a similar manner; it uses the frequency of rare species (those with ten or fewer individuals) to estimate species richness (Chao et al. 2005). To evaluate differences in alpha diversity between the sites, Fisher's alpha was calculated. EstimateS 8.2.0 (R.K. Colwell, http://viceroy.eeb.uconn.edu/estimates) was used for all calculations of species richness and diversity. A linear regression was used to test for any relationship between temperature and estimated species richness or temperature and Fisher's alpha. For data that shows a break in the trend, as is seen in unimodal humps, a single linear regression will not give any significant results. Breakpoint regression allows for the regression before and after the breakpoint to be calculated separately (Beck and Chey 2008). Therefore, breakpoint regression was used to analyze the relationship between (log)elevation and estimated species richness following Beck and Chey (2008). For all regression analyses, r^2 and p-values were calculated using Minitab.

The percent contribution of each subfamily to species richness and total abundance at each site was calculated and used to analyze community composition and structure. Species were classified using the method of Beck and Kitching (2009), who determined species' ranges by assuming they occupied all areas between their highest and lowest altitudinal record. Species were classified as 1) cosmopolitan (those collected at least at Los Amigos and Wayquecha, and so assumed to occupy the entire gradient), 2) lowland (those only collected at the three lowest sites), 3) montane (those where at least 80% of the specimens were collected at the two highest sites), or 4) endemic (those only collected at a single site). Endemics were further divided into 1) regular endemics (those with more than three individuals), and 2) rare endemics (those with fewer than three individuals). These classifications, from Ignatov et al. (2011), were used to make comparisons between the two data sets easier.

RESULTS

Species richness

Among the five sites, 134 species in 23 genera were collected (data not corrected for sampling intensity). Table 1 provides a full list of species and their abundances at each site and Table 2 gives a list of observed richness, estimated richness, and Fisher's alpha for each site. The most speciose site was Vitobamba, followed by Atalaya, Los Amigos, and Culebrayoc; Wayquecha was the least speciose. A clear unimodal pattern is present, though it is skewed toward the lower elevations (Fig. 2). Species richness decreases faster at the lower elevations than higher elevations. This is evidenced by the fact that Los Amigos and Culebrayoc have nearly equal estimated species richness values, but Culebrayoc is 900 m up slope from Vitobamba

Table 1. Species abundances at each site. Data for Los Amigos, Atalaya, and Wayquecha is not corrected for sampling intensity.

No	Species	Los Amigos	Atalaya	Vitobamba	Culebrayoc	Wayquecha
1	*Adhemarius dentoni* (Clark, 1916)	0	2	0	0	0
2	*Adhemarius gagarini* (Zikan, 1935)	1	0	0	0	0
3	*Adhemarius palmeri* (Boisduval, 1870)	43	15	17	0	0
4	*Adhemarius sexoculata* (Grote, 1865)	0	0	0	25	108
5	*Adhemarius tigrina* (Felder, 1874)	0	0	16	34	0
6	*Adhemarius ypsilon* (Rothschild & Jordon, 1903)	0	1	8	0	0
7	*Adhemarius* sp. 1 (likely *A. gannascus* or *A. daphne*)	21	26	30	42	0
8	*Agrius cingulate* (Fabricius, 1775)	3	8	2	6	54
9	*Amphimoea walker* (Boisduval, 1875)	18	1	1	0	0
10	*Callionima acuta* (Rothschild & Jordon, 1910)	44	24	6	0	1
11	*Callionima denticulate* (Schaus, 1895)	0	6	5	0	0
12	*Callionima falcifera* (Gehlen, 1943)	30	4	0	0	1
13	*Callionima inuus* (Rothschild & Jordon, 1903)	27	18	1	4	0
14	*Callionima nomius* (Walker, 1856)	8	4	1	1	0
15	*Callionima pan* (Cramer, 1779)	21	6	0	0	0
16	*Callionima parce* (Fabricius, 1775)	12	28	13	7	0
17	*Cocytius antaeus* (Drury, 1773)	0	3	1	0	1
18	*Cocytius belzebuth* (Boisduval, 1875)	9	0	0	0	0
19	*Cocytius duponchel* (Poey, 1832)	121	27	36	31	11
20	*Cocytius lucifer* (Rothschild & Jordon, 1903)	7	2	19	1	0
21	*Enyo bathus* (Rothschild, 1904)	7	5	0	1	0
22	*Enyo cavifer* (Rothschild & Jordon, 1903)	0	1	0	3	0
23	*Enyo gorgon* (Cramer, 1777)	1	0	0	0	0
24	*Enyo lugubris* (Linnaeus, 1771)	14	8	3	16	40
25	*Enyo ocypete* (Linnaeus, 1758)	62	29	17	3	1
26	*Erinnyis alope* (Drury, 1773)	12	56	23	2	23
27	*Erinnyis crameri* (Schaus, 1898)	0	0	0	2	2
28	*Erinnyis ello* (Linnaeus, 1758)	43	93	8	4	217
29	*Erinnyis impunctata* Rothschild & Jordon, 1903	0	0	3	0	0
30	*Erinnyis lassauxii* (Boisduval, 1859)	3	25	0	0	0
31	*Erinnyis obscura* (Fabricius, 1775)	3	13	2	3	15
32	*Erinnyis oenotrus* (Cramer, 1780)	13	51	27	4	8
33	*Eumorpha anchemolus* (Cramer, 1779)	13	11	6	2	1
34	*Eumorpha capronnieri* (Boisduval, 1875)	8	11	16	0	0
35	*Eumorpha cissi* (Schaufuss, 1870)	0	1	2	39	0
36	*Eumorpha fasciatus* (Sulzer, 1776)	1	1	0	0	4
37	*Eumorpha labruscae* (Linnaeus, 1758)	3	0	1	1	2
38	*Eumorpha megaeacus* (Hübner, 1816)	0	1	0	0	0
39	*Eumorpha obliquus* (Rothschild & Jordon, 1903)	0	0	2	0	0
40	*Eumorpha phorbas* (Cramer, 1775)	17	33	12	0	1
41	*Eumorpha satellitia* (Linnaeus, 1771)	4	17	3	0	0
42	*Eumorpha triangulum* (Rothschild & Jordon, 1903)	2	13	81	37	0
43	*Eumorpha vitis* (Linnaeus, 1758)	5	8	0	7	0
44	*Euryglottis aper* (Walker, 1856)	0	0	0	1	18
45	*Euryglottis dognini* Rothschild, 1869	0	0	0	88	29
46	*Euryglottis guttiventris* (Rothschild & Jordon, 1903)	0	0	0	128	8
47	*Hemeroplanes ornatus* Rothschild, 1894	1	2	1	0	0
48	*Hemeroplanes triptolemus* (Cramer, 1779)	1	4	0	0	0
49	*Isognathus caricae* (Linnaeus, 1785)	0	1	0	0	0
50	*Isognathus excelsior* (Boisduval, 1875)	0	1	1	0	0
51	*Isognathus leachii* (Swainson, 1823)	59	31	0	0	2
52	*Lintneria aurigutta* (Rothschild & Jordon, 1903)	0	0	0	11	0
53	*Madoryx bubastus* (Cramer, 1777)	5	1	0	0	0
54	*Madoryx plutonius* (Hübner, 1819)	5	3	4	0	0
55	*Madoryx* sp. 1	2	0	0	0	0
56	*Madoryx* sp. 2	1	0	0	0	0
57	*Manduca albiplaga* (Walker, 1856)	19	16	13	0	0
58	*Manduca andicola* (Rothschild & Jordon, 1916)	5	1	13	1	0
59	*Manduca clarki* (Rothschild & Jordon, 1916)	2	1	0	0	0
60	*Manduca dalica* (Kirby, 1877)	5	0	7	0	0
61	*Manduca diffissa* (Butler, 1871)	15	22	14	0	0
62	*Manduca extrema* (Gehlen, 1926)	0	27	52	0	0
63	*Manduca florestan* (Stoll, 1782)	0	9	3	0	0
64	*Manduca hannibal* (Cramer, 1779)	10	3	1	0	1
65	*Manduca lamasi* Eitschberger & Haxaire, 2007	47	9	4	0	0
66	*Manduca lefeburii* (Guérin-Méneville, 1844)	0	3	8	0	0

Continues

No	Species	Los Amigos	Atalaya	Vitobamba	Culebrayoc	Wayquecha
67	*Manduca rustica* (Fabricius, 1775)	9	34	4	0	0
68	*Manduca schausi* (Clark, 1919)	0	0	11	0	1
69	*Manduca scutata* (Rothschild & Jordon, 1903)	0	3	0	0	0
70	*Manduca sexta* (Linnaeus, 1763)	12	4	0	0	0
71	*Manduca trimacula* (Rothschild & Jordon, 1903)	0	0	12	1	0
72	*Manduca* sp. 1	2	0	0	0	0
73	*Manduca* sp. 2	0	2	0	0	0
74	*Manduca* sp. 3	0	2	0	0	0
75	*Manduca* sp. 4	22	2	51	32	1
76	*Manduca* sp. 5	3	0	0	0	0
77	*Neococytius cluentius* (Cramer, 1776)	45	6	9	3	18
78	*Nyceryx coffaeae* (Walker, 1856)	8	9	1	0	0
79	*Nyceryx hyposticta* (Felder, 1874)	0	1	1	20	45
80	*Nyceryx maxwelli* (Rothschild, 1896)	0	1	1	0	0
81	*Nyceryx nictitans* (Boisduval, 1875)	0	0	4	0	0
82	*Nyceryx stuarti* (Rothschild, 1894)	15	18	0	0	0
83	*Nyceryx tacita* (Druce, 1888)	0	0	2	0	0
84	*Oryba achemenides* (Cramer, 1779)	3	3	1	0	0
85	*Oryba kadeni* (Schaufuss, 1870)	2	2	4	0	0
86	*Pachylia darceta* Druce, 1881	338	54	37	0	0
87	*Pachylia ficus* (Linnaeus, 1758)	84	28	24	9	11
88	*Pachylia syces* (Hübner, 1819)	1	2	0	0	1
89	*Pachylioides resumens* (Walker, 1856)	31	19	15	20	2
90	*Perigonia grisea* Rothschild & Jordon, 1903	0	2	1	12	0
91	*Perigonia lusca* (Fabricius, 1777)	18	6	1	3	0
92	*Perigonia stulta* Herrich-Schaffer, 1854	0	3	18	9	9
93	*Perigonia* sp. 1	0	0	0	0	1
94	*Perigonia* sp. 2	0	0	0	0	2
95	*Protambulyx astygonus* (Boisduval, 1875)	0	1	0	0	0
96	*Protambulyx eurycles* Rothschild & Jordon, 1903	22	2	5	0	0
97	*Protambulyx goeldii* Rothschild & Jordon, 1903	27	9	0	0	0
98	*Protambulyx ockendeni* Rothschild & Jordon, 1903	5	0	0	1	0
99	*Protambulyx strigilis* (Linnaeus, 1771)	99	17	33	5	0
100	*Pseudosphinx tetrio* Rothschild & Jordon, 1903	38	215	2	0	37
101	*Xylophanes amadis* (Stoll, 1872)	7	1	0	0	0
102	*Xylophanes anubus* (Crammer, 1777)	23	4	29	1	1
103	*Xylophanes chiron* (Drury, 1773)	100	61	15	2	39
104	*Xylophanes cyrene* (Druce, 1881)	0	3	0	0	0
105	*Xylophanes cosmius* Rothschild & Jordon, 1903	41	22	10	0	0
106	*Xylophanes docilis* (Butler, 1875)	0	0	19	259	4
107	*Xylophanes dolius* (Rothschild & Jordon, 1906)	18	40	28	0	0
108	*Xylophanes elara* (Druce, 1878)	6	0	1	0	0
109	*Xylophanes fassli* Gehlen, 1928	0	0	8	1	0
110	*Xylophanes fusimacula* (Felder, 1874)	0	15	30	0	0
111	*Xylophanes guianensis* (Rothschild, 1894)	5	3	0	0	1
112	*Xylophanes germen* (Schaus, 1890)	0	0	8	3	0
113	*Xylophanes hannemanni* Closs, 1917	0	17	13	0	0
114	*Xylophanes lamontagnei* Vagilia & Haxaire, 2003	0	7	32	51	0
115	*Xylophanes libya* (Druce, 1878)	45	89	61	1	0
116	*Xylophanes loelia* (Druce, 1878)	0	2	0	0	1
117	*Xylophanes mariae* Haxaire, 2013	0	0	0	39	123
118	*Xylophanes media* Rothschild & Jordon, 1903	0	14	30	0	0
119	*Xylophanes ockendeni* Rothschild, 1904	0	0	0	4	0
120	*Xylophanes pluto* (Fabricius, 1777)	4	27	2	1	0
121	*Xylophanes porcus* (Hübner, 1823)	0	2	30	0	0
122	*Xylophanes* Pyrrhus Rothschild & Jordon, 1906	0	0	1	79	9
123	*Xylophanes resta* Rothschild & Jordon, 1903	0	0	0	198	7
124	*Xylophanes rhodotus* Rothschild, 1904	0	0	0	1	0
125	*Xylophanes schausi* (Rothschild, 1894)	3	0	0	1	0
126	*Xylophanes tersa* ((Linnaeus, 1771)	2	13	31	71	17
127	*Xylophanes thyelia* (Linnaeus, 1758)	8	43	66	0	0
128	*Xylophanes titana* (Druce, 1787)	16	40	25	7	0
129	*Xylophanes undata* Rothschild & Jordon, 1903	2	43	82	0	0
130	*Xylophanes* sp. 1	0	0	1	0	0
131	*Xylophanes* sp. 2	0	0	23	0	0
132	*Xylophanes* sp. 3	0	0	99	507	36
133	*Xylophanes* sp. 4	3	1	0	0	0
134	*Xylophanes* sp. 5	0	3	21	0	0
	Total: 7,545	1,820	1,581	1,385	1,845	914

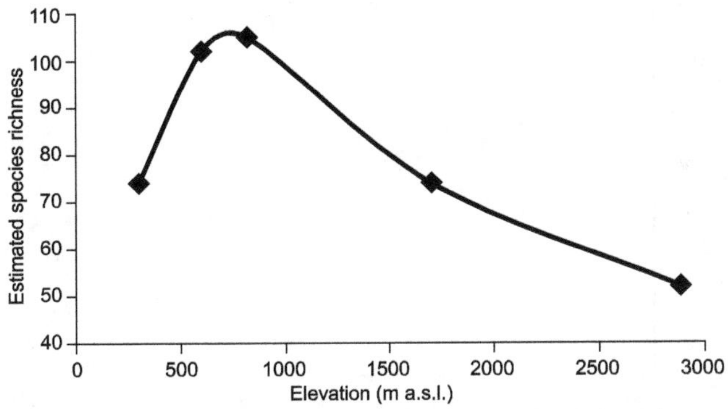

Figure 2. The relationship between elevation and species richness. There is a clear low-elevation skewed unimodal pattern, which peaks around 800 m.

Table 2. Observed and estimated species richness, observed number of genera, and Fisher's alpha for Los Amigos, Atalaya, Vitobamba, Culebrayoc, and Wayquecha. S_{obs} = observed number of species; S_{est} = estimated number of species; completeness = percentage of S_{est} collected at each site; $Genera_{obs}$ = observed number of genera.

Site	S_{obs}	S_{est}	Completeness (%)	$Genera_{obs}$	Fisher's alpha
Los Amigos	65	74	87.84	19	17.77
Atalaya	85	103	82.52	20	23.88
Vitobamba	86	105	81.90	21	20.29
Culebrayoc	56	74	75.68	17	10.90
Wayquecha	39	52	75.00	17	9.61

while Los Amigos is only 500 m down slope. Both species richness estimations and Fisher's alpha positively correlated with temperature. The relationship was weak and not significant when considered for the whole elevational range for estimated species richness ($r^2 = 0.552$, $p = 0.150$) but the relationship had a higher correlation for Fisher's alpha ($r^2 = 0.726$, $p = 0.067$) (Figs 3–6). When the relationship was only considered for the three highest elevations it was relatively highly correlated, but still non-significant, for species richness ($r^2 = 0.939$, $p = 0.159$), and not significant for Fisher's alpha ($r^2 = 0.736$, $p = 0.343$) (Fig. 7). Species richness was positively, but not significantly, correlated with (log)elevation below the breakpoint (300–818 m a.s.l.; $r^2 = 0.953$, $p = 0.139$) and was negatively, significantly correlated with elevation above the breakpoint (818–2,886 m a.s.l.; $r^2 = 0.999$, $p = 0.003$); Fisher's alpha did not significantly correlate with (log)elevation above or below the breakpoint, and was only strongly correlated with (log)elevation above the breakpoint (300–818 m a.s.l.; $r^2 = 0.357$, $p = 0.593$; 818–2,886 m a.s.l.; $r^2 = 0.901$, $p = 0.203$).

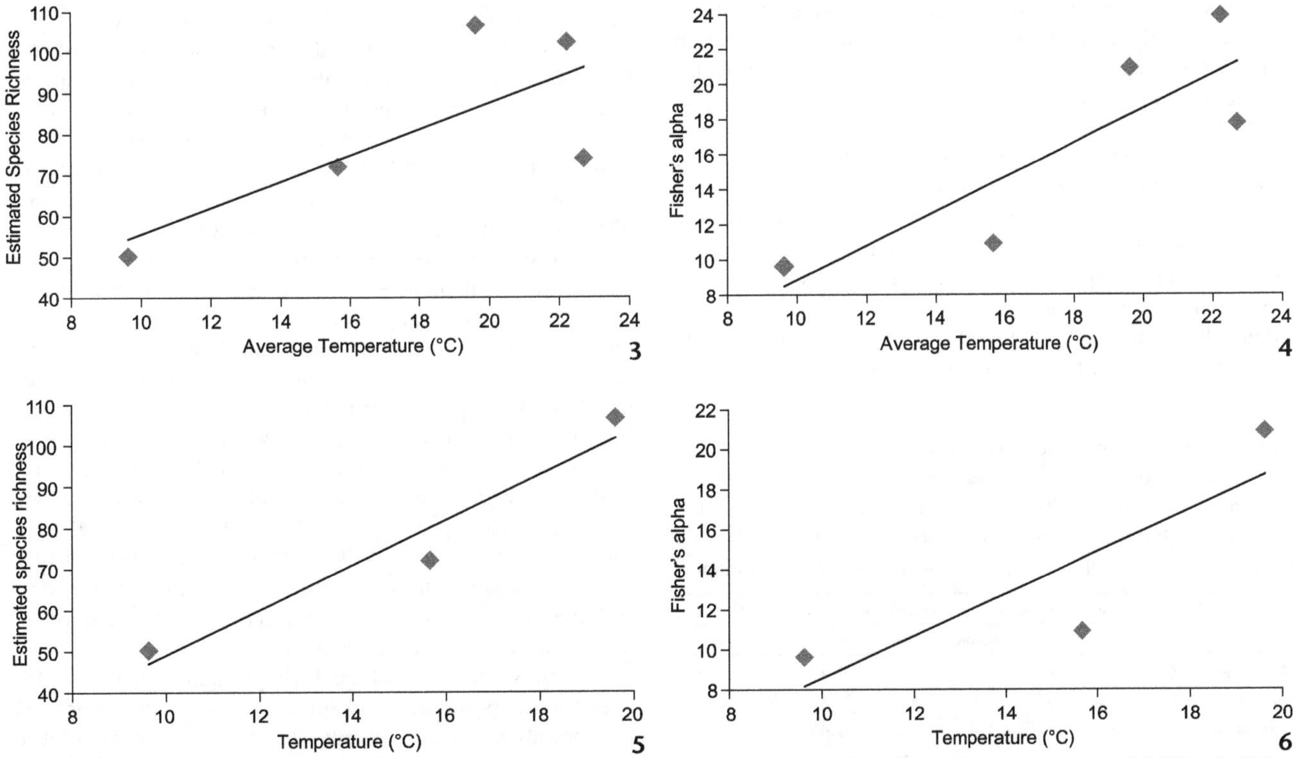

Figures 3–6. Regression analysis across the entire gradient for average temperature with (3) estimated species richness, (4) Fisher's alpha, and regression analysis at the three highest elevations for average temperature with (5) estimated species richness, and (6) Fisher's alpha. (3) $r^2 = 0.5518$, $p = 0.150$; (4) $r^2 = 0.7264$, $p = 0.067$; (5) $r^2 = 0.9391$, $p = 0.159$; (6) $r^2 = 0.7364$, $p = 0.343$.

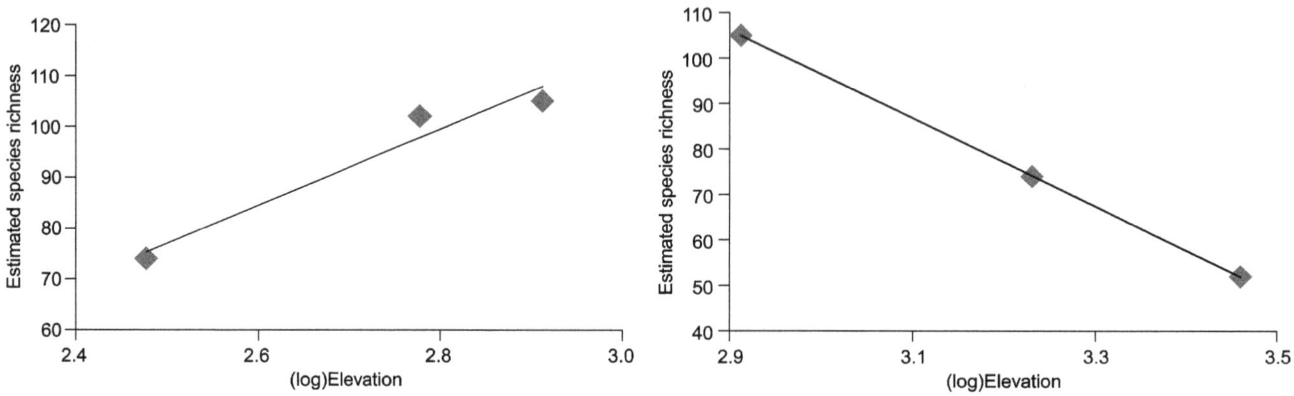

Figure 7. Breakpoint regression analysis for (log)elevation with estimated species richness. Vitobamba (800 m), which was the break point, is included in both graphs. The characteristic mid-elevation unimodal richness pattern can be seen with the increase in species richness with elevation up to Vitobamba ($r^2 = 0.953$, $p = 0.139$), and the decrease in species richness with elevation above Vitobamba ($r^2 = 0.999$, $p = 0.003$).

There were only 26 cosmopolitan species, which accounted for 21.0% of the total species richness and 34.2% of the total abundance (Table 3). Lowland species comprised 34.7% of total species richness and 23.2% of total abundance. Low land species had similar contributions to total species richness and total abundance at Los Amigos and Vitobamba. At Culebrayoc, montane species only accounted for 17.9% of the species richness but 71.9% of the total abundance, which is far greater than at either Vitobamba or Wayquecha (Table 3). This is primarily due to three species: *Xylophanes docilis* (Butler, 1875), *X. resta* Rothschild & Jordan, 1903, and *X.* sp. 3, whose combined abundances accounted for 964 of the 1,845 moths collected at Culebrayoc (Table 1). Endemic species contributed 21.0% to total species richness, but much less to total abundance (1.2%), and very little to the species richness or abundance at any one site (Table 3). Only four endemic species: *Cocytius belzebuht* (Boisduval, 1875), *Lintneria aurigutta* (Roghschild & Jordan, 1903), *Nyceryx nictitans saturate* Rothschild & Jordan, 1903, and *Xylophanes* sp. 2 were represented by more than three specimens (Table 1).

Table 3. Percent contribution of cosmopolitan, montane, lowland, and endemic species to species richness and total abundance at individual sites and all sites combined. Cosmo = cosmopolitan species; LL = lowland species; MO = montane species; endemic = endemic species; total = total species richness or total abundance.

Site	Contribution to species richness (%)				Contribution to total abundance (%)			
	Cosmo	LL	MO	Endemic	Cosmo	LL	MO	Endemic
Los Amigos	32.5	38.8	0	8.8	43.9	39.5	0	1.0
Atalaya	26.0	42.7	0	8.3	47.1	32.7	0	0.9
Vitobamba	24.4	38.4	3.5	7.0	23.1	37.6	8.6	2.5
Culebrayoc	30.4	0	17.9	5.4	11.4	0	71.9	0.9
Wayquecha	60.5	0	23.3	4.7	55.8	0	37.6	0.3
Total	21.0	34.7	8.1	21.0	34.2	23.3	23.7	1.2

Community composition

On a species level, there appears to be a division between lowland species, those collected at Los Amigos, Atalaya, and Vitobamba, and montane species, those with populations centered at Culebrayoc and Wayquecha (Table 1). Even cosmopolitan species and species found at both lowland and montane sites can be divided into two groups: those with Los Amigos-Atalaya-Vitobamba centered populations or those with Vitobamba-Culebrayoc-Wayquecha centered populations. These species had individuals extending beyond their population centers, but only in small numbers. *Protambulyx strigilis* (Linnaeus, 1771), for example, had 73 individuals collected at lowland sites but only five individuals from Culebrayoc. Few species showed uniform distributions among high and low elevations.

Most genera were cosmopolitan, with species along the entire altitudinal gradient. Some genera, however, appeared to be more restricted in range, and most of these were lowland centered. The genera *Callionima, Eumorpha, Hemeroplanes, Madoryx, Manduca, Nyceryx, Oriba*, and *Protambulyx* all contain predominantly lowland species, with high elevation representatives being rare. Three genera, *Hemeroplanes, Madoryx*, and *Oriba*, appear to be lowland endemics, with all individuals coming from Los Amigos, Atalaya or Vitobamba. Only one genus, *Euryglottis*, was completely restricted to montane forests. Individuals of all three species, *Euryglottis aper* (Walker, 1856), *E. dognini* Rothschild, 1896, and *E. guttiventris* (Rothschild & Jordan, 1903), were only collected in Culebrayoc and Wayquecha, with the majority of individuals collected in Culebrayoc.

The contributions of the three sphingid subfamilies, Macroglossinae, Sphinginae, and Smerinthinae, to species richness and total abundance were similar at all five sites (Figs 8–9). Moreover, at each site the contribution of each subfamily to species richness was similar to its contribution to abundance. There were no sites in which a subfamily made up a large proportion

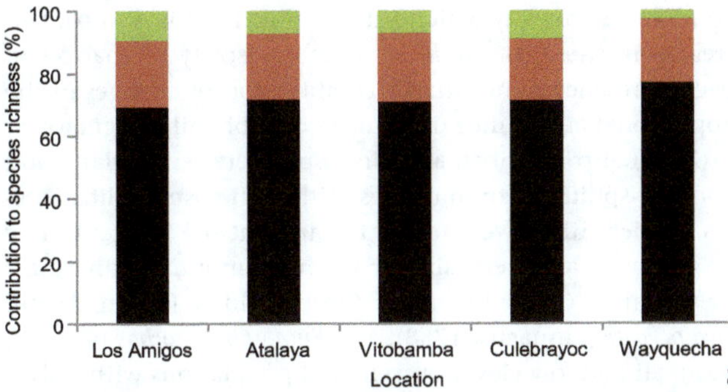

8

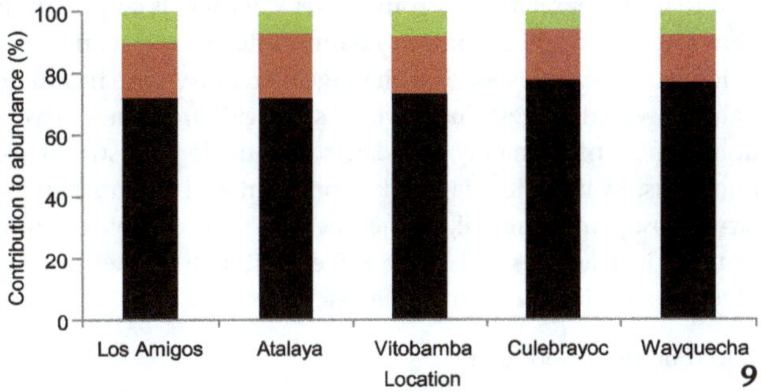

9

Figures 8–9. Contribution of sphingid subfamilies to (8) species richness and (9) total abundance. Blue bars = Macroglossinae; orange bars = Sphinginae; grey bars = Smerinthinae. Contributions of each subfamily are similar among sites, and contributions of any particular subfamily to richness and abundance are similar within a site.

of species richness but a small proportion of the abundance, or vice versa. For example, at Los Amigos Macroglossinae made up 69.2% of the species richness and 71.9% of the abundance.

DISCUSSION

Factors affecting species richness

Hawkmoth diversity did not correlate well with temperature across the entire elevational range. However, Hawkins et al. (2003) and McCain (2007) noted that with water-energy dynamics the importance of energy depended on the latitude or altitude, and Beck and Chey (2008) stated that temperature could be expected *a priori* to only be correlated with richness at higher elevations. When the temperature-species richness relationship is only considered for the three highest elevations the predictive power of temperature increases dramatically, which is consistent with other studies. Temperature should, then, only be part of what drives species richness. Longino and Colwell (2011) found that although temperature was the best predictor of elevational species richness, no single environmental variable matched the unimodal, low-elevation skewed richness pattern they observed for litter ants in Costa Rica.

Species richness peaked at ~800 m. This is consistent with the findings of other studies on altitudinal insect diversity patterns (e.g., Samson et al. 1997 for ants, Beck and Chey 2008 for geometrids, Beck and Kitching 2009 for sphingids, Longino and Colwell 2011 for litter ants). Transitional plant communities at the sites of peak richness can potentially explain a unimodal pattern. The two sites with the highest species richness (Atalaya and Vitobabma) both contain multiple plant communities. The vegetation at Atalaya is characterized by a combination of moist lowland tropical forest and pluvial submontane tropical forest, while Vitobamba is characterized by a transition from lowland to montane forest. These transitional plant communities likely contain food plants for montane and lowland forest hawkmoths. Beck et al. (2002) and Beck and Chey (2008) reported positive correlations between moth and plant species richness, though Beck and Chey only found this at higher elevations; Schulze et al. (2001) found evidence that, for some butterflies, distribution was influenced by host plant distribution. Atalaya and Vitobamba may represent the lower boundary limit for the range of some montane plants and the upper boundary limit for lowland plant species, and therefore the range limit for the moths that depend on those plants. Overlap between lowland and montane faunas could explain the mid-elevational peak in species richness (Holloway et al. 1990 referenced in Beck et al. 2006b, but see Beck and Chey 2008). To test this specifically, plant species richness and sphingid host plant data are needed for these sites.

The question remains as to what factors drive species richness. As mentioned previously, hypotheses of indirect, trophic cascade effects (host plant diversity) (Beck and Chey 2008) and a combination of direct and indirect effects (McCain 2007) have been proposed as means by which water-energy dynamics limit species richness. Addressing the determinants of species' ranges may also prove insightful to understanding diversity patterns. Longino and Colwell (2011) raise the question as to why mobile species are not constantly expanding their range. They consider a hypothesis in which the upper limits of a species' range is constrained by its thermoregulatory adaptations while lower range limits are determined by competitive interactions (references within Longino and Colwell (2011) provide more information and support for this hypothesis). Gaston and Chown (1999) showed that the thermal minimum temperature of beetles decreased strongly with elevation while the thermal maximum only decreased slightly. This suggests that high elevation species are physiologically able to inhabit lower elevations, but some other factor, such as competition, is restricting their range, further supporting the hypothesis in Longino and Colwell (2011). The extent to which this can explain sphingid ranges depends on the intensity of interspecific competition. Kitching and Cadiou (2000) describe hawkmoth larvae in the Guanacaste Conservation Area, Costa Rica as being host specific, only feeding on a specific genus or several genera within a single family. Depending on the amount of host plant overlap, this would suggest interspecific competition to be either relatively strong or weak. If there is competition between lowland and montane communities, the larger number of lowland centered species with ranges extending into montane habitats would suggest that lowland sphingids are competitively dominant. Unfortunately, this data set is unable to answer these questions. Even with detailed data on plant and moth richness, larval host plants, rainfall, and temperature, the complex nature of these interactions will likely make it difficult to identify the mechanisms driving moth species richness.

Community composition

Community composition changed between the five sites on a species but not on a subfamily level. Other researchers (Kemp et al. 1990 for grasshoppers, Boggs and Murphy 1997 for butterflies, Kumar et al. 2009 for paper wasps, Longino and Colwell 2011 for ants) have also observed similar species level community composition changes. Boggs and Murphy (1997) noted that weather conditions, such as temperature and precipitation, affect the availability of food resources and therefore the distribution of species. Kemp et al. (1990) found that habitat type largely influenced species presence, and habitat type, precipitation, and elevation were important in structuring community composition. This explanation fits with the hypothesis that vegetation type and water-energy dynamics are important determinants of species richness.

The analysis of community composition at the subfamily level shows this to be too broad a taxonomic level for describing sphingid communities at a small scale. Even at the generic level it is likely that many of the community composition changes would have been missed, since most genera are found throughout the altitudinal gradient. Analyzing data at the species level seems to be the most accurate means of describing community composition dynamics. Analysis of sphingids at the subfamily level does reveal, however, that although species, and even some genera, are restricted to particular habitats, none of the subfamilies appear to specialize in any one habitat or elevational range. Beck et al. (2006a) investigated the relative contributions of sphingid subfamilies to faunas of Southeast Asia and Malesia, and found that forest cover was a strong, though not significant, predictor of Smerinthinae and Macroglossinae. This rainforest effect was strongest for Smerinthinae, which they attributed to life history characteristics. Beck et al. (2006b) showed disturbance had a significant effect on subfamily frequency (negative for Smerinthinae and positive for Macroglossinae) while there was only a trend between elevation and subfamily frequency. These studies suggest that disturbance is the main driving force behind proportional abundances of sphingid subfamilies. The small proportion of Smerinthinae, large proportion of Macroglossinae, and similar subfamily frequencies between all five sites in this study suggests that the five sampling locations have faced comparable levels of disturbance in the past.

There appear to be two distinct moth communities: a lowland community which extends up to Vitobamba, and a montane community which extends down to Vitobamba. Differences in these communities were due solely to changes in species presence rather than a combination of changes in the proportional abundance or richness of subfamilies. Ignatov et al. (2011) also described a distinct split between lowland and montane sphingid communities. Although cosmopolitan species, by definition, were found throughout the gradient, they too followed a pattern similar to the community divisions. For example, *Callionima acuta* (Rothschild & Jordan, 1910), *Enyo ocypete* (Linnaeus, 1758), and *Eumorpha phorbas* (Cramer, 1775), all had low elevation centered populations with only a few individuals collected in the montane region, while *Enyo lugubris* (Linnaeus, 1771) showed a montane-centered population with few individuals from the lowlands. A difference in plant communities is the most intuitive explanation for this pattern. Below Vitobamba the dominant vegetation type is lowland forest, while areas above Vitobamba are dominated by montane forest. Ignatov et al. (2011) reported that many of the lowland genera feed on plants that are most abundant and diverse at lower elevations; species of *Eumorpha*, for example, specialize on "Vitaceae, represented by the genus *Cissus* in the study region, which is largely absent or less diverse at the cloud forest site [Wayquecha]." Since hawkmoths tend to be host plant specialists (Kitching and Cadiou 2000), it should be expected that these two forest types support different moth communities.

Endemic species contributed little to single site species richness, or single site abundance (Table 3). This seems contradictory to the findings of Ignatov et al. (2011), who reported a high number of endemic species (up to 23.4% at one site). This decrease is mainly a result of many of the endemics being reclassified after they were collected at a second site. For example, *Adhemarius sexoculata* (Grote, 1865), *Erinnyis crameri* (Schaus, 1898), *Euryglottis aper*, *Euryglottis dognini* Rothschild, 1896, and *Xylophanes mariae* Haxaire, 2013 were classified as endemics from Wayquecha by Ignatov et al. (2011), but were collected at Culebrayoc in this study, placing them in the montane classification. If montane ecosystems are thought to be hotspots for endemism (Meyers et al. 2000, Orme et al. 2005), why is the number of endemic species so low? Endemic as used in this paper, to describe species that were only collected at a single site, is somewhat misleading. This terminology was adopted to allow more accurate comparisons with the results from Ignatov et al. (2011). In their study on overall species richness, threatened species, and endemic species, Orme et al. (2005) defined endemic as the "25% of species with the smallest geographical breeding ranges." The narrow definition of "endemic" as used in this paper likely underestimates the true number of endemic hawkmoth species. In future papers it may be more constructive to use a more traditional definition of the term.

Sources of potential errors

Estimated species richness at Los Amigos, calculated from the corrected data, was lower than observed species richness from all 11 months of sampling. Reexamination of the raw data revealed that all the species which were lost in the correction process were only collected between November and May, the rainy season. Data used for analysis only came from the dry season. This raises the potential concern that, when data from only one season is used, non-parametric richness estimators cannot account for species turnover between seasons. Gotelli et al. (2010), using different methods from this paper, commented that few existing models account for imperfect detection, such as only sampling during the wet or dry season. This suggests that species richness was underestimated at all five sites. If it is the case that non-parametric estimators do not account for temporal turnover, researchers may face problems when trying to estimate species richness from single season samples.

One other potential problem is the small number of sample sites and the lack of replicates. Such a coarse-grained view makes it difficult to determine the exact elevation of peak species richness. Longino and Colwell (2011) faced a similar problem when studying litter ants in Costa Rica. A true mid-elevation peak (~1,400 m) seems unlikely, and in fact may not be expected (Beck and Chey 2008). A finer altitudinal resolution with replicates at some elevations will give a clearer picture of the elevation-species richness relationship, and allow for more rigorous hypothesis testing of determinants of species richness.

ACKNOWLEDGEMENTS

We thank George Diggs, Steve Goldsmith, and Peter Schulze of Austin College for providing feedback on the initial manuscript. Funding was provided by the Austin College Environmental Studies Department, the Botanical Research Institute of Texas' Andes to Amazon Biodiversity Program, and the Entomological Society of America's SysEB Student Travel Grant.

LITERATURE CITED

Amorim FW, de Avila Jr RS, de Camargo JA, Vieira AL, Oliveira PE (2009) A hawkmoth crossroads? Species richness, seasonality and biogeographical affinities of Sphingidae in a Brazilian Cerrado. Journal of Biogeography 36: 662–674. https://doi.org/10.1111/j.1365-2699.2008.02043.x

Axmacher JC, Fiedler K (2004) Manual versus automatic moths sampling at equal light sources – A comparison of catches from Mt. Kilimanjaro. Journal of the Lepidopterists' Society 58: 196–202.

Barlow HS, Woiwod IP (1989) Moth diversity of a tropical forest in Peninsular Malaysia. Journal of Tropical Ecology 5: 37–50. https://doi.org/10.1017/S0266467400003205

Beck J, Chey VK (2008) Explaining the elevation diversity pattern of geometrid moths from Borneo: a test of five hypotheses. Journal of Biogeography 35: 1452–1464. https://doi.org/10.1111/j.1365-2699.2008.01886.x

Beck J, Kitching IJ (2009) Drivers of moth species richness on tropical altitudinal gradients: a cross-regional comparison. Global Ecology and Biogeography 18: 361–371. https://doi.org/10.1111/j.1365-2699.2005.01418.x

Beck J, Kitching IJ, Linsenmair KE (2006a) Determinants of regional species richness: an empirical analysis of the number of hawkmoth species (Lepidoptera: Sphingidae) on the Malesian archipelago. Journal of Biogeography 33: 694–706. https://doi.org/10.1111/j.1365-2699.2005.01418.x

Beck J, Kitching IJ, Linsenmair KE (2006b) Effects of habitat disturbance can be subtle yet significant: biodiversity of hawkmoth-assemblages (Lepidoptera: Sphingidae) in Southeast-Asia. Biodiversity and Conservation 15: 465–486.

Beck J, Schulze CH, Linsenmair KE, Fiedler K (2002) From forest to farmland: Diversity of geometrid moths along two habitat gradients on Borneo. Journal of Tropical Ecology 18: 33–51. https://doi.org/10.1017/S026646740200202X

Boggs CL, Murphy DD (1997) Community composition in mountain ecosystems: climatic determinants of montane butterfly distributions. Global Ecology and Biogeography Letters 6: 39–48.

Brehm G, Colwell R, Kluge J (2007) The role of environment and mid-domain effect on moth species richness along a tropical elevational gradient. Global Ecology and Biogeography 16: 205–219. https://doi.org/10.1111/j.1466-8238.2006.00281.x

Chao A, Chazdon RL, Colwell RK, Shen T (2005) A new statistical approach for assessing similarity of species composition with incidence and abundance data. Ecology Letters 8: 148–159. https://doi.org/10.1111/j.1461-0248.2004.00707.x

Gaston KJ, Chown SL (1999) Elevation and climatic tolerance: a test using dung beetles. Oikos 86: 584–590.

Gotelli NJ, Dorazio RM, Ellison AM, Grossman GD (2010) Detecting temporal trends in species assemblages with bootstrapping procedures and hierarchical models. Philosophical Transactions of the Royal Society of London, Series B, Biological Sciences 356: 3621–3631.

Haber WA, Frankie GW (1989) A tropical hawkmoth community: Costa Rican dry forest Sphingidae. Biotropica 21: 155–172.

Hawkins BA, Field R, Cornell HV, Currie DJ, Gue'gan JF, Kaufman DM, Kerr JT, Mittelbach GG, Oberdorff T, O'Brien EM, Porter EE, Turner JRG (2003) Energy, water, and broad-scale geographic patterns of species richness. Ecology 84: 3105–3117. https://doi.org/10.1890/03-8006

Holloway JD, Robinson GS, Tuck KR (1990) Zonation in the Lepidoptera of northern Sulawesi. In: Knight WJ, Holloway JD (Eds) Insects and the rain forests of South East Asia (Wallaceae). London, Proceedings of the Royal Entomological Society of London, 153–156.

Ignatov II, Centeno P, Janovec JP, Tobler M.W, Grados J, Lamas G (2011) Patterns of diversity, composition, and distribution of sphingid moths along an elevational gradient in the Andes-Amazon region of southeastern Peru. Annals of the Entomological Society of America 104: 68–76. https://doi.org/10.1603/AN09083

Jackson JK, Flowers RW (2006) Measuring watershed health – Madre de Dios River basin – Appendix 6. Macroinvertebrate Diversity/Ecology. Avondale, Stroud Water Research Center, 114 pp.

Kemp WP, Harvey SJ, O'Neill KM (1990) Patterns of vegetation and grasshopper community composition. Oecologia 83: 299–308. https://doi.org/10.1007/BF00317552

Kitching IJ, Cadiou JM (2000) Hawkmoths of the world: an annotated and illustrated revisionary checklist (Lepidoptera: Sphingidae). Ithaca, Natural History Museum, London & Cornell University Press, 240 pp.

Kumar A, Longino JT, Colwell RK, O'Donnell S (2009) Elevational patterns of diversity and abundance of eusocial paper wasps (Vespidae) in Costa Rica. Biotropica 41: 338–346. https://doi.org/10.1111/j.1744-7429.2008.00477.x

Longino JT, Colwell RK (2011) Density compensation, species composition, and richness of ants on a neotropical elevational gradient. Ecosphere 2: 1–20. https://doi.org/10.1890/ES10-00200.1

Malhi Y, Silman M, Salinas N, Bush M, Meir P, Saatchi S (2010) Introduction: elevation gradients in the tropics: laboratories for ecosystem ecology and global change research. Global Change Biology 16: 3171–3175.

McCain MC (2007) Could temperature and water availability drive elevational species richness patterns? A global case study for bats. Global Ecology and Biogeography 16: 1–13. https://doi.org/10.1111/j.1365-2486.2010.02323.x

McCoy ED (1990) The distribution of insects along elevational gradients. Oikos 58: 313–322.

Myers N, Mittermeier RA, Mittermeier CG, da Fonseca GAB, Kent J (2000) Biodiversity hotspots for conservation priorities. Nature 403: 853–858.

Nabli H, Bailey WC, Necibi S (1999) Responses of Lepidoptera in Central Missouri to traps with different light sources. Journal of the Kansas Entomological Society 72: 82–90.

Orme CDL, Davies RG, Burgess M, Eigenbrod F, Pickup N, Olson VA, Webster AJ, Ding T-S, Rasmussen PC, Ridgely RS, Stattersfield AJ, Bennett PM, Blackburn TM, Gaston KJ, Owens IPF (2005) Global hotspots of species richness are not congruent with endemism or threat. Nature 436: 1016–1019.

Palmer MW (1991) Estimating species richness: the second-order jackknife reconsidered. Ecology 72: 1512–1513.

Powell JA, Brown JW (1990) Concentrations of lowland sphingid and noctuid moths at high mountain passes in eastern Mexico. Biotropica 22: 316–319.

Price PW, Diniz IR, Morais HC, Marques ESA (1995) The abundance of insect herbivore species in the tropics: the high local richness of rare species. Biotropica 27: 468–478.

Ricklefs RE (2004) A comprehensive framework for global patterns in biodiversity. Ecology Letters 7: 1–15. https://doi.org/10.1046/j.1461-0248.2003.00554.x

Samson DA, Richart EA, Gonzales PC (1997) Ant diversity and abundance along an elevational gradient in the Philippines. Biotropica 29: 349–363. https://doi.org/10.1111/j.1744-7429.1997.tb00436.x

Schulze CH, Fiedler K (2003) Hawkmoth diversity in northern Borneo does not reflect the influence of anthropogenic habitat disturbance. Ecotropica 9: 99–102.

Schulze CH, Linsenmair KE, Fiedler K (2001) Understorey versus canopy: patterns of vertical stratification and diversity among Lepidoptera in a Bornean rain forest. Plant Ecology 153: 133–152. https://doi.org/10.1023/A:1017589711553

Spector S (2002) Biogeographic crossroads as priority areas for biodiversity conservation. Conservation Biology 16: 1480–1487. https://doi.org/10.1046/j.1523-1739.2002.00573.x

Walther BA, Morand S (1998) Comparative performance of species richness estimation methods. Parasitology 116: 395–405. https://doi.org/10.1017/S0031182097002230

Wei S, Li L, Walther BA, Ye W, Huang Z, Cao H, Lian J, Wang Z, Chen Y (2010) Comparative performance of species-richness estimators using data from a subtropical forest tree community. Ecological Research 25: 93–101. https://doi.org/10.1007/s11284-009-0633-2

Author Contributions: CAS and JPJ collected specimen data in Peru. CAS and JLC provided the analysis of data and writing of the paper.

Competing Interests: The authors have declared that no competing interests exist.

PERMISSIONS

All chapters in this book were first published in ZOOLOGIA, by Pensoft Publishers; hereby published with permission under the Creative Commons Attribution License or equivalent. Every chapter published in this book has been scrutinized by our experts. Their significance has been extensively debated. The topics covered herein carry significant findings which will fuel the growth of the discipline. They may even be implemented as practical applications or may be referred to as a beginning point for another development.

The contributors of this book come from diverse backgrounds, making this book a truly international effort. This book will bring forth new frontiers with its revolutionizing research information and detailed analysis of the nascent developments around the world.

We would like to thank all the contributing authors for lending their expertise to make the book truly unique. They have played a crucial role in the development of this book. Without their invaluable contributions this book wouldn't have been possible. They have made vital efforts to compile up to date information on the varied aspects of this subject to make this book a valuable addition to the collection of many professionals and students.

This book was conceptualized with the vision of imparting up-to-date information and advanced data in this field. To ensure the same, a matchless editorial board was set up. Every individual on the board went through rigorous rounds of assessment to prove their worth. After which they invested a large part of their time researching and compiling the most relevant data for our readers.

The editorial board has been involved in producing this book since its inception. They have spent rigorous hours researching and exploring the diverse topics which have resulted in the successful publishing of this book. They have passed on their knowledge of decades through this book. To expedite this challenging task, the publisher supported the team at every step. A small team of assistant editors was also appointed to further simplify the editing procedure and attain best results for the readers.

Apart from the editorial board, the designing team has also invested a significant amount of their time in understanding the subject and creating the most relevant covers. They scrutinized every image to scout for the most suitable representation of the subject and create an appropriate cover for the book.

The publishing team has been an ardent support to the editorial, designing and production team. Their endless efforts to recruit the best for this project, has resulted in the accomplishment of this book. They are a veteran in the field of academics and their pool of knowledge is as vast as their experience in printing. Their expertise and guidance has proved useful at every step. Their uncompromising quality standards have made this book an exceptional effort. Their encouragement from time to time has been an inspiration for everyone.

The publisher and the editorial board hope that this book will prove to be a valuable piece of knowledge for researchers, students, practitioners and scholars across the globe.

LIST OF CONTRIBUTORS

Teresa C. S. Ávila-Pires
Museu Paraense Emílio Goeldi, Coordenação de Zoologia, Caixa Postal 399, 66017-970 Belém, PA Brazil

Cristiano de C. Nogueira and Marcio Martins
Departamento de Ecologia, Instituto de Biociências, Universidade de São Paulo. 05508-090 São Paulo, SP, Brazil

Mirella B. Alonso, Débora R. de Carvalho and Paulo S. Pompeu
Laboratório de Ecologia de Peixes, Setor de Ecologia, Departamento de Biologia, Universidade Federal de Lavras. Campus Universitário, 37200-000 Lavras, MG, Brazil

Carlos B. M. Alves
Laboratório Nuvelhas, Projeto Manuelzão, Universidade Federal de Minas Gerais. Avenida Antônio Carlos 6627, 31270-901 Belo Horizonte, MG, Brazil

Marcelo Z. Moreira
Laboratório de Ecologia Isotópica, Centro de Energia Nuclear na Agricultura, Universidade de São Paulo. Avenida Centenário 303, 13416-000 Piracicaba, SP, Brazil

Rodrigo Morais, Luiz Carlos de Araújo and Gleidson Ramos da Silva
Programa de Pós-graduação em Ecologia de Ecossistemas, Universidade Vila Velha. Campus Boa Vista, 29102-920 Vila Velha, ES, Brazil

Charles Duca
Laboratório de Ecologia de Populações e Conservação, Universidade Vila Velha. Campus Nossa Senhora da Penha, 29102-623 Vila Velha, ES, Brazil

José Eriberto De Assis
Departamento de Educação, Secretaria Municipal de Educação, Prefeitura Municipal de Bayeux. 50670-420 Bayeux, PB, Brazil

José Roberto B. de Souza, Manuela M. de Lima, Gislaine V. de Lima and Ralf T.S. Cordeiro
Departamento de Zoologia, Centro de Biociências, Universidade Federal de Pernambuco. 50670-420 Recife, PE, Brazil

Carlos D. Pérez
Centro Acadêmico de Vitória, Universidade Federal de Pernambuco. 55608-680 Vitória de Santo Antão, PE, Brazil

Leroy Soria-Díaz and Lorena Garrido-Olvera
Instituto de Ecología Aplicada, Universidad Autónoma de Tamaulipas. División del Golfo 356, La Libertad, 87019, Ciudad Victoria, Tamaulipas, Mexico

Jaime Luis Rábago-Castro
Facultad de Medicina Veterinaria y Zootecnia, Universidad Autónoma de Tamaulipas. Carretera Victoria-Mante km 5, 87000, Ciudad Victoria, Tamaulipas, Mexico

Hublester Domínguez-Vega and Yuriana Gómez-Ortiz
División de Desarrollo Sustentable, Universidad Intercultural del Estado de México. Libramiento Francisco Villa SN, 50640, San Felipe del Progreso, Estado de México, Mexico

Javier Manjarrez
Centro de Investigación en Recursos Bióticos, Facultad de Ciencias, Universidad Autónoma del Estado de México. Instituto Literario 100, 50000, Toluca, Estado de México, Mexico

Sandra M. Hartz and Fernando G. Becker
Departamento de Ecologia, Universidade Federal do Rio Grande do Sul. Avenida Bento Gonçalves 9500, 91501-970 Porto Alegre, RS, Brazil
Programa de Pós-Graduação em Ecologia, Universidade Federal do Rio Grande do Sul. Avenida Bento Gonçalves 9500, 91501-970 Porto Alegre, RS, Brazil

Elise A. Rocha, André L. Luza and Taís de F.R. Guimarães
Programa de Pós-Graduação em Ecologia, Universidade Federal do Rio Grande do Sul. Avenida Bento Gonçalves 9500, 91501-970 Porto Alegre, RS, Brazil

Fernanda T. Brum
Programa de Pós-Graduação em Ecologia e Conservação, Setor de Ciências Biológicas, Universidade Federal do Paraná. Avenida Coronel Francisco H. dos Santos 210, Jardim das Américas, 81531-980 Curitiba, PR, Brazil

Giulianne Sampaio Ferreira and Danilo Augusto Almeida dos Santos
Programa de Pós-Graduação em Recursos Aquáticos Continentais Amazônicos, Universidade Federal do Oeste do Pará. Rua Vera Paz, 68035-110 Santarém, PA, Brazil

Edson Varga Lopes
Programa de Pós-Graduação em Recursos Aquáticos Continentais Amazônicos, Universidade Federal do Oeste do Pará. Rua Vera Paz, 68035-110 Santarém, PA, Brazil
Instituto de Biodiversidade e Florestas, Universidade Federal do Oeste do Pará. Rua Vera Paz, 68035-110 Santarém, PA, Brazil

Pedro Brito and Walquíria Arruda
Laboratório de Estudos Morfológicos, Departamento de Histologia, Embriologia e Biologia Celular, Instituto de Ciências Biológicas, Universidade Federal de Goiás. Avenida Esperança, Campus Samambaia, Setor Itatiaia, 74690-900 Goiânia, GO, Brazil

Cíntia P. Targueta
Laboratório de Genética e Biodiversidade, Departamento de Genética, Instituto de Ciências Biológicas, Universidade Federal de Goiás. Avenida Esperança, Campus Samambaia, Setor Itatiaia, 74690-900 Goiânia, GO, Brazil

Fernanda Santos
Laboratório de Histofisiologia, Departamento de Histologia, Embriologia e Biologia Celular, Instituto de Ciências Biológicas, Universidade Federal de Goiás. Avenida Esperança, Campus Samambaia, Setor Itatiaia, 74690-900 Goiânia, GO, Brazil

Rogério Bastos
Laboratório de Herpetologia e Comportamento Animal, Departamento de Ecologia, Instituto de Ciências Biológicas, Universidade Federal de Goiás. Avenida Esperança, Campus Samambaia, Setor Itatiaia, 74690-900 Goiânia, GO, Brazil

Francisco G. Araújo and Iracema D. Gomes
Laboratório de Ecologia de Peixes, Universidade Federal Rural do Rio de Janeiro. Rodovia BR 465, km 7, 23890-000 Seropédica, Rio de Janeiro, Brazil

Aparecida A. Nascimento, Armando Sales and Beatriz A. Chagas de Oliveira
Laboratório de Histologia e Embriologia Animal, Universidade Federal Rural do Rio de Janeiro. Rodovia BR 465, km 7, 23890-000 Seropédica, Rio de Janeiro, Brazil

Wesley O. de Sousa, Rodrigo Aranda and Fátima R. J. da Silva
Departamento de Biologia, Universidade Federal de Mato Grosso, Campus Universitário de Rondonópolis. Avenida dos Estudantes 5055, Cidade Universitária, 78736-900 Rondonópolis, MT, Brazil

Lincey E. Sousa
Instituto Federal de Mato Grosso. Rodovia BR 158, Radial José Maurício Zampa, Industrial, 78600-000 Barra do Garças, MT, Brazil

Wildio I. da Graça Santos
Departamento de Zoologia, Setor de Ciências Biológicas, Universidade Federal do Paraná. Jardim das Américas, Caixa Postal 19020, 81531-980 Curitiba, PR, Brazil

Ana Carolina da Rocha Neves, Maria Cleide de Mendonça and Gabriel Costa Queiroz
Departamento de Entomologia, Museu Nacional, Universidade Federal do Rio de Janeiro. Quinta da Boa Vista, São Cristóvão, 20940-040 Rio de Janeiro, RJ, Brazil

Lisiane Dilli Wendt and Luciane Marinoni
Departamento de Zoologia, Universidade Federal do Paraná. Caixa Postal 19020, 81531-980 Curitiba, PR, Brazil

Luane Maria Melo Azeredo and Luiz Carlos Serramo Lopez
Laboratório de Ecologia Comportamental e Psicobiologia, Departamento de Sistemática e Ecologia, Centro de Ciências Exatas e da Natureza, Universidade Federal da Paraíba. Campus I, Cidade Universitária, Castelo Branco, 58051-900 João Pessoa, PB, Brazil

Monique Silva Ximenes
Programa de Pós-Graduação em Ecologia e Monitoramento Ambiental, Universidade Federal da Paraíba. Campus IV, Rio Tinto, 58297-000 PB, Brazil

Kleytone Alves Pereira
Programa de Pós-Graduação em Psicobiologia, Universidade Federal do Rio Grande do Norte. Campus Lagoa Nova, 59078-970 Natal, RN, Brazil

Maria Paula Aguiar Fracasso
Centro Universitário de João Pessoa. Rodovia BR 230, km 22, Água Fria, 58053-000 João Pessoa, PB, Brazil

Flavia D. F. Sampaio
Instituto Federal do Paraná. Rua João Negrão 1285, 80230-150 Curitiba, PR, Brazil

Helena C. Silva-de-Assis and Franciele L. Bettim
Departamento de Farmacologia, Setor de Ciências Biológicas, Universidade Federal do Paraná. 81531-980 Curitiba, PR, Brazil

Luís F. Fávaro
Departamento de Biologia Celular, Setor de Ciências Biológicas, Universidade Federal do Paraná. 81531-980 Curitiba, PR, Brazil

Carolina A. Freire
Departamento de Fisiologia, Setor de Ciências Biológicas, Universidade Federal do Paraná. 81531-980 Curitiba, PR, Brazil

Zhumagazy Kurzhykayev and Viktor Fefelov
Fisheries Research and Production Center, North Kazakhstan Branch. Armandostar Street 2B, Nur-Sultan, Kazakhstan

Kuanysh Syzdykov, Ainur Assylbekova and Dinara Sabdinova
S. Seifullin Kazakh Agro Technical University. Zhenis Avenue 62, Nur-Sultan, Kazakhstan

Sarah Mângia and Diego José Santana
Universidade Federal de Mato Grosso do Sul, Instituto de Biociências, Laboratório de Zoologia, Av. Costa e Silva, s/n, Cidade Universitária, 79070-900, Campo Grande, MS, Brazil

Lucas Ribeiro Mariotto
Universidade de São Paulo, Faculdade de Filosofia Ciências e Letras de Ribeirão Preto, Departamento de Biologia. Avenida Bandeirantes, 3900, Monte Alegre, 14040-901, Ribeirão Preto, SP, Brazil

Mohd Y. Bhat and Ashok Channa
Department of Zoology, University of Kashmir. Srinagar 190006, India

Bilal A. Paray and Mohammed K. Al-Sadoon
Department of Zoology, College of Science, King Saud University, Riyadh, 11451, Saudi Arabia

Irfan A. Rather
Department of Biological Sciences, Faculty of Science, King Abdulaziz University. Jeddah, 21589, Saudi Arabia
Centre of Excellence in Bionanoscience Research, King Abdulaziz University. Jeddah, 21589, Saudi Arabia

Gilson R. P. Moreira
Departamento de Zoologia, Instituto de Biociências, Universidade Federal do Rio Grande do Sul. Avenida Bento Gonçalves 9500, 91501-970 Porto Alegre, RS, Brazil

Cristiano M. Pereira
Programa de Pós-graduação em Biologia Animal, Departamento de Zoologia, Instituto de Biociências, Universidade Federal do Rio Grande do Sul. Avenida Bento Gonçalves 9500, 91501-970 Porto Alegre, RS, Brazil

Vitor O. Becker
Reserva Serra Bonita, Instituto Uiraçú. Caixa Postal 001, 45880-970 Camacan, BA, Brazil

Alexandre Specht
Embrapa Cerrados. Rodovia BR 020, km 18, Caixa Postal 08223, 73310-970 Planaltina, DF, Brazil

Gislene L. Gonçalves
Departamento de Recursos Ambientales, Facultad de Ciencias Agronómicas, Universidad de Tarapacá. Casilla 6-D, Arica, Chile

Cristiane Apolinario and Luís Fábio Silveira
Museu de Zoologia da Universidade de São Paulo. Avenida Nazaré 481, Ipiranga, 04263-000 São Paulo, SP, Brazil

Reinaldo T. Medeiros
Graduação em Ciências Biológicas, Universidade Veiga de Almeida. Rua Ibituruna 108, 20271-020 Rio de Janeiro, RJ, Brazil

Flávia G. Chaves
Programa de Pós-graduação em Ecologia e Evolução, Instituto de Biologia Roberto Alcantara Gomes, Universidade do Estado do Rio de Janeiro. Rua São Francisco Xavier 524, 20550-011 Rio de Janeiro, RJ, Brazil
Instituto Nacional da Mata Atlântica. Avenida José Ruschi 4, 29650-000 Santa Teresa, ES, Brazil

Maurício B. Vecchi and Maria Alice S. Alves
Departamento de Ecologia, Instituto de Biologia Roberto Alcantara Gomes, Universidade do Estado do Rio de Janeiro. Rua São Francisco Xavier 524, 20550-011 Rio de Janeiro, RJ, Brazil

Denise M. Nogueira
Departamento de Genética, Instituto de Ciências Biológicas e da Saúde, Universidade Federal Rural do Rio de Janeiro. Rodovia BR 465, km 7, 23897-000 Seropédica, RJ, Brazil

Hassan Al-Razi and Marjan Maria
Faculty of Life and Earth Science, Department of Zoology, Jagannath University. Dhaka, Bangladesh
Bangladesh Slow Loris Research and Conservation Project. 531/2, Shahidbagh, Mosque Road, Dhaka, Bangladesh

Sabir Bin Muzaffar
Department of Biology, United Arab Emirates University. Al Ain, United Arab Emirates

Luísa L. Mota, Eduardo P. Barbosa and Junia Y. O. Carreira
Departamento de Biologia Animal, Instituto de Biologia, Universidade Estadual de Campinas. Rua Monteiro Lobato 255, 13083-862 Campinas, SP, Brazil

André V. L. Freitas
Departamento de Biologia Animal, Instituto de Biologia, Universidade Estadual de Campinas. Rua Monteiro Lobato 255, 13083-862 Campinas, SP, Brazil
Museu de Zoologia, Instituto de Biologia, Universidade Estadual de Campinas. Rua Charles Darwin, Bloco N, Barão Geraldo, 13083-863 Campinas, SP, Brazil

Gabriel A. R. Melo
Laboratório de Biologia Comparada de Hymenoptera, Departamento de Zoologia, Universidade Federal do Paraná. Caixa Postal 19020, 81531-980 Curitiba, PR, Brazil

Luiz R. R. Faria
Instituto Latino-Americano de Ciências da Vida e da Natureza, Universidade Federal da Integração Latino-Americana. Avenida Tarquínio Joslin dos Santos 1000, Jardim Universitário, 85870-901 Foz do Iguaçu, PR, Brazil

Leandro M. Santos
Instituto de Formación Docente Salomé Ureña (ISFOODSU). Santo Domingo, República Dominicana

Clayton A. Sublett, Jerry L. Cook and John P. Janovec
Sam Houston State University Natural History Museum, Sam Houston State University, Box 2116, Huntsville, TX, USA

Index

www.ingramcontent.com/pod-product-compliance
Lightning Source LLC
LaVergne TN
LVHW080917210926
848485LV00004B/663